AF777282

P88-2905
10/26/88

GEOLOGICAL SOCIETY SPECIAL PUBLICATION NO. 36

128065

QE
262
A1
G56
#36
c. 2

Diagenesis of Sedimentary Sequences

EDITED BY
J. D. MARSHALL
Department of Geological Sciences,
University of Liverpool,
Liverpool L69 3BX, UK

SHELL WESTERN E&P INC.
WOODCREEK LIBRARY
P. O. BOX 4423
HOUSTON, TX 77210

Services Integration Gr
Bellaire Technology Cente
E P Library-Houston
P.O. Box 481
Houston, TX 77001-04

1987
Published for
the Geological Society by
Blackwell Scientific Publications
OXFORD LONDON EDINBURGH
BOSTON PALO ALTO MELBOURNE

Geological Society Special Publications

Series Editor K. COE

Published for
The Geological Society by
Blackwell Scientific Publications
Osney Mead, Oxford OX2 0EL
(*Orders*: Tel. 0865-240201)
8 John Street, London WC1 2ES
23 Ainslie Place, Edinburgh EH3 6AJ
52 Beacon Place, Boston
Massachusetts 02108, USA
667 Lytton Avenue, Palo Alto
California 94301, USA
107 Barry Street, Carlton, Victoria 3053
Australia

First published 1987

© 1987 The Geological Society. Authorization to photocopy items for internal or personal use, or the internal or personal use of specific clients, is granted by the Geological Society for libraries and other users registered with the Copyright Clearance Center (CCC) Transactional Reporting Service, providing that a base fee of $03.00 per copy is paid directly to CCC, 27 Congress Street, Salem, MA 01970, USA. 0305-8719/87/$03.00

Printed and bound in Great Britain by
William Clowes Limited, Beccles and London

DISTRIBUTORS

USA and Canada
Blackwell Scientific Publications Inc.
PO Box 50009, Palo Alto
California 94303
(*Orders*: Tel. (415) 965-4081)

Australia
Blackwell Scientific Publications (Australia) Pty Ltd.
107 Barry Street, Carlton, Victoria 3053
(*Orders*: Tel. (03) 347-0300)

British Library Cataloguing in Publication Data

Diagenesis of sedimentary sequences.—
(Geological Society special publication;
ISSN 0305-8719; no. 36).
1. Diagenesis
I. Marshall, J. D. (James D.) II. Series
552′.5 QE471

ISBN 0-632-01939-5

Library of Congress
Cataloging-in-Publication Data
Diagenesis of sedimentary sequences/edited by
James D. Marshall.
p. cm. — (Geological Society special publication; no. 36)
Papers of a meeting held in Liverpool, Sept. 30–Oct. 1, 1986 under the auspices of the British Sedimentological Research Group (BSRG).
Includes bibliographies.
ISBN 0-632-01939-5
1. Diagenesis—Congresses. 2. Sedimentation and deposition—Congresses. I. Marshall, J. D. (James D.) II. Geological Society of London. III. British Sedimentological Research Group. IV. Series.
QE471.D53 1988
552′.5—dc19

Contents

Diagenesis and Sedimentary Sequences—Introduction

The meeting on 'The Diagenesis of Sedimentary Sequences' was held in Liverpool on 30 September and 1 October 1986 under the auspices of the British Sedimentological Research Group (BSRG), a Specialist Group of the Geological Society.

The aim of the meeting was to bring together an international group of research workers on carbonate and clastic sediments, to discuss the major controls on sediment diagenesis from deposition to deep-burial. This book contains a number of the papers presented at the meeting, together with one that was offered but which the authors were unable to present. I believe that these papers are a good reflection of the proceedings of a lively and diverse meeting, attended by around 150 representatives of universities, governmental organizations and industry from nine different countries.

I leave readers to delve into the papers for themselves but would like to take this opportunity to reiterate some of the points that I raised at the meeting concerning diagenesis and the way in which it is approached.

Diagenesis is an integral part of the history of the fill of a sedimentary basin and needs to be treated as such. In order to understand the post-depositional history of a sedimentary rock or, in an industrial context, to understand the evolution of its reservoir properties, we must use information from more general geological studies. A knowledge of depositional setting, facies architecture and burial history are all invaluable in the deduction of a well-constrained diagenetic history. Similarly, quantitative diagenetic investigations may contribute information, particularly about pore-fluid evolution and temperature changes, which needs to be taken into account by those concerned with broader syntheses of basin history and hydrocarbon prospectivity.

Many published works on diagenesis and indeed most of those in this book, are concerned with the post-depositional history of a single lithological unit, or even a particular phase within a unit. As the author of a paper on the diagenetic adventures of just three ammonites (Marshall 1981) I can scarcely be too critical! Such studies concentrate attention on the interesting but perhaps atypical features of the sediment. They are undoubtedly useful in determining the local controls on diagenesis: indeed they often reveal just how complex the interaction of processes (cementation, neomorphism, compaction and dissolution) can be. However, if we only work on the small scale, it becomes difficult to determine what is typical of a sequence as a whole and indeed what is regionally rather than just locally significant.

It is extremely difficult to be objective in sampling for any geological study and this is particularly true for diagenetic investigations. Material is often only available from a restricted area of the basin (determined by outcrop pattern or on a structure that has been drilled for hydrocarbons) and our attention is automatically drawn to the 'different'. Even with careful sampling we can only ever hope to look at a minute proportion of what is there; a colleague once estimated that he was being asked to assess the reservoir potential of a North Sea reservoir unit on something like $10^{-15}\%$ of the rock in the area! We should always be aware of this! Having said that, it is interesting, and I think particularly welcome, to see that in a number of the papers in this book authors have been able to pool sufficient data to treat diagenetic evolution in a regional context.

In organizing the meeting one of the main aims was to enable specialists from different areas to talk to each other. The ways in which rocks of one type influence the diagenetic evolution of other sediments within the same sedimentary sequence are far from clear. We might, for example, expect chemical and isotopic buffering to preclude extremes of acidity, alkalinity or even geochemical values as fluids pass from one rock type to another. All too often however, in diagenetic studies, authors invoke a source, for ions, acids, fluids or whatever, from outside the rocks that they are studying. Too often, in the past, they have not considered the feasibility of reactions or the problems of mass balance or transfer.

Mudrocks are a fine example of an oft-quoted source or sink that we have yet to fully understand: they are, after all, fine-grained, and full of reactive organic and inorganic chemical species which can, through compaction, be expelled into adjacent sandstones and limestones. Several recent studies have shown that mudrocks have complex diagenetic histories and have indicated that many reaction products may stay very close to where they started. It is a shame that for a number of reasons more of the 'mudrock' papers presented at the meeting are not included in this volume.

Conduits also tend to be poorly understood by workers in our field. Faults, for example, are commonly invoked as carriers for fluids, which travel up (and occasionally down) the sedimentary pile, yet the same fractures are known to seal hydrocarbons in place. Clearly, then, we need a more integrated approach to constrain our grander conclusions. There is certainly scope for detailed, quantitative petrographic studies of mudrocks (now possible with modern back-scatter electron microscopy) and a need for us to talk to structural geologists. Organic geochemists and thermal modellers also have a lot to offer the diagenetic investigator.

In planning the meeting and the layout of this book, to be consistent with the ideas expressed above, I have tried to avoid grouping papers on purely lithological grounds. In the book therefore the papers are, perhaps somewhat arbitrarily, arranged into three sections. The first contains a collection of papers on 'Diagenetic Processes', the second on 'Early Diagenesis' and the third on 'Regional Studies and Burial Diagenesis'; each contains papers dealing with both clastic and carbonate rocks. I hope that readers will have a look at papers that lie outside their direct field of specialization.

Finally I would like to express my thanks to all the people who have helped in the organization of the meeting and the preparation of the book. Generous financial support for the meeting was provided by grants from the Geological Society and the Royal Society, and donations from BP (London), BP Research, Britoil, Esso and Shell; the funds to enable colour printing were provided by the authors' employers, the authors themselves and by additional donations from Esso and BP. The success of the meeting was due in large part to the Liverpool postgraduates and technical staff who took the whole thing over and ran it very smoothly! (Thanks go to Steve, John, Greg, Greg, Jim, Paul and Hilary.) My gratitude, for all their work, goes to the contributors, both to the meeting and the book, as it does to the forty or so reviewers who have put so much effort into manuscripts and maintaining scientific standards. Nick Parsons of Blackwells helped ensure rapid publication and Hilary Davies prepared the index. I am especially grateful to my wife, Lesley, who has shown great patience and given enormous support, practical and moral, throughout the whole enterprise.

JIM MARSHALL
Liverpool, Easter, 1987

Reference

MARSHALL, J. D. 1981. Zoned calcites in Jurassic ammonite chambers; isotopes, trace elements and neomorphic origin. *Sedimentology*, **28**, 867–87.

Diagenetic processes

Hydrodynamic modelling of cementation patterns in modern reefs

Ian R. Goldsmith & Peter King

SUMMARY: Cementation patterns in modern reefs are inhomogeneous, even on the thin section scale. The concept of the microenvironment has been developed in order to explain this irregularity, this microenvironment being determined by both chemistry and permeability. Identification of the rate limiting step in carbonate cement growth has led to the suggestion that cementation patterns are controlled principally by the microenvironmental permeability. In order to test this hypothesis, modelling of the permeability characteristics of biogenic frameworks has been attempted. Representative matrix geometries were created mathematically and Poiseuille flow through these has produced cementation patterns similar to those observed in thin section. Visual comparison of the features of the real system and model results indicates that the conceptual model is realistic and permeability may indeed be the major control on cementation.

In the future, modelling of the matrix geometry will be controlled using the measured pore throat radius distribution of the rock. At that stage the model will be applicable not only to the prediction of cementation patterns but to many other problems involving fluid flow through porous media.

Introduction

Cementation patterns in reef limestones are extremely inhomogeneous on a range of scales, down to and including thin-section level. This feature can be clearly observed in the photomicrographs shown in Figs 1 and 2.

Cores taken through reefs of the SE Florida carbonate platform show an irregular variation of cementation with depth, with cements comprising from zero to 15% of the total rock volume.

In the past this inhomogeneity has been explained using the concept of the microenvironment (Schroeder 1972). It has been generally considered that the major factor controlling this microenvironment is the seawater chemistry; spatial variation in the chemical nature of the pore waters therefore produces the observed variation in the cement products.

Mass balance calculations show that many thousands of volumes of porewater need to pass through a pore in order to fill it with cement (Bathurst 1975, Berner 1980); therefore the diagenetic systems must be physically 'open' on all levels, so that effective chemical exchange is possible (Pingitore 1982). However, this appears to contradict the concept of a spatially variable chemical microenvironment which would require at least a partly closed diagenetic system.

The precipitation and growth of carbonate cements

A fundamental understanding of the controls on cementation must be based on the identification of the rate limiting step in the process. Crystal growth in the hydrodynamic regime present in reef sediments is almost certainly transport and not surface controlled (Berner 1980). Hence the rate limiting step in the growth of calcium carbonate cements is the supply of solutes (Ca^{++} and HCO_3^-) to the crystal surface. It is therefore apparent that the variation in degree of cementation may be governed simply by the fluid flow rates, these in turn being determined by local permeabilities.

The suggestion is, therefore, that the microenvironments are not chemically controlled (in space) but are permeability controlled. One could, of course, argue that the microenvironments are then chemically controlled in time.

Modelling of permeability characteristics

In order to analyse the above hypothesis, computer programs developed at the BP Research Centre have been used to model the permeability characteristics of carbonate framestones. The modelling to date has concentrated exclusively on coral framework as this is the

From: MARSHALL, J. D. (ed.), 1987, *Diagenesis of Sedimentary Sequences*, Geological Society Special Publication No. 36, pp. 1–13.

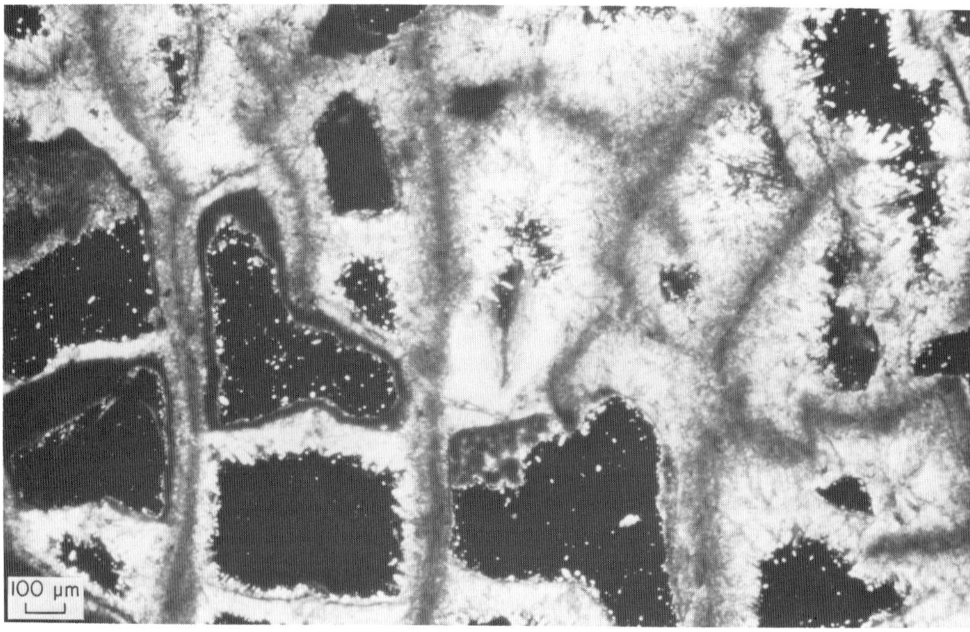

FIG. 1. The inhomogeneous nature of cementation in a Holocene coral framestone from a reef of the SE Florida Shelf. Cements are acicular aragonite and peloidal high magnesium calcite. Photomicrograph, crossed polars.

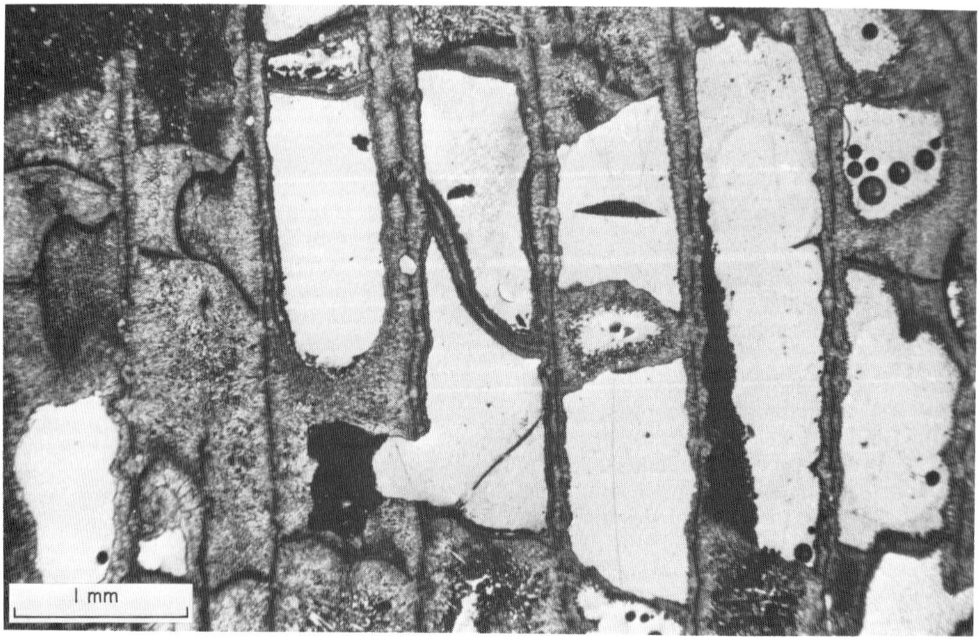

FIG. 2. The inhomogeneous nature of cementation in a coral framestone from the Belize barrier reef. Note that some pores are devoid of cement, some have a thin isopachous coating and others are full of aragonite fibres. Photomicrograph, crossed polars.

major component of the Florida Holocene reefs. The results described below represent volumes of cement only; no attempt has been made to explain the distribution of different morphologies and mineralogies. Most importantly, the models attempt to show that the distribution of the cements is also determined by the permeability characteristics of the rock matrix.

A number of assumptions have been made in order to be able to describe the *real* system in simple terms:

(a) The real system is in a laminar flow regime, i.e. flow through the pores of the system is non-turbulent. This assumption can be verified by calculating the dimensionless number—the Reynolds number—as follows:

$$N_{Re} = \frac{v.r.\rho}{\mu} \tag{1}$$

where v is the fluid velocity in a pore of radius r, ρ is the fluid density and μ is the fluid viscosity. This is a standard fluid mechanical term which is used to determine the nature of the flow through any system. The flow becomes turbulent at Reynolds numbers between 1200 and 2000. Whilst this range is strictly only valid for flow through a uniform, smooth cylindrical tube, flow in rougher tubes is still laminar up to about $N_{Re} = 1$.

Calculation of the Reynolds number using geologically reasonable values for the flow of pore fluids (1–10 m day^{-1}), through pores of radius 1–1000 μm, and standard data for the viscosity and density of water, reveals a range for N_{Re} of 10^{-1}–10^{-6}. This clearly indicates that the real system is in a laminar flow regime.

(b) Solute transport is considered to be by bulk flow only; there is no significant ionic diffusion. Another dimensionless number, the Peclet number (Lerman 1979) can be used to confirm this assumption:

$$N_P = \frac{Q.w}{D} \tag{2}$$

where Q is the fluid flux (related to fluid velocity, v), w is a particular length scale and D is the diffusion coefficient for the solutes of interest, in this case Ca^{++} and HCO_3^-. For Peclet numbers greater than 1, bulk flow is the most important. Using values of the fluid velocity of 1–10 m day^{-1}, a length scale (w), of 1 mm to 1 m and standard data for the diffusion coefficients (both approximately 10^{-5} cm^2 s^{-1}) the calculated Peclet number has a range of 10–10^5 for the real system. Therefore, diffusive mass transport is negligible when compared to bulk mass transport, so no diffusion term is included and transport of solutes is considered to be by bulk flow only.

Theoretical basis of the model

The rock matrix

Network codes describe a regular square grid of capillary tubes (Fig. 3) which represents the rock matrix (rock in grey, porosity in white). In all of the examples in this paper the grid is 10 by 10 square, though theoretically any size can be computed.

Fluid flow

There is considered to be a fixed pressure gradient (P), across the grid; this drives the pore fluid from bottom to top. Laminar flow is controlled by Poiseuille's law;

$$Q = \frac{\pi.r^4.\rho.\Delta p}{8.\mu.l} \tag{3}$$

where Q is the fluid flux through the capillary, r is the capillary radius, ρ is the pore fluid density, Δp is the pressure difference across the capillary, μ is the viscosity of the fluid and l is the length of the capillary.

The model then calculates the fluid velocities in each segment of the grid in the following manner. At each node in the grid, fluid must be conserved, so the sum of the velocities at each node must equal zero. A standard numerical iterative technique is then used to calculate the velocities in each segment. An initial guess at the pressure difference between each node is made, and a velocity calculated from that using equation (3). The iterative technique is pursued to the point when the sum of all the velocities at every node in the grid is zero. Note that the inlet and outlet boundaries (the top and bottom of the diagrams), are at constant pressure, and that the edges of the cell are connected together as though the diagrams were wrapped around a cylinder.

Cementation process

We are assuming that the rate controlling step in the growth of the cements is the rate of supply of solutes to the growing surfaces. As we have shown that the main transport of solutes is by bulk flow, then the rate of growth is directly proportional to the flow velocity.

The cementation process is described as a decrease in the radius of the capillary tubes. This decrease is calculated to be directly proportional to the flow velocity;

$$\frac{dr_i}{dt} = \beta.v_i \tag{4}$$

where r_i is the radius of the ith capillary, v_i is the fluid velocity in the ith capillary and β is a proportionality constant. Note that this constant acts as a time-scale and relates the model results to geological reality. In the later examples this constant is called a cementation factor.

In the reef system, as cementation progresses, the pore radii will decrease. From equation (3), the fluid flux or velocity will also decrease and so will the rate of precipitation (from equation (4)). This negative feedback is built into the model by calculating the overall result via a number of stages or time steps (usually 10). After each step new flow velocities are calculated from the capillary radii as modified in the previous step.

In the figures below, the cement precipitated is shown in black.

Matrix geometry

The simple example in Fig. 3 is perhaps representative of a supermature sandstone or a carbonate grainstone, but it is obviously not a good description of a coral framework as shown in Figs 4 and 5.

There are several parameters which may be used to alter the matrix geometry so that it is more representative of carbonate framestones. These parameters are input by the user at the start of the computer program and are listed at the edges of the figures (Figs 3 and 7 to 15). The parameters are described below:

(i) λ_x, λ_y These are two parameters which control the capillary radius distribution (in the x and y directions respectively) according to the following condition:

The probability of there being a capillary tube of radius r is $\frac{1}{2}\lambda$ for;

$$1-\lambda<r<1+\lambda \qquad (5)$$

and zero otherwise (Fig. 6).

A value for each parameter is input at the start of the program. This can be any number between 0 and 1.

In the 10 by 10 matrix used throughout this study, there are 100 capillary segments in both the x- and y-axis directions. The use of a probability distribution, such as that above, allows a non-uniform capillary radii distribution to be created, simply by the input of a single value for the x and y directions. As the value of λ is increased, the variety in the capillary radii also increases.

If $\lambda=0$, from equation (5), r varies from $1-0$, to $1+0$, i.e. all the radii are equal and of unit size. Similarly, if $\lambda=0.5$, r varies from 0.5 to 1.5 units. Increasing λ to 0.9 produces a wider distribution in the radii, from 0.1 to 1.9 units. The probability distribution (Fig 6), shows that, for any value of λ, there is an equal probability

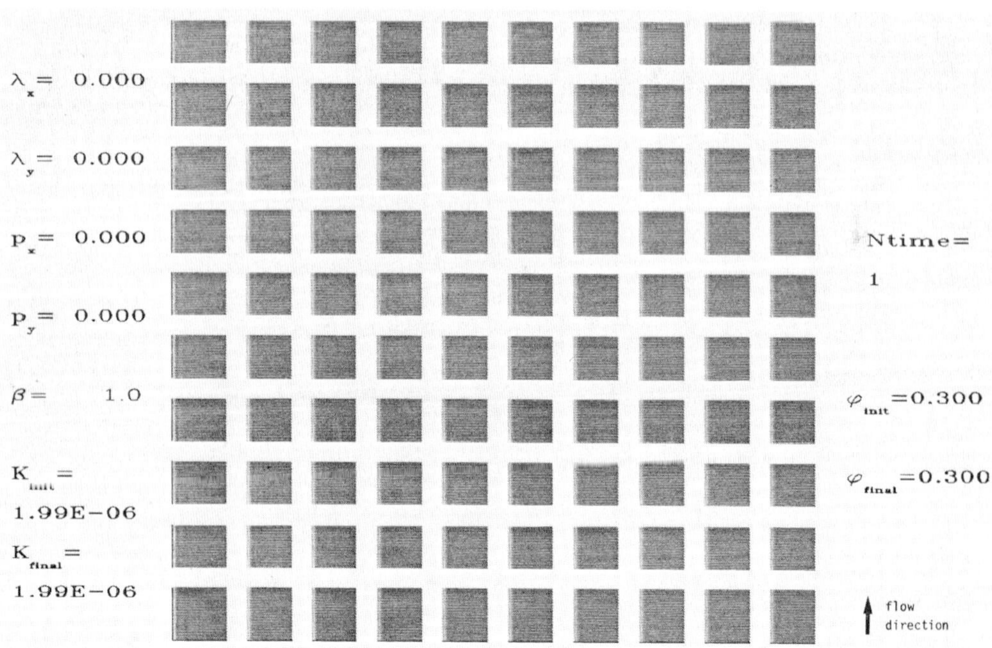

FIG. 3. The regular square grid of capillary tubes which forms the basis of the model matrix. In these examples the grid is 10 × 10 square, though any size can be computed. The rock is in grey and the porosity in white.

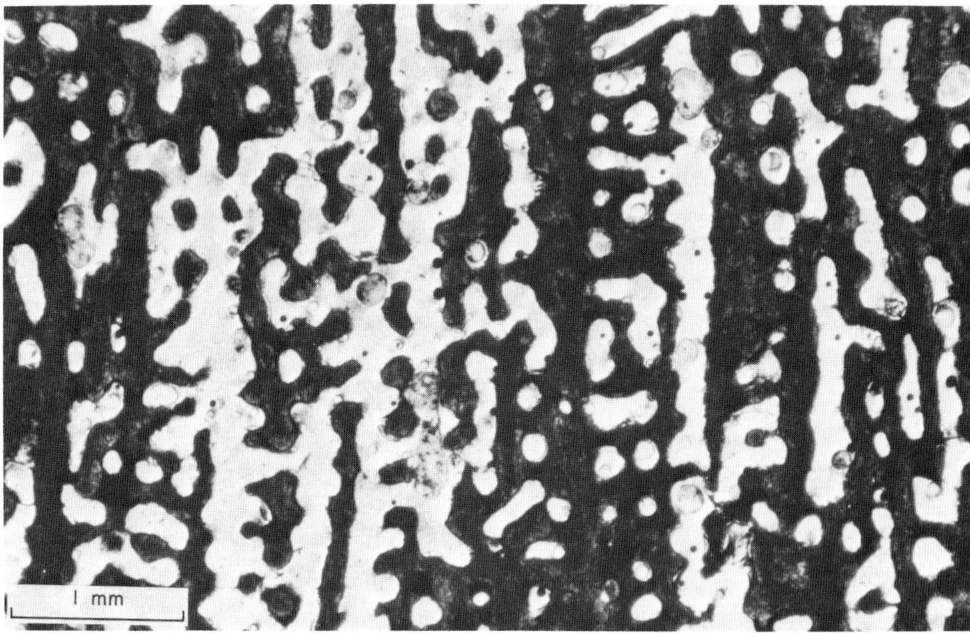

FIGS 4 and 5. Typical appearance of Holocene coral frameworks in thin section. This is the type of carbonate framestone whose permeability characteristics are being modelled here. The sections are parallel and perpendicular to the corallites. Photomicrographs, plane polarized light.

($\frac{1}{2}\lambda$) that any radius within the range of $1 - \lambda$, to $1 + \lambda$, will occur in the grid.

(ii) P_x, P_y These two parameters represent the fraction of the capillary segments (in the x and y directions respectively) which are non-conducting. This therefore introduces a degree of matrix connectivity.

The parameters can have any value from 0 to 1. Hence, $P_x = 0.5$ means that half of the capillary

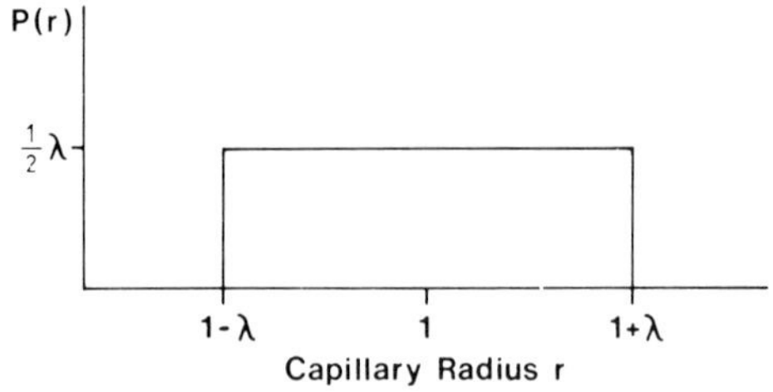

FIG. 6. The capillary radius distribution function.

segments in the x direction are non-conducting. The visible effect on the matrix is that half of the x-axis segments are sealed.

The program calculates the initial and final permeabilities (K init. and K final) of the grid from a simplified version of Poiseuille's law;

$$Q = \frac{K.\Delta p}{\mu.L} \tag{6}$$

where Q = fluid flux out of the top of the grid, Δp is the pressure difference across the grid, μ is the fluid viscosity and L is the length of the grid. These permeabilities are of course only relative.

The initial porosity can also be chosen by the user; the program will then calculate the final porosity (φ init. and φ final).

Variation of these input parameters leads to a wide range of results, some of which are shown in Figs 7 to 15.

Results

Presentation of the results of this modelling is in two parts:

(i) the effects of the matrix geometry on the cementation patterns;
(ii) the effects of varying β, the cementation rate constant.

Each example below is the result of an individual experiment created using the parameters listed at the edges of the relevant figure.

Matrix geometry

Figure 7 shows the cementation pattern created using the regular square grid as in Fig. 3. The precipitated cement (shown in black) occurs in a uniform pattern along the y-axis surfaces only. This is because there is no pressure difference across the capillaries in the x direction, and so flow is in the y direction only.

This result is in itself geologically unreasonable, though mathematically valid, as we are assuming the system to be in a laminar flow regime. In reality, eddies and edge effects are bound to deflect the flow of pore waters through the rock matrix. However, introduction of irregularities into the matrix in the following examples does indeed cause precipitation of cements on the x-axis surfaces.

This example serves as a standard with which to compare the effect of changing the matrix parameters. The uncemented 'gap' at the base of all the figures is a function of the graphics program and should be ignored.

Variation in P_x, P_y

Figures 8 and 9 show the effect of introducing a proportion of pores which are non-conducting. Sealing half of the pores in the x direction has very little effect on the cementation pattern as it does not alter the flow path or the initial permeability (K init. for Figs 7 and 8 is 1.99×10^{-6}).

However, it is interesting to note that sealing half the pores in the y direction has a marked effect (Fig. 9). In this case the matrix tortuosity is greatly increased and the flow is forced along capillaries in the x direction. An appreciable degree of inhomogeneity has been created. Note the concomitant reduction in initial permeability. (K init. is 8×10^{-7} in Fig. 9 compared with 1.99×10^{-6} in Fig. 7.)

Variation in λ_x, λ_y

In the examples discussed above (Figs 7 to 9), λ_x and λ_y are both zero, so the capillary radius distribution function (Fig. 6) reduces to a single value and all the pore radii are equal. Varying λ introduces a distribution of pore radii as shown in Fig. 10, where $\lambda_x = 0$, $\lambda_y = 0.5$. Note that for the analogous situation, $\lambda_x = 0.5$, $\lambda_y = 0$ there would be very little effect on the cementation pattern (for the same reason as P_x has no effect) as this has no effect on the flow paths. In Fig. 10 there is therefore a distribution of y-axis capillary radii, with all x-axis radii equal and all pores conducting ($P_x, P_y = 0$).

The Poiseuille flow dependence on the capillary radius raised to the fourth power is evident here ($Q \propto r^4$, equation (3)). A substantial degree of variation in the cement thickness has been created, with the larger pores initially receiving a greater volume of cement.

Representation of coral framework

A combined variation of λ_x, P_x and P_y, creates a matrix geometry that is a reasonable representation of coral frameworks (Figs 11–15).

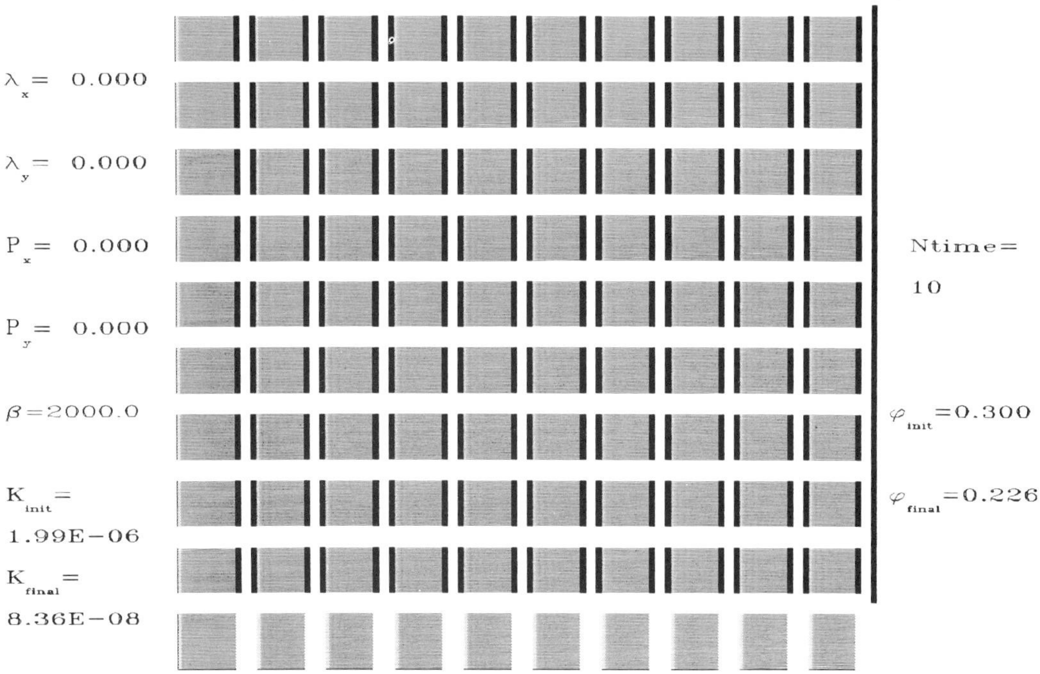

FIG. 7. Regular square grid. Cementation (shown in black) occurs along the y-axes only because flow is in that direction only. This example serves as a standard with which to compare the effects of changing the matrix dscription.

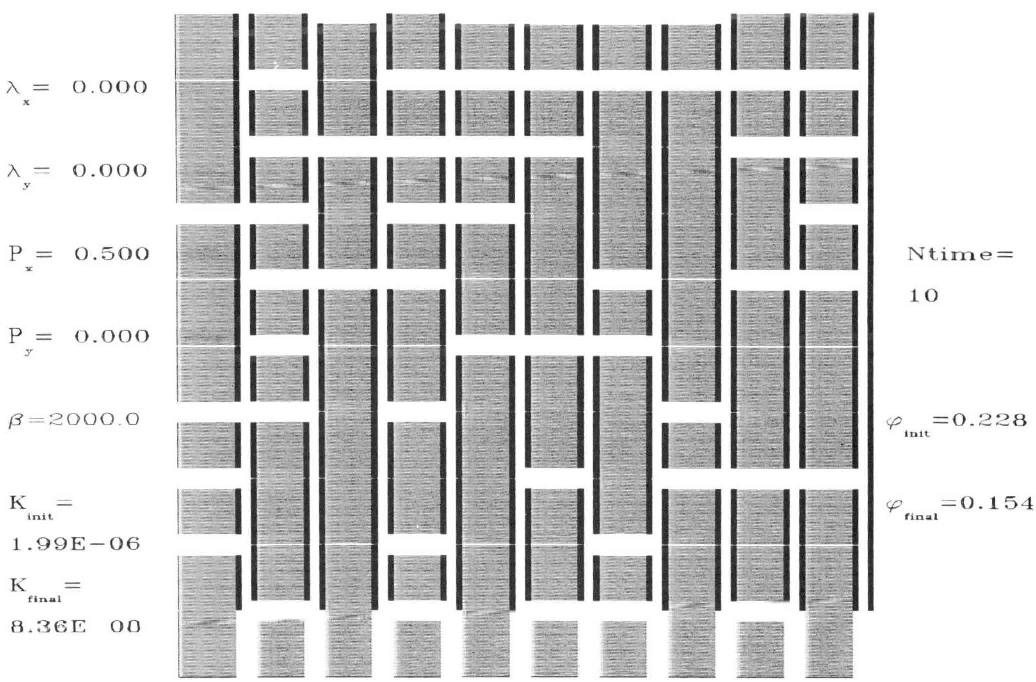

FIG. 8. In this example, half of the pores in the x direction are non-conducting. The effect on the cementation pattern is minimal and flow is still only in the y direction.

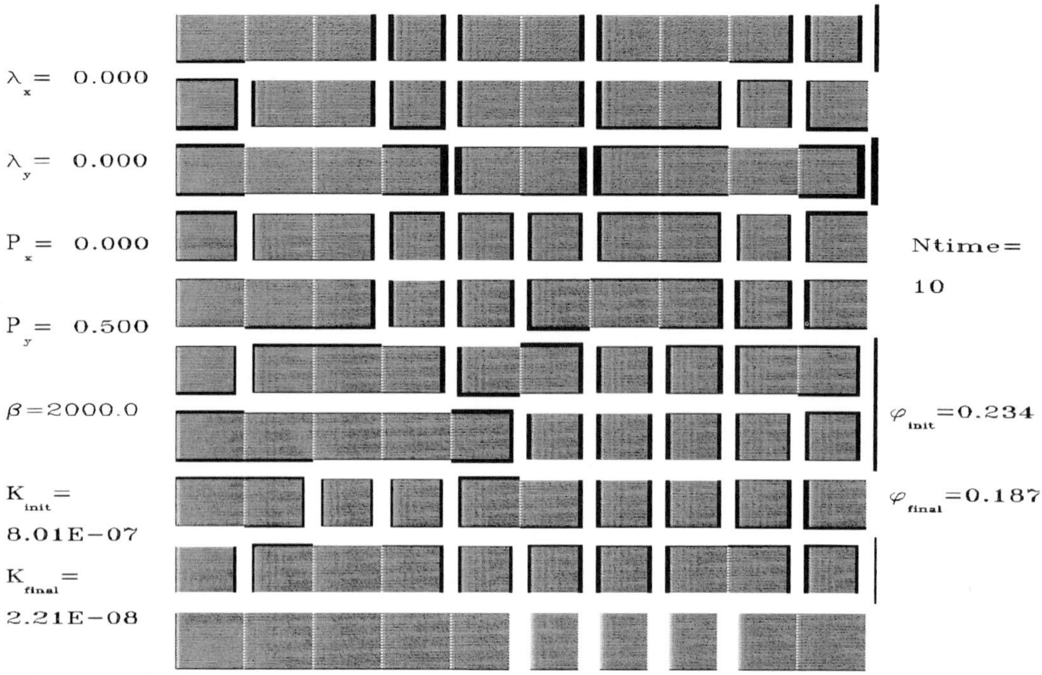

FIG. 9. Sealing half of the pores in the y direction has a more pronounced effect on the cementation pattern. The matrix tortuosity is increased and flow is forced along capillaries in the x direction. Note the concomitant reduction in initial permeability (cf. Fig. 7).

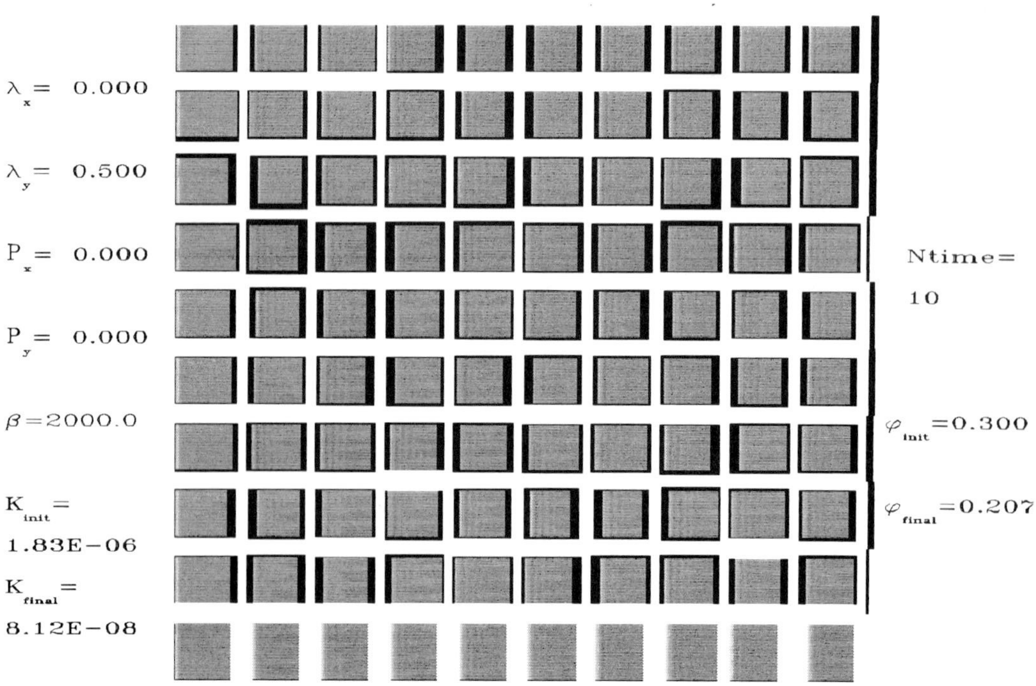

FIG. 10. $\lambda_y = 0.5$ creates an irregular distribution of y-axis capillary radii. As the rate of cementation is directly proportional to the flow velocity, which is itself dependent on the fourth power of the radius, larger pores receive substantially more cement.

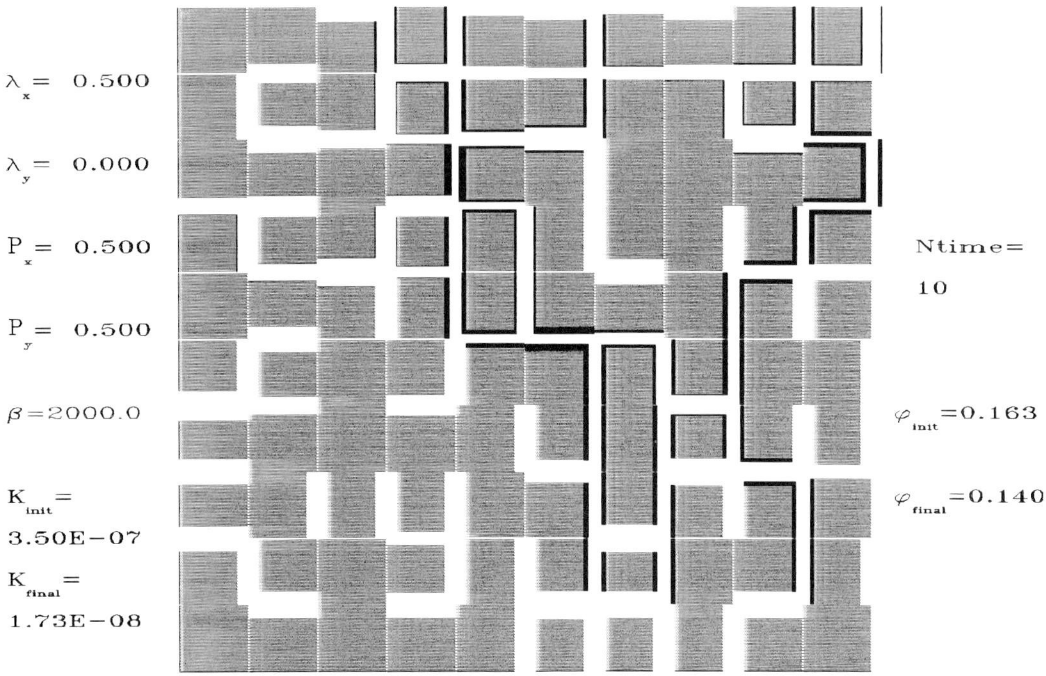

FIG. 11. Results more representative of the reality (cf. Fig. 2) can be produced by increasing the matrix connectivity as in this example. Both dead end and non-effective porosity are apparent here, and there is a considerable variation in the degree of cementation.

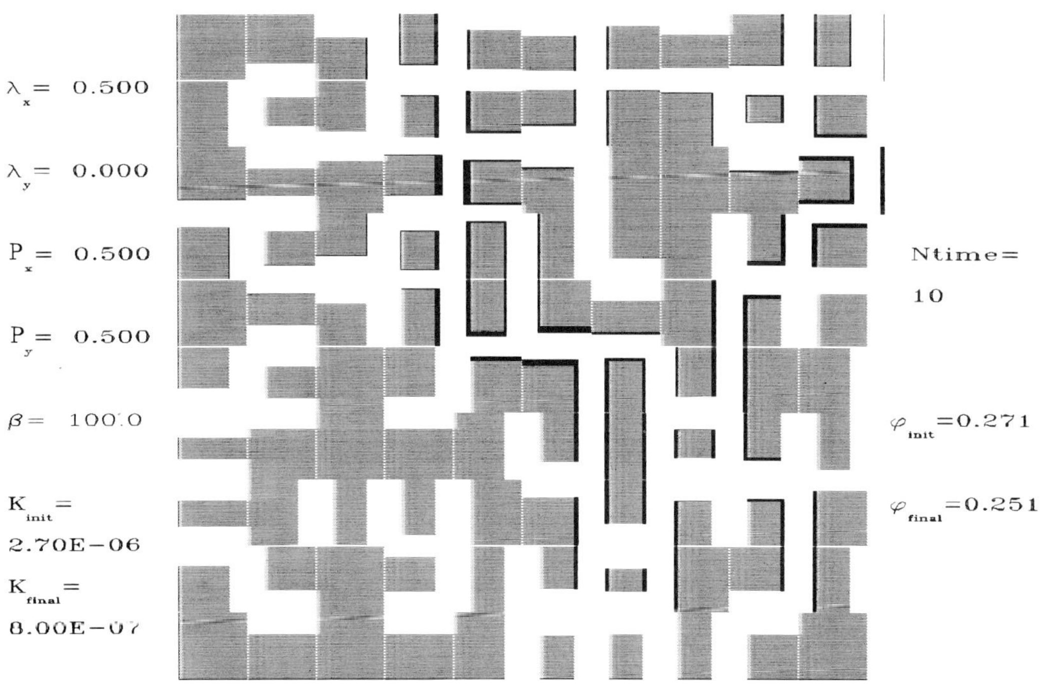

FIG. 12. Using the same matrix description as in Fig. 11, the porosity has been increased. Due to the r^4 dependence, the cementation factor has to be reduced by an order of magnitude.

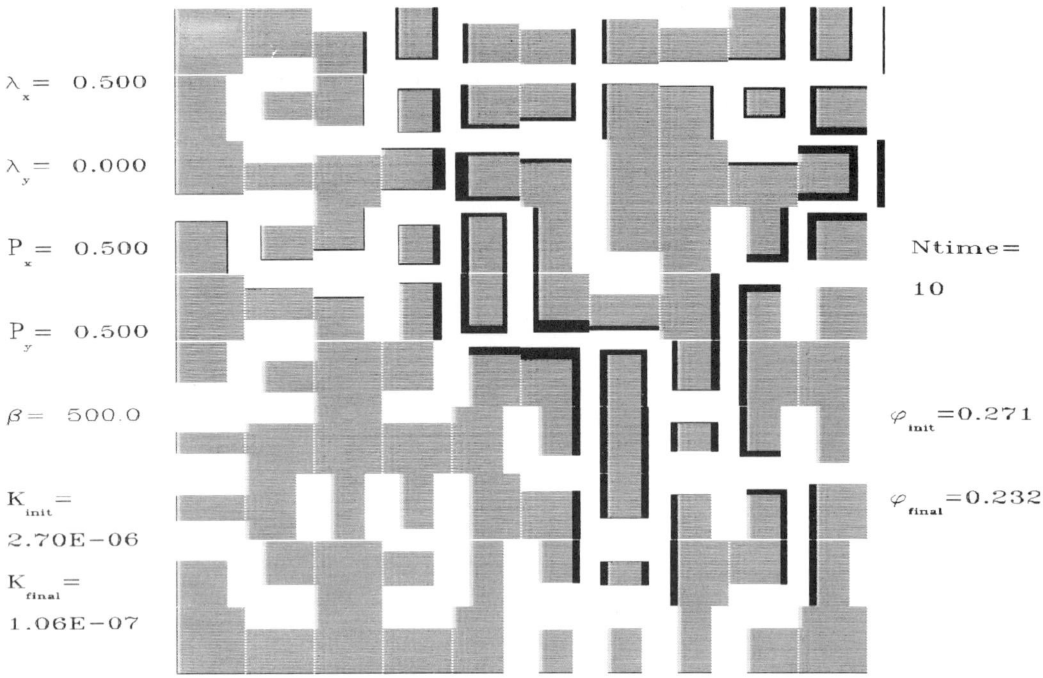

FIG. 13. It is interesting to observe the effects of progressively increasing β. In this example $\beta = 500$ (cf. Fig. 12, $\beta = 100$).

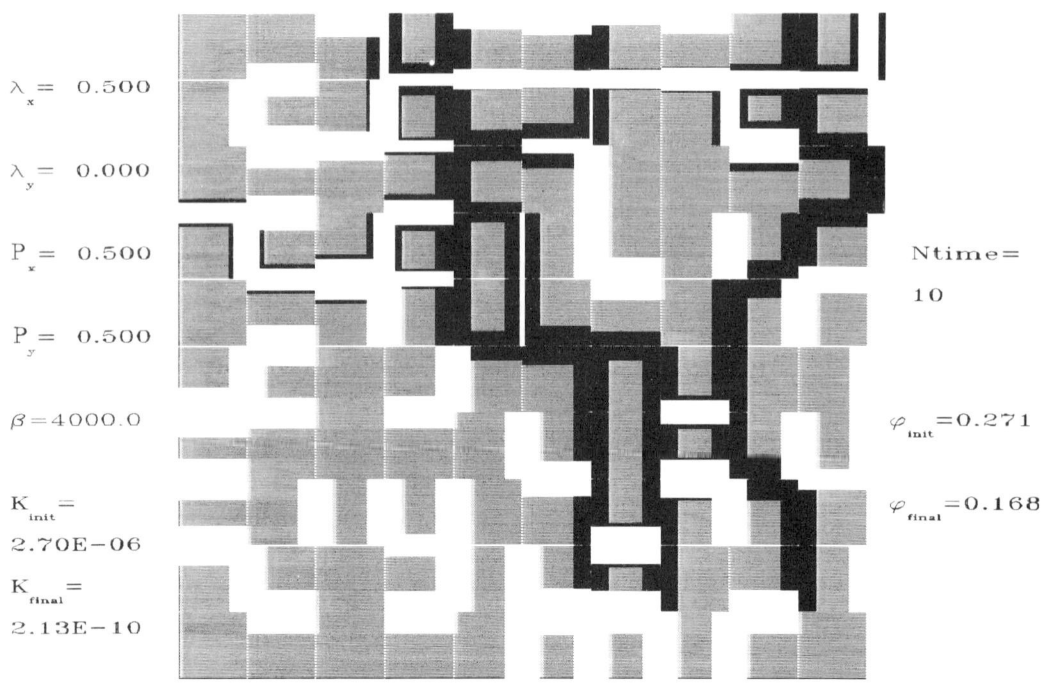

FIG. 14. In this example $\beta = 4000$ and much of the effective porosity has been filled by cement.

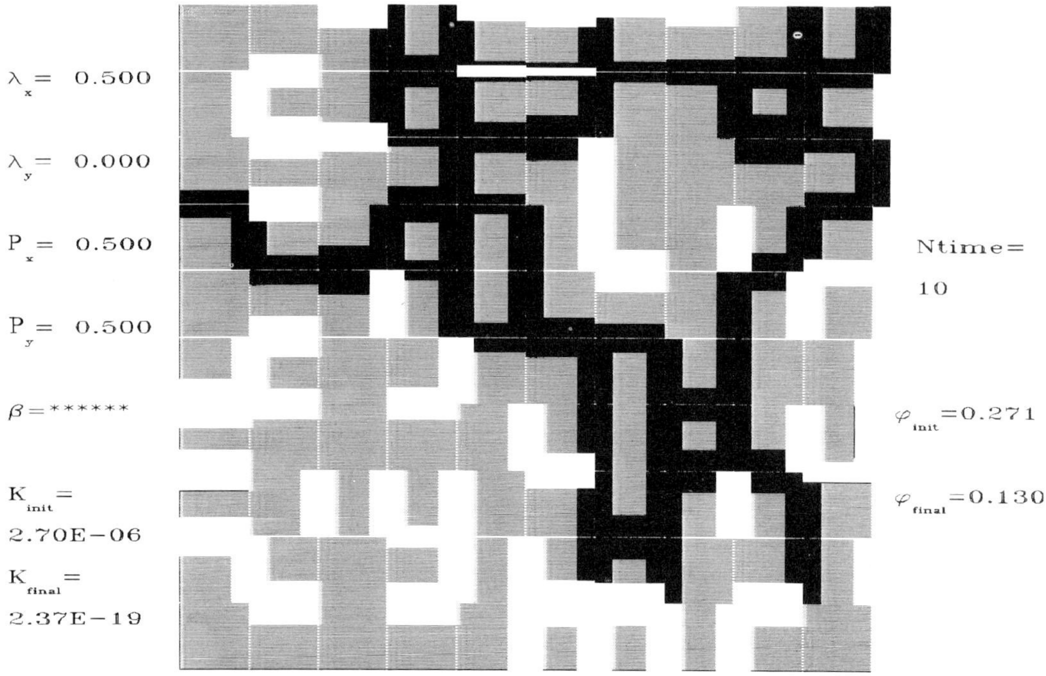

FIG. 15. Here β is very large and the grid has been fully cemented. Note that as flow is laminar, dead end porosity remains uncemented. Non-effective porosity receives no fluid flow and so is also unaltered.

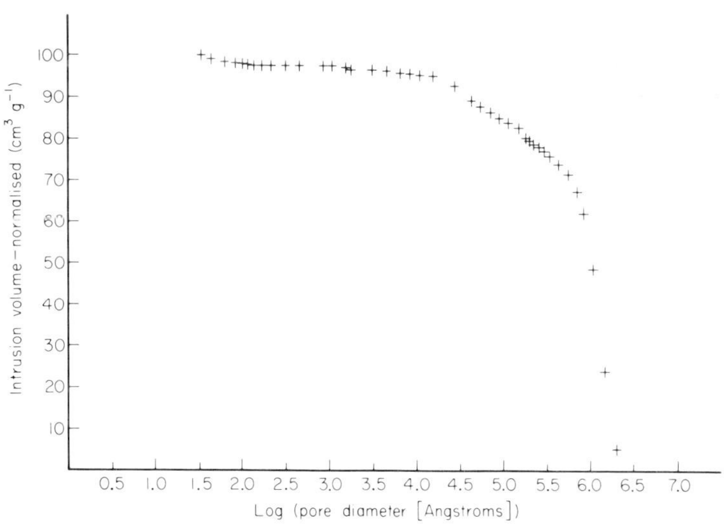

FIG. 16. The pore throat radius distribution for an unaltered Holocene coral framestone. This is a capillary pressure curve derived by mercury injection porosimetry. In future modelling these types of data will be used to control the capillary radius distribution of the model matrix.

In Fig. 11, P_x and P_y are both 0.5, thereby increasing the matrix connectivity. Many of the features apparent in the real system are now present in the model results. Both dead-end and non-effective porosity are present and a wide variation in the degree of cementation is apparent. The lower left corner of Fig. 11 has received no cement whatsoever as all the porosity is either non-effective or dead-end. Some of the pores in the upper left side have received small volumes of cement, but the majority of the cement has been 'precipitated' in the centre and right side of the grid.

Using this matrix description, the following examples show the effects of varying the porosity and the cementation factor β.

In Fig. 12 the input parameters are the same as for Fig. 11, but the initial porosity figure has been increased. The capillary radii are therefore increased and again the Poiseuille dependence on r^4 is evident: to produce roughly the same volume of cement as in Fig. 11, the cementation factor β has to be reduced from 2000 to 100.

The cementation factor β

The parameters used in Fig. 12 produce the most accurate representation of coral framestone geometry as observed in thin section. Using this basis it is interesting to observe the effects of increasing β on the cementation patterns (Figs 13, 14 and 15). In Fig. 15, β is very large and the grid has been fully cemented, with the exception of non-effective and dead-end porosity. Comparison of this result with the photomicrograph in Fig. 2 suggests that those empty pores adjacent to fully cemented ones are probably non-effective porosity only revealed in the two-dimensional section. Such pores have very little or no fluid flow through them.

Whilst the last four examples are separate results created by increasing the cementation parameter β, it is possible to use the model in an incremental manner such that successive cement zones are created. Figures 7 to 15 were created using 10 time steps, with only the final result displayed. By setting the program to display the results of each time step, a series of 10 cement zones is produced. The thickness of each zone progressively decreases according to the dependance on the fourth power of the radius.

Discussion

The modelling to date has been on the thin section level only. Our aim is to be able to model the cementation patterns on a variety of larger scales. In order to do this, some of the limitations of the present network codes need to be modified.

(1) The use of a regular square grid restricts the variation in capillary radii essentially to one order of magnitude. Modification of the network codes will eventually allow control of the capillary radius distribution of the model using the measured pore throat radius distribution of the rock. An example of the pore throat radius distribution is given for a Holocene coral framestone in Fig. 16. This modification will permit modelling of the permeability characteristics of any rock given its pore throat distribution.

(2) At present the model operates in two dimensions only which clearly involves some fundamental approximations. Expansion of the capillary grids to three dimensions would be advantageous. This will necessitate changes in the method of visualization of the results, either by observation of successive 2-D sections or by producing a capillary radius distribution after cementation.

(3) Modelling on a larger scale requires a detailed knowledge of the porosity distribution on that scale followed by use of those data for the description of the model matrix.

(4) Cementation interacts dynamically in both space and time with other processes of reef formation (Schroeder & Zankl 1974). Reef construction, reef destruction and sedimentation alter the initial depositional permeability characteristics of the reef limestones, so affecting the continuing cementation patterns. Some aspects of these processes will be included in the modelling in the future.

Conclusions

The results of this first stage model are very promising. Many of the features apparent in the real system have been reproduced using a fairly simple mathematical description of the framestone matrices.

The creation of the two-dimensional matrix has indicated that a significant proportion of the porosity observed in thin section may in fact be non-effective. This therefore explains why areas devoid of cement are often observed in an otherwise well cemented sample. In addition, much of the porosity may be poorly interconnected with the major flow conduits (dead-end porosity), and will therefore receive substantially less cement than the well connected pores.

The results of the cementation experiments indicate that larger pores initially receive a greater volume of cement than the smaller pores and confirm that the rate of cementation will be reduced as the pores become restricted.

The most important conclusion of this study is that the model indicates that the variation in the volumes and the spatial distribution of precipitated cement is controlled by the microenvironmental permeability alone. We have not yet developed a statistical method to compare rigorously the results and the reality, but visual comparison suggests that the conceptual model is successful.

ACKNOWLEDGMENTS: We gratefully acknowledge the permission of the British Petroleum Company to publish this work. One of us (IRG) would like to thank NERC and British Petroleum for financial assistance and Dr M. E. Tucker for supervision throughout this research. We are grateful to Professors R. N. Ginsburg and E. A. Shinn for kindly donating the samples on which this research has been carried out.

References

BATHURST, R. G. C. 1975. *Carbonate Sediments and their Diagenesis*. Elsevier, Amsterdam.

BERNER, R. A. 1980. *Early Diagenesis: a Theoretical Approach*. Princeton University Press, New Jersey.

LERMAN, A. 1979. *Geochemical Processes, Water and Sediment*, pp. 64–65. Wiley, New York.

PINGITORE, N. E. 1982. The role of diffusion during carbonate diagenesis. *Journal of Sedimentary Petrology*, **52**, 27–39.

SCHROEDER, J. H. 1972. Fabrics and sequences of submarine carbonate cements. *Geologische Rundschau*, **61**, 2, 708–30.

—— & ZANKL, H. 1974. Dynamic reef formation: a sedimentological concept based on studies of Recent Bermuda and Bahama Reefs. *Proceedings of the 2nd International Coral Reef Symposium*, **2**, 314–28.

IAN R. GOLDSMITH, Department of Geological Sciences, South Road, University of Durham, Durham DH1 3LE, UK.

PETER KING, Reservoir Technology Branch, BP Research Centre, Chertsey Road, Sunbury on Thames, Middlesex TW16 7LN, UK.

Some aspects of diagenetic sulphate-hydrocarbon redox reactions

H. G. Machel

SUMMARY: Sulphate-hydrocarbon redox-reactions occur at two specific diagenetic temperature/thermal maturity levels: less than about 75–85°C (0.2–0.3% R_0), and more than 100–140°C (>1.5% R_0), respectively. In low-temperature/maturity environments these redox reactions take place only with the mediation of bacteria. In high-temperature/maturity environments these reactions take place thermochemically, and certain catalysts must interact in order to overcome the high activation energies and to sustain the reactions at geologically significant rates.

The reaction products and by-products may be identical for both temperature/maturity levels: altered and oxidized hydrocarbons (including bitumen), hydrogen sulphide, metal sulphides (including Mississippi Valley Type deposits), elemental sulphur, carbonates (mainly calcite and dolomite), and other minerals. An important by-product of these redox reactions may be porosity resulting from the dissolution of solid sulphates and/or the carbonate host rock. The net reaction is exothermic, and the released heat may generate a geothermal hot-spot in some cases.

Introduction

The association of dissolved sulphates and hydrocarbons is thermodynamically unstable in diagenetic environments (for the purpose of this paper, the term hydrocarbons includes any type of organic matter, carbohydrates, kerogen, crude oil, bitumens, dissolved and gaseous organic compounds). Hence, redox reactions occur, either with or without the mediation of bacteria. For simplicity, these reactions are discussed from the vantage point of sulphate reduction, which always implies concomitant hydrocarbon oxidation. Bacterially mediated sulphate reduction is called BSR (bacterial sulphate reduction), and abiologically mediated sulphate reduction is called TSR (thermochemical sulphate reduction). TSR is also called non-biogenic (Barton 1967), non-microbial (Orr 1974), organic (Ohmoto & Rye 1979), thermal (Siebert 1985), and abiological (Trudinger *et al.* 1985) sulphate reduction.

BSR was discovered by a bacteriologist who published his findings in a journal of bacteriology, parasitology, infectious diseases, and hygiene (Beijerinck 1895). BSR occurs in a large variety of sedimentary and low-temperature (less than about 85°C) diagenetic environments including ground water aquifers (e.g. Champ *et al.* 1979), marine sediments (e.g. Berner 1980), reefal carbonates and layered or diapiric evaporites (e.g. Feely & Kulp 1957, Sassen 1980), and clastic rocks (e.g. Coleman 1985). First data on TSR were presented by Toland (1960) who performed hydrothermal experiments with a variety of dissolved sulphates and hydrocarbons. Later experiments showed that TSR can take place at temperatures at least as low as 175°C (summarized in Trudinger *et al.* 1985). Geological evidence, however, suggests lower minimum temperatures, perhaps 100–140°C (e.g. Powell & Macqueen 1984, Siebert 1985, Machel & Krouse, in prep. a). In either case, TSR is restricted to high-temperature diagenetic and hydrothermal environments and probably is the main process generating large quantities of hydrogen sulphide in numerous deep subsurface sour gas provinces of the world (Orr 1977). Presumably TSR is also involved in the formation of sulphide ore deposits, i.e. some Mississippi Valley Type deposits (Powell & Macqueen 1984).

Despite their frequent occurrence, some aspects of sulphate-hydrocarbon redox reactions are not well understood. Evidence for reaction paths, products, and conditions is scattered in the literature and partly contradictory, especially regarding the associated changes in pH, carbonate alkalinity, and precipitation or dissolution of carbonates. A major objective of the present paper is to clarify some of these problems. Secondly, it is not clear under what circumstances sulphate-hydrocarbon redox reactions proceed without the mediation of bacteria. In particular, Trudinger *et al.* (1985) suggested that TSR may not occur at geologically significant rates at diagenetic temperatures because of kinetic barriers and the paucity of reactive hydrocarbons. Therefore, another objective of the present paper is to discuss reaction kinetics and catalysts for TSR. Elemental and isotopic

From: MARSHALL, J. D. (ed.), 1987, *Diagenesis of Sedimentary Sequences*, Geological Society Special Publication No. 36, pp. 15–28.

TABLE 1. *Reaction scheme for bacterial and thermochemical sulphate reduction. See text for explanation*

Reaction 1:	paraffinic hydrocarbons	→	biodegraded hydrocarbons
Reaction 2:	crude oil	→	light crude oil + H_2S + CH_4
Reaction 3 a:	$4R\text{-}CH_3 + 3SO_4^{2-} + 6H^+$	→	$4R\text{-}COOH + 4H_2O + 3H_2S$
b:	$R\text{-}CH_3 + 2R{=}CH_2 + CH_4 + 3SO_4^{2-} + 5H^+$	→	$3R\text{-}COOH + HCO_3^- + 3H_2O + 3H_2S$
c:	$2CH_2O + SO_4^{2-}$	→	$2HCO_3^- + H_2S$
Reaction 4 a:	$2H_2S + O_2$	↔	$2S^\circ + 2H_2O$
b-1:	$3H_2S + SO_4^{2-} + 2H^+$	↔	$4S^\circ + 4H_2O$
b-2:	$H_2S + SO_4^{2-} + 2H^+$	↔	$S^\circ + 2H_2O + SO_2$
c:	H_2S + hydrocarbons	→	S° + altered hydrocarbons
d:	S^{2-}	→	S°
Reaction 5:	$4S^\circ + 1.33(\text{-}CH_2\text{-}) + 2.66\ H_2O + 1.33\ OH^-$	→	$4H_2S + 1.33\ HCO_3^-$
Net Reaction:	hydrocarbons + SO_4^{2-}	→	altered hydrocarbons + bitumen HCO_3^- + H_2S (+CO_2?) + heat

characteristics of the reaction products of BSR and TSR will be discussed in a forthcoming paper (Machel & Krouse, in prep. b). Some of these characteristics are mentioned below where necessary.

Important reactions

Many reactions have been proposed for BSR and TSR. However, no previous reaction scheme provides a satisfactory explanation for all the important reactants and products occurring in diagenetic environments. Therefore, a reaction scheme consisting of previously published and of modified reactions is suggested that are believed to be a better representation of sulphate-hydrocarbon redox reactions in diagenetic environments (Table 1). The reactions in Table 1 have been (a) demonstrated experimentally, (b) observed in natural environments, (c) calculated thermodynamically, or (d) inferred on the basis of elemental or isotopic evidence, as indicated in each case for verification. The redox-steps in reactions 3, 4, and 5 (Table 1) are valid for bacterial *and* for abiological sulphate-hydrocarbon redox reactions, as are the major reaction paths of many redox reactions (e.g. the formation of pyrite: Morse *et al.* 1987). BSR and TSR are therefore discussed together for each of these redox-steps. However, the reaction scheme in Table 1 is valid only for systems that are initially free of base and transition metals (the addition of these metals is discussed separately), and metastable or transient reaction products such as polysulphides are not included. It should also be kept in mind that each reaction in Table 1 is a net mass and charge balance reaction consisting of several sub-reactions. These sub-reactions are omitted from Table 1 for simplicity.

Reaction 1: biodegradation of hydrocarbons

Biodegradation of hydrocarbons is not a prerequisite where methane is the main carbon source, or where sulphate reduction proceeds thermochemically. However, biodegradation of hydrocarbons is a prerequisite for BSR in environments devoid of methane for two reasons: (1) sulphate-reducing bacteria generally cannot digest *n*-paraffins, which are common in non-biodegraded hydrocarbons such as crude oil; and (2) sulphate-reducing bacteria depend on the metabolic residues of hydrocarbon biodegradation as nutrients e.g. various organic acids (experimentally demonstrated by Bailey *et al.* 1973, Nazina *et al.* 1985, Jobsen *et al.* 1979). This is supported by evidence from natural environments of BSR where most hydrocarbons are biodegraded oils, i.e. typically heavy to medium gravity naphthenic oils with most *n*-paraffins and isoprenoids removed (Davis & Kirkland 1970, Philippi 1977). Accordingly, reaction 1 (Table 1) represents the sequential and partially overlapping oxidation of (a) *n*-paraffins, (b) isoprenoids, and (c) lower-ring naphthenes and aromatics with increasing extent of biodegradation. As used in this context, the term biodegradation designates hydrocarbon oxidation/decomposition by *aerobic* bacteria. Hydrocarbon decomposition by fermenting, sulphate reducing, or methanogenic bacteria is *not* called biodegradation.

Reaction 2: thermal cracking of crude oil

Reaction 2 (Table 1) is thermal cracking of crude oil which, among other reaction products, results in the formation of methane and (a few per cent) H_2S (e.g. Orr 1977). Reaction 2 may be involved in TSR because the resulting H_2S is a catalyst for reactions 3 and 5 via reaction 4b (experimentally shown: e.g. Toland 1960). Reaction 2 is not necessary for BSR.

Reaction 3: initial S–O bond rupture and reduction of S^{6+}

Reaction 3 consists of several sub-reactions between sulphates and hydrocarbons involving the initial S–O bond rupture and reduction of S^{6+} to lower valence states (most S^{6+} is reduced to S^{2-}). Reaction rates vary for different sulphate and hydrocarbon species, and not all species are reactive. Firstly, only *dissolved* sulphate can be utilized during BSR and TSR. Low-temperature experiments demonstrate that sulphate-reducing bacteria cannot directly decompose solid sulphates, although they can solubilize solid sulphates via removal (reduction) of sulphate from solution (experimentally shown by Bolze *et al.* 1974 and McCready & Krouse 1980). Similarly, TSR of solid sulphates is not possible in diagenetic environments: non-aqueous mixtures of hydrocarbons and solid sulphates remain unreactive even at temperatures of 180–315°C (Toland 1960). Furthermore, the reaction rate depends on the specific sulphate species in solution, and some species (i.e. $NaSO_4^-$) appear to be non-reactive even at temperatures up to 300°C (Kiyosu 1980). Regarding the hydrocarbons, a clear distinction has to be made between BSR and TSR. Sulphate reducing bacteria can utilize a large variety of low-molecular weight organic compounds including many products of aerobic biodegradation, but generally not *n*-paraffins (e.g. Jobson *et al.* 1979, Peck 1984, Nazina *et al.* 1985). On the other hand, relatively few hydrocarbons are known to react abiologically with sulphate, but they include *n*-paraffins: low-*n*-alkanes (i.e. methane, ethane), low-*n*-alkenes, *n*-octadecane, carbohydrates (i.e. sugar), alkylated aromatic compounds, and other, partly oxygenated compounds (experimentally shown by Toland 1960, Davis & Yarbrough 1966, Kiyosu 1980, Trudinger *et al.* 1985 and references therein). As in the case of the sulphate species, reaction rates vary for different hydrocarbons, and certain compounds are non-reactive (Toland 1960, Trudinger *et al.* 1985).

For these reasons, and because there are conflicting data in the literature regarding the reaction products, reaction 3 has been written in three alternative ways (Table 1). Reactive organic compounds are: alkanes/paraffins in reaction 3a; alkanes/paraffins, alkenes/olefins, and methane in reaction 3b; and carbohydrate in reaction 3c. Obviously, reactions 3a, 3b and 3c are not the only possibilities. For example, simultaneous and intermediate reactions are possible, or methane could be the only organic reactant. Excepting the last possibility, the main reaction products of reaction 3 are organic acids (i.e. carboxylic acids for paraffins, olefins, and alkylated aromatic compounds as reactants), bicarbonate ions, and hydrogen sulphide (experimentally and theoretically shown for BSR: e.g. Rickard 1969, Berner *et al.* 1970; for TSR: e.g. Toland 1960). One important reaction product is not included in Table 1: some TSR experiments produced 'black precipitates' (e.g. Toland 1960, Kiyosu 1980). The equivalent to these precipitates in natural environments is bitumen. Subordinate reaction products, also omitted from Table 1 for simplicity, are liquid and gaseous inorganic and organic sulphur compounds, including polysulphides (e.g. Toland 1960, Burnie 1979). Also, small amounts of CO_2 and CO gas evolved during BSR and TSR laboratory experiments (e.g. Toland 1960, Ward & Brock 1978, Kiyosu 1980). Where methane is the only organic reactant, all carbon is oxidized to carbonate species and some CO_2 and CO, and no bitumen is formed.

There is abundant geological evidence for an increase in carbonate alkalinity in solution, and for precipitation of carbonates, as a result of sulphate reduction in a variety of diagenetic environments (e.g. Berner *et al.* 1970, Curtis 1977, Berner 1980, Jeffries & Krouse 1984). Therefore, reactions 3a, 3b and 3c (Table 1) have been written in such a way as to allow for carbonate precipitation. For example, reaction 3a involves a net increase in pH because the organic acid released is very weak. This would lead to carbonate precipitation utilizing carbonate (and metal) ions that are already in solution (these carbonates would *not* contain organic carbon, as observed in some geological environments: Machel & Krouse, in prep. b). Alternatively, carbonates would precipitate as a result of an increase in carbonate alkalinity (release of

HCO_3^- with organic carbon) in the cases of reactions 3b and 3c. It should be kept in mind, however, that there is considerable controversy regarding the changes of pH and carbonate alkalinity, and regarding carbonate precipitation or dissolution resulting from sulphate reduction. For example, experimental evidence suggests that the pH may increase or decrease during BSR depending on the types of hydrocarbons utilized by the bacteria (e.g. Birnbaum & Wireman 1984). Chemical models are partly contradictory and suggest carbonate dissolution or precipitation depending on the involvement of other phases and components, i.e. silicates, carbonates, sulphides, and ammonia. For example, Berner *et al.* (1970) predicted an increase in carbonate alkalinity and concomitant carbonate precipitation in organic-rich silicate muds. The data of Berner *et al.* (1970) also suggest that carbonate precipitation occurs in chemically 'inert' and carbonate sediments. On the other hand, Gardner's (1973) model predicts carbonate dissolution in 'inert' and carbonate sediments as a result of sulphate reduction. It also predicts carbonate precipitation along with sulphide precipitation in sediments containing iron oxides. This is at variance with the models and observations of Anderson (1983) and Anderson & Garven (1987) who predict carbonate dissolution if sulphides precipitate, unless the acid released during sulphide precipitation is consumed by other reactions, i.e. by sulphate reduction or reactions involving silicates. Reactions 3a, 3b and 3c (Table 1) conform with the models of Berner *et al.* (1970), Anderson (1983) and Anderson & Garven (1987), and with the evidence from most natural environments of BSR/TSR, in that carbonates precipitate where base and transition metals are scarce or absent.

Reaction(s) 3 represent BSR and TSR. In the case of BSR, the temperature must be cool enough to permit bacterial metabolism, that is less than about 45°C for most sulphate-reducers, although some are known to metabolize between about 60 and 92°C (Peck 1984, Trudinger *et al.* 1985, Nazina *et al.* 1985, Stetter *et al.* 1987). Furthermore, the environment must be anaerobic because sulphate-reducing bacteria cannot grow under aerobic conditions (experimentally shown: e.g. Jobson *et al.* 1979). On the other hand, sulphate-reducing bacteria are a taxonomically and physiologically extremely diverse group of microorganisms consisting of many genera (Desulfo-x) that can grow over a wide range of temperatures (about +0–92°C), salinities, and water compositions (Peck 1984), which explains their common occurrence in low-temperature diagenetic environments. All sulphate-reducing bacteria use sulphate as terminal electron acceptor to yield sulphide, but some reduce nitrate, nitrite, and other compounds and can grown even in the absence of detectable sulphate (Peck 1984, Postgate 1984). For reaction(s) 3 to proceed thermochemically, catalysts and temperatures in excess of 100–140°C are necessary (see discussion of kinetics, below).

Reaction 4: partial oxidation of S^{2-} to S°

Sulphide-sulphur oxidation may take place in several alternative ways. Reactions 4a and 4b represent reversible reactions for abiological oxidation of sulphide, step 4c represents the non-reversible abiological oxidation of sulphide by hydrocarbons, and step 4d represents a variety of non-reversible bacterial reactions.

Reaction 4a: Abiological oxidation of sulphide with molecular oxygen is possible where H_2S escapes into the atmosphere, and where H_2S comes into diffusive or hydrological contact with dissolved oxygen (Berner 1980, Morse *et al.* 1987). Abiological sulphide oxidation with dissolved oxygen was experimentally demonstrated for temperatures as low as 70°C, and it is thermodynamically likely for temperatures in excess of 25°C (Davies *et al.* 1970). Importantly, elemental sulphur should be formed only in environments of relatively low partial pressure of oxygen, or where oxygen is supplied relatively slowly to the reaction site (reaction 4a). Where oxygen is abundant, hydrogen sulphide is oxidized to sulphuric acid according to

$$H_2S + 2O_2 \leftrightarrow 2H^+ + SO_4^{2-}.$$

This reaction is omitted from Table 1 because elemental sulphur is not a reaction product.

Reaction 4b: Abiological oxidation with excess dissolved sulphate at low pH is not likely at low temperatures but possible during TSR (experimentally and theoretically shown by Husain 1967, Davies *et al.* 1970, Husain & Krouse 1978). Water and elemental sulphur are the only reaction products where the hydrogen sulphide concentration is high (reaction 4b-1). Sulphur dioxide is produced along with elemental sulphur where little hydrogen sulphide is available (reaction 4b-2) (experimentally shown by Husain & Krouse 1978).

Reaction 4c: Above about 100°C, H_2S reacts abiologically with (mainly saturated) hydrocarbons to produce elemental sulphur and NSO-compounds (organic compounds containing N, S, O, e.g. resins), if catalysts such as silica gel and/or clays are present (experimentally shown: e.g. Rudakova & Velikovskii 1947, Ho *et al.* 1974).

Reaction 4d: Bacterial oxidation of sulphide was experimentally shown and observed in natural environments (Chen & Morris 1972, Pawlowski *et al.* 1979, Sieburth 1979, Paull *et al.* 1984). These bacteria may be anaerobes (e.g. photosynthetic bacteria, *Chromatium*) or aerobes (*Thiobacillus*, *Beggiatoa*, and other genera). However, even bacteria that need free oxygen live as microaerophiles or anaerobes, because a thin and almost anaerobic aqueous boundary layer surrounds the colonies even in turbulent water (Jorgensen & Revsbech 1983). Oxygen reaches the bacteria only via diffusion through this boundary layer. Another important factor for bacterial sulphide-sulphur oxidation is pH. Chen & Morris (1972) suggested that the reaction rate during aerobic bacterial oxidation of sulphide is highest at intermediate pH (about 8). However, the optimum pH depends on the type of bacteria carrying out oxidation. For example, *Beggiatoa* and *Thiotrix* thrive best at slightly alkaline pH, but *Thiobacillus* thrives best in rather acidic environments (Larkin, pers. comm. 1986).

The electron acceptors during bacterial sulphide-sulphur oxidation may be different for the various genera and environments, and usually involve several enzyme-catalysed steps. Therefore, the simplest representation of the reaction path during bacterial sulphide-sulphur oxidation is the valence change of sulphur (reaction 4d). Regarding mass balance, however, reaction 4a may be taken to represent bacterial sulphide-sulphur oxidation in natural oxygenated environments, and reaction 4c may be taken to represent anaerobic bacterial sulphide-sulphur oxidation. Reaction 4b is probably not an alternative in this context, because no bacteria are known that can couple sulphide oxidation with sulphate reduction (Larkin, pers. comm. 1986).

Reaction 5: reduction of S° to S^{2-}

At temperatures in excess of about 100°C, elemental sulphur is an active oxidizing agent for many organic compounds (represented by the methylene functional group in Table 1), resins and asphaltenes generally being the most reactive (experimentally shown with bacteria: e.g. Stetter & Gaag 1983, Peck 1984; and abiologically/thermochemically: Pryor 1962, Douglas & Mair 1965, Valitov 1974). Therefore, reaction 5 (Table 1) is valid for BSR and TSR. As in the case of reaction 3, reaction 5 is a very simplified representation of the involved redox-steps. Major reaction products are H_2S and bicarbonate ions (and some CO_2 and CO). Experiments also produced 'viscous black tar' and 'dark viscous gum' with up to 11% S (Douglas & Mair 1965). Similarly, sulphurized bitumens and carbonates with isotopically light organic carbon are found in natural environments of BSR and TSR (e.g. Powell & Macqueen 1984, Sassen in press). Furthermore, reaction 5 could destroy light saturated hydrocarbons and generate naphthenic acid, aromatic compounds, and numerous inorganic and organic sulphur compounds (polysulphides, mercaptans, thialkenes, thiopenes, and others: e.g. Ho *et al.* 1974, Burnie 1979).

Net mass balance

Using a reaction such as reaction 3c (Table 1) as a net mass balance reaction is correct (e.g. Berner *et al.* 1970), but it does not account for all important reaction products. Reactions 4 and 5 do not seem to be necessary to account for net sulphate reduction, but these reactions are likely to occur simply because the oxidizing bacteria or oxidizing agents are present. Once instigated, reactions 3, 4 and 5 proceed simultaneously because some of the reaction products (i.e. hydrogen sulphide, elemental sulphur) are used up as they are generated. Therefore, reactions 3, 4 and 5 can be summarized in a net mass balance reaction (Table 1) representing the reaction products if all sulphur generated during reaction 4 is used up in reaction 5. Hydrocarbons are expressed with words rather than formulae because there are numerous possibilities to write and balance the net reaction. Important reaction products are altered (non-polymerized) hydrocarbons, bitumen, HCO_3^-, and H_2S. No bitumen is formed where methane is used as the only carbon source. Some carbon may evolve as carbon dioxide gas. The net reaction is exothermic (discussed below).

For simplicity, polysulphides and other compounds with intermediate valence states of sulphur have not been included in the reaction scheme of Table 1 although they are undoubtedly involved (e.g. Davis & Kirkland 1970, Orr 1974, Krouse 1977, Morse *et al.* 1987). For example, elemental sulphur and excess sulphide may react abiologically to form polysulphides (experimentally shown by Chen & Morris 1972). Polysulphides, generated in this way, or during reactions 3 or 5, may react with bicarbonate to form elemental sulphur (Davis & Kirkland 1970), and polysulphides may be used in reaction 5 instead of S° (experimentally shown: e.g. Orr 1977). Furthermore, various bacteria may mediate redox-steps that are not shown in Table 1. For example, bacteria oxidizing sulphide to sulphur (reaction 4) may be accompanied by other bacteria (e.g. certain species of *Thiobacillus*, *Thiomicrospora*, and other genera: Sieburth 1979) which further oxidize the elemental sulphur to sulphate if the environment obtains some dissolved oxygen, and should there be an appropriate balance between oxygen, sulphide, carbon dioxide, organic compounds, and pH (experimentally shown by Iwatsuka & Mori 1960, Nakai & Jensen 1964, Ivanov 1968, Larkin 1981). Of course, oxidation of sulphur to sulphate by oxygen may also proceed abiologically, but this takes place much more slowly (Chen & Morris 1972).

Geological and geochemical implications

The reactions listed in Table 1 have a number of geological and geochemical implications, depending on the types of reactive hydrocarbons, presence or absence of base and transition metals, presence of catalysts, and temperature (mainly controlling bacterial metabolism and kinetics of TSR).

Initial order of reactions for non-gaseous hydrocarbons

Regarding BSR with non-gaseous hydrocarbons as a carbon source (i.e. crude oil, but also solid and dissolved hydrocarbons), Table 1 is a simplification of complex associations of bacteria performing specific redox-steps (including intermediate redox-steps that are not shown for simplicity). Most bacteria mediating reactions 1 and 4 are aerobes, whereas those mediating reactions 3 and 5 are anaerobes. Initially, the normal sequence of reactions for BSR would be 1–3. If the environment obtains some oxygen (i.e. at the groundwater/oil interface), reaction 4a will proceed to form S°, which will accumulate as a net reaction product only if reaction 5 is retarded or inhibited. This may take place if the bacteria necessary for reaction 5 are not present, or if the environment is not conducive to growth. This, in turn, depends mainly on a sufficient supply of suitable nutrients (including specific hydrocarbons that can be metabolized). If the environmental conditions are appropriate, reactions 3, 4 and 5 proceed simultaneously at or near the aerobic/anaerobic interface, which may be an oil/water contact or a diffusion zone within anaerobic sediments overlain by aerobic water. Should the whole environment become closed and totally anaerobic, reactions 3 and 5 will take over until the nutrients and reactants are depleted.

The initial sequence of reactions is different and possibly variable for TSR of non-gaseous hydrocarbons. This depends mainly on the initial availability of 'catalytic' H_2S, which may form a number of partially reduced sulphur compounds (S°, polysulphides, sulphite, thiosulphate) by reaction with dissolved sulphate (similar to reaction 4b, Table 1; experimentally shown: e.g. Toland 1960). These compounds appear to be more reactive than SO_4^{2-} or H_2S, forming H_2S, carboxylic acids, CO_2, NSO-compounds, and other polysulphides via reactions 3 and 5. Hence, two possibilities exist for the initial sequence of reactions during TSR, assuming that reaction 2 is the most likely process to provide 'catalytic' H_2S in deep subsurface diagenetic environments: 2, 4b and/or 4c, 5, 3; or 2, 4b and/or 4c, 3, 5 (note: reduced sulphur other than H_2S, such as NSO-compounds from early diagenesis, may also act as catalysts: Burnie 1979). Whatever the initial sequence, reactions 3, 4 and 5 proceed simultaneously once they are instigated because H_2S generated in reactions 3 and 5 is (at least partially) recycled in reaction 4. As in the case of BSR, S° accumulates as a net reaction product only if reaction 5 does not proceed or is retarded. This happens if the hydrocarbons suitable for TSR are no longer available, or if they are not supplied fast enough (see discussion of reaction kinetics, below).

Methane

In natural environments, methane is the only gaseous hydrocarbon abundant enough for large-scale sulphate reduction. Methane could be derived from methanogenic bacteria (in low temperature/maturity environments), or from thermal cracking of hydrocarbons (reaction 2: in high-maturity environments). The first case was observed or inferred in natural BSR environments on the basis of isotopic and circumstantial evidence (e.g. Barnes & Goldberg 1976, Kirkland & Evans 1976). On the other hand, thermodynamic calculations indicate that TSR should take place between methane and dissolved sulphate at temperatures lower than 200°C (Barton 1967), and geological examples occur in several sour gas provinces in the world (e.g. Orr 1977). In both cases, reactions 3 to 5 can be simplified because bitumen would not be formed: all methane would be oxidized to HCO_3^- (and/or CO_2, CO).

Stable reaction products and by-products

H_2S evolves as a separate gas phase if the system does not contain, or has used up, base and transition metals. During BSR, H_2S could be generated as long as reactants and nutrients for the bacteria are available, and as long as the H_2S concentration is below the toxic level for bacterial metabolism. Hence, for large quantities of H_2S to be formed, the system must be open and allow for continuous inward diffusion of sulphate as well as a sink for hydrogen sulphide. In fact, H_2S in most natural environments of BSR is bonded as metal sulphides, organic compounds, or escapes as gas from the reaction site, and does not accumulate as reservoir gas (e.g. Orr 1977, Krouse 1980). However, the possibility of biogenic H_2S migrating into a shallow trap and forming a sour gas reservoir cannot be ruled out. On the other hand, deep reservoirs with more than a few per cent H_2S (in reservoir gas) are suspected to have undergone TSR. Using crude oil with about 3% sulphur as representative of sulphur-rich crude, Orr (1977) calculated that no more than 2–3 vol-% H_2S (in reservoir gas) can evolve from thermal cracking of crude oil. Crude oil may contain more than 3% sulphur (Sassen, pers. comm. 1986) and perhaps release more than 3% H_2S, but deep reservoirs with 20–80% H_2S are invariably due to TSR (Orr 1977).

Elemental sulphur accumulates as a net reaction product if reaction 5 is inhibited. This happens if the system runs out of reactive hydrocarbons (BSR/TSR), or if reactive hydrocarbons are not supplied fast enough (TSR).

BSR and TSR form many distinctive NSO-compounds (discussed in Ho *et al.* 1974, Burnie 1979). Some of these compounds are contained in the solid organic precipitates (bitumens) which should be formed where hydrocarbons other than methane are utilized. These bitumens must be differentiated from those generated by biodegradation or thermal maturation on the basis of elemental and isotopic criteria (which has been attempted in isolated cases: Macqueen & Powell 1983, Powell & Macqueen 1984, Sassen 1986 and in press, Machel & Krouse, in prep. b).

Several minerals could precipitate as the result of BSR/TSR if the respective metal ions are (a) present or (b) transported to the reaction site, or (c) if the reaction products of BSR/TSR are transported into an environment containing metal ions. Firstly, the presence of alkali earth metals will result in precipitation of carbonates (mainly calcite and dolomite), either as cement or as replacement of the dissolving sulphates (mainly gypsum and anhydrite), as a direct result of reactions 3 and 5 because of the generated bicarbonate. Additionally, Davis & Kirkland (1970) suggested that the reaction of polysulphides with bicarbonate also leads to calcite precipitation, and Schneider & Nielson (1965) suggested that bicarbonate may form as a result of bacterial oxidation of sulphide to sulphur (reaction 4), with subsequent precipitation of this bicarbonate as calcite. Other carbonates (ankerite, siderite, witherite, strontianite) are also formed as a result of BSR/TSR where the respective metals are available.

If transition or base metals are present, disseminated or stratiform base metal or Mississippi Valley type deposits could be formed. Minerals such as pyrite, galena, and sphalerite are precipitated as a result of sulphide generation during reactions 3 and 5. In these cases, partial dissolution of the host rock should occur (at least initially), because the precipitation of sulphides generally are strongly acid-generating reactions (several related phenomena are discussed in Anderson 1983, Coleman 1985, Anderson & Garven 1987, Morse *et al.* 1987). In addition, BSR/TSR themselves could generate acidity due to the release of CO_2, perhaps in reaction 3. However, the low acidity resulting from this carbon dioxide is probably negligible in most natural environments. On the other hand, much larger quantities of CO_2 evolve from thermal cracking of kerogen or crude oil (e.g. Krouse 1983). Hence, dissolution by thermogenic carbonic acid in deeply buried hydrocarbon reservoirs may compound potential dissolution caused by precipitation of sulphides.

Subordinate reaction by-products that form in the vicinity of BSR and TSR as a result of released and/or consumed phases are cerrusite, barite, fluorite, and gases such as nitrogen and helium (e.g. Barton 1967, Dunsmore 1971, Anderson 1983, Siebert 1985).

Naturally, one or several of these reaction products and by-products may be absent. The availability and mass proportions of the reactants determine which products and by-products are formed, and in which proportions.

Temperature ranges and reaction kinetics of BSR and TSR

The major reaction pathways are similar for biological and abiological sulphate-hydrocarbon redox reactions. However, BSR and TSR appear to be mutually exclusive processes in natural environments. One of the best indications for this phenomenon are the natural occurrences and isotopic compositions of H_2S, suggesting that BSR and TSR take place at two particular temperature/thermal maturity levels: at low levels of less than about 75–85°C (equivalent to about 0.2–0.3% R_0), and at high levels in excess of 100–140°C (>1.5% R_0) (Burnie 1979, Krouse 1980, Sassen 1985, see Fig. 1; note: these temperature-maturation correlations are only generalized approximations because of the time-dependence of thermal maturation).

The lower temperature/maturity level coincides with BSR, because sulphate reducing bacteria cannot thrive at temperatures of the upper level. Bacteria can survive up to temperatures at which their cytoplasm does not boil

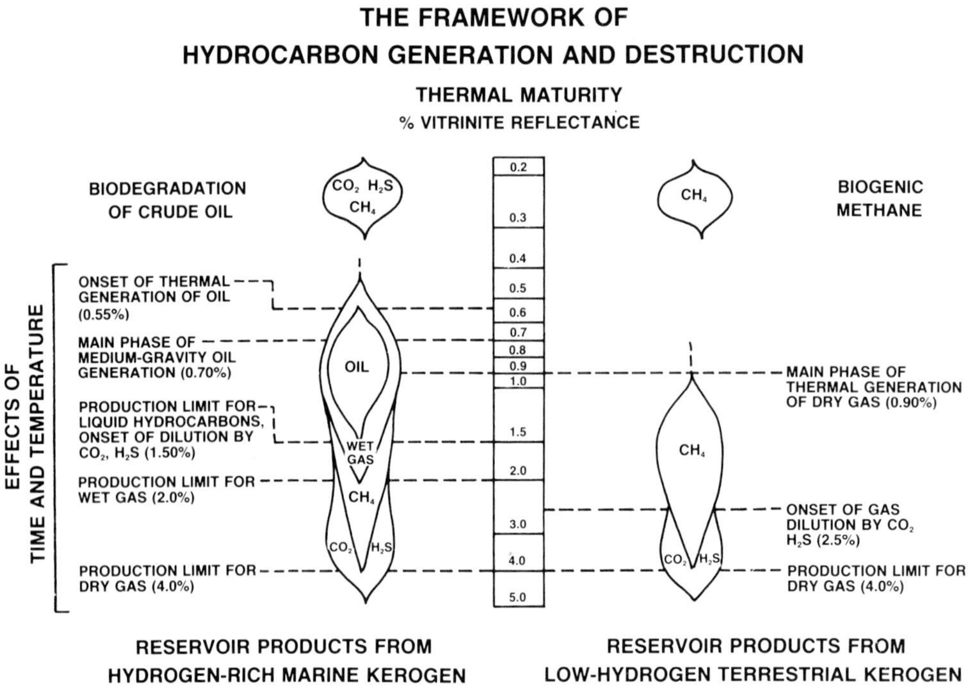

FIG. 1. Generalized relationships between thermal maturity and hydrocarbon generation and destruction. Key vitrinite reflectance values useful in hydrocarbon exploration are shown in parentheses. Biogenic methane and the gases resulting from oil biodegradation are also included because they are significant reservoir constituents. Note that H_2S is formed mainly in association with marine kerogen at low maturity levels (0.2–0.3% vitrinite reflectance), and at high maturity levels (greater than about 1.5% vitrinite reflectance). Reproduced from Sassen (1985).

(which may be in excess of 100°C if the pressure is high enough), but they become dormant as a certain temperature is exceeded. No sulphate-reducing bacteria are known that metabolize at temperatures in excess of about 92°C, and most sulphate-reducing bacteria do not metabolize at temperatures in excess of about 45°C (at 1 atm: Peck 1984, Trudinger *et al.* 1985, Nazina *et al.* 1985, Stetter *et al.* 1987). Hence, BSR can take place only in near-surface and shallow subsurface environments. In these environments, BSR proceeds almost instantaneously on a geological time-scale and does not depend on any catalysts (other than enzymes within the cells). This was demonstrated experimentally in the laboratory (e.g. Nakai & Jensen 1964), and in the field. For example, water injection for enhanced petroleum recovery in several shallow Russian and Canadian oil fields resulted in rapid H_2S production. Within less than 10 years, up to about 10% H_2S was generated in previously sweet reservoirs because of the introduction of bacteria into hitherto sterile reservoirs (Ashirov 1962, Krouse 1980).

The upper temperature/thermal maturity level ($>$100–140°C, $>$1.5% R_0) coincides with TSR which is very slow or inhibited at low temperatures, even though the reactions may have a large negative free energy change of reaction. The activation energy is also very large and differs for various hydrocarbons (experimentally shown: e.g. Toland 1960). For TSR of an 'average' hydrocarbon, the activation energy has been estimated to be about 50 kcal (Dhannoun & Fyfe 1972). Acordingly, TSR has been performed in the laboratory only at high temperatures (in excess of 175°C), and geologically significant reaction rates were measured only above about 250°C. On the basis of this evidence, and the apparent paucity of reactive hydrocarbons, Trudinger *et al.* (1985) concluded that TSR may not be possible at geologically significant rates in natural diagenetic environments. On the other hand, Dunsmore (1971), Powell & Macqueen (1984), Machel (1985), Machel & Krouse (in prep. a), and Sassen (in press), as well as a number of other studies (Kartsev *et al.* 1959, Gutsalo & Krivosheya 1966, Barton 1967, Andreev *et al.* 1968, Dunsmore 1971, Reznikov 1971, Orr 1974, 1977, 1982, Drean 1978, Ohmoto & Rye 1979, Anderson, 1983, Macqueen & Powell 1983, Krebs & Macqueen 1984, Powell & Macqueen 1984), have provided cumulative geological, theoretical, and circumstantial evidence which is considered sufficient proof by the present author for the occurrence of TSR at diagenetic temperatures as low as at least 135–140°C (three examples are discussed in the section on 'Selected geological examples', below). The discrepancy between this writer's and Trudinger *et al.*'s (1985) assessment is based on the role catalysts and kinetic factors play that are not present or involved in most laboratory experiments. There is no shortage of potential catalysts in natural environments and some have been identified experimentally for TSR. Also, the reaction chain in Table 1 contains several reaction loops that provide for 'autocatalysis', in the sense that reaction products of one step may increase the reaction rate and degree of completion of another step. The main kinetic factors and catalysts can be summarized as follows (this list is certainly incomplete).

(1) Salts (soaps) of organic acids are known to catalyse TSR in an unknown manner (Kartsev *et al.* 1959). Also, the formation of soaps could lower the pH and thus drive reaction 3 to the right.
(2) The formation of compounds using up reaction products of reversible reactions would drive these steps to the right. For example, S° is used up by polysulphide formation, favouring reactions 4a and 4b. Similarly, S° is used up in reaction 5 and by bitumen, and by the resin and asphaltene fractions in crude oil ('sulphurization': Kaplan *et al.* 1963, Orr 1974, Milner *et al.* 1977, Powell & Macqueen 1984), again favouring reactions 4a and 4b. Of course, sulphurization also occurs with sulphur compounds of valence states other than zero, but this does not favour any of the reactions in Table 1.
(3) Formation of certain compounds by one reaction increases the rate and degree of completion of another reaction. For example, formation of elemental sulphur in reaction 4 increases the rate and completion of reaction 5 which uses up sulphur.
(4) Formation of complexes, including polysulphides, is known to catalyse redox-reactions (Stumm & Morgan 1970, Berner 1971, Morse *et al.* 1987).
(5) Organic acids are involved in a variety of processes, i.e. complexing of metals, decarboxylation, and oxidation by metals (e.g. Hanor & Workman 1986), some of which may catalyse TSR (and perhaps BSR).
(6) The oxidation of H_2S is known to be catalysed by metal ions in water in the sequence $Ni^{2+} > Co^{2+} > Mn^{2+} > Cu^{2+} > Fe^{2+} > Ca^{2+} = Mg^{2+}$, and by organic compounds such as phenols, aldehydes, aniline, urea, and vanillin (Morse *et al.* 1987).
(7) Clay minerals (particularly montmorillonite) and silica gel have been shown experimentally to catalyse TSR (Rudakova & Velikovskii 1947, Dhannoun & Fyfe 1972).
(8) H_2S from thermal decomposition of hydrocarbon NSO-compounds (reaction 2) drives step 4 to the right (also invoked by Orr 1974, Ohmoto & Rye 1979, Powell & Macqueen 1984).
(9) Reactions 3 and 4b are driven to the right by low pH (represented by H^+ on the left sides; experimentally shown by Toland 1960; references in Trudinger *et al.* 1985). Therefore, precipitation of carbonates and particularly sulphides, which are associated with a decrease in pH (e.g. Anderson 1983, Coleman 1985, Siebert 1985) and often accompany TSR, would drive reactions 3 and 4b to the right. Of course, simultaneous thermal cracking of organic matter during maturation also generates acidity, as do other reactions not related to the redox-scheme of Table 1 (i.e. clay mineral diagenesis, and other processes: Giles & Marshall 1986). Acids derived from these processes may be generated within the system of BSR/TSR, or invade from elsewhere, and would also drive reactions 3 and 4b to the right.
(10) Reaction 5 is driven to the right at high pH, represented by OH^- on the left side (experimentally shown by Toland 1960).
(11) Oxidation of crude oil should proceed faster than oxidation of methane because crude oil is thermodynamically less stable than methane (Barton 1967).
(12) The rate of reaction 3 depends on the sulphate species (i.e. SO_4^{2-}, HSO_4^-, $NaSO_4^-$) and temperature (Kiyosu 1980), as well as on the type of hydrocarbons (e.g. Toland 1960, Trudinger *et al.* 1985). The rate of reaction 5 probably also depends on the types of hydrocarbons utilized.

Given the presence of at least some of the above catalysts and kinetic factors, and an intermediate pH, the reaction scheme could be triggered thermochemically. A minimum temperature of about 135–140°C (Machel 1985, Siebert 1985), perhaps only about 100°C (Rye & Williams 1981,

Macqueen & Powell 1983, Sassen 1986) is suggested for the onset of TSR even under the most favourable conditions (as inferred by these authors from fluid inclusion and thermal maturation data from carbonate rocks and Mississippi Valley type deposits of Canada, the US Gulf coast, and Australia). Once the reactions are instigated, several reaction loops provide for 'autocatalysis'. The most important loops are probably: sulphur generated in reactions 4b and 4c appears as reactant in reaction 5 driving this reaction to the right; the removal of sulphur from reaction 4b drives this reaction to the right; H_2S generated in reaction 5 appears as reactant in reactions 4b and 4c, again driving these reactions to the right (reaction 4a is not an alternative in this context because natural TSR environments are anoxic). Additionally, polysulphides form similar reaction loops. Therefore, the reaction rate of this scheme may increase considerably, provided the system begins with a surplus of the initial reactants: reactive sulphates and hydrocarbons. In fact, shortage of these reactants is the only factor that may inhibit exponential increase of the reaction rate. For example, if hydrocarbons seep to the reaction site only slowly, this could be the rate-limiting step. Similarly, if the sulphates dissolve or are supplied only very slowly, this will be the rate-limiting step.

Heat

The net reaction (Table 1) is exothermic. The heat released was estimated to be about 30 or 10 kcal mol^{-1} calcium sulphate by Feely & Kulp (1957) and Dhannoun & Fyfe (1972), respectively. It is not clear, however, whether these authors took into account the heat released during reactions 4 and 5, which may also be exothermic. This is suggested by observations that the bacterial oxidation of sulphur to sulphate (in quarry dumps of rocks containing elemental sulphur) is accompanied by a marked production of heat (Pawlowski *et al.* 1979). Hence, 10 kcal mol^{-1} calcium sulphate is a conservative estimate of the minimum heat released during TSR. This heat may be sufficient to generate a geothermal hot-spot at the reaction site, if the reactions proceed fast enough. This was invoked by several authors to explain present or past positive geothermal anomalies (e.g. Bush 1970, for Pine Point; Dunsmore 1971, for Keg River and Pine Point). These interpretations are supported by calculations (Dhannoun & Fyfe 1972) indicating that heat production in a 1 km thick rock column containing 30% calcium sulphate would be of the same order of magnitude as normal geothermal heat flow, if TSR occurred in the order of 1 Ma (taking 10 kcal mol^{-1} calcium sulphate as the heat released during TSR). Naturally, the faster the reaction rate, the greater the potential geothermal anomaly.

Selected geological examples

Because of the limited temperature tolerance of most bacteria, BSR has a relatively shallow depth limit in natural environments. The maximum depth for BSR is probably close to 3000 m. Scarce living microflora occurs down to depths of 3290 m (equivalent to about 85°C and a pressure of about 350 bar: Ashirov 1962, Rosanova & Khudyakova 1974), but these deep environments are anoxic, and large-scale BSR is not likely to occur because of the paucity of nutrients for the sulphate-reducers (from aerobic biodegradation of hydrocarbons). Hence sulphate reduction, the associated aerobic biodegradation of hydrocarbons, and aerobic sulphide-sulphur oxidation, are generally bracketed between the 'aeration zone' of exposed rocks (0–200 m depth) and the bottom of the oxygenated groundwater zone (generally about 600 m) (Andreev *et al.* 1968). Accordingly, carbonates, bitumen, and elemental sulphur resulting from BSR and partial reoxidation of sulphide occur generally in near-surface environments (i.e. in salt dome cap rocks of the United States and Europe: Feely & Kulp 1957, Dessau *et al.* 1962, Pawlowski *et al.* 1979) and along/below the oil-water interface in shallow oil reservoirs (i.e. in Russia: Ashirov 1962). Meteoric water is believed to be the vehicle to transport microbes into contact with subsurface hydrocarbon pools, i.e. via faults, fractures, and other conduits (e.g. Milner *et al.* 1977), and oxygenated groundwater probably is the cause of H_2S oxidation to S° in most shallow sulphur deposits. Where bitumen is absent, and the carbon isotope ratios of precipitated carbonates are lower than those of oil, methane is indicated as the main reactant hydrocarbon (e.g. in the limestone buttes of west Texas: Kirkland & Evans 1976). Perhaps both crude oil and methane may be utilized in some cases.

The natural occurrence of TSR is probably restricted to environments that have a minimum temperature of about 100–140°C. For mass balance reasons, any deeply buried sour gas reservoir with more than a few per cent H_2S is suspected to be formed by TSR. On the other hand, uplifted rocks that underwent sulphate-hydrocarbon redox reactions may have undergone BSR or TSR. Perhaps the most convincing examples for these two possibilities have been described from western Canada.

Examples for TSR at deep burial are the subsurface Keg River and Nisku Formations. Dunsmore (1971) showed for the Middle Devonian Keg River Formation that (a) the present reservoir temperatures of about 70–100°C, corresponding to depths of about 1500–2300 m, represent abnormally high temperatures (these temperatures and depths also represent the minima since uplifting and erosion began in the early Tertiary); (b) these abnormally high temperatures are not due to elevated heat flow from below; (c) most reaction products of TSR are present; and (d) the composition of present Keg River oil field brines and hydrocarbons, as well as sulphur isotope data, are consistent with TSR. On the basis of these results, Dunsmore (1971, p. 55) suggested that, at present, 'heat may be originating at the site of petroleum occurrence'. Dunsmore also concluded that the present geothermal anomalies may be inherited from TSR that has gone to completion at greater depths.

Regarding the Upper Devonian Nisku Formation, Machel (1985) and Machel & Krouse (in prep. a) demonstrated that sour gas (up to 30% H_2S), elemental sulphur at a former oil/water contact, and saddle dolomites with organic carbon and isotopically depleted oxygen, must have resulted from TSR. Importantly, these reservoirs are presently at depths in excess of about 3400 m, which are their shallowest depths since oil emplacement. Isotopic, fluid inclusion, and thermal maturation data indicate that (1) the maximum burial temperature was between 135 and 170°C; (2) these reservoirs have been hydrodynamically isolated since oil emplacement at similar depths in the late Cretaceous; and (3) no meteoric water intruded since then. Hence, BSR is not an alternative because these reservoirs were sterile at the time of sulphate reduction.

An example for BSR and TSR in an uplifted rock sequence is the Devonian Pine Point Mississippi Valley Type district in northwestern Canada. Macqueen & Powell (1983) and Powell & Macqueen (1984) presented isotopic and compositional data on reservoir bitumens, integrated into thermal maturation, petrographic, isotopic, and fluid inclusion data of the mineralic precipitates. Among other details, they could demonstrate that (a) the maximum temperature of the mineralized Pine Point district was about 100°C (perhaps 10–30°C higher: Macqueen, pers. comm. 1987), which represents a thermal palaeo-anomaly because the maximum temperature of the surrounding rocks was only about 60°C and (b) insoluble bitumens are enriched in isotopically heavy sulphur relative to their precursor bitumen. Considering all the data, these authors interpreted TSR to have taken place at a temperature of about 100°C (or somewhat higher, should the fluid inclusion temperatures be too low; Macqueen, pers. comm. 1987). The heat anomaly in the mineralized zone was not interpreted to be the result of TSR. Rather, heat was thought to be imported by the basinal fluids carrying the metals.

Conclusions

The preceding discussion sheds some light on the environmental conditions and kinetics of sulphate-hydrocarbon redox reactions. It is concluded that both bacterial and thermochemical sulphate reduction are common phenomena in diagenetic environments, and that these processes occur at mutually exclusive temperature/thermal maturity levels. Bacterially mediated redox reactions generally proceed much faster than thermochemical redox reactions. Given the presence of certain catalysts, however, thermochemical sulphate reduction may proceed at geologically significant rates and may even generate a geothermal hot spot.

Future work should focus on clarification of the specific roles of various catalysts for thermochemical sulphate reduction. Also, elemental and isotopic criteria should be developed that discriminate products of bacterial and thermochemical sulphate reduction from one another and from other processes (as discussed in Machel & Krouse, in prep. b).

ACKNOWLEDGMENTS: H. R. Krouse introduced me to the subject of sulphate-hydrocarbon redox reactions. Several enthusiastic diurnal and nocturnal discussions with him paved the way for this paper. A. Aplin, J. M. Larkin, R. W. Macqueen, R. Raiswell, R. Sassen and Z. Sofer critically read earlier versions of the manuscript, and their numerous suggestions greatly improved the final version. G. M. Anderson and J. W. Morse kindly provided unpublished manuscripts. The permission for reproduction of Fig. 1 by R. Sassen is greatly appreciated. Last but not least, C. H. Moore was a patient and constant source of encouragement without which this study would not have been possible. This research was supported by the Basin Research Institute of Louisiana State University.

References

ANDERSON, G. M. 1983. Some geochemical aspects of sulfide precipitation in carbonate rocks. *In*: KISVARSANYI, G., GRANT, S. K., PRATT, W. P. & KOENIG, J. W. (eds). *International Conference on Mississippi Valley Type lead-zinc deposits*. University of Missouri-Rolla, Rolla, Missouri, pp. 61–76.

—— & GARVEN, G. 1987. Sulfate-sulfide-carbonate associations in Mississippi-Valley-type lead-zinc deposits. *Economic Geology*, in press.

ANDREEV, P. F., BOGOMOLOV, A. I., DOBRYANSKII, A. F. & KARTSEV, A. A. 1968. *Transformation of Petroleum in Nature*. Pergamon Press, Oxford.

ASHIROV, K. B. 1962. Life activity of formational microflora as an index of geologic environment and processes obtaining in petroliferous formations. *In*: KUZNETSOV, S. I. (ed.). *Geologic Activity of Microorganisms*. Transactions of the Institute of Microbiology, **IX**, 84–91.

BAILEY, N. J. L., JOBSON, A. M. & ROGERS, M. A. 1973. Bacterial degradation of crude oil: comparison of field and experimental data. *Chemical Geology*, **11**, 203–21.

BARNES, R. O. & GOLDBERG, E. D. 1976. Methane production and consumption in anoxic marine sediments. *Geology*, **4**, 297–300.

BARTON, P. B. 1967. Possible role of organic matter in the precipitation of the Mississippi Valley ores. *In*: BROWN, J. S. (ed.). *Genesis of Stratiform Lead-Zinc-Barite-Fluorite Deposits*. Economic Geology, Monograph 3, 371–8. The Economic Publishing Company, Lancaster, PA.

BEIJERINCK, M. W. 1895. Über *Spirillum desulfuricans* als Ursache von Sulfat-reduktion. *Centralblatt für Bakteriologie, Parasitenkunde, Infektionskrankheiten und Hygiene, Abteilung I, Originale*, **1**, 1–9, 49–59, 104–14.

BERNER, R. A. 1971. *Principles of Chemical Sedimentology*. McGraw-Hill, New York.

—— 1980. *Early Diagenesis*. Princeton University Press, New Jersey.

——, SCOTT, M. R. & THOMLINSON, C. 1970. Carbonate alkalinity in the pore waters of anoxic marine sediments. *Limnology and Oceanography*, **15**, 544–9.

BIRNBAUM, S. J. & WIREMAN, J. N. 1984. Bacterial sulfate reduction and pH; implications for diagenesis. *Chemical Geology*, **43**, 143–9.

BOLZE, C. E., MALONE, P. G. & SMITH, M. J. 1974. Microbial mobilization of barite. *Chemical Geology*, **13**, 141–3.

BURNIE, S. W. 1979. A sulphur and carbon isotope study of hydrocarbons from the Devonian of Alberta, Canada. *PhD Thesis*, University of Alberta, Edmonton.

BUSH, P. R. 1970. Chloride brines from sabkha sediments and their possible role in ore formation. *Institute of Mining and Metallurgy Transactions Section B*, **79**, 137–44.

CHAMP, P. R., GULENS, J. & JACUSON, R. E. 1979. Oxidation-reduction sequences in ground water flow systems. *Canadian Journal of Earth Science*, **16**, 12–23.

CHEN, K. Y. & MORRIS, J. C. 1972. Kinetics of oxidation of aqueous sulfide by O_2. *Environmental Science and Technology*, **6**, 529–37.

COLEMAN, M. L. 1985. Geochemistry of diagenetic non-silicate minerals: kinetic considerations. *Philosophical Transactions of the Royal Society of London, A*, **315**, 39–56.

CURTIS, C. D. 1977. Sedimentary geochemistry: environments and processes dominated by involvement of an aqueous phase. *Philosophical Transactions of the Royal Society of London*, **286**, 353–72.

DAVIS, J. B. & KIRKLAND, D. W. 1970. Native sulfur deposition in the Castile formation, Culberson County, Texas. *Economic Geology*, **65**, 107–21.

——, STANLEY, J. P. & CUSTARD, H. C. 1970. Evidence against oxidation of hydrogen sulfide by sulfate ions to produce elemental sulfur in salt domes. *American Association of Petroleum Geologists Bulletin*, **54**, 2444–7.

—— & YARBROUGH, H. F. 1966. Anaerobic oxidation of hydrocarbons by Desulfovibrio desulfuricans. *Chemical Geology*, **1**, 137–44.

DESSAU, G., JENSEN, M. L. & NAKAI, N. 1962. Geology and isotopic studies of Sicilian sulfur deposits. *Economic Geology*, **57**, 410–38.

DHANNOUN, H. Y. & FYFE, W. S. 1972. Reaction rates of hydrocarbons with anhydrite. *Progress in Experimental Petrology*, **2**, 69–71.

DOUGLAS, A. G. & MAIR, B. J. 1965. Sulfur: role in genesis of petroleum. *Science*, **147**, 499–501.

DREAN, T. A. 1978. Reduction of sulfate by methane, xylene, and iron at temperatures of 175 to 350°C. *MS Thesis*, Pennsylvania State University.

DUNSMORE, H. E. 1971. Diagenetic model for middle Devonian Keg River Formation, Rainbow area, northwestern Alberta. *Unpublished BSc Thesis*, University of Calgary.

FEELY, H. W. & KULP, J. L. 1957. Origin of gulf coast salt-dome sulphur deposits. *American Association of Petroleum Geologists Bulletin*, **41**, 1802–53.

GARDNER, L. R. 1973. Chemical models for sulfate reduction in closed anaerobic marine environments. *Geochimica et Cosmochimica Acta*, **37**, 53–68.

GILES, M. R. & MARSHALL, J. D. 1986. Constraints on the development of secondary porosity in the subsurface: re-evaluation of processes. *Marine and Petroleum Geology*, **3**, 243–55.

GUTSALO, L. K. & KRIVOSHEYA, V. A. 1966. Importance of sulfates from subsurface waters in prospecting for oil and gas in Dnepr-Donets trough. *International Geological Reviews*, **9**, 1144–9.

HANOR, J. S. & WORKMAN, A. L. 1986. Distribution of dissolved volatile fatty acids in some Louisiana oil field brines. *Applied Geochemistry*, **1**, 37–46.

HO, T. Y., ROGERS, M. A., DRUSHEL, H. V. & KROONS, C. B 1974. Evolution of sulphur compounds in crude oils. *American Association of Petroleum Geologists Bulletin*, **58**, 2338–48.

HUSAIN, S. A. 1967. Sulphur isotope exchange reactions. *PhD Dissertation*, University of Alberta, Edmonton, Alberta.

—— & KROUSE, H. R. 1978. Sulphur isotope effects during the reaction of sulphate with hydrogen sulphide. *In*: ROBINSON, B. W. (ed.). *Stable Isotopes in the Earth Sciences*. Department of Scientific and Industrial Research Bulletin, **220**, 207–10.

IVANOV, M. V. 1968. Microbiological processes in the genesis of native sulfur deposits. *Institute for Physical Science and Technology, Maryland, No. 1850*.

IWATSUKA, H. & MORI, T. 1960. Studies on the metabolism of a sulfur-oxidizing bacterium. I. Oxidation of sulfur. *Plant Cell Physiology*, **1**, 163–72.

JEFFRIES, M. O. & KROUSE, H. R., 1984. Isotope geochemistry of stratified water bodies on northern Ellesmere Island, Canadian arctic. *Zentralinstitut für Isotopen- und Strahlenforschung Mitteilungen, Leipzig*, **84**, 159–69.

JOBSON, A. M., COOK, F. D. & WESTLAKE, D. W. S. 1979. Interaction of aerobic and anaerobic bacteria in petroleum biodegradation. *Chemical Geology*, **24**, 355–65.

JORGENSEN, B. & REVSBECH, N. P. 1983. Colourless sulfur bacteria *Beggiatoa* sp.. and *Thiovulum* spp. in O_2 and H_2S microgradients. *Applied and Environmental Microbiology*, **45**, 1261–70.

KAPLAN, I. R., EMERY, K. O. & RITTENBERG, S. C. 1963. The distribution and isotopic abundance of sulphur in recent marine sediments off southern California. *Geochimica et Cosmochimica Acta*, **27**, 297–331.

KARTSEV, A. A., TABASARANSKII, Z. A., SUBBOTA, M. I. & MOGILEVSKII, G. A. 1959. *Geochemical Methods of Prospecting and Exploration for Petroleum and Natural Gas*. University of California Press, Berkeley.

KIRKLAND, D. W. & EVANS, R. 1976. Origin of limestone buttes, gypsum plain, Culberson County, Texas. *American Association of Petroleum Geologists Bulletin*, **60**, 2005–18.

KIYOSU, Y. 1980. Chemical reduction and sulfur-isotope effects of sulfate by organic matter under hydrothermal conditions. *Chemical Geology*, **30**, 47–56.

KREBS, W. & MACQUEEN, R. 1984. Sequence of diagenetic and mineralization events, Pine Point lead-zinc property, Northwest Territories, Canada. *Bulletin Canadian Petroleum Geology*, **32**, 434–64.

KROUSE, H. R. 1977. Sulfur isotope studies and their role in petroleum exploration. *Journal of Geochemical Exploration*, **7**, 189–211.

—— 1980. Stable isotope geochemistry of non-hydrocarbon constituents of natural gas. *Proceedings 10th World Petroleum Congress*, pp. 85–92, Bucharest, Romania.

—— 1983. Stable isotope research in support of more effective utilization of gas fields in Alberta. *Alberta — Canada Energy Resource Research Fund Agreement U-30*.

LARKIN, J. M. 1981. Isolation of *Thiothrix* in pure culture and observation of a filamentous epiphyte on *Thiothrix*. *Current Microbiology*, **4**, 341–67.

MACHEL, H. G. 1985. Facies and diagenesis of the upper Devonian Nisku Formation in the subsurface of central Alberta. *Unpublished PhD Thesis*, McGill University, Montreal.

—— & KROUSE, H. R. in prep. a. Thermochemical sulphate reduction in deeply buried Upper Devonian Nisku carbonates of the Alberta subsurface.

—— & —— in prep. b. Products and distinguishing criteria of bacterial and thermochemical sulphate-hydrocarbon redox-reactions in diagenetic environments.

MACQUEEN, R. W. & POWELL, T. G. 1983. Organic geochemistry of the Pine Point lead-zinc ore field and region, Northwest Territories, Canada. *Economic Geology*, **78**, 1–25.

MCCREADY, R. G. L. & KROUSE, H. R. 1980. Sulfur isotope fractionation by *Desulfovibrio vulgaris* during metabolism of $BaSO_4$. *Geomicrobiology Journal*, **2**, 55–62.

MIGDISOV, A. A., CHERKOVSKIY, S. L. & GRINENKO, V. A. 1974. The effects of formation conditions on the sulfur isotopes of aquatic sediments. *Translations from Geokhimiya No. 10*, 1482–502.

MILNER, C. W. D., ROGERS, M. A. & EVANS, M. A. 1977. Petroleum transformation in reservoirs. *Journal of Geochemical Exploration*, **7**, 101–53.

MORSE, J. W., MILLERO, F. J., CORNWELL, J. C. & RICKARD, D. T. 1987. The chemistry of the hydrogen sulfide and iron sulfide systems in natural waters. *Earth Science Reviews*, **24**, 1–42.

NAKAI, N. & JENSEN, M. L. 1964. The kinetic isotope effect in the bacterial reduction and oxidation of sulfur. *Geochimica et Cosmochimica Acta*, **28**, 1893–912.

NAZINA, T. N., ROZANOVA, E. P. & KNZETSOV, S. I. 1985. Microbial oil transformation processes accompanied by methane and hydrogen-sulfide formation. *Geomicrobiology Journal*, **4**, 103–30.

OHMOTO, H. & RYE, R. O. 1979. Isotopes of sulfur and carbon. *In*: BARNES, H. L. (ed.). *Geochemistry of Hydrothermal Ore Deposits*, pp. 509–65. Wiley, New York.

ORR, W. L. 1974. Changes in sulfur content and isotopic ratios of sulfur during petroleum maturation—study of Big Horn Basin Paleozoic oils. *American Association of Petroleum Geologists Bulletin*, **58**, 2295–318.

—— 1977. Geologic and geochemical controls on the distribution of hydrogen sulfide in natural gas. *In*: CAMPOS, R. & GONI, J. (ed.). *Advances in Organic Geochemistry*, pp. 571–97. Enadisma, Madrid, Spain.

—— 1982. Rate and mechanism of non-microbial sulfate reduction. *Geological Society of America, Annual Meeting, Abstracts with Programs*, **14**, 580.

PAULL, C. K., HECKER, B., COMMEAU, R., FREEMAN-LYNDE, R. P., NEUMANN, C., CORSO, W. P., GOLUBIC, S., HOOK, J. E. & CURRAY, J. 1984. Biological communities at the Florida Escarpment resemble hydrothermal vent taxa. *Science*, **226**, 965–7.

PAWLOWSKI, S., PAWLOWSKA, K. & KUBICA, B. 1979. Geology and genesis of the Polish sulfur deposits. *Economic Geology*, **74**, 475–83.

PECK, H. D. 1984. Physiological diversity of the sulfate-reducing bacteria. *In*: STROHL, W. R. & TUOVINEN, O. H. (eds). *Microbial Chemoautotrophy*, pp. 231–335. Ohio State University Press, Columbus.

PHILIPPI, G. T. 1977. On the depth, time and mechanism of origin of the heavy to medium-gravity naphthenic crude oils. *Geochimica et Cosmochimica Acta*, **41**, 33–52.

POSTGATE, J. R. 1984. *The Sulphate-reducing Bacteria*. Cambridge University Press.

POWELL, T. G. & MACQUEEN, R. W. 1984. Precipitation of sulfide ores and organic matter: sulfate reactions at Pine Point, Canada. *Science*, **224**, 63–6.

PRYOR, W. A. 1962. *Mechanisms of Sulfur Reactions.* McGraw-Hill, New York.

REZNIKOV, A. N. 1971. The conversion of petroleum gases in a high-temperature environment. *Geologiya Nefti i Gaza*, **4**, 44–8.

RICKARD, D. T. 1969. The chemistry of iron sulfide formation at low temperatures. *Stockholm Contributions to Geology*, **20**, 67–95.

ROSANOVA, E. P. & KHUDYAKOVA, A. I. 1974. A new non-sporeforming thermophilic sulfate-reducing organism. *Desulfovibrio thermophilus* nov. sp. *Microbiology*, **43**, 908–12.

RUDAKOVA, E. F. & VELIKOVSKII, A. S. 1947. Conditions for the formation of sulphur compounds and sulphur in crude oils. *Neftyanoe Khozyaistro*, **25**, 49–54.

RYE, D. M. & WILLIAMS, N. 1981. Studies of the base metal sulfide deposits at McArthur River, Northern Territory, Australia: III. The stable isotope geochemistry of the H.Y.C. Ridge, and Cooley deposits. *Economic Geology*, **76**, 1–26.

SASSEN, R. 1980. Biodegradation of crude oil and mineral deposition in a shallow Gulf Coast salt dome. *Organic Geochemistry*, **2**, 153–66.

—— 1985. Basic geochemical strategies. *Oil and Gas Journal* 1985.

—— 1986. Crude oil destruction and bitumen precipitation in deep carbonate reservoirs of the Smackover Formation. *Third Annual Meeting, Society for Organic Petrology, Abstract and Program*, pp. 24–6. Lexington, Kentucky.

—— in press. Geochemical and carbon isotope studies of crude oil destruction, bitumen precipitation, and sulfate reduction in the deep Smackover Formation. *Organic Geochemistry*.

SCHNEIDER, A. & NIELSON, H. 1965. Zur Genese des elementaren Schwefels im Gips von Weenzen (Hils). *Beiträge zur Mineralogie und Petrographie*, **11**, 705–18.

SIEBERT, R. M. 1985. The origin of hydrogen sulfide, elemental sulfur, carbon dioxide, and nitrogen in reservoirs. Timing of siliciclastic diagenesis: relationship to hydrocarbon migration. *Sixth Annual Research Conference*, Gulf Coast Section, Society of Economic Paleontologists and Mineralogists Foundation, pp. 30–31.

SIEBURTH, J. McN. 1979. *Sea Microbes.* Oxford University Press, New York.

STETTER, K. O. & GAAG, G. 1983. Reduction of molecular sulphur by methanogenic bacteria. *Nature*, **305**, 309–11.

STETTER, K. O., LAUERER, G., THOMM, M. & MEUNER, A. 1987. Isolation of extremely thermophilic sulfate reducers: evidence for a novel branch of archaebacteria. *Science,* **236**, 822–4.

STUMM, W. & MORGAN, J. J. 1970. *Aquatic Chemistry.* Wiley-Interscience, New York.

TOLAND, W. G. 1960. Oxidation of organic compounds with aqueous sulfate. *Journal of the American Chemical Society*, **82** 1911–6.

TRUDINGER, P. A., CHAMBERS, L. A. & SMITH, J. W. 1985. Low-temperature sulphate reduction: biological versus abiological. *Canadian Journal of Earth Science*, **22**, 1910–8.

VALITOV, N. B. 1974. Elemental sulphur as a factor in the generation of hydrogen sulphide in deep-lying carbonate reservoir rocks. *Doklandy Akademii Nauk SSSR*, **219**, 206–8.

WARD, D. M. & BROCK, T. D. 1978. Anaerobic metabolism of hexadecane in sediments. *Geomicrobiology Journal*, **1**, 1–9.

H. G. MACHEL, Basin Research Institute, Louisiana State University, Baton Rouge, Louisiana 70803-4101, USA. Present address: Department of Geology, University of Alberta, Edmonton, Alberta T6G 2E3, Canada.

Porosity reduction, microfabric and resultant lithification in UK uncemented sands

S. N. Palmer & M. E. Barton

SUMMARY: Studies of the extent of diagenetic change in matrix-free, uncemented, quartzose sands ranging in age from the Jurassic to the Recent in the UK have been carried out as part of a geotechnical research programme. All the sands studied are thought to have experienced only a relatively small depth of burial and the extent of diagenetic change is consequently small. Previous studies of the *in situ* fabric of sands in this category have been limited owing to the sampling difficulties created by their very friable nature. Careful sampling, however, has succeeded in obtaining undisturbed material and has facilitated studies of the porosity, microfabric and degree of lithification. Distinctive changes, progressive with age, include reduction of porosity, an increase in the numbers and complexity of grain contacts and an increasing degree of lithification. The cause of these diagenetic changes are discussed and it is concluded that the evidence strongly favours pressure solution of the detrital quartz grains as the dominant process.

This is a study of very clean, virtually matrix-free, uncemented, mature quartzose sands of ages ranging from Recent to Jurassic in the UK. The primary motivation for the research is geotechnical. Surprisingly, despite other areas of progress in soil and rock mechanics, very little geotechnical work has been done on the transition from loosely compacted, unlithified sands to compact, indurated sandstones. Dusseault & Morgenstern (1979) coined the phrase 'locked sand' to denote the intermediate state where the original depositional porosity is reduced, the grains are inter-locked (or 'locked') and the sand has acquired a small but measurable degree of lithification without true cementation. The authors have investigated the extent to which these characteristics are present in UK sands (Barton *et al.* 1986b, Barton *et al.* in prep.) The study has included observations on the porosity, microfabric and lithification and has revealed the extent of diagenetic changes in these uncemented sands. Because of the friable nature of such sediments, studies of the *in situ* fabric of weakly lithified sands have tended to be neglected and it is therefore considered that the results are of significant sedimentological interest.

Sands studied

Samples of matrix-free, uncemented, quartzose sands have been obtained from numerous quarries and natural exposures. The Mesozoic and Tertiary sands, although readily disaggregated by manipulation, with care, can be block-sampled (Barton *et al.* 1986a) thereby preserving the *in situ* fabric and allowing laboratory impregnation (Palmer & Barton 1986). The Quaternary sands are impregnated *in situ*. Eight of these sands (Table 1) have a similar median grain size, similar sorting (Fig. 1) and quartz dominated mineralogy as shown in Table 2. The particle shapes and roundness of the various sands show some variation, but all eight sands have similar minimum remoulded (see below) porosities. It is important to note that all the sands have minor matrix (not greater than 2.0% in any case) and negligible cement content (not greater than 0.5%). It can be considered therefore that these sands have sufficient similarities in the sedimentary characteristics influencing depositional and diagenetic fabrics (Meade 1966, Berner 1971, Wolf & Chilingarian 1976, Chilingarian 1983) to permit comparison between them in respect of their porosity reduction and microfabric.

The uncemented nature of these sands, particularly those of Mesozoic age, is distinctive since there are other sand deposits of equivalent burial histories which are considerably more lithified due to the development of an authigenic mineral phase. Indeed, as noted in Table 1, the petrographic features discussed here are not necessarily present throughout the complete stratigraphic thickness of the strata detailed. The Mesozoic strata, for example, contain horizons possessing the textural features described in this paper, but also frequently contain zones of cemented and/or matrix-rich (>5% in this context) sands. The reason for the lack of interstitial cement is unknown and requires further investigation. There is no evidence of secondary porosity or cement dissolution. The

From: MARSHALL, J. D. (ed.), 1987, *Diagenesis of Sedimentary Sequences*, Geological Society Special Publication No. 36, pp. 29–40.

TABLE 1.

Sands studied	Age: Stratigraphic	Age: Absolute Ma (bp)	Estimated maximum depth of burial (m)	Approx. thickness of sand unit* (m)	Location: Site	Location: NGR
Modern-day beach sand	Holocene	0.0	0.2	>2	Holkham Beach, Wells-next-the-Sea, Norfolk	TF 892 455
Norwich Crag sand	Pleistocene	*c.* 1.65	20	*c.* 3	Eastern Bavents, Southwold, Suffolk	TM 518 780
Barton sand, Barton Beds (Bed K), Barton Formation	Late Eocene	*c.* 42	169	*c.* 5	Becton Bunny, Barton-on-Sea, Essex	SZ 425 092
Thanet sand, Thanet Formation	Late Palaeocene	*c.* 57	295	*c.* 30	Linford Pit, Thurrock, Essex	TQ 667 799
Woburn Sands	Albian/Aptian L. Cretaceous	*c.* 100	302	*c.* 60	Nine Acres Pit, Leighton Buzzard, Bedfordshire	SP 939 277
Folkestone sand, Folkestone Beds	Albian L. Cretaceous	*c.* 100	474	*c.* 78	West Heath Commons, West Harting, West Sussex	SU 785 227
Kellaways Sand Kellaway Beds	Callovian, M. Jurassic	*c.* 139	700	*c.* 10	South Cave Pit, South Cave, Humberside	SE 920 330
Grantham sand, Grantham Formation	Aalenian, M. Jurassic	*c.* 170	780	*c.* 7	Wittering Grange Pit, Wittering, Cambridgeshire	TL 048 101

* The petrographic features detailed for each sand in this paper do not necessarily occur throughout the sand's total thickness.

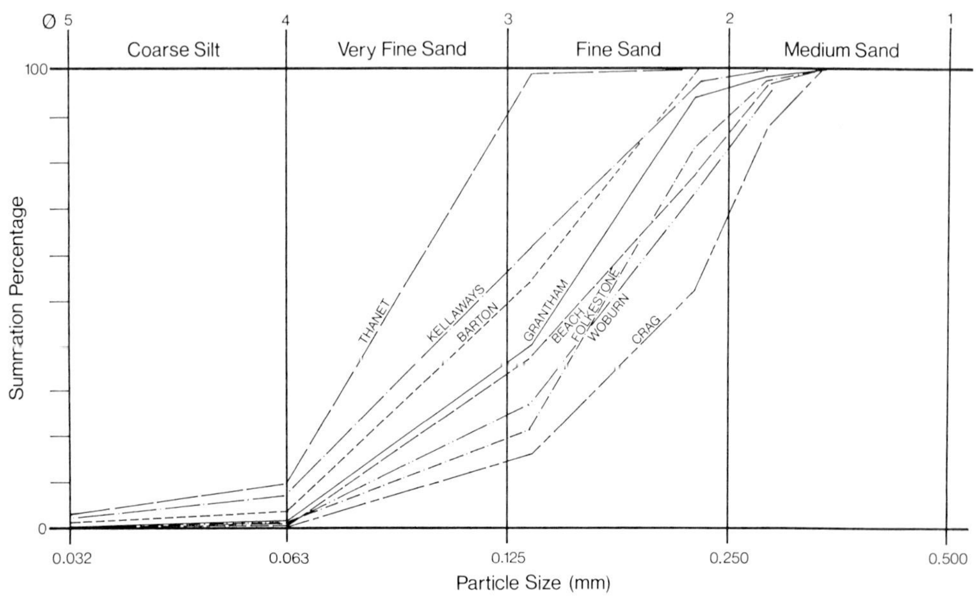

FIG. 1. Particle size distribution of the sands, illustrating their very similar grain sizes and sorting.

TABLE 2.

Sands studied \ Index properties	Median particle size (mm)	Sorting	Petrographic components (%): Clasts*	Matrix†	Cement‡	Porosity %	Min. lab. remoulded porosity %	Mineralogical composition (%): Quartz	Lithics	Feldspar	Mica	Other
Modern-day beach sand	0.16	0.58	99.2	0.8	0.0	47.2	38.8	97	3	0	0	T
Norwich Crag sand	0.19	0.63	97.7	1.8	0.2	43.1	38.5	93	5	1	T	1
Barton sand	0.15	0.54	98.0	1.7	0.3	35.6	39.0	96	3	1	T	T
Thanet sand	0.10	0.47	97.0	1.9	0.5	35.1	40.7	93	3	2	1	1
Woburn sand	0.14	0.55	98.4	1.5	0.1	35.0	38.7	94	3	1	1	1
Folkestone sand	0.18	0.56	98.3	1.2	0.5	34.5	37.9	96	2	T	2	T
Kellaways sand	0.12	0.51	97.3	1.7	0.4	34.1	39.6	95	3	1	T	1
Grantham sand	0.14	0.57	98.6	1.1	0.3	33.6	38.6	94	2	2	1	1

* Detrital grains of diameter greater than 20 μm.
† Detrital grains of diameter less than 20 μm.
‡ Authentic mineral development.

porosity and microfabric are therefore considered to be primary features.

Ideally, it would be most appropriate to make comparisons between the sands' porosities and microfabrics in terms of factors such as depth of burial, maximum past temperature and other conditions likely to influence their diagenesis. Unfortunately, the majority of these factors remain speculative or unknown. It is considered necessary, however, in such a study as this to make some estimate of the burial depth for each deposit to serve as a rough guide for comparative purposes.

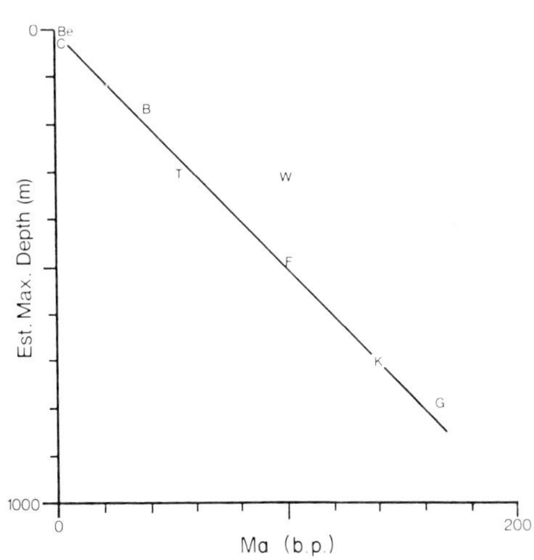

FIG. 2. Estimate of the maximum depth of burial (based on regional stratigraphy) versus chronological age. The plot suggests that there is an approximately linear relationship between maximum burial depth and age for these deposits.

With the exception of the modern beach material, the sands are shallow water deposits which, in the majority of cases, have been preserved as the deposits of stable, shelf areas (Whittaker 1985) and, while the Tertiary deposits are associated with basins, the samples studied are either from the upper stratigraphic levels or from the marginal areas and are similarly unlikely to have experienced any great depth of burial. Estimates of the maximum burial depths are given in Table 1. These figures are maximum thicknesses based on the lithostratigraphy presented in the relevant Geological Society Report volumes (Mitchell *et al.* 1973, Curry *et al.* 1978, Rawson *et al.* 1978, Cope *et al.* 1980). No attempt has been made to allow for possible thicknesses of sediments eroded prior to subsequent periods of deposition and hence the figures must be treated as speculative estimates of the regional burial depths for each sand.

Since the chronological age of the sands is well documented (Geological Society Special Report volumes cited above and Anderton *et al.* 1979), it is considered a more satisfactory parameter for detailed comparison between the sands. The chronological age does, of course, represent a time-scale and, in terms of slow diagenetic processes requiring extended time periods, such comparisons will be germane. It is also possible,

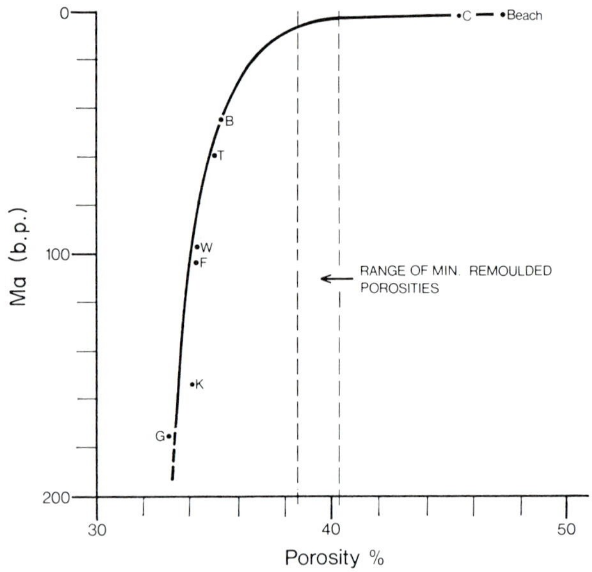

FIG. 3. Mean porosities of sands versus chronological age. The curve shows an initially rapid decrease in porosity with time, followed by a continuing, slow reduction over a long time period. The Folkestone (F) and Woburn (W) sands both have an approximate age of 100 Ma and have been separated purely for clarity.

as suggested by Fig. 2, that the chronological age of the sediments may represent a scale of generally increasing depth of burial.

Porosity

The natural intact or *in situ* porosities of the sands, as measured by standard petrographic and soil engineering techniques, are given in Table 2. They are plotted versus age in Fig. 3 which shows a rapid decrease in porosity from the recent beach sand (with a porosity equal to the deposited value) to the Mesozoic sands, where significant porosity reduction has occurred. With sands such as these (well sorted, quartz dominated, matrix-free) the *in situ* porosities can usefully be compared with (i) the likely original deposited porosity and (ii) the minimum porosity obtained in the laboratory by re-compacting the disaggregated grains (producing a remoulded, dense sand deposit).

Original depositional porosities of sands are generally in the range of 40–50% and more specifically 46–50% for fine sands such as those of this study (Pryor 1973, Friedman & Sanders 1978, North 1985). There are apparently no records of denser packing being achieved in sands by the normal, natural processes of sedimentation and thus no reason to believe that the depositional porosities of the sands detailed here were any different to the figures quoted above. Thus, as seen in Fig. 3, apart from the modern Beach and the Pleistocene Sand, the existing *in situ* porosities are considerably reduced (by about 12–13% on average).

Comparisons of *in situ* porosities with recompacted (or 'remoulded') porosities are common practice in soil engineering (Kolbuszewski 1948). Various techniques can be used to obtain a range of porosities from a maximum possible at one end of the scale to a minimum (without grain fracturing or crushing) at the other. In this study the minimum porosity was achieved using a device applying a lateral vibration during sedimentation (Barton & Brookes, in press); giving values in the range from 38 to 41%. The existing *in situ* porosities, as shown in Fig. 3, are significantly less than these values (by about 4–5% on average).

Associated with these reductions in porosity are changes in the microfabric and the degree of lithification.

Microfabric

Standard petrographic techniques show all the sands to have remarkably similar microfabrics, i.e. high intergranular porosities, grain-supporting, matrix-free, uncemented micro-textures.

Similarly to the reduction in porosity, the numbers of grain contacts per grain (Taylor 1950) increase with age, as shown in Fig. 4.

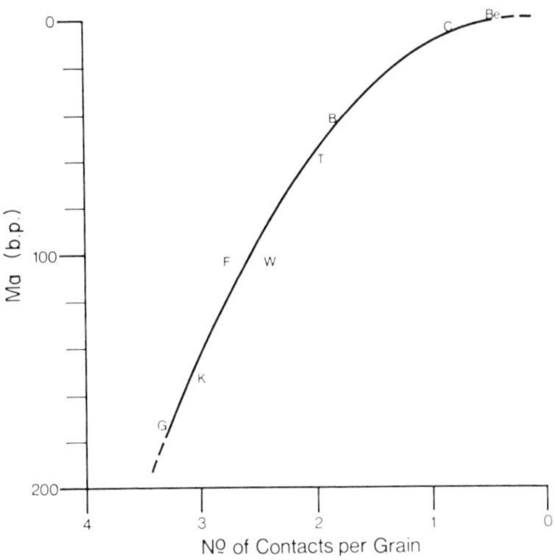

FIG. 4. Number of grain contacts per grain (contact index) versus age, illustrating an increase in the state of packing with time.

Concomitantly, a change in the type of contacts occurs, principally as a reduction in the numbers of tangential contacts with an increase in straight and concavo/convex contacts (Fig. 5). The microfabrics are illustrated in the photomicrographs (Fig. 6) and SEM photographs (Fig. 7). A small number of sutured contacts are seen in the Cretaceous and Jurassic sands and although a few of these are clearly inherited (in polycrystalline grains) most appear to be primary diagenetic features. A search has been made, using cathodoluminescence petrography, SEM and XRD techniques, for evidence of cementation but none has been found by the authors in any of these sands. Confirmation of a lack of quartz overgrowths in the Grantham, Folkestone and Barton Sands has been made by GAPS Geological Services. Grain contact areas and grain surface textures seen under the SEM in the older sands (Fig. 6b) show features strongly suggestive of pressure solution. Both the reduced porosities and the types of

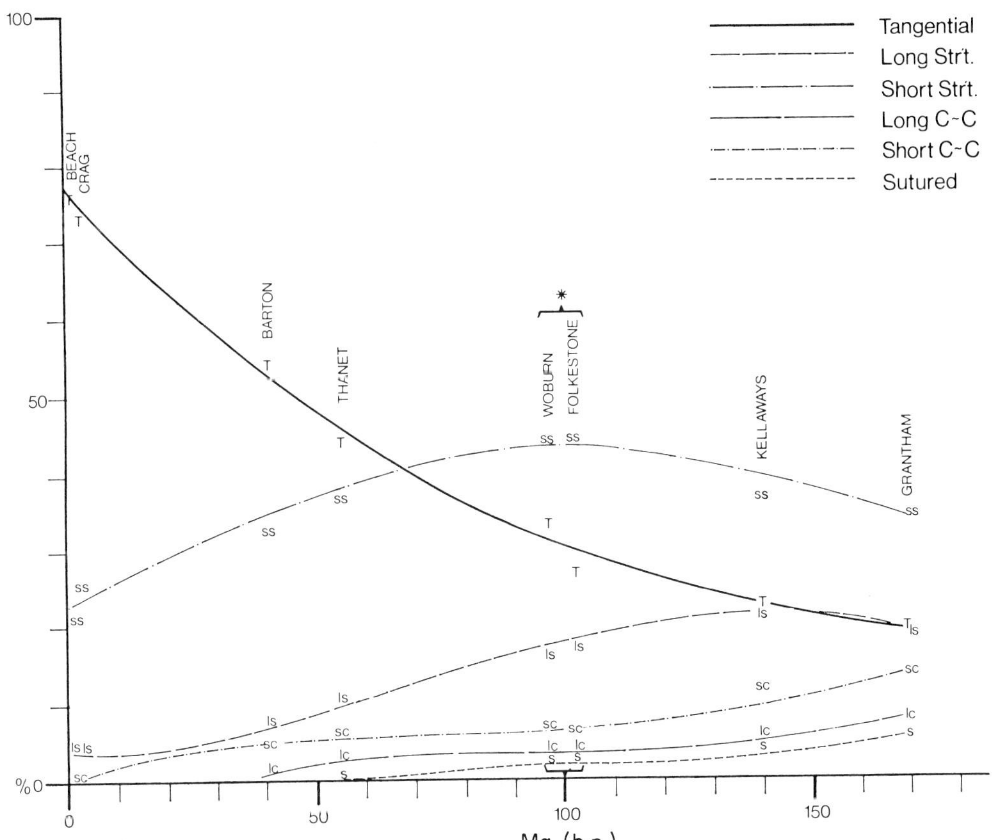

FIG. 5. Grain contact morphology versus chronological age. The plot illustrates a progressive decrease of tangential contacts and increase in planar-type contacts with age. The latter causes gradual increase in the grain-grain contact areas and the state of fabric interlocking with age. *The Woburn and Folkestone sands both have an approximate age of 100 Ma, and have been separated purely for clarity.

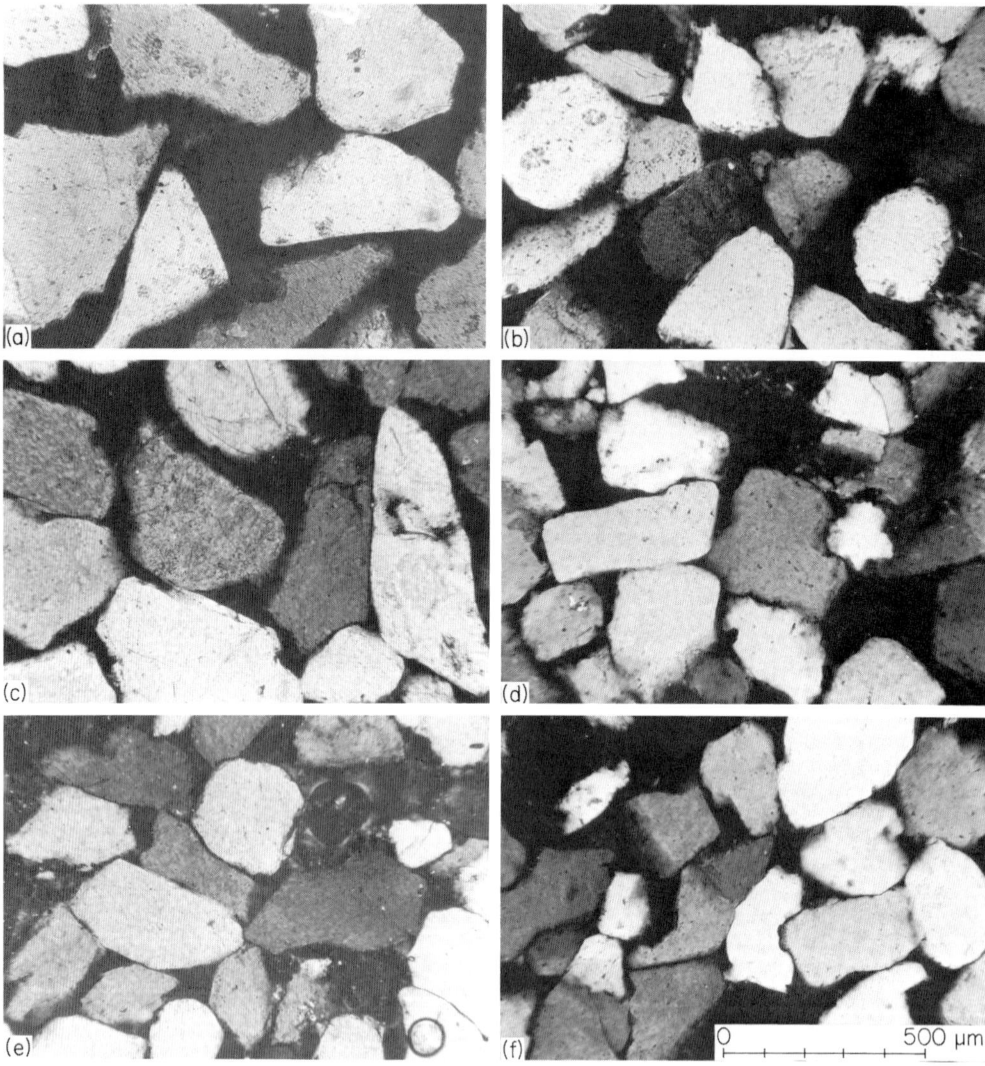

FIG. 6. Photomicrographs of five of the sands to illustrate the progressive porosity reduction, increase in the interlocking of the fabric (tightness of packing) and change in contact type morphologies: (a) modern-day beach sand, with high porosity, very low number of grain contacts per grain, tangential contacts dominating; (b) Barton Sand; (c) Folkestone Sand, tightness of packing increasing, planar contacts important; (d) Kellaways Sand; (e) and (f) Grantham Sand, marked porosity reduction, high number of grain contacts per grain, interlocked fabric, predominantly straight and concavo/convex contacts giving high grain-grain contact areas. With the exception of the modern-day beach sand, which was impregnated *in situ*, all the sands were block-sampled and impregnated in the laboratory using the Drip Method and the epoxy Epo-tec 301, spiked with the dye Waxoline Blue AP-FW in a 1% wt concentration (Palmer & Barton 1986). (Scale in (f) applies to all photomicrographs.)

grain contacts are consistent with features known to be produced by pressure solution.

Lithification

Lithification is thought of in geological terms as a change from unconsolidated to indurated rock. A quantitative measure of lithification can be obtained either from the shear strength measured on dry samples in the direct shear box (Fig. 8) or by taking values of unconfined compressive strength (Fig. 9). Samples for these tests were prepared by hand carving (Fig. 10). The increase in age, and concomitant decrease in porosity, is associated with an increase in

FIG. 7. SEM photomicrographs of (a) Barton (scale bar = 40 μm) and (b) Grantham Sand (scale bar = 100 μm) showing the interlocked fabric, absence of cement and matrix. Features suggestive of pressure solution are seen in the Grantham Sand.

lithification as measured by these tests of strength.

The direct shear strength tests were conducted as slow, drained tests at normal stresses ranging from 50 to 900 kPa. The samples gave approximately linear envelopes over this stress range and show increasing values of shearing resistance angle (ϕ'), and increasing values of the cohesive intercept (c'), with age of the sand, correlating with porosity reduction and an increase in the number of grain contacts per grain. The beach sand, which could not be prepared for shear box testing of the *in situ* fabric, and the Pleistocene sand have values of cohesive intercept (c') equal to zero. There is some evidence to suggest that at very low normal stresses the envelopes may become curvilinear but, owing to the usual difficulty of obtaining reliable results in this range, further work will be required to substantiate this.

Measurement of unconfined compressive strength, using dry cylindrical samples in a triaxial testing machine, are given in Fig. 9 against values of grain-grain contact areas (Dobereiner & de Freitas 1986) and numbers of tangential contacts (a measure of fabric 'locking'). A further confirmation of the increasing lithification with age is obtained together with the relationship with the other diagenetic features—a decrease in tangential contacts and an increase in grain contact areas. The most 'lithified' formation, the Grantham Sand, remains geologically 'a sand' and geotechnically it is best classed as an engineering soil. Although none of these sands are sufficiently indurated to withstand coring, with the exception of the beach sand they are compact enough to be sampled and transported as intact blocks in which the porosity and micro-fabric remain in the *in situ* state (Barton *et al.* 1986a).

Discussion

The sands studied clearly show a gradual series of diagenetic changes. Admittedly the degree of diagenetic change can be regarded as small—thus we may compare, for instance, the considerably altered but more deeply buried Brent Sand of the North Sea with the comparably aged Grantham Sand as shown here. Comparisons of porosity with depth have been made by Atwater & Miller (1965), Hsü (1977), Selley (1978), Zieglar & Spotts (1978), Magara (1980) and Loucks *et al.* (1984) among others, but in all these cases the depth of burial is considerable and the porosity much less (especially at the deeper levels) than found in the sands of this study. Even so the porosities recorded in Table 2, and as shown in Fig. 3, are reduced below the original deposited values for such sands and, therefore, indicate a measure of diagenetic change.

Similarly we can compare the gradual changes in packing and particle contact types (as shown in Figs 3 and 4 respectively) with the well-known and referenced study by Taylor (1950) of the deeply buried Jurassic and Cretaceous sands in Wyoming. In this case, however, the changes in contact index and contact types are considerably more intense than observed in this study. Taylor concluded that porosity reduction, increase in contact index, decrease in tangential contacts and increase in concavo/convex and sutured contacts relate to an increase in compaction. Selley (1982) and Burley *et al.* (1985) suggest that the features in the Wyoming Sands are due to the development of quartz overgrowths. In this latter case the observed changes would be clearly of a different origin from the uncemented fabrics studied here. Other studies of grain contacts and packing parameters versus depth of burial include Beaudry (1950) and Hays (1951),

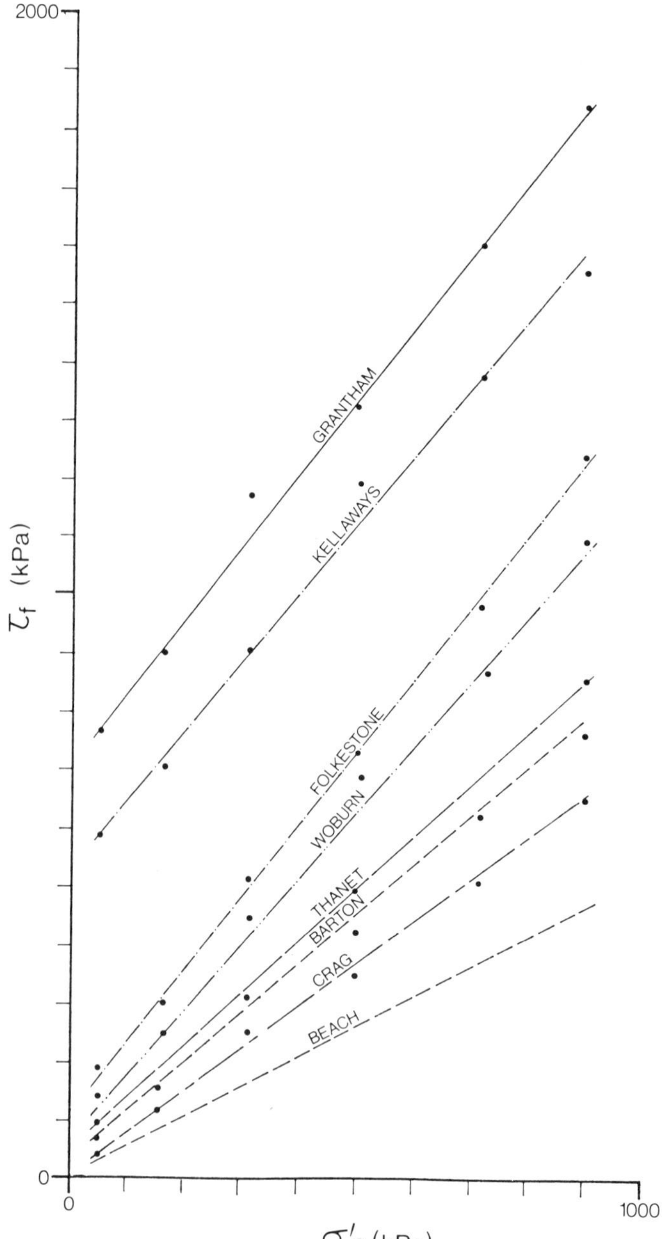

FIG. 8. Shear strength envelopes obtained from direct shear tests on dry samples using a standard 60 mm square shear box. The envelopes are drawn as best-fit straight lines through the experimental points: there is evidence that at low stresses the envelopes become markedly curvilinear. The strength of the sands increases with the age of the deposits. The envelope of the beach sand is obtained from the maximum angle of repose in the laboratory, since undisturbed samples for shear box testing could not be produced for this sand, it being a truly unlithified deposit.

as reported by Wolf & Chilingarian (1976), of deeply buried sandstones showing cementation, unlike the sands of this study. It would appear there is an absence of studies in the literature where porosity reduction in sand is unequivocally due to compaction (mechanical and/or chemical) alone.

It is not intended to discuss here in detail the reasons for the diagenetic changes observed in the sands studied, but simply to comment

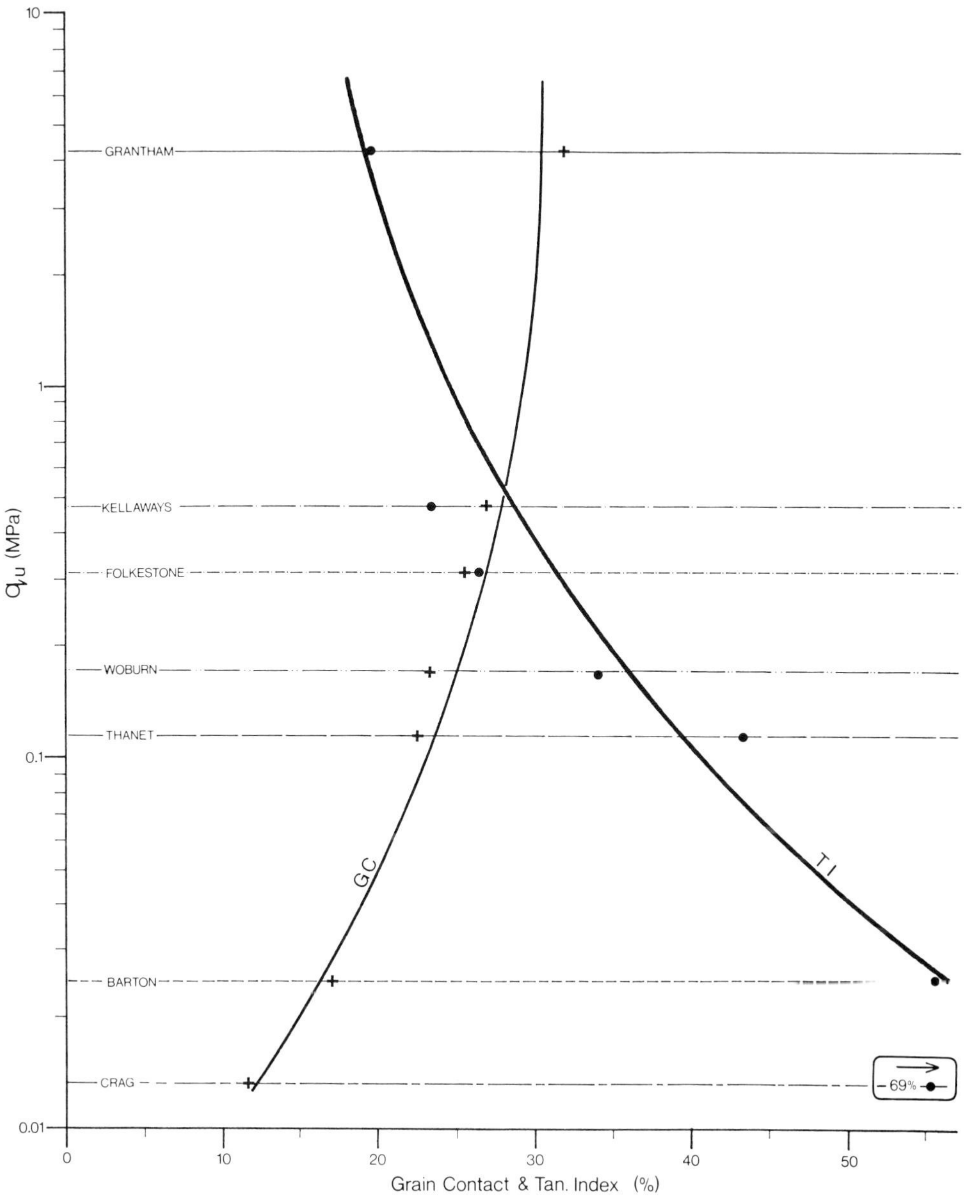

FIG. 9. Unconfined or uniaxial compressive strength versus number of tangential contacts (tan. index) and grain contact areas (grain contact). The plot illustrates that the unconfined compressive strengths of the sands increase with age, and correlates to decreasing numbers of tangential contacts and increasing grain contact areas, i.e. an increase in the interlocking.

upon the fundamental processes that may have been operative. The main diagenetic processes which primarily control porosity reduction in sediments, and therefore also the development of microfabric in these sands (as discussed by Fuchtbauer 1967, Bjørlykke 1983 and Houseknecht 1984) are:

(i) Mechanical compaction due to overburden pressure—causing the rotation, reorientation and fracturing of competent grains, and the plastic deformation of ductile components.

(ii) Mechanical compaction due to seismic disturbances over a geological time-scale—causing the geometrical rearrangement of framework components.

(iii) Chemical compaction—the intergranular pressure solution (dissolution) at contacts between particles, resulting in material mass transfer.

FIG. 10. Preparation of prismatic samples for direct shear box testing (left) and cylindrical samples (centre and right) for confined and unconfined compressive tests. These samples are prepared by hand carving from block samples such as that in the left background.

(iv) Cementation—the precipitation of authigenic mineral matter in interstitial pore space.

Various studies indicate that the observed porosities and microfabrics in the sands of this type could not have been obtained by simple mechanical compaction. The fabrics produced by laboratory compaction do not produce features consistent with the microfabrics observed in natural sands and sandstones (Maxwell 1960). Berner (1971), Pettijohn *et al.* (1972), North (1985) and preliminary studies by the authors all indicate that minimal porosity reduction will occur by the mechanical mutual adjustment of grains under load. This is particularly so since the sands' compositions are dominated by pure, stable quartz, there being only very small proportions of incompetent ductile clasts, thus very much limiting porosity reduction by plastic deformation of grains (Nagtegaal 1978, Dodge & Loucks 1979). There is no petrographic evidence to suggest that grain fracturing (caused by mechanical compaction) significantly contributed to porosity reduction. All the sands show textural parameters, such as good or moderate sorting, which are similarly known to promote the retention of porosity under load (Rittenhouse 1971). The textural and mineralogical maturity are widely recognized as major factors in determining the mechanical compaction of a sediment (e.g. North 1985 and Selley 1985).

Mechanical compaction due to seismic disturbance over geological time-scales is also not considered as likely to affect the observed porosity reductions. The work of Selig (1963) and the authors shows that the presence of an overburden load inhibits the porosity reduction resulting from vibration.

It is suggested that sands of this type, with depositional porosities of 47–48%, have undergone only a small porosity reduction resulting directly from mechanical compaction owing to the increasing stability produced in the structure of the sand as porosity reduces. Furthermore it is considered that porosities equivalent to the laboratory minimum values are unlikely to be achieved diagenetically by mechanical compaction alone at the relatively small burial depths of these sands. Thus other diagenetic processes must have been operative to produce the dense state of packing observed, particularly in the Mesozoic sands for example. Chemical compaction and cementation are the other main processes which could be invoked to explain the remaining porosity reduction in the sands.

Petrographic, SEM and CL analyses reveal insignificant amounts of authigenic mineral phases (Table 2), so cementation does not account for the observed reductions in porosity. There is no evidence to suggest previous higher grade diagenetic changes: the observed porosities must therefore be considered as primary porosities.

The simplest and most reasonable explanation of the porosity reduction and microfabric changes therefore is that they are the products of the effects of pressure solution. The geometries of the planar grain contacts, especially in the older sands, are those widely known to result from congruent quartz dissolution. Although quantitatively the sands show relatively small amounts of dissolution, minor pressure solution can cause marked porosity reduction (Füchtbauer 1967). SEM analysis of the Cretaceous and Jurassic sands shows micro-pitting and contact-hollows consistent with pressure dissolution. It is therefore tentatively proposed that, with reference to Fig. 3, initial rapid but small porosity reduction is obtained by mechanical compaction. Slower but more significant porosity loss is obtained later by chemical compaction. A similar conclusion for more deeply buried sands was presented by Füchtbauer (1967).

It has been shown how the decrease in porosity and increase in grain contact areas relates to an increase in lithification, and it follows from the discussion above that this also results predominantly from pressure solution. Any contribution to the lithification by the cohesion provided by simple clay bonding from the minor matrix (as discussed by Waldschmidst 1941 and Dapples 1972) and the negligible amounts of cement are thought to be minimal. Evidence for this is also provided by the lack of any correlation between strength and the percentage of either matrix or cement. Pressure cohesion or intergranular welding of the detrital grains (Dapples 1967, 1972, Pettijohn *et al.* 1972, Fairbridge 1983, North 1985) probably contributes to the lithification, possibly significantly, but further work is required to assess its relative importance.

The decrease in porosity, but more significantly the progressive change of contact types and increase of grain contact areas with age, is thought to explain the increase in the degree of induration. It is suggested, therefore, that these sands are primarily lithified by the physical inter-locking of the detrital grains, mainly due to the high states of packing obtained through pressure solution.

ACKNOWLEDGMENTS: The authors wish to express their thanks to the following companies for giving permission to visit and to sample at their sand-pits: Joseph Arnold & Sons Ltd; Barker-Mill Estates; S. E. Borrow Ltd; Ready Mixed Concrete (UK) Ltd; St Albans Sand and Gravel Co. Ltd and in particular British Industrial Sand Ltd.

We would sincerely like to thank Dr P. J. Gregson, Department of Engineering Materials, Southampton University for the use of the SEM; Drs I. M. West and T. Clayton, Department of Geology, Southampton University, for the use of the XRD and CL apparatus and their helpful comments; Dr G. M. Power, Department of Geology, Portsmouth Polytechnic for the use of the photomicroscope, and Mr P. G. Cambridge, Geological Society of Norfolk. Special thanks are due to Messrs A. Brookes and Y. L. Wong for technical assistance with sampling and testing and for drawing the figures.

Financial support for this research was provided by the Science and Engineering Research Council.

References

ANDERTON, R., BRIDGES, P. H., LEEDER, M. R. & SELLWOOD, B. W. 1979. *A Dynamic Stratigraphy of the British Isles*. Allen & Unwin, London.

ATWATER, G. I. & MILLER, E. E. 1965. The effects of decrease in porosity with depth on future development of oil and gas reserves in south Louisiana (abs.) *Bulletin of the American Association of Petroleum Geologists*, **49**, 334.

BARTON, M. E. & BROOKES, A. (in press). Lateral shaking during sedimentation—a new technique for obtaining the minimum porosities of granular soils. *Ground Engineering*.

—— BROOKES, A., PALMER, S. N. & WONG, Y. L. 1986a. A collapsible sampling box for the collection and transport of intact block samples of friable uncemented sands. *Journal of Sedimentary Petrology*, **56**, 540–41.

—— PALMER, S. N. & WONG, Y. L. 1986b. A geotechnical investigation of two Hampshire Tertiary sand beds: are they locked sands? *Quarterly Journal of Engineering Geology*, **19**, 399–412.

——, —— & —— (in prep.) Studies of a locked sand of Jurassic age in the U.K.: the Grantham Formation sand at Wittering, Cambridgeshire.

BEAUDRY, D. A. 1950. Pore-space reduction in some deeply buried sandstones. *Unpublished Thesis*, University of Cincinnatti.

BERNER, R. A. 1971. *Principles of Chemical Sedimentology*. McGraw-Hill, New York.

BJØRLYKKE, K. 1983. Diagenetic reactions in sandstones. *In*: PARKER, A. & SELLWOOD, B. W. (eds) *Sediment Diagenesis*, pp. 169–213. Reidel, Dordrecht, Holland.

BURLEY, S. D., KANTOROWICZ, J. D. & WAUGH, B. 1985. Clastic diagenesis. *In*: BRENCHLEY, P. J. & WILLIAMS, P. J. W. (eds) *Sedimentology. Recent Developments and Applied Aspects*, pp. 189–226. Geological Society Special Publication No. 18. Blackwell Scientific Publications, Oxford.

CHILINGARIAN, G. V. 1983. Compactional diagenesis. *In*: PARKER, A. & SELLWOOD, B. W. (eds) *Sediment Diagenesis*, pp. 57–167. Reidel, Dordrecht, Holland.

COPE, J. C. W., DUFF, L. L., PARSONS, C. K., TORRENS, H. S., WIMBLEDON, W. A. & WRIGHT, J. K. 1980. A correlation of the Jurassic rocks in the British Isles. Part Two: Middle and Upper Jurassic. *Geological Society of London, Special Report No. 15*.

CURRY, D., ADAMS, C. G., BOULTER, M. C., DILLEY, F. C., EAMES, F. E., FUNNELL, B. M. & WELLS, M. K. 1978. A correlation of Tertiary rocks in the British Isles. *Geological Society of London, Special Report No. 12*.

DAPPLES, E. C. 1967. Diagenesis of sandstones. *In*: LARSEN, G. & CHILINGAR, G. V. (eds). *Diagenesis in Sediments*, pp. 91–125. Elsevier, Amsterdam.

—— 1972. Some concepts of cementation and lithification of sandstones. *Bulletin of the American Association of Petroleum Geologists*, **56**, 3–25.

DOBEREINER, L. & DE FREITAS, M. H. 1986. Geotechnical properties of weak sandstones. *Géotechnique*, **36**, 79–94.

DODGE, M. M. & LOUKS, R. G. 1979. Mineralogic composition and diagenesis of Tertiary sandstones along Texas Gulf Coast. *Bulletin of the American Association of Petroleum Geologists*, **63**, 440.

DUSSEAULT, M. B. & MORGENSTERN, N. R. 1979. Locked sands. *Quarterly Journal of Engineering Geology*, **12**, 117–31.

FAIRBRIDGE, R. W. 1983. Syndiagenesos-anadiagenesis-epidiagenesis: phases in lithogenesis. *In*: LARSEN, G. & CHILINGAR, G. V. (eds) *Diagenesis in Sediments and Sedimentary Rocks*, **2**, 17–113. Elsevier, Amsterdam.

FRIEDMAN, G. M. & SANDERS, J. E. 1978. *Principles of Sedimentology*, Wiley, London.

FÜCHTBAUER, H. 1967. Influence of different types of diagenesis on sandstone porosity. *Proceedings of the 7th World Petroleum Congress, Mexico*, **2**, 353–69.

HAYS, F. R. 1951. Petrographic analysis of deep well cores. *Unpublished Thesis*, University of Cincinnatti.

HOUSEKNECHT, D. W. 1984. Influence of grain size and temperature on intergranular pressure solution, quartz cementation, and porosity in a quartzose sandstone. *Journal of Sedimentary Petrology*, **54**, 348–61.

HSÜ, K. J. 1977. Studies of Ventura Field, California II: Lithology, compaction, and permeability of sands. *Bulletin of the American Association of Petroleum Geologists*, **61**, 169–91.

KOLBUSZEWSKI, J. 1948. An experimental study of the maximum and minimum porosities of sands. *Proceedings of the 2nd International Conference of Soil Mechanics and Foundation Engineering*, Rotterdam, **1**, 158-65.

LOUCKS, R. G., MARIANNE, M. D. & GALLOWAY, W. E. 1984. Regional controls on diagenesis and reservoir quality in Lower Tertiary Sandstones along the Texas Gulf coast. *In*: MCDONALD, D. A. & SURDAM, R. C. (eds) *Clastic Diagenesis*, pp. 15–45. The American Association of Petroleum Geologists.

MAXWELL, J. C. 1960. Experiments on compaction and cementation of sand. *In*: GRIGGS, D. & HANDIN, J. (eds) *Rock Deformation*. Geological Society of America Memoir, **79**, 105–32.

MAGARA, K. 1980. Comparison of porosity-depth relationships of shale and sandstones. *Journal of Petroleum Geology*, **3**, 175–85.

MEADE, R. H. 1966. Factors influencing the early stages of the compaction of clays and sand—review. *Journal of Sedimentary Petrology*, **36**, 1085–101.

MITCHELL, G. F., PENNY, L. F., SHOTTON, F. W. & WEST, R. G. 1973. A correlation of Quaternary deposits in the British Isles. *Geological Society of London, Special Report No. 4*.

NAGTEGAAL, P. J. C. 1978. Sandstone-framework instability as a function of burial depth. *Journal of the Geological Society of London*, **135**, 101–6.

NORTH, F. K. 1985. *Petroleum Geology*. Allen & Unwin, London.

PALMER, S. N. & BARTON, M. E. 1986. Avoiding microfabric disruption during the impregnation of friable, uncemented sands with dyed epoxy. *Journal of Sedimentary Petrology*, **56**, 556–7.

PETTIJOHN, F. J., POTTER, P. E. & SIEVER, R. 1972. *Sand and Sandstone*. Springer-Verlag, Berlin.

PRYOR, W. A. 1973. Permeability-porosity patterns and variations in some Holocene sand bodies. *Bulletin of the American Association of Petroleum Geologists*, **57**, 162–89.

RAWSON, P. F., CURRY, D., DILLEY, F. C., HANCOCK, J. M., KENNEDY, W. J., NEALE, J. W., WOOD, C. J. & WORSSAM, B. C. 1978. A correlation of Cretaceous rocks in the British Isles. *Geological Society of London, Special Report No. 9*.

RITTENHOUSE, G. 1971. Mechanical compaction of sands containing different percentages of ductile grains: a theoretical approach. *Bulletin of the American Association of Petroleum Geologists*, **52**, 92–6.

SELIG, E. T. 1963. *Effect of Vibration on Density of Sand*. 2nd Pan-American Conference on Soil Mechanics and Foundation Engineering, Rio de Janeiro, **1**, 129–44.

SELLEY, R. C. 1978. Porosity gradients in North Sea oil-bearing sandstones. *Journal of the Geological Society of London*, **135**, 119–31.

—— 1982. *An Introduction to Sedimentology* (2nd ed.). Academic Press, London.

—— 1985. *Elements of Petroleum Geology*. Freeman, San Francisco.

TAYLOR, J. M. 1950. Pore-space reduction in sandstones. *Bulletin of the American Association of Petroleum Geologists*, **34**, 701–16.

WALDSCHMIDT, W. A. 1941. Cementing materials in sandstones and their probable influence on migration and accumulation of oil and gas. *Bulletin of the American Association of Petroleum Geologists*, **25**, 1839–79.

WHITTAKER, A. (ed.) 1985. *Atlas of Onshore Sedimentary Basins in England and Wales: Post-Carboniferous Tectonics and Stratigraphy*. Blackie, London.

WOLF, K. H. & CHILINGARIAN, G. V. 1976. Diagenesis of sandstones and compaction. *In*: CHILINGARIAN, G. V. & WOLF, K. H. (eds) *Compaction of Coarse-grained Sediments*, II, 69–444. Elsevier, Amsterdam.

ZIEGLAR, D. L. & SPOTTS, J. H. 1978. Reservoir and source-bed history of Great Valley, California. *Bulletin of the American Association of Petroleum Geologists*, **62**, 813–26.

S. N. PALMER & M. E. BARTON, Department of Civil Engineering, University of Southampton, Southampton SO9 5NH, UK.

Non-steady state microbiological diagenesis and the origin of concretions and nodular limestones

R. Raiswell

SUMMARY: Morphological, chemical, isotopic and textural studies of pyritiferous carbonate concretions from the Jet Rock (Lower Jurassic, England) are all consistent in indicating an origin by anaerobic methane oxidation. Concretionary growth occurred in a stratigraphically thin, sub-surface zone of uncompacted sediment, where methane (diffusing from below) was consumed to stimulate a late phase of locally intense sulphate reduction, which caused the precipitation of $CaCO_3$ and isotopically heavy FeS_2. The carbon isotope composition of the concretionary carbonate was indistinguishable from that of sulphate reduction (using sedimentary organic matter), due to the simultaneous diffusion of isotopically light CH_4 and isotopically heavy dissolved carbonate from the underlying methanogenic zone. The alkalinity generated by anaerobic methane oxidation resulted in precipitation being restricted to a thin, vertically confined zone beneath the sediment/water interface. A pause in deposition then allowed continued precipitation in the form of concretionary growth. Different stages in the evolution of isolated concretions into nodular limestones depend on the magnitude of carbonate supersaturation in the zone of anaerobic methane oxidation and the duration of the pause in sedimentation. Vertically confined zones of carbonate precipitation may also be developed deeper in the sediment due to iron reduction, a further pause in deposition can then produce isotopically heavy carbonate concretions in the zone of methanogenesis.

The formation of carbonate concretions is one of the most ubiquitous but enigmatic features in the diagenesis of ancient organic carbon-rich sediments. Substantial progress has been made in identifying the timing and nature of the concretionary growth process and stable isotopes have proved particularly valuable in unambiguously showing that the principal authigenic carbonate source is from the microbiological decay of organic matter, either by sulphate reduction or methanogenesis (Curtis *et al.* 1972, Hudson 1978, Pearson 1979, Coleman & Raiswell 1981, Gautier 1982 and others). Concretionary growth, however, is a localized phenomenon despite the fact that every horizon in an ancient organic-rich marine sediment must have passed through the successive burial zones of sulphate reduction and/or methanogenesis. This poses a fundamental question: precisely what factors localized microbial decay to produce the stratigraphic confinement of concretionary horizons, and the siting of individual concretions?

The present contribution will address these questions specifically for the occurrence of two concretionary horizons in the Jet Rock (Lower Jurassic) of NE England. A model of concretionary growth in the Jet Rock will be proposed which should be generally applicable to other early diagenetic carbonate concretions formed by sulphate reduction (those comprising iron-poor calcite with a light $\delta^{13}C$ signature, often associated with pyrite). Finally a tentative extrapolation will be made to explain how alternative diagenetic pathways lead either to early diagenetic concretions (as above) or to later diagenetic concretions (formed by methanogenesis, giving a characteristically heavy $\delta^{13}C$ signature often associated with iron-rich carbonates).

Previous studies

Localized cementation as carbonate concretions has been envisaged as resulting either from isolated nucleation within an environment which is uniformly supersaturated with respect to carbonates, or from the generation of localized supersaturation by the decay of concentrations of organic matter at the concretionary site (Raiswell 1976). For example the excellent preservation of the fossils found inside concretions can be interpreted as suggesting that fossil cores are an essential feature of concretionary growth, either because the fossil originally contained organic body material, or because the shell nucleated carbonate precipitation.

However a nucleation control is difficult to reconcile with the extreme variability in the relative proportions of barren and fossiliferous concretions which occur in some sediments (Weeks 1957) and with the coexistence of barren

From: MARSHALL, J. D. (ed.), 1987, *Diagenesis of Sedimentary Sequences*, Geological Society Special Publication No. 36, pp. 41–54.

and fossiliferous concretions within the same horizons in which non-concretionary fossils also occur (Waage 1964). Furthemore in many cases non-concretionary fossil nuclei are sufficiently abundant in concretionary sediments such that the distribution of macrofossils is essentially unrelated to concretions (e.g. Dickson & Barber 1976, Raiswell 1976). These observations demonstrate that fossil nuclei cannot always have been the principal control on concretion siting.

Instead, siting control is usually attributed to the microbiological decay of local concentrations of organic matter (Weeks 1957, Zangerl *et al.* 1969, Sass & Kolodny 1972, Dickson & Barber 1976, Berner 1980). Experimental work (Berner 1968) has shown that the anaerobic decay of fish and clam body material can generate supersaturation with respect to carbonates, although precipitation in fact occurs as Ca soap (which may later be converted to $CaCO_3$). A well-documented example of siting control through local concentrations of organic matter occurs in the Fayetteville Shale (Zangerl *et al.* 1969), where concretionary growth occurred around coprolitic material and shells containing organic carcasses. Here the presence of organic matter was clearly indicated by several lines of evidence; namely, the textural similarities with coprolite, the absence of sediment from inside shell body chambers and, crucially, by the explosive brecciation of shells from the pressure of gases generated during microbiological decay (whose escape was prevented by concretionary cementation). Similarly, concretions which encompass intact skeletons of organisms such as fish (Weeks 1957) are also observed to preserve delicate features in tissue, body and other non-skeletal material. Such concretions clearly owe their origin to the microbiological decay of organic matter associated with the residual skeleton.

However there are considerable difficulties in accepting the decay of organic carcasses as the main control on the siting of concretions like those in the Jet Rock, because there is simply no evidence to indicate the presence of any organic carcasses. Large, intact skeletal cores are uncommon in the Jet Rock concretions, nor are there any apparent associations with an organic carcass. Instead the concretions consist of an intimate mixture of intact and broken skeletal debris, authigenic calcite and pyrite, and small proportions (<15%) of the host sediment (Raiswell 1976). There is no textural evidence for the presence of organic tissues nor is there any field evidence for a local concentration of organic matter.

Collectively, these observations can only be reconciled by proposing the existence of a mass of organic soft-bodied material which could decompose entirely without leaving any residual tissue or skeletal material, and which was also thoroughly disseminated and intimately mixed with the host sediment. Here these circumstances are regarded as unlikely and instead it is proposed that the Jet Rock concretions grew in thin zones below the sediment surface where alkalinity (and hence carbonate supersaturation) was locally enhanced by the anaerobic oxidation of dissolved methane.

Approach

At Port Mulgrave, 11 km NW of Whitby (Yorkshire), the Jet Rock consists of fine-grained, well-laminated sediment containing up to 12% organic carbon (Raiswell & Berner 1985). Two laterally extensive concretionary horizons (Howarth 1962) contain pyritiferous carbonate concretions and are known as the Curling Stones and the Cannonball Doggers.

The Cannonball Doggers range from flattened elongate bodies to regular ellipsoids (maximum diameter 30 cm) whereas the Curling Stones are commonly oblate spheroids up to 70 cm maximum diameter. In each case the host sediment is differentially compacted around the concretion, demonstrating growth and cementation in poorly compacted sediment (Raiswell 1976).

A single concretion was collected from each of the horizons bearing the Cannonball Doggers (concretion UA) and the Curling Stones (concretion UB). Each concretion was sampled by removing a slice from the centre, parallel to the bedding plane. This slice was further sub-divided to provide a representative edge-centre-edge traverse. Further details of the sampling scheme and analytical methodologies are given elsewhere; here we are concerned with a new interpretation, integrating earlier chemical (Raiswell 1976), isotopic (Coleman & Raiswell 1981) and textural (Raiswell 1982) data in the context of recent studies of anaerobic pore water chemistry.

Chemical, isotopic and textural features of the Jet Rock concretions

The two most extensively studied concretionary horizons in the Jet Rock (the Curling Stones, the Cannonball Doggers) can be traced at the same stratigraphic level over nearly 20 km distance through which there is little apparent variation in the size, shape or frequency of the concretions. Clearly the horizons are laterally persistent but stratigraphically confined. Evidence for their early diagenetic origin in uncompacted sediment is conclusive. Field observations include the presence of uncrushed skeletal debris and the compaction of host sediment laminae around the concretions (Raiswell 1976). An early origin is further confirmed by estimates of the host sediment porosity at the time of concretionary growth, based on the assumption that the concretion cement replaces water-filled pore space (Raiswell 1971). Porosity estimates of 77–87% were obtained from the Jet Rock concretions which indicates an origin in the top few metres of sediment (Raiswell 1976).

The carbon isotope composition of the concretionary carbonate is −13 to −16‰ (Coleman & Raiswell 1981), which has been interpreted as mainly indicating a mixture of marine-derived dissolved carbonate (at about 0‰) and carbonate generated by sulphate reduction from organic matter (at −25 to −30‰). However, organic carbon contents inside the concretions show no clear evidence for any local concentrations of organic carbon, on an authigenic mineral (i.e. carbonate- and pyrite-free) basis. Thus the concretions (UA 8.3 ± 2.2%C, UB 8.6 ± 3.7%C) are more variable, but not statistically dissimilar, in organic carbon content to the host sediments (UA 6.3 ± 0.1%C, UB 9.0 ± 0.6%C). Moreover the isotopic compositions of the residual organic matter inside the concretions (UA −32 to −33‰, UB −26 to −27‰) and outside in the sediments (UA −31 to −35‰, UB −30 to −31‰) are similarly light, probably due to the selective preservation of a lipid fraction (Coleman & Raiswell 1981). These observations indicate that microbial concretionary growth required an organic matter source which could decompose entirely to dissolved carbonate, without leaving any chemical or isotopic signature.

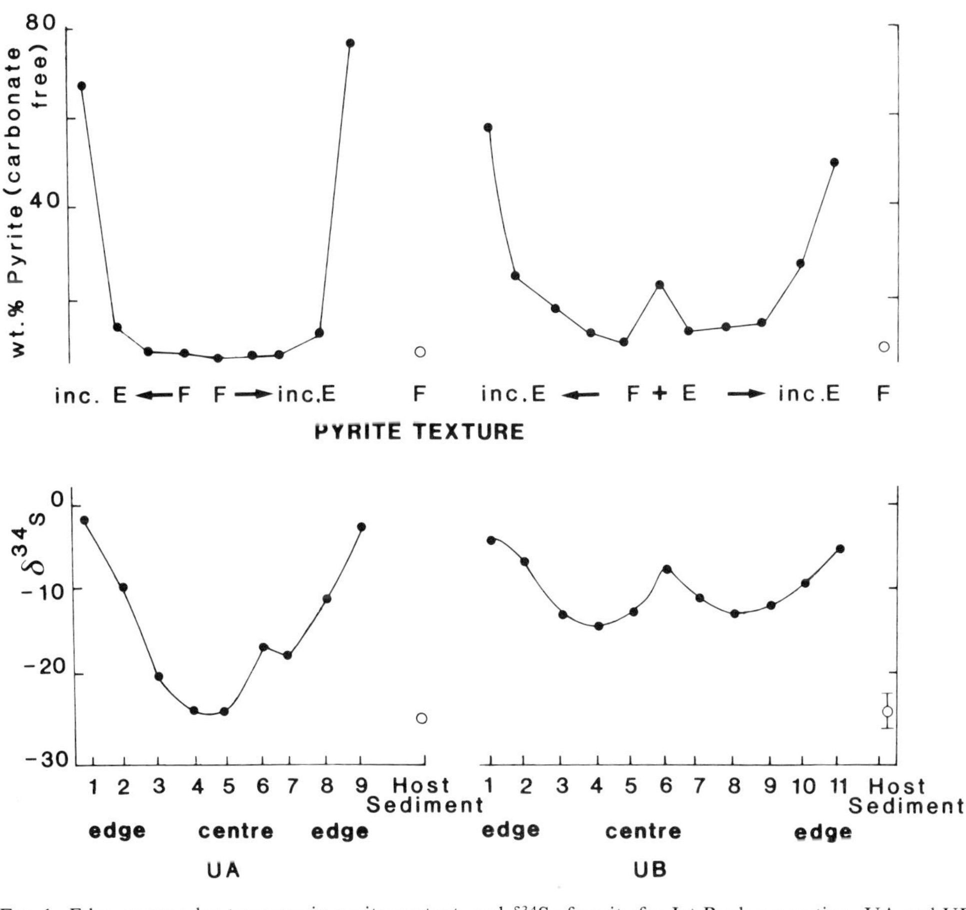

FIG. 1. Edge-centre-edge traverses in pyrite content, and $\delta^{34}S$ of pyrite for Jet Rock concretions UA and UB. Host sediments contain mainly framboidal pyrite (F), as also does the central part of concretion UA. The centre of concretion UB contains framboids overgrown by euhedral pyrite (E). Both concretions show a trend of increasing euhedral pyrite towards their margin (after Coleman & Raiswell 1981).

Figure 1 shows values for the masses and isotopic composition of pyrite inside the concretions and host sediments (on a carbonate-free basis). Pyrite in both host sediments occurs mainly as framboids, which have $\delta^{34}S = -24$ to $-26‰$. Framboids appear similarly abundant inside both concretions, where they are enveloped by a later phase of equigranular, euhedral pyrite which is associated with concretionary growth. This pyrite nucleates on the framboids and steadily increases in abundance towards the concretion margins (Raiswell 1982), as shown in Fig. 1.

The pyrite at the centre of concretion UA is exclusively framboidal and, after correction to a carbonate-free basis, its abundance is similar to that in the host sediment. However with increasing amounts of the syn-concretionary euhedral pyrite, the bulk isotopic composition becomes steadily heavier and reaches -2 to $-3‰$ at the margin of the concretion. By contrast the samples at the centre of concretion UB have some euhedral overgrowths on the framboids, and also contain more pyrite than the host sediments on a carbonate-free basis (Fig. 1). The presence of the euhedral pyrite causes the central samples of concretion UB to be isotopically heavier ($\sim -8‰$) and more pyritiferous than the host sediments. Apart from these central samples, concretion UB shows essentially the same trend towards isotopically heavy pyrites at the concretion margin (-5 to $-6‰$) as the euhedral pyrite becomes more abundant. The margins of both concretions contain 5–10 times more pyrite than their host sediments.

The trends in both concretions can be explained on the same basis, where an initial phase of framboids formed uniformly throughout the sediment prior to concretionary growth. The isotopic composition of these pyrites indicates that they formed at shallow depths, in a pore system with a free diffusive connection to overlying seawater. These pre-concretionary pyrites were passively occluded by the growing concretions; thus framboids of the same isotopic composition and abundance can be recognized at the centre of concretion UA and in the host sediments. Concretionary growth itself was associated with formation of the euhedral pyrite ($\delta^{34}S$ ranging up to $-2‰$), which increased the total pyrite in the concretions to levels significantly above those in the host sediments. The isotopically heavy pyrite formed in a pore system from which sulphate was depleted.

In summary, these data indicate that concretionary growth:

(i) Was confined to a thin, subsurface zone of uncompacted sediment within a few metres of the sediment surface.

(ii) Required an organic matter source which could decompose entirely to leave no chemical or textural trace.

(iii) Occurred at a depth where the pore system contained high concentrations of isotopically light carbonate, sufficient to produce carbonate supersaturation.

(iv) Was uniquely characterized by the occurrence of a renewed phase of local sulphate reduction, at a depth within the sediment where sulphate was isotopically heavy and hence depleted.

This combination of circumstances has been recognized in studies of modern pore waters, and provides a unique fingerprint for the involvement of anaerobic methane oxidation in concretionary growth.

Anaerobic methane oxidation in modern sediments

The presence of anaerobic methane oxidation has been inferred from studies of pore water chemistry in a variety of modern sediments (Barnes & Goldberg 1976, Reeburgh 1976, 1980, Martens & Berner 1977, Devol & Ahmed 1981, Devol 1983). These studies show the following generalized features in the depth distributions of dissolved carbon and sulphur species and their isotopic compositions. Distinct zones of steady state diagenesis can be recognized in modern anaerobic sediments (Fig. 2), arising from the influences of sulphate reduction, methane oxidation and methanogenesis (Reeburgh 1980, Rice & Claypool 1981) as described below.

Sulphate reduction

In the uppermost zone sulphate reduction

$$2CH_2O + SO_4^{2-} \rightarrow 2HCO_3^- + H_2S \quad (1)$$

causes a steady decrease in dissolved sulphate with depth. The instantaneous isotopic fractionation between seawater sulphate ($+20‰$) and sulphide is commonly about -40 to $-60‰$, so that the pyrite being formed by reaction between dissolved H_2S and detrital iron oxides is always lighter than the residual sulphate (which itself progressively evolves towards heavier values). Thus later pyrites formed from a depleted sulphate reservoir are isotopically heavy. Dissolved carbonate ($\sum CO_2$) concentrations increase as sulphate decreases and, because the carbonate is derived from the oxidation of organic matter ($\delta^{13}C = -25$ to $-30‰$), its isotopic composition decreases rapidly with depth from surface values of $0‰$ to approach the

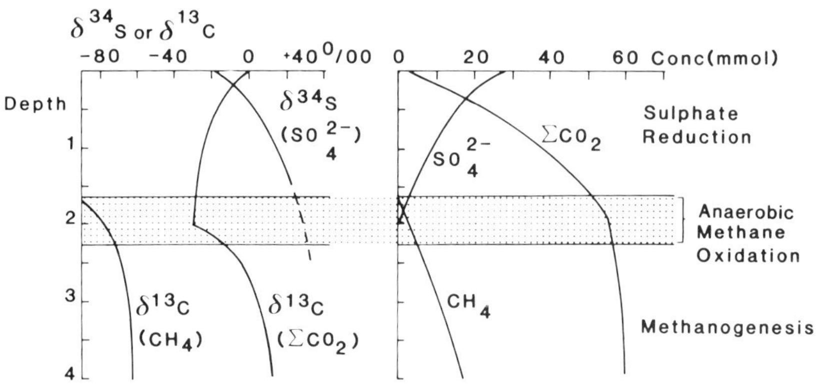

FIG. 2. Steady state diagenetic zones in modern anaerobic sediments (depth scale in arbitrary units). Sulphate Reduction Zone is characterized by $[SO_4{}^{2-}] \sim 0$ to 28 mmol with $\delta^{34}S$ at ~ 0 to $-20‰$; $[\Sigma CO_2] > 2$ mmol with $\delta^{13}C$ at 0 to $-25‰$. Anaerobic Methane Oxidation Zone (shaded) has $[SO_4{}^{2-}] \sim 0$ mmol with $\delta^{34}S \sim 0‰$; $[\Sigma CO_2] \gg 2$ mmol with $\delta^{13}C$ at -15 to $-25‰$; $[CH_4]$ low with $\delta^{13}C$ at -60 to $-80‰$. Methanogenic Zone has $[\Sigma CO_2] \gg 2$ mmol with $\delta^{13}C \sim 0‰$; $[CH_4]$ increasing with $\delta^{13}C$ at -60 to $-80‰$. After Reeburgh (1980) and Rice & Claypool (1981).

value of the organic phase. Note that sulphate reduction adds dissolved carbonate as $HCO_3{}^-$ and the consequent increases in alkalinity (together with that due to ammonia formation) produce oversaturation with respect to $CaCO_3$ (Berner *et al.* 1970, Sholkovitz 1973). The precipitation of the pre-concretionary pyrite by the reaction between iron oxides and H_2S,

$$2Fe_2O_3 + 8SO_4{}^{2-} + 15CH_2O \rightarrow 4FeS_2 + 15HCO_3{}^- + 7H_2O + OH^- \quad (2)$$

also generates alkalinity (Coleman 1985). Clearly the processes occurring in the sulphate reduction zone are important in producing oversaturation of $CaCO_3$, but most Ca^{2+} removal from porewaters by precipitation appears to occur at the base of the sulphate reduction zone, where methane oxidation becomes significant, e.g. 5 cm and below in Saanich Inlet sediments (Murray *et al.* 1978, Devol 1983).

Methanogenesis

Below the sulphate reduction zone, the main microbiological process is methanogenesis which is often referred to as CO_2 reduction (Claypool & Kaplan 1974). The reducing agent is usually assumed to be H_2, which must be derived from organic matter.

$$2CH_2O + H_2O \rightarrow 2CO_2 + 4H_2 \quad (3)$$

$$4H_2 + CO_2 \rightarrow CH_4 + 2H_2O. \quad (4)$$

Methanogenic bacteria can also use other substrates but the above mechanism probably accounts for most methane produced in deep water sediments (Rice & Claypool 1981). Methane production does not apparently become significant until pore water sulphate is nearly depleted although the reasons for this are still under discussion (Claypool & Kvenvolden 1983). In any event the pore-water profiles show that dissolved CH_4 becomes significant only when sulphate is nearly depleted, also that concentrations of dissolved carbonate subsequently increase less rapidly, and are close to their maximum levels (Claypool & Kaplan 1974, Reeburgh & Heggie 1977, Reeburgh 1980, Claypool & Kvenvolden 1980, Devol *et al.* 1984), because CH_4 is now being produced by CO_2 reduction. The isotopic trends in dissolved carbonate confirm these processes. The CO_2 removed to form methane is typically $\sim 70‰$ lighter than the $\delta^{13}C$ of the bulk dissolved carbonate, due to a kinetic isotope effect (Claypool & Kaplan 1974). Thus continuing methanogenic CO_2 reduction is indicated by increasing $\delta^{13}C$ values of the residual dissolved carbonate. The relative rates of metabolic CO_2 addition and methanogenic removal will determine the precise nature of the depth changes in the isotopic composition and concentrations of dissolved carbonate, but dissolved carbonate usually reaches its minimum $\delta^{13}C$ value at the start of methanogenesis, although concentrations of dissolved carbonate continue to increase (albeit now more slowly), as shown in Fig. 2. Note that dissolved carbonate is added as CO_2 in the methanogenic zone (rather than as $HCO_3{}^-$ in the sulphate reduction zone) and hence alkalinity will now decrease (Rice & Claypool 1981) unless significant iron reduction accompanies methanogenesis (Coleman 1985), as shown below.

$$13CH_2O + 2Fe_2O_3 + 3H_2O \rightarrow 6CH_4 + 7HCO_3^- + 4Fe^{2+} + OH^-. \quad (5)$$

In the Jet Rock, iron is almost completely consumed to form the pre-concretionary pyrite (Raiswell 1982) and hence insufficient alkalinity can be generated by iron reduction to offset the production of CO_2 by methanogenesis alone.

Anaerobic methane oxidation

A significant feature of Fig. 2 is that dissolved methane concentrations are low or absent in near-surface sediments but increase with depth. It is argued that this type of profile can only result from CH_4 consumption reactions which act as a barrier to the transfer of methane across the sediment/water interface (Martens & Berner 1977, Reeburgh 1976, 1980, Devol 1983). In sediments deposited from anoxic bottom waters (as is believed to be the case for the Jet Rock; Raiswell & Berner 1985), removal cannot occur by bioturbation or aerobic oxidation and anaerobic consumption by sulphate reduction has been proposed.

$$CH_4 + SO_4^{2-} \rightarrow HS^- + HCO_3^- + H_2O. \quad (6)$$

The reaction is thermodynamically favourable and sulphate reducers capable of oxidizing CH_4 have been cultured in the laboratory, but not isolated from modern sediments (Reeburgh 1980). Observations of pore water chemistry provide the strongest evidence for anaerobic methane oxidation and these have recently been successful in documenting the existence of a thin zone of locally enhanced sulphate reduction, occurring at the base of the sulphate reduction zone (Fig. 3). The zone where sulphate reduction rates are enhanced also corresponds with maximum rates of methane oxidation and occurs within the top metre of sediment (Devol & Ahmed 1981, Devol 1983, Devol *et al.* 1984, Iverson & Jorgensen 1985).

The contributions of anaerobic methane oxidation to the isotopic composition of the dissolved carbonate in the porewater will initially be exceedingly light, since the CH_4 utilized will be ~70‰ lighter than the CO_2 from which it was derived. However the removal of light CO_2 to form CH_4 is accompanied by a rapid increase in the $\delta^{13}C$ of the residual carbonate (see earlier). The overall isotopic composition of the pore waters in the zone of anaerobic methane oxidation will initially reflect the size and isotopic composition of the pool of dissolved carbonate initially added from skeletal debris and seawater (at 0‰) and from sulphate reduction (at −25 to −30‰), which is modified by the rate of addition of light CO_2 from anaerobic methane oxidation, and by the rates of diffusion of dissolved carbonate to, or from, adjacent porewaters.

A more detailed attempt is made in the following section to model the $\delta^{13}C$ changes in the zone of anaerobic methane oxidation. At this point it is noted that despite the complexity of the system the generalized sequences of Reeburgh (1980) and Claypool & Kvenvolden (1983) show that the boundary between sulphate reduction and methanogenesis is marked by minimum $\delta^{13}C$ values for dissolved carbonate (−15 to −25‰) at maximum or near-maximum concentrations of dissolved carbonate. This range of $\delta^{13}C$ values is typical of those found in carbonate concretions, probably because anaerobic methane oxidation increases carbonate supersaturation to the point of precipitation more rapidly than the $\delta^{13}C$ of the dissolved carbonate pool can

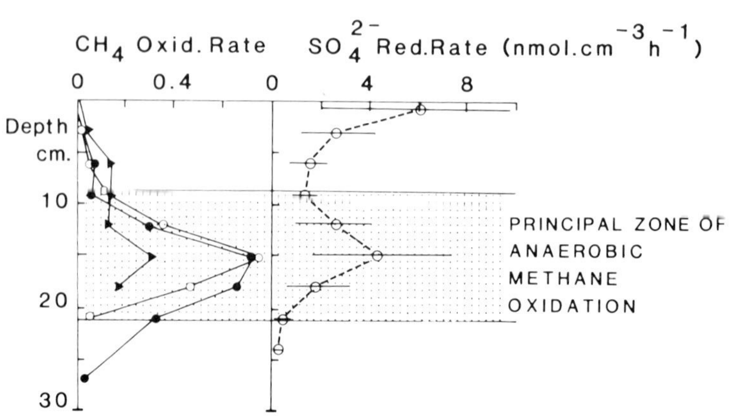

FIG. 3. Correspondence between the depth profiles of rates of methane oxidation and rates of sulphate reduction, indicating methane consumption by sulphate reducing bacteria (after Devol 1983).

be changed (see later). Atypical circumstances where the $\delta^{13}C$ of the dissolved carbonate can be more rapidly altered may be responsible for the very light $\delta^{13}C$ values recorded in some concretions (e.g. $-42‰$, Hudson & Friedman 1978). Anaerobic methane oxidation clearly increases alkalinity (equation 6), and thus may produce further sufficient increases in carbonate supersaturation (over and above that resulting from the earlier phase of sulphate reduction and pyrite formation; see equations 1 and 2) to induce precipitation.

In summary, these studies of modern sediments indicate that anaerobic methane oxidation:

(i) Is confined to a thin subsurface zone of uncompacted sediment, within a metre of the sediment surface. Zone thickness are 10–50 cm (Reeburgh 1983) and the upper surface lies about 10–100 cm below the sediment/water interface (Reeburgh 1976, 1980, Devol & Ahmed 1981, Devol 1983, Iverson & Jorgensen 1985).

(ii) Utilizes dissolved or gaseous methane, which could be decomposed without leaving any residual phases in the sediment.

(iii) Occurs at a depth where the pore system contains high concentrations of isotopically light carbonate, and where the further increases in alkalinity could be sufficient to cause carbonate precipitation.

(iv) Is uniquely characterized by the occurrence of a renewed phase of sulphate reduction, at a depth within the sediment where sulphate is substantially depleted and isotopically heavier.

Carbon isotope sources to the zone of methane oxidation

The previous section demonstrated that pore waters in the zone of anaerobic methane oxidation typically have carbon isotope compositions in the range -15 to $-25‰$. Under steady state diagenesis this signature potentially reflects an initial reservoir composition inherited from shallow diagenetic sources (marine-derived material and sulphate reduction carbonate) modified by the addition of light dissolved carbonate from anaerobic methane oxidation and by diffusive fluxes. The usual treatment of carbon isotopes in diagenesis has been to assume that the carbonates precipitated reflect the isotopic signature of the input processes (i.e are independent of initial reservoir composition). In general this is only true when the initial isotopic reservoir is small, relative to the mass of carbonate precipitated. That this is also the case for the Jet Rock concretions is demonstrated below.

The Curling Stones are separated by a mean nearest neighbour distance of approximately 0.47 m (Raiswell & White 1978). The Curling Stone analysed by Raiswell (1976) is of average size and contains approximately 153 moles of $CaCO_3$. Typical anaerobic pore waters at the base of the sulphate reduction zone contain about 50 meq l^{-1} of alkalinity (e.g. Berner *et al.* 1970, Sholkovitz 1973). Assuming that all alkalinity is HCO_3^- and that its complete precipitation can be achieved, a supply zone of radius 23.5 cm in a sediment of 85% porosity is estimated to contain the following amounts of carbonate.

$$\frac{4}{3}\pi \frac{(23.5)^3}{1000} \times 0.85 \times 50 \times 10^{-3}$$
$$= 2.3 \text{ moles } HCO_3^-.$$

This is one to two orders of magnitude less than that required for concretionary growth. Clearly most carbonate must be derived from inputs uniquely associated with concretionary growth, furthermore the isotopic signature of the concretions will reflect these inputs, rather than that from the pool of dissolved carbonate arising from the pre-concretionary diagenetic processes.

There are two kinetic fractions which are of potential importance in the methane formation and oxidation cycle. The first is that associated with methanogenesis, where the CO_2 removed to form CH_4 is around 70‰ lighter than the CO_2 from which it was derived. Continued removal of CO_2 thus leads to the residual dissolved carbonate becoming heavier (Fig. 2), and hence the methane also evolves to heavier values. The initial methane often ranges from -70 to $-85‰$ and porewaters found in the methanogenic zone (Claypool & Kaplan 1974) typically show that methanogenesis has continued to give CH_4 at -60 to $-70‰$, which is associated with dissolved carbonate at 0 to $+10‰$. Note that whatever the concentrations and isotopic compositions of the separate CH_4 and CO_2 phases, the bulk carbon phase ($CO_2 + CH_4$) will still retain the approximate isotopic signature of the organic matter from which the CO_2 and CH_4 were both initially derived (i.e. -25 to $-30‰$).

A further kinetic fractionation is associated with anaerobic methane oxidation, such that the residual methane becomes heavier by 2–14‰; Whiticar & Faber (1986). Thus the bicarbonate produced by oxidation should be this amount lighter than the CH_4 from which it was derived. In proposing methane as the organic carbon source for concretionary growth, it is clearly necessary to explain why exceedingly light $\delta^{13}C$

values are not commonly found in carbonate concretions. It is suggested here that both CH_4 and dissolved carbonate diffuse upwards to the zone of methane oxidation in such a way that their net isotopic signature (-25 to $-30‰$) is not grossly altered. Once in the zone of methane oxidation, the light HCO_3^- derived from CH_4 mixes with the heavier transported HCO_3^- and thus a light diagenetic phase is precipitated.

Flux measurements of dissolved carbonate and CH_4 suggest that this explanation is reasonable. Sediments from the Cariaco Trench have equal or nearly equal fluxes of CH_4 and dissolved carbonate from the methanogenic zone into the zone of methane oxidation (Reeburgh 1976). In Saanich Inlet, methanogenesis produces CH_4 and CO_2 at similar rates (Murray *et al.* 1978), however, although the existing carbonate pool (mainly from sulphate reduction) enhances carbonate fluxes over CH_4, the overall $\delta^{13}C$ value must still be light. In these cases the extent of fractionation within the methanogenic zone (and the oxidation zone) will be unimportant and the net isotopic signature of the carbon phases may be close to that of organic matter, provided methane is completely oxidized to dissolved carbonate. The absence of CH_4 in near surface sediments (Fig. 2) indicates that this is generally the case, except perhaps where rapid methane ebullition is caused by tidal hydrostatic pressure changes in shallow marine sediments (Martens & Klump 1984). This flux evidence provides strong support for the transportation of dissolved carbonate plus CH_4 without isotopic separation, as is also indicated by the $\delta^{13}C$ values of dissolved carbonate in the zone of methane oxidation (Fig. 2). In these circumstances the isotopic signature of sulphate reduction using sedimentary organic matter and using CH_4 will be indistinguishable.

Synthesis

The two summaries demonstrate a remarkable similarity between the chemical and physical conditions required for concretionary growth and caused by anaerobic methane oxidation. The carbon isotope evidence has also been shown to be consistent with the involvement of methane oxidation in concretionary growth, and a causal link between the two processes is strongly indicated. Further studies of modern porewaters could add still more support if full porewater anlyses were made which would allow carbonate saturation indices, and rates of carbonate precipitation, to be estimated and to be linked to rates of anaerobic methane oxidation. No such data have yet been published, although the data of Murray *et al.* (1978) do appear to demonstrate carbonate removal in Saanich Inlet sediments at depths both before and during anaerobic methane oxidation. In the following section a model of concretionary growth is proposed which explains how the involvement of methane oxidation imposes spatial constraints on the distribution of concretionary growth.

Chemical and physical constraints on the distribution of concretionary growth

The features of anaerobic methane oxidation already discussed are summarized in the following model of concretionary growth (Fig. 4). The growth process is initiated at depth, in the methanogenic zone, where isotopically light CH_4 and heavy dissolved carbonate are produced. Both diffuse upwards at rates which in most cases are not sufficiently different to produce a distinct isotopic separation. Thus the carbon phases overall more or less retain their original, organic matter, isotopic signature. In the zone of anaerobic methane oxidation, the HCO_3^- derived from CH_4 mixes with the transported dissolved carbonate and the locally enhanced alkalinity generates carbonate precipitation. Iron may also diffuse upwards from the methanogenic zone to react with the sulphide produced by methane oxidation (Raiswell 1982).

The vertical confinement of concretionary horizons arises in the first instance from the chemical controls on the methane oxidation process, which is constrained upwards by lack of CH_4 (because CH_4 generation does not begin until sulphate is nearly depleted) and downwards by lack of sulphate. Thus increased carbonate supersaturation is generated only in a vertically localized zone.

The preceding phase of sulphate reduction may also be important in increasing alkalinity, and hence carbonate supersaturation, to the point where precipitation is close. Whether or not concretionary growth then occurs would depend on sufficient further supersaturation being induced by anaerobic methane oxidation, despite the inhibiting effects of dissolved organics and Mg^{2+} on $CaCO_3$ precipitation. Anaerobic methane oxidation appears to be more important at some locations than others (Devol *et al.* 1984) and in some cases its magnitude may be insufficient to cause carbonate precipitation. The reasons for variations in the importance of anaerobic methane oxidation between locations are not at present understood (Devol *et al.* 1984).

Within the zone of methane oxidation, rates of sulphate reduction may be seasonally and spatially variable (Reeburgh 1980, Devol & Ahmed 1981, Crill & Martens 1983, Iverson & Jorgensen 1985). Seasonal variations are probably unimportant on the time-scale of concretionary

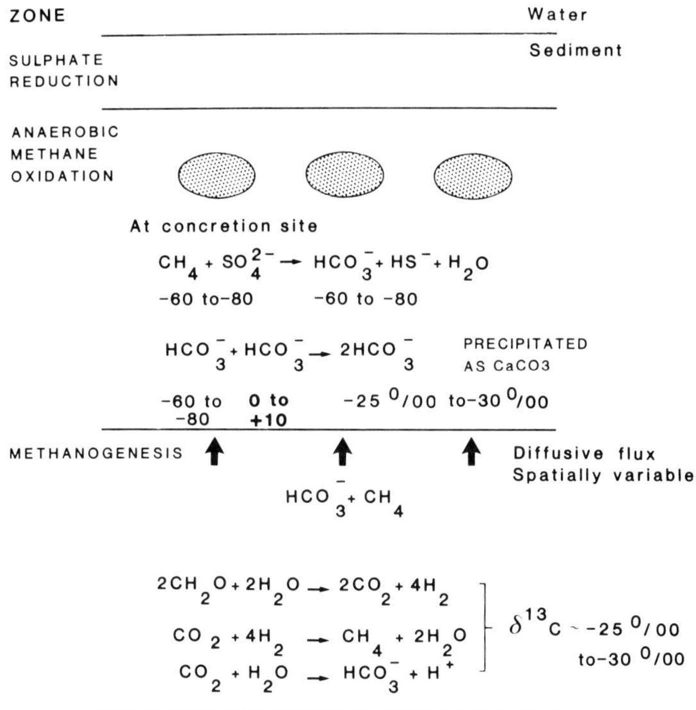

FIG. 4. A schematic representation of the chemical processes occurring at the site of a growing concretion, where a relatively high flux of methane sustains high methane oxidation rates, and hence produces locally enhanced carbonate supersaturation. Equal HCO_3^- and CH_4 fluxes from methanogenesis, unaccompanied by other carbonate sources.

growth, but the spatial variations may determine the siting of individual concretions. The most permeable vertical pathways will have the highest rates of CH_4 transport from the methanogenic zone and into the zone of methane oxidation, and these are likely to produce the highest rates of sulphate reduction and are the most probable sites of carbonate precipitation and concretionary growth. Such sites will usually be randomly distributed, as indeed are the Jet Rock concentrations (Raiswell & White 1978). The presence of favourable nuclei (e.g. macrofossils) near a potential site of concretionary growth may initiate carbonate precipitation at a somewhat lower degree of supersaturation, but siting is mainly controlled by the generation of local carbonate supersaturation through enhanced rates of sulphate reduction (in turn determined by minor vertical permeability variations which produce point-to-point variations in the CH_4 flux from below). Any fine-grained sediment containing skeletal debris should have at least some grains capable of nucleating carbonate precipitation at a potential concretion site, thus no consistent relationship should be expected between concretionary growth and macrofossil distribution.

However the occurrence of anaerobic methane oxidation is not, in itself, sufficient to cause concretionary growth. An idealized model of anaerobic methane oxidation can demonstrate that a further crucial requirement is a pause in sedimentation (or an abrupt decrease in sedimentation rates). For example, assume that the depth variation in the carbonate saturation index roughly mimics the depth variations in the rates of methane oxidation (Fig. 5), and that the shallower pore waters are on the threshold of carbonate precipitation. Thus further supersaturation resulting from methane oxidation is represented by the area of the shaded triangle. Under continuous sedimentation the area swept by the triangle is always the same throughout the sediment column and a diffuse phase of carbonate cementation would result. By contrast, any sedimentological break stabilizes the potential zone of carbonate precipitation at a fixed distance below the sediment/water interface.

Concretion morphology (Fig. 5) would then be a function of:

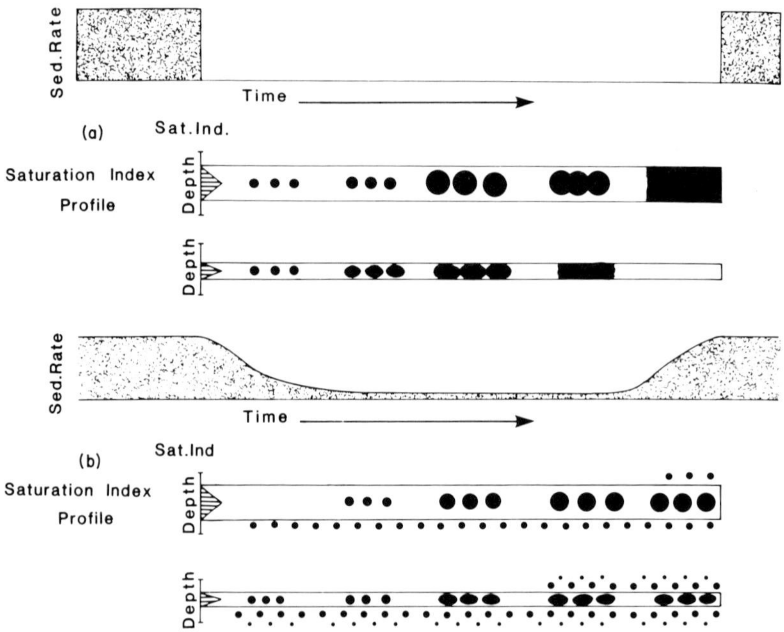

FIG. 5. Schematic representation of the progressive variations in concretion morphology caused by changes in sedimentation rates with time, and carbonate saturation index with depth. A complete break in sedimentation (a) fixes the zone of anaerobic methane oxidation (AMO), and hence carbonate supersaturation, a few tens of centimetres below the sediment/water interface. The vertical extent of the concretionary horizon is then determined by the thickness of the zone of AMO and coalescence is encouraged by longer breaks in sedimentation and high saturation indices. During periods of reduced sedimentation (b), the zone of AMO may instigate brief periods of growth at several closely spaced horizons, with the size and frequency of concretionary growth reaching its maximum extent when sedimentation rates are at a minimum.

(a) Relative changes in sedimentation rates. A complete absence of deposition would give the minimal possible vertical range between the highest and lowest concretions in that horizon, slow deposition would give a longer vertical range.
(b) Time. Longer sedimentological breaks would encourage the growth of large concretions, and ultimately would tend to favour the coalescence of adjacent concretions into diagenetic limestone beds.
(c) Shape of the saturation index versus depth profile. Sharp, narrow profiles would favour a limited vertical extent to the concretionary bed and would tend to produce coalescence into thin limestones. A wider profile would disperse concretionary growth over a thicker sequence of sediments.

The non-steady state conditions arising from a depositional hiatus would, of course, change the nature of the isotopic gradients in Fig. 2. In the methanogenic zone, both methane and dissolved carbonate evolve to heavier values with time. In this model it has been assumed that the diffusive fluxes of methane and dissolved carbonate are nearly equal, such that their net isotopic signature is approximately that of sedimentary organic matter. The net signature of the residual carbon pool then also retains the same value. However, if the diffusive fluxes differ, then the carbonate precipitated in the zone of anaerobic methane oxidation will become either lighter (if the methane flux is larger) or heavier (if the carbonate flux is larger) than sediment organic matter. Furthermore the cements will then also evolve with time. In fact the Jet Rock concretions have edges which are little more than 1–2‰ heavier than their centres (Coleman & Raiswell 1981), which implies that methane and dissolved carbonate fluxes were little different.

Clearly representative horizons of concretionary growth will tend to arise where sedimentation is episodic, and the closer steady state conditions are approached the less likely are concretionary phenomena. A model of this type can explain cyclic limestone-shale rhythms, as in the Lower Lias of southern England (Hallam 1964) and the Ordovician of Norway (Gluyas 1984). The limestone horizons in these sequences are isotopically attributable to mixtures of marine-derived and sulphate reduction carbonate (Campos & Hallam 1979, Gluyas 1984) and they differ from the Jet Rock concretions only in two important respects; their morphology and the smaller amounts of a pyrite phase specifically associated with anaerobic methane oxidation. The differences in morphology can be explained by a suitable choice of the variables illustrated in Fig. 5, such that concretionary growth occurred in thin zones and was of sufficient persistence to result in coalescence of the individual growth centres. Less later pyrite is probably the result of low amounts of reactive iron within the sediment, after the end of the initial phase of sulphate reduction (Raiswell 1982, Coleman & Raiswell in preparation).

Alternative diagenetic pathways to methane-derived or methanogenic concretions

The above conclusions, it should be emphasized, apply in detail only to the early diagenetic calcareous concretions with an isotopically light cement, as distinct from the later concretions discussed by Curtis *et al.* (1972), Irwin *et al.* (1977), Pearson (1979) and Gautier (1982), where the carbonate phase is often iron-rich (sideritic or ankeritic) and invariably isotopically heavy. However the same characteristics of non-steady state sedimentation may also favour the formation of these methanogenic carbonates, where the carbonate saturation index instead develops a deeper maximum in the methanogenic zone. The following tentative model outlines how this

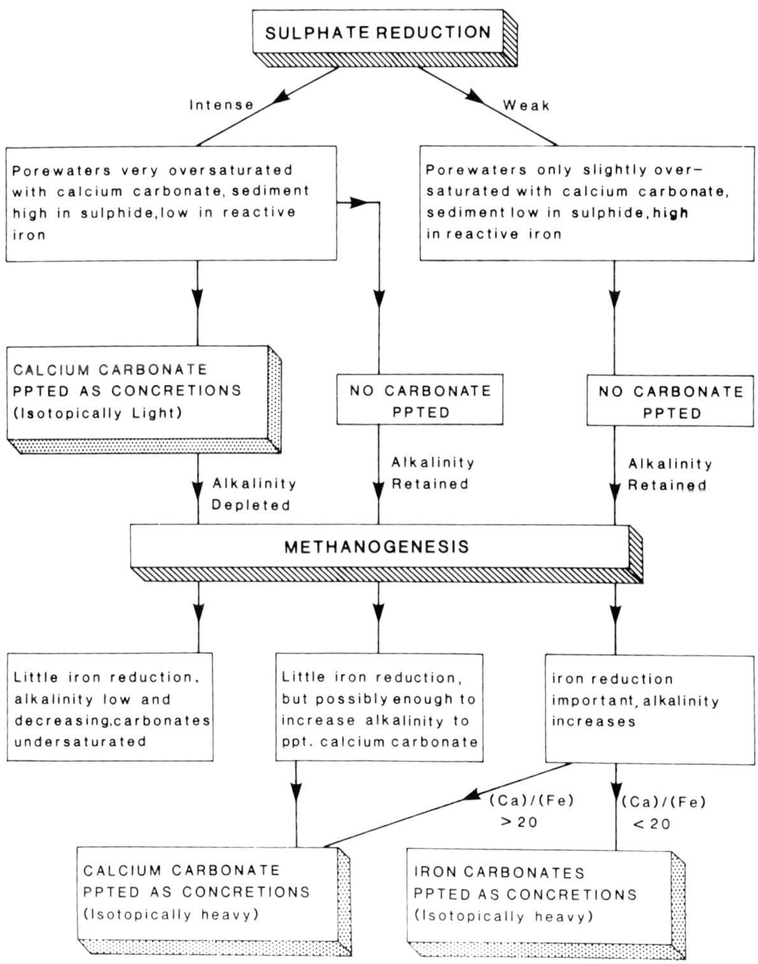

FIG. 6. Alternative diagenetic pathways leading to the location of a zone of carbonate precipitation either in the zone of anaerobic methane oxidation, or the zone of methanogenesis. The role of iron is crucial in determining whether, and where, sufficient alkalinity can be developed to give carbonate supersaturation. The criteria of Berner (1971) is used as a general indication of how Ca/Fe ratios affect carbonate mineralogy.

can occur, by showing alternative diagenetic pathways which could lead to early, isotopically light $CaCO_3$ formed by methane oxidation and later, isotopically heavy carbonates formed by methanogenesis (Fig. 6). The role of reactive iron is crucial (Coleman 1985) and two end-member cases can be so identified depending on whether or not reactive iron is substantially depleted prior to burial into the methanogenic zone.

Both sulphate reduction and anaerobic methane oxidation generate alkalinity as HCO_3^- (equations 1 and 6) and hence produce porewaters which become oversaturated with respect to $CaCO_3$ (e.g. Berner *et al.* 1970, Murray *et al.* 1978). The greater the intensity of these processes, the more alkalinity is generated and the more likely is $CaCO_3$ precipitation. However the concomitant generation of H_2S also consumes iron to form pyrite, such that little reactive iron survives an intense phase of sulphate reduction and methane oxidation. The extent of iron depletion can be estimated by the parameter Degree of Pyritization (DOP) defined by Berner (1971) as

$$\mathrm{DOP} = \frac{\text{Pyrite Fe}}{(\text{Pyrite Fe} + \text{HCl-soluble Fe})}.$$

Calculation of DOP values assumes that all pyrite Fe was originally present in a form reactive towards H_2S, also that HCl-soluble Fe would so react, given sufficent time. In the Jet Rock, DOP values in excess of 0.85 occur and hence little reactive iron survived passage through the zones of sulphate reduction which have the pre-concretionary pyrite. If $CaCO_3$ concretions are subsequently precipitated as a result of methane oxidation, then the porewaters entering the methanogenic zone will be substantially depleted in alkalinity and Ca^{2+}, because

$$Ca^{2+} + 2HCO_3^- \rightarrow CaCO_3 + CO_2 + H_2O. \quad (7)$$

Furthermore methanogenesis generates CO_2, rather than alkalinity, hence the porewaters will approach undersaturation with respect to $CaCO_3$, unless a new alkalinity-generating mechanism occurs. This cannot be iron reduction (Coleman 1985)

$$3H_2O + 2Fe_2O_3 + CH_2O \rightarrow HCO_3^- + 4Fe^{2+} + 7OH^- \quad (8)$$

since insufficient reactive iron is now available to produce the alkalinity needed to offset methanogenic CO_2 production. Thus the very intensity of the early sulphate reduction and methane oxidation phases of diagenesis needed to produce $CaCO_3$ precipitation will, in general, preclude further carbonate precipitation by producing porewaters of low alkalinity and sediments with little alkalinity-generating potential. This diagenetic pathway is exemplified by the Jet Rock concretions.

The alternative pathway is where sulphate reduction is less intense (and anaerobic methane oxidation weak or non-existent). Here insufficient alkalinity is generated to result in early precipitation, and relatively large amounts of reactive iron (low DOP values) survive entry into the methanogenic zone. Porewaters entering this zone are perhaps over-saturated with respect to $CaCO_3$, but insufficiently so to induce precipitation. Concretionary growth can now occur if iron reduction (see above) generates alkalinity sufficient to offset the CO_2 production by methanogenesis. Iron-rich carbonate will result, with the iron content (either iron-rich calcites, siderites or ankerites) reflecting the relative proportions of Ca^{2+} and Fe^{2+} in the porewaters (values in Fig. 6 from Berner 1971), and the occurrence of calcite or dolomitic-type minerals reflecting the abundance in sulphate (Coleman 1985). This pathway is very probable in freshwater sediments, where low sulphate concentrations act to limit the extent of pyrite formation, with the result that proportionally more iron survives burial into the methanogenic zone. However the same pathway can result in marine sediments, where isotopically heavy siderite concretions (Gautier 1982) are found because rapid burial limited the time spent in the sulphate reduction zone, and hence iron was subsequently still available to generate a methanogenic alkalinity maximum.

Conclusions

Morphological features of pyritiferous carbonate concretions in the Jet Rock show that growth had to occur during early diagenesis and was confined to a thin, subsurface zone of uncompacted sediment, within a few metres of the sediment surface. These concretions contain no evidence of formation around local concentrations of organic matter, although their carbon isotope composition (-13 to $-16‰$) clearly indicates a biogenic source. Crucial evidence as to the nature of that source is revealed by the concretionary pyrite, which is both isotopically heavier than the host sediment pyrite (-24 to $-26‰$) and also texturally distinct. Thus an early phase of pyrite formation occurred throughout the sediment prior to concretionary growth and the concretion site was subsequently distinguished from the surrounding sediments by the occurrence of locally intense, later phase of sulphate reduction and pyrite formation which occurred in a pore system now depleted in sulphate.

Studies of modern sediments show that anaerobic methane oxidation

$$CH_4 + SO_4{}^{2-} \rightarrow HCO_3{}^- + HS^- + H_2O$$

exhibits all these necessary features of a concretionary growth process. It occurs in a thin subsurface zone of uncompacted sediment and stimulates a late phase of renewed sulphate reduction in pore systems which are sulphate depleted. Anaerobic methane oxidation also generates alkalinity which could result in carbonate precipitation. The carbon isotope signals of cements precipitated by methane oxidation would be indistinguishable from those of simple sulphate reduction, provided the methane diffusing up into the oxidation zone is accompanied by the methanogenic-derived dissolved carbonate.

Anaerobic methane oxidation produces carbonate from a dissolved organic source (which leaves no fossil trace within the concretion) and is a stratigraphically-confined process (which explains the preferential development of concretions along certain bedding planes). A complete absence of sedimentation (or a sharp reduction in the sedimentation rate) is necessary for anaerobic methane oxidation to localize carbonate precipitation for long enough to produce a concretionary horizon. Concretions represent an initial stage in carbonate precipitation, and their evolution into nodular limestones is related to the duration of the break in sedimentation, and the rapidity of cementation. Similar non-steady state sedimentation conditions may also cause the origin of later, methanogenic concretions, where a deeper maximum in carbonate supersaturation is developed due to alkalinity derived from iron reduction.

ACKNOWLEDGMENTS: Charles Curtis, John Hudson, Don Gautier and Joe Macquaker are thanked for their contributions on a recent visit to the Jet Rock. Joe Macquaker and John Gluyas also provided most encouraging reviews.

References

BARNES, R. O. & GOLDBERG, E. D. 1976. Methane production and consumption in anoxic marine sediments. *Geology*, **4**, 297–300.

BERNER, R. A. 1968. Calcium carbonate concretions formed by the decomposition of organic matter. *Science*, **159**, 195–7.

—— 1971. *Principles of Chemical Sedimentology*. McGraw-Hill, New York.

—— 1980. *Early Diagenesis, a Theoretical Approach*. Princeton University Press.

——, SCOTT, M. R. & THOMLINSON, C. 1970. Carbonate alkalinity in the pore waters of anoxic marine sediments. *Limnology and Oceanography*, **15**, 544–9.

CAMPOS, H. S. & HALLAM, A. 1979. Diagenesis of English Lower Jurassic limestones as inferred from oxygen and carbon isotope analysis. *Earth and Planetary Science Letters*, **45**, 23–31.

CLAYPOOL, G. E. & KAPLAN, I. R. 1974. The origin and distribution of methane in marine sediments. *In*: KAPLAN, I. R. (ed.) *Natural Gases in Marine Sediments*, pp. 97–139. Plenum Press, New York.

—— & KVENVOLDEN, K. A. 1983. Methane and other hydrocarbon gases in marine sediment. *Annual Reviews in Earth and Planetary Sciences*, **11**, 299–327.

COLEMAN, M. L. 1985. Geochemistry of diagenetic non-silicate minerals: kinetic considerations. *Philosophical Transactions of the Royal Society of London*, **315A**, 39–56.

—— & RAISWELL, R. 1981. Carbon oxygen and sulphur isotope variations in concretions from the Upper Lias of N.E. England. *Geochimica et Cosmochimica Acta*, **45**, 329–40.

CRILL, P. M. & MARTENS, C. S. 1983. Spatial and temporal fluctuations of methane production in anoxic coastal marine sediments. *Limnology and Oceanography*, **28**, 1117–30.

CURTIS, C. D., PETROWSKI, C. & OERTEL, G. 1972. Stable carbon isotopes ratios within carbonate concretions: a clue to place and time of formation. *Nature*, **235**, 98–100.

DEVOL, A. H. 1983. Methane oxidation rates in the anaerobic sediments of Saanich Inlet. *Limnology and Oceanography*, **28**, 738–42.

—— & AHMED, S. I. 1981. Are higher rates of sulphate reduction associated with anaerobic methane oxidation? *Nature*, **291**, 407–8.

——, ANDERSON, J. J., KUIVILA, K. & MURRAY, J. W. 1984. A model for coupled sulfate reduction and methane oxidation in the sediments of Saanich Inlet. *Geochimica et Cosmochimica Acta*, **48**, 993–1004.

DICKSON, J. A. D. & BARBER, C. 1976. Petrography, chemistry and origin of early diagenetic concretions in the Lower Carboniferous of the Isle of Man. *Sedimentology*, **23**, 189–211.

GAUTIER, D. L. 1982. Siderite concretions—indications of early diagenesis in the Gammon Shale (Cretaceous). *Journal of Sedimentary Petrology*, **52**, 859–71.

GLUYAS, J. G. 1984. Early carbonate diagenesis within Phanerozoic shales and sandstones of the N.W. European Shelf. *Clay Minerals*, **19**, 309–21.

HALLAM, A. 1964. Origin of the limestone-shale rhythm in the Blue Lias of England: a composite theory. *Journal of Geology*, **72**, 157–69.

HOWARTH, M. K. 1962. The Jet Rock Series and Alum Shale Series of the Yorkshire Coast. *Proceedings of the Yorkshire Geological Society*, **33**, 381–422.

HUDSON, J. D. 1978. Concretions, isotopes, and the diagenetic history of the Oxford Clay (Jurassic) of central England. *Sedimentology*, **25**, 339–70.

—— & FRIEDMAN, I. 1978. Carbon and oxygen isotopes in concretions: relationship to pore-water changes during diagenesis. *In*: CADEK, J. & PACES, T. (eds). *Proceedings, Symposium on Water-Rock Interaction, Czechoslovakia*, pp. 331–9.

IRWIN, M., COLEMAN, M. L. & CURTIS, C. D. 1977. Isotopic evidence for source of diagenetic carbonates formed during burial of organic-rich sediments. *Nature*, **269**, 209–13.

IVERSON, N. & JORGENSEN, B. B. 1985. Anaerobic methane oxidation rates at the sulfate-methane transition in marine sediments from Kattegat and Skagerrak (Denmark). *Limnology and Oceanography*, **30**, 944–55.

MARTENS, C. S. & BERNER, R. A. 1977. Interstitial water chemistry of anoxic Long Island Sound sediments. 1. Dissolved gases. *Limnology and Oceanography*, **22**, 10–25.

—— & KLUMP, J. V. 1984. Biogeochemical cycling in an organic-rich coastal marine basin: an organic carbon budget for sediments dominated by sulphate reduction and methanogenesis. *Geochimica et Cosmochimica Acta*, **48**, 1987–2004.

MURRAY, J. W., GRUNDMANIS, V. & SMETHIC, M S. JR. 1978. Interstitial water chemistry in the sediments of Saanich Inlet. *Geochmica et Cosmochimica Acta*, **42**, 1011–26.

PEARSON, R. 1979. Geochemistry of the Hepworth Carboniferous sediment sequence and origin of the diagenetic iron minerals and concretions. *Geochimica et Cosmochimica Acta*, **43**, 927–41.

RAISWELL, R. 1971. The growth of Cambrian and Liassic concretions. *Sedimentology*, **12**, 147–71.

—— 1976. The microbiological formation of carbonate concretions in the Upper Lias of N.E. England. *Chemical Geology*, **18**, 227–44.

—— 1982. Pyrite texture, isotopic composition and the availability of iron. *American Journal of Science*, **282**, 1244–63.

—— & BERNER, R. A. 1985. Pyrite formation in euxinic and semi-euxinic sediments. *American Journal of Science*, **285**, 710–24.

—— & WHITE, N. J. M. 1978. Spatial aspects of concretionary growth in the Upper Lias of N.E. England. *Sedimentary Geology*, **20**, 291–300.

REEBURGH, W. S. 1976. Methane consumption in Cariaco Trench waters and sediments. *Earth and Planetary Science Letters*, **28**, 337–44.

—— 1980. Anaerobic methane oxidation rate depth distribution in Skan Bay sediments. *Earth and Planetary Science Letters*, **47**, 345–52.

—— 1983. Rates of biogeochemical processes in anoxic sediments. *Annual Reviews in Earth and Planetary Sciences*, **11**, 269–98.

—— & HEGGIE, D. T. 1977. Microbial methane consumption reactions and their effect on methane distributions in freshwater and marine environments. *Limnology and Oceanography*, **22**, 1–9.

RICE, D. D. & CLAYPOOL, G. W. 1981. Generation, accumulation and resource potential of biogenic gas. *American Association of Petroleum Geologists Bulletin*, **65**, 5–25.

SASS, E. & KOLODNY, Y. 1972. Stable isotopes, chemistry and petrology of carbonate concretions (Mishash Formation, Israel). *Chemical Geology*, **10**, 261–86.

SHOLKOVITZ, E. 1973. Interstitial water chemistry of the Santa Barbara Basin sediments. *Geochimica et Cosmochimica Acta*, **37**, 2043–73.

WAAGE, K. M. 1964. Origin of repeated fossiliferous concretion layers in the Fox Hills Formation. *Kansas Geological Survey Bulletin*, **169**, 541–63.

WEEKS, L. G. 1957. Origin of carbonate concretions in shales, Magdalena Valley, Columbia. *Geological Society of America Bulletin*, **68**, 95–102.

WHITICAR, M. J. & FABER, E. 1986. Methane oxidation in sediment and water column environments—isotope evidence. *Organic Geochemistry*, **10**, 759–68.

ZANGERL, R., WOODLAND, B. G., RICHARDSON, E. S. JR & ZACHRY, D. L. JR. 1969. Early diagenetic phenomena in the Fayetteville Black Shale (Mississippian) of Arkansas. *Sedimentary Geology*, **3**, 87–119.

R. RAISWELL, Department of Earth Sciences, University of Leeds, Leeds LS2 9JT, UK.

The application of a solution-mineral equilibrium model to the diagenesis of Carboniferous sandstones, Bothamsall oilfield, East Midlands, England

E. A. Warren

SUMMARY: Within the fluviatile reservoir sandstones of the Bothamsall oilfield, authigenic cements of early quartz, kaolinite and illite predominate. Detrital feldspars show varying degrees of alteration and muscovite micas are commonly altered to intergrowths of kaolinite and illite.

Solution-mineral equilibria in the idealized system of K_2O-Al_2O_3-SiO_2-H_2O are used to relate the simple silicate mineral assemblage observed to possible pore-fluid compositions. Previous studies considered aluminium to be immobile in this system. Petrographic evidence of the paragenetic sequence at Bothamsall indicates that aluminium is mobile, albeit over short distances. This study therefore considers aluminium mobility in the solution-mineral equilibria. Graphical plots of total aluminium versus pH for given potassium ion activity, at a temperature of 298 K, are constructed to illustrate the stability fields of the phases identified. These show that aluminium can be an important component, together with pH and potassium-ion activity, in affecting mineral stabilities in this system.

The existence of several co-stable phases at Bothamsall considerably reduces the possible permutations for pore-fluid chemistry; kaolinite-illite alteration of mica proves particularly sensitive in this respect. The relative merits of two different pore-fluid models are compared: one, where the pore-fluid composition is considered to be constant in an open system, and another where the pore-fluid composition is assumed to vary in response to *in situ* mineralogy in a closed system. The former would require a very restricted range of pore-fluid compositions to result in the diagenetic modifications observed, whereas a wide range of pore fluids could evolve in a closed system driven by feldspar dissolution. This pore-fluid evolutionary model is preferred.

Introduction

One useful aim of the study of diagenesis is to relate the authigenic minerals observed in the rock to the chemistry of the pore fluid. Mineral precipitation sequences can then be examined in terms of the chemical composition and evolution of the pore fluid.

The use of thermodynamic phase diagrams *in the interpretation of mineral systems* at low temperatures has long been recognized. When combined with formation water data they can form a powerful tool in the elucidation of diagenetic sequences (e.g. Kaiser 1984). However, where reliable chemical compositional information for formation water is non-existent, or known to be unrelated to the minerals considered, interpretation of the pore-fluid system from which diagenetic minerals precipitated can be rather difficult. This paper describes the construction of a solution-mineral equilibrium model relevant to diagenetic systems, in which all components are assumed to be mobile. It is used to analyse possible pore-fluid evolutionary pathways of a simple paragenetic sequence described from the Bothamsall oilfield (National Grid Reference SK6674), Nottinghamshire, England. The example taken concentrates on the early part of the paragenetic sequence, that primarily involving silicates, which was followed by oil emplacement and dissolution. These subsequent events indicate a change in pore-fluid chemistry, so modern formation water data (Downing & Howitt 1969) are unlikely to be accurate analogues of the early pore fluid.

Petrography

The Bothamsall oilfield reservoir rocks studied comprise the Crawshaw and sub-Alton sandstones (Westphalian A, Upper Carboniferous). These have been interpreted as representing delta-front and low-sinuousity fluviatile deposits (Hawkins 1972, 1978, Guion 1971).

The results presented here form part of a detailed petrographic and chemical study of 100 samples from 10 wells within the oilfield. The interpretations are based on extensive examination of thin-sections, secondary and backscatter electron imaging (SEI and BEI) of rock chips and polished thin-sections and hot- and cold-cathodoluminescence. Semi-quantitative and quantitative chemical analyses of mineral phases

From: MARSHALL, J. D. (ed.), 1987, *Diagenesis of Sedimentary Sequences*, Geological Society Special Publication No. 36, pp. 55–69.

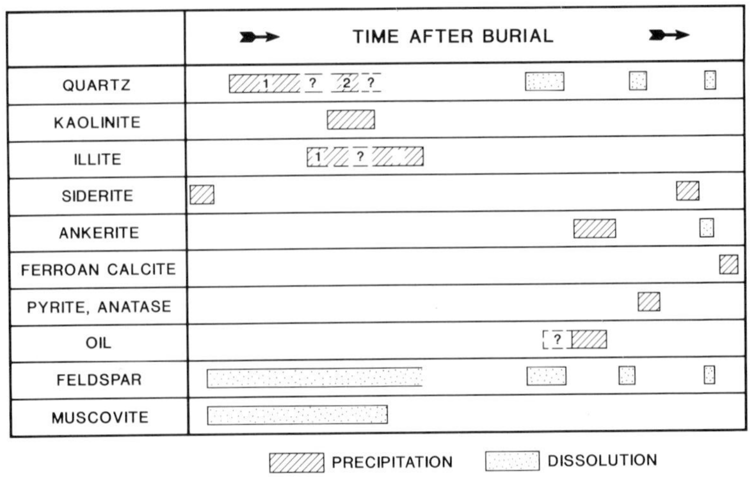

FIG. 1. A diagenetic sequence for the Bothamsall oilfield. The relative timing of precipitation and dissolution of the more important diagenetic phases is shown. Note the sharp division between silicate precipitation (early), and subsequent precipitation of carbonates and oil emplacement. Two possible phases of quartz overgrowths and illite precipitation relative to kaolinite are indicated.

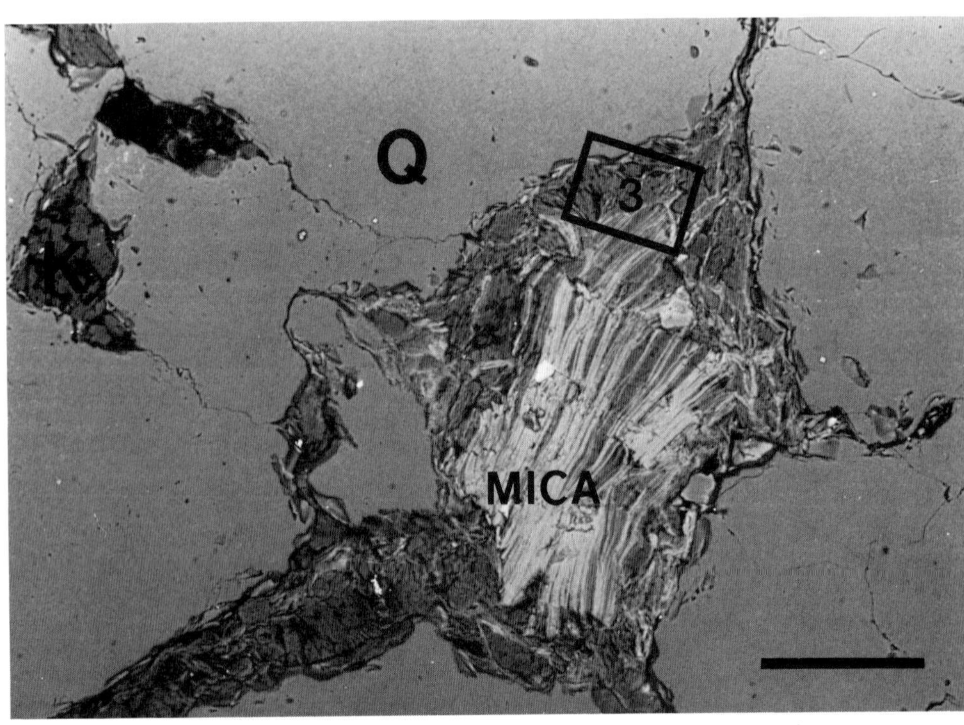

FIG. 2. Muscovite mica. Alteration of detrital muscovite mica to laths of illite (I, light) and kaolinite (K, dark). Authigenic kaolinite and illite infill remanent porosity. Note only kaolinite infills the adjacent, but non-connecting pore. This may indicate the existence of different pore-fluid chemical compositions in adjacent pores resulting in different cement sequences. Illite formation may be controlled by mica dissolution. Back-scatter SEM micrograph (Oxford Polytechnic). Scale bar = 100 μm.

were obtained using Energy Dispersive Spectral (EDS) X-ray analysis systems on scanning electron microscopes and Wavelength Dispersive Spectral (WDS) X-ray analysis on a microprobe. The mineralogy and chemical composition of clay minerals were studied using an Analytical Transmission Electron Microscope (ATEM) in ultra-thin (*c.* 300 nm) sections, and powder X-ray diffractometry (XRD) of fine separates from whole rock samples.

The sandstone samples examined range from fine- to coarse-grained arkoses to quartz-arenites. Present porosity varies significantly both between, and within, samples, but can be as high as 15%. The main detrital phases are quartz, alkali feldspar and muscovite.

The most abundant authigenic phases identified are quartz overgrowths, kaolinite, illite and zoned ankerite. All authigenic phases are distributed throughout the reservoir; none appears to be controlled by the presence of unstable detrital phases, such as feldspar or muscovite. A general diagenetic sequence of these minerals (Fig. 1) involves early silicates, followed either by oil emplacement or, in the water zone, ankerite (cf. Kantorowicz 1985). This event was accompanied, or preceded, by substantial dissolution of both quartz and feldspar creating secondary porosity. Although no phyllosilicate dissolution was recognized in this event, no subsequent precipitation of silicates has been observed. In addition several other minerals, most notably siderite, pyrite and anatase were observed but found to be of only local volumetric significance.

Potassium feldspar grains show varying degrees of alteration and dissolution; partially altered feldspar grains are often associated with kaolinite and/or illite. The presence of abundant overgrowths tends to obscure grain textures, but generally angular grains with point contacts were identified, indicating minimal pressure solution prior to quartz cementation. Muscovite micas are frequently altered to complex intergrowths of kaolinite and illite (Figs 2 and 3) (cf. Huggett 1984). Kaolinite appears to have grown by the displacement of the mica parallel to cleavage (Fig. 3).

The precipitation sequences within the silicates are complex. Kaolinite commonly abuts (Fig. 4), or has been partially enclosed by (Figs 6 and 7), quartz overgrowths. Quartz thus both

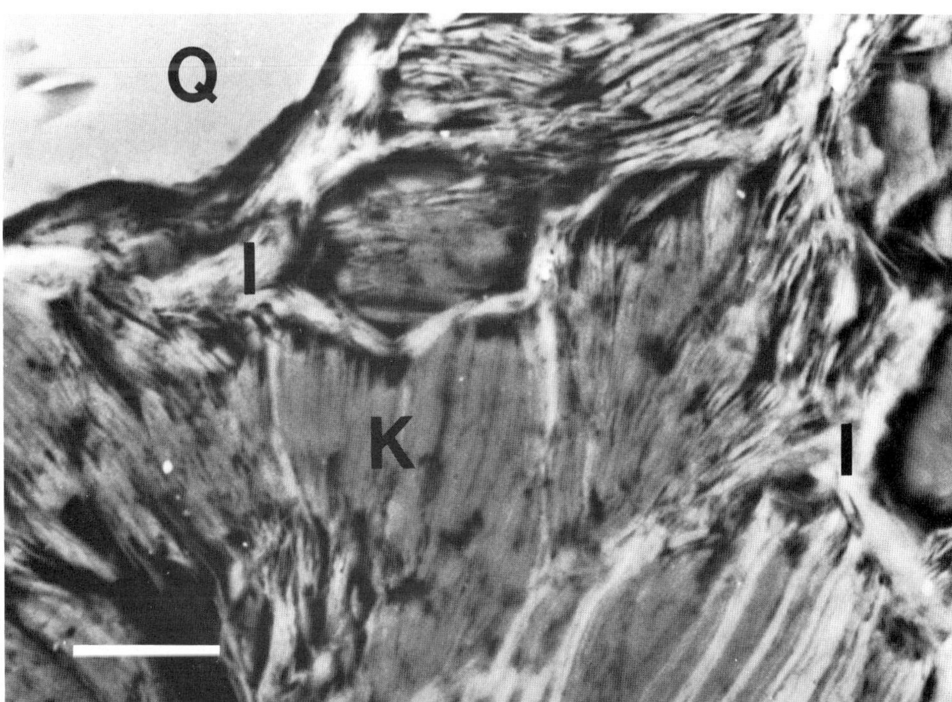

FIG. 3. Detail of Fig. 2. Inset of mica intergrowth and authigenic cements. Illite lines mica and quartz grains. Authigenic kaolinite rouleaux appear to have displaced earlier illite resulting in the 'mesh-work' texture seen. Back-scatter SEM micrograph (Oxford Polytechnic). Scale bar = 10 μm.

FIG. 4. Authigenic kaolinite and quartz.The quartz overgrowth partially encloses some kaolinite booklets, whilst other booklets lie upon the overgrowth. This is interpreted as either two generations of quartz, with kaolinite precipitation in between, or as continual precipitation of quartz, before and during kaolinite formation. Lack of signs of dissolution (e.g. pitting) suggests that both phases were stable, i.e. fluid was saturated with respect to both. SEM. Scale bar = 10 μm.

FIG. 5. Authigenic illite on kaolinite. Kaolinite rouleaux are coated by later illite of both platy and hairy morphologies. No evidence of dissolution indicates that kaolinite was stable when illite precipitated. SEM. Scale bar = 10 μm.

FIG. 6. Illite and kaolinite on quartz overgrowth. Note the hairy habit of illite. No evidence of dissolution is apparent, which would indicate that the accompanying solution was saturated with respect to all three phases. SEM. Scale bar = 10 μm.

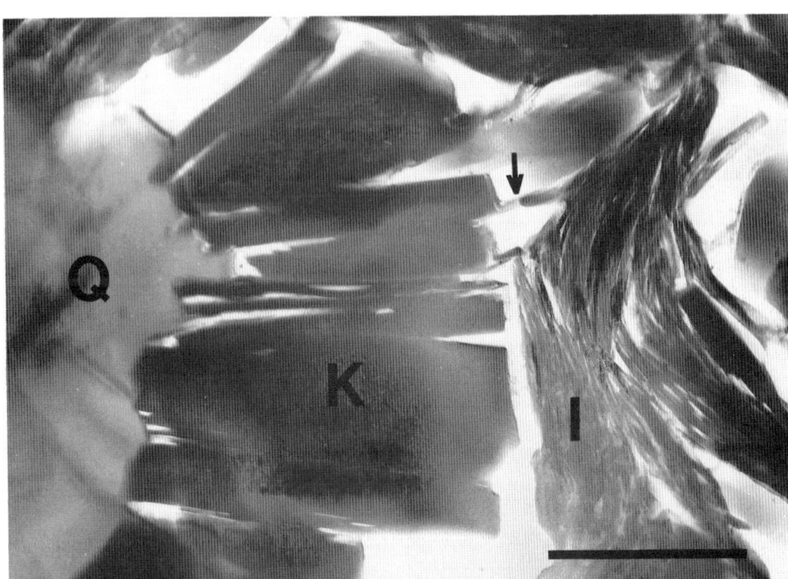

FIG. 7. Ultra-thin section of authigenic silicates. Kaolinite plates (K) abutting, and partially enclosed by, quartz overgrowth (Q). Later pore-filling illite (I) is delicately attached to kaolinite (arrowed). A solution saturated in quartz, then quartz plus kaolinite, then quartz plus kaolinite plus illite (i.e. three co-stable phases) is interpreted. TEM micrograph. Scale bar = 500 nm.

preceded and followed the start of kaolinite precipitation. This could be interpreted as either two distinct generations of quartz precipitation, or as a period during which two silicates, kaolinite and quartz, precipitated. Cathodoluminescence did not resolve this aspect. Neither phase shows evidence of dissolution. This could imply that the pore-fluid present was at least continually saturated with respect to both kaolinite and quartz, although kinetic factors, such as dissolution rate, may be important. Illite has been interpreted as both preceding (Fig. 3) and post-dating kaolinite (Figs 5 and 7) in different areas of the oilfield. It is often intimately associated with kaolinite, to which delicate attachments have been observed (Fig. 7), indicating illite post-dating kaolinite. Elsewhere, illite replacement of muscovite is often followed by the precipitation of pore-lining illite, which is displaced by later pore-filling kaolinite (Fig. 3). This indicates that supersaturation of the pore fluid with respect to illite and kaolinite occurred at different times in different parts of the reservoir. Lack of dissolution textures (e.g. etch pits), on quartz grains and overgrowths (Fig. 6) indicates that the solution precipitating illite was also saturated with respect to quartz.

A very different relationship is observed between the authigenic silicates and ankerite. Authigenic ankerite rhombs commonly appear to displace clay minerals, and quartz overgrowths are often corroded to the grain. Pore fluids precipitating ankerite thus must have been undersaturated with respect to quartz.

A pore-fluid history can be constructed by considering the assemblages of authigenic phases and their textures. The initial pore water was probably a Carboniferous meteoric water, in accordance with the depositional interpretation. This was perhaps quartz saturated, but with burial became supersaturated, resulting in quartz precipitation as grain overgrowths. Progressive supersaturation in kaolinite and illite followed during which the pore fluid remained saturated, or supersaturated, in quartz. The same fluid was also undersaturated with respect to K-feldspar and muscovite, resulting in their dissolution or replacement by illite and/or kaolinite. Subsequently, the pore fluid became undersaturated with respect to quartz, resulting in dissolution of both grain and overgrowths. This was accompanied by the emplacement of oil or the precipitation of ankerite. Large-scale fluid movement must have occurred, as oil generation within the reservoir is unlikely. This indicates that the present-day pore fluid is of a different chemical composition to the pore fluids preceding oil emplacement. This study attempts to examine the early pore-fluid chemistry which resulted in the precipitation of the silicates by considering a chemical model based on the mineral solubility equilibria of the phases involved. The distribution of authigenic aluminosilicates within the reservoir also indicates that aluminium should be treated as a mobile species.

A thermodynamic model

The Gibb's phase rule considers phase, component and resulting degrees of freedom (or variance), F, within any given system through the relationship:

$$P + F = C + 2.$$

Studies of real diagenetic systems (e.g. Kaiser 1984) often suffer from the problem that the number of components greatly exceeds the number of phases present, thus resulting in non-unique solutions for the variance. This is true for the Bothamsall sandstones, for which a rigorous treatment must consider 10 or more components but far fewer phases. To overcome this problem the number of components must be reduced in relation to the number of phases present. In this case study only the early part of the diagenetic sequence was considered—that restricted to silicate diagenesis. Furthermore, the consideration of 'ideal' stoichiometric compositions for the phases enabled the components to be limited to four: K_2O, Al_2O_3, SiO_2, H_2O. This assumption would appear to be reasonable from published analyses of quartz, kaolinite, muscovite and K-feldspar (e.g. Deer *et al.* 1962), but less so for illite. A composition of $K_{1.5}Al_{5.5}Si_{6.5}O_{20}(OH)_4$ was used for illite, which approximates to the average analysis determined directly for the Bothamsall material using Analytical TEM—$K_{1.2}Al_{4.8}Mg_{0.2}Fe_{0.2}Si_{6.8}O_{20}(OH)_4$ (details to be published separately).

A variety of methods for modelling diagenetic systems has appeared in the literature. Experimental investigations of solution-mineral equilibria (e.g. Surdam *et al.* 1984) have the obvious advantage of enabling the sensitivity of phases to components to be determined directly, but suffer from problems of kinetics and the metastability of intermediate phases (May *et al.* 1986). Computer programs of solution models based on thermodynamics, e.g. MINEQL (Westall *et al.* 1976) and PHREEQE (Parkhurst *et al.* 1980), provide an elegant method of predicting fluid evolution but require an initial estimate of the pore fluid composition. In the case of Bothamsall this presents difficulties for it is unlikely that the present-day formation water (data in Downing & Howitt 1969) should reflect the physicochemical composition of the possible Carboniferous waters from which the authigenic silicate phases

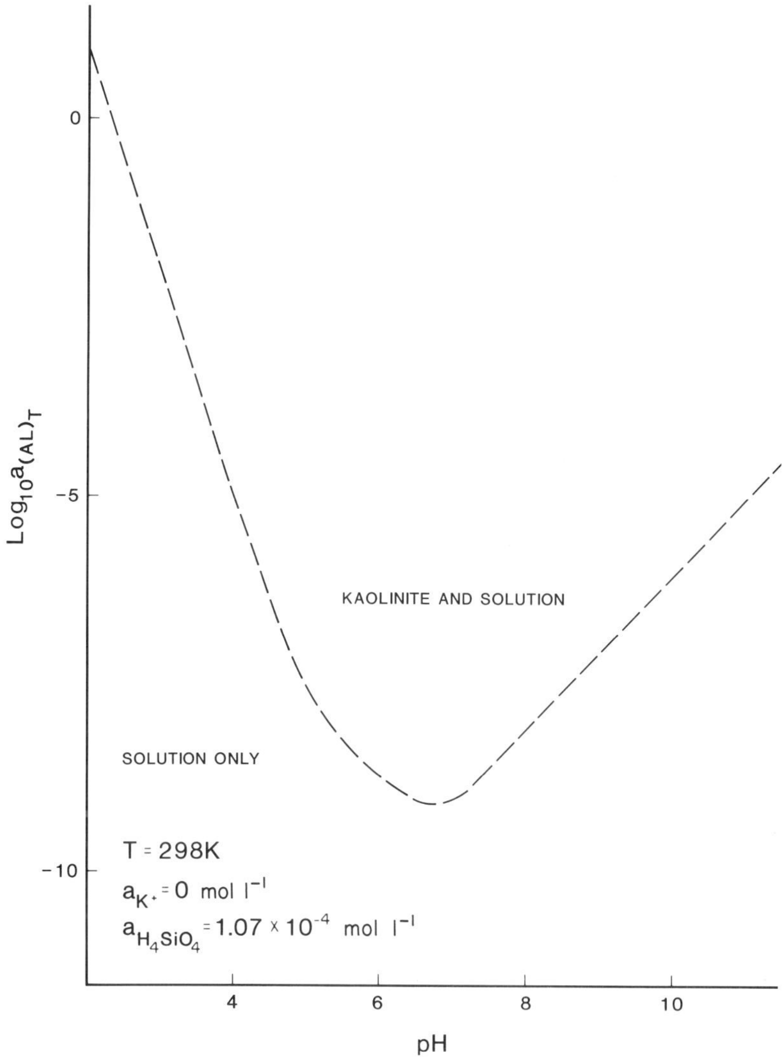

FIG. 8. Solution-mineral equilibria in the system: Al_2O_3-SiO_2-H_2O, $T = 298$ K, $P = 1$ bar, quartz-saturated (after Curtis 1983). Activity of potassium ion is zero. Kaolinite will precipitate from any solution whose chemical composition lies within the saturation curve. All solutions should precipitate quartz (quartz-saturated). N.B. $a(Al)_T$ is the sum of the activities of the individual aluminium species calculated from the free energy data of the solution-mineral equilibria.

precipitated, when subsequent diagenetic events (e.g. dissolution and ankerite formation) indicate a change in fluid chemistry.

The approach used here follows that described in Garrels & Christ (1965) and Kaiser (1984). Solution-mineral equilibria for the phases and compositions considered were determined from basic thermodynamic principles (Atkins 1978) and plotted in the system K_2O-Al_2O_3-SiO_2-H_2O. Thermodynamic data for the species were taken from Robie *et al.* (1978). A problem existed in the case of illite for available thermodynamic data were not applicable to the composition considered. Instead, a model (Tardy & Garrels 1974) was used to derive an empirical value for the Gibb's free energy of formation of -1320 kcal mol^{-1}. For each solution-mineral reaction the free energy for reaction, and hence the equilibrium constant, was calculated by summation of the free energy of formation data of the species considered.

In the treatment of similar systems previous authors (Garrels & Christ 1965, Kaiser 1984) have considered the aluminium species to be

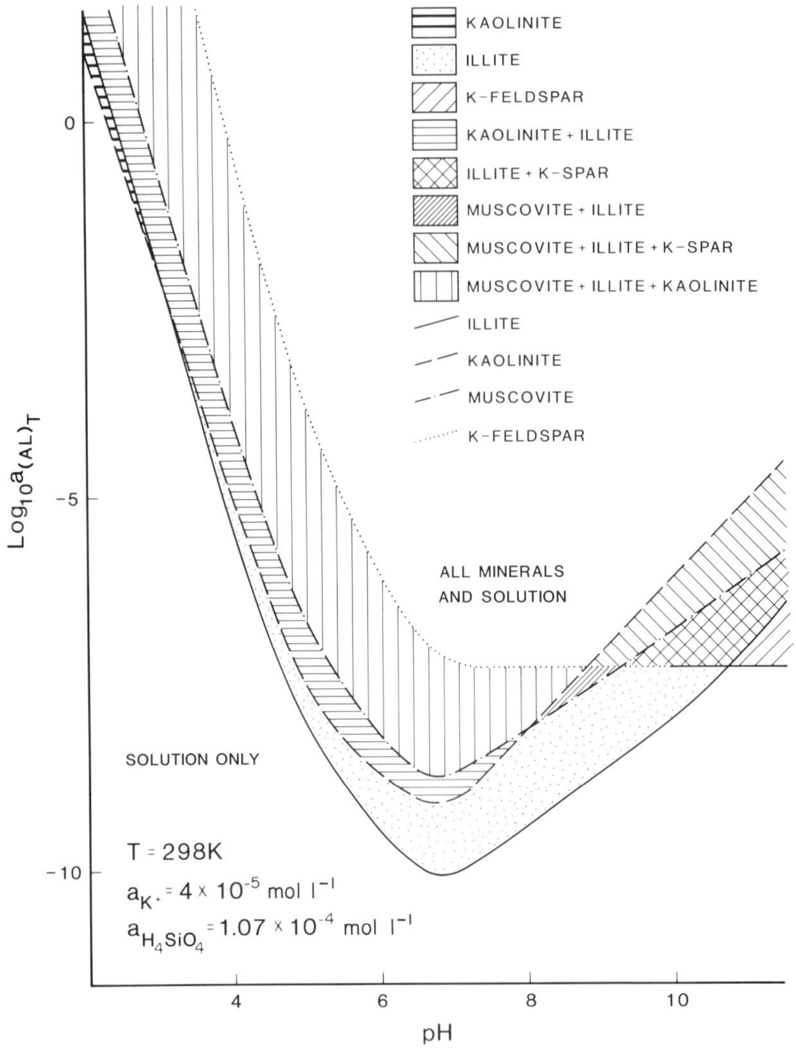

FIGS 9–12. Solution-mineral equilibria in the system: K_2O-Al_2O_3-SiO_2-H_2O, $T = 298$ K, $P = 1$ bar, quartz-saturated, and increasing potassium ion activity. Individual mineral saturation curves (lines) and stability fields (shaded areas) are indicated. Note sensitivity of K-feldspar field with K^+. See text for discussion. N.B. $a(Al)_T$ is the sum of the activities of the individual aluminium species calculated from the free energy data of the solution-mineral equilibria.

effectively immobile. This has the advantage of eliminating aluminium, so reducing the number of components. It also means that the only charged species are potassium ions, hydrogen ions and hydroxyl; as charge must be conserved the potassium activity and pH are thus interdependent. This effectively reduces the components one further and enables a four oxide system to be plotted on a two-dimensional diagram (e.g. K^+/H^+ versus H_4SiO_4).

At Bothamsall, however, the presence of authigenic pore-filling kaolinite and illite distributed throughout the reservoir suggests that aluminium is mobile, if only over short distances. Aluminium, together with potassium ions and pH, have thus been considered as independent components in this system. This presents problems representing the system graphically as a single dimension for each of the four components is required. The presence of several phases co-stable with quartz enabled the solution-mineral equilibria to be considered at quartz saturation. Several alternative graphical plots were possible. A plot of total aluminium versus pH for different activities of potassium ion typical of analysed porewaters (e.g. Downing & Howitt 1969, Stumm & Morgan 1981) has been considered here in Figs 8–12.

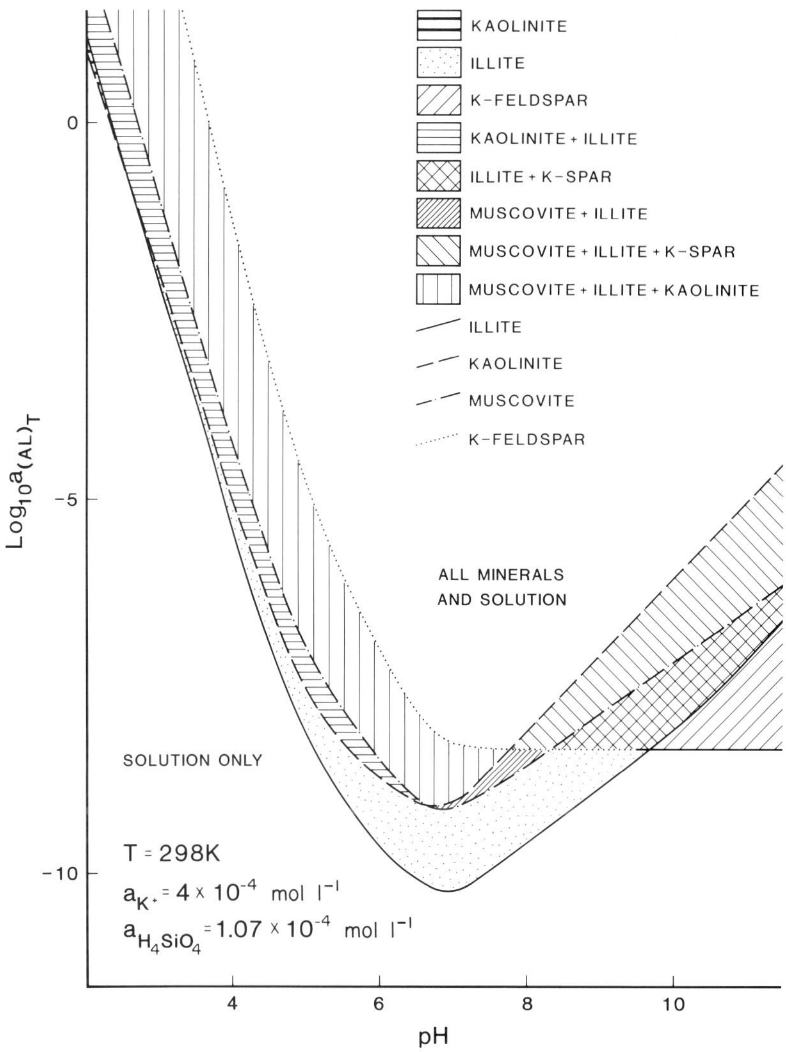

FIG. 10.

One important component not considered is temperature. The temperature increase associated with burial can affect mineral stability. The solubility of many phases increases with temperature which could alter the juxtaposition of the solubility curves in Figs 8–13. Unfortunately, modelling the pore-fluid system considered here at elevated temperatures (e.g. 100°C) has been hampered by the sparse data base available for many species, aluminium in particular (Hovey & Tremaine 1986). The diagrams presented here therefore only indicate the relative stability of minerals and solution at 25°C. They cannot be used to obtain absolute values for pore-fluid compositions. Values cited in the discussion are merely used to highlight the areas of the diagrams being referred to.

Discussion

Amphoteric aluminium

Aluminium forms many ionic species in solution of which three are dominant: Al^{3+}, $Al(OH)_2^+$, $Al(OH)_4^-$. These are responsible for the amphoteric nature of aluminium in solution; aluminium being soluble at both high and low pH but markedly insoluble at neutral and near neutral pH. A graphical plot of total aluminium species versus pH thus results in a curve (Fig. 8). This effect gives rise to the curve for kaolinite (Curtis 1983), and also illite, muscovite and K-feldspar. At high pH the solubility of all silicates is increased further, due to the formation of charged ionic species of silica.

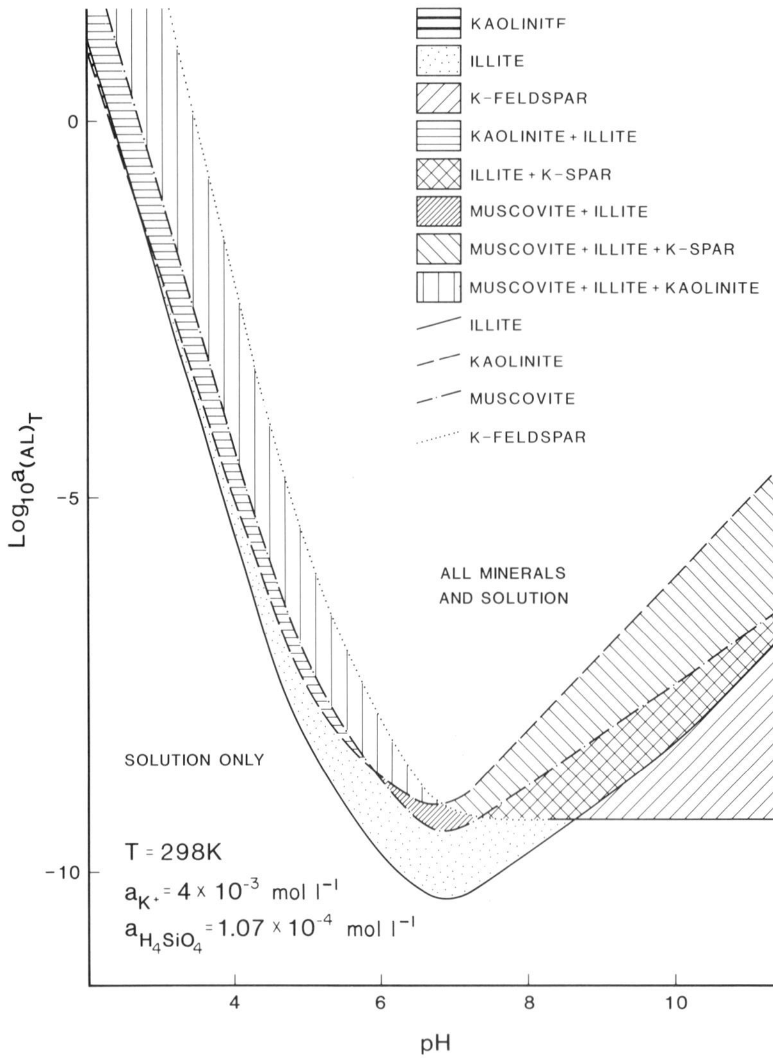

FIG. 11.

Phases are only stable in solutions whose compositions lie within the solubility boundary, or saturation curves derived from the solution-mineral equilibria. If a pore fluid lies outside the solubility boundary of a phase, then that phase should dissolve as the solution is undersaturated. Similarly a mineral should precipitate only if the pore-fluid composition lies within the solubility boundary of that phase—the solution is then supersaturated in that phase. Kinetic factors, such as nucleation and growth, increase the degree of supersaturation required from the phase diagram for precipitation to commence. Dissolution is affected by activation energies in a similar way. Thus the solution-mineral equilibria only indicate where a particular phase is thermodynamically stable. Kinetics will determine whether it actually dissolves or precipitates.

Mineral precipitation, of kaolinite for example, would be favoured in neutral pH solutions where solubility is lowest (Fig. 8). Conversely, mineral dissolution would be most favourable in solutions of either high or low pH. A subtle change in pH could thus drastically affect the stability of the minerals, causing either dissolution of pre-existing phases or precipitation of existing and additional phases, thus resulting in a change in the bulk assemblage.

Potassium ion activity

The solubility of the potassium-bearing phases is also sensitive to variations of potassium activity in solution (Figs 9–12). None of these phases is stable in a solution of zero K^+ activity, leaving kaolinite and quartz as the only stable phases

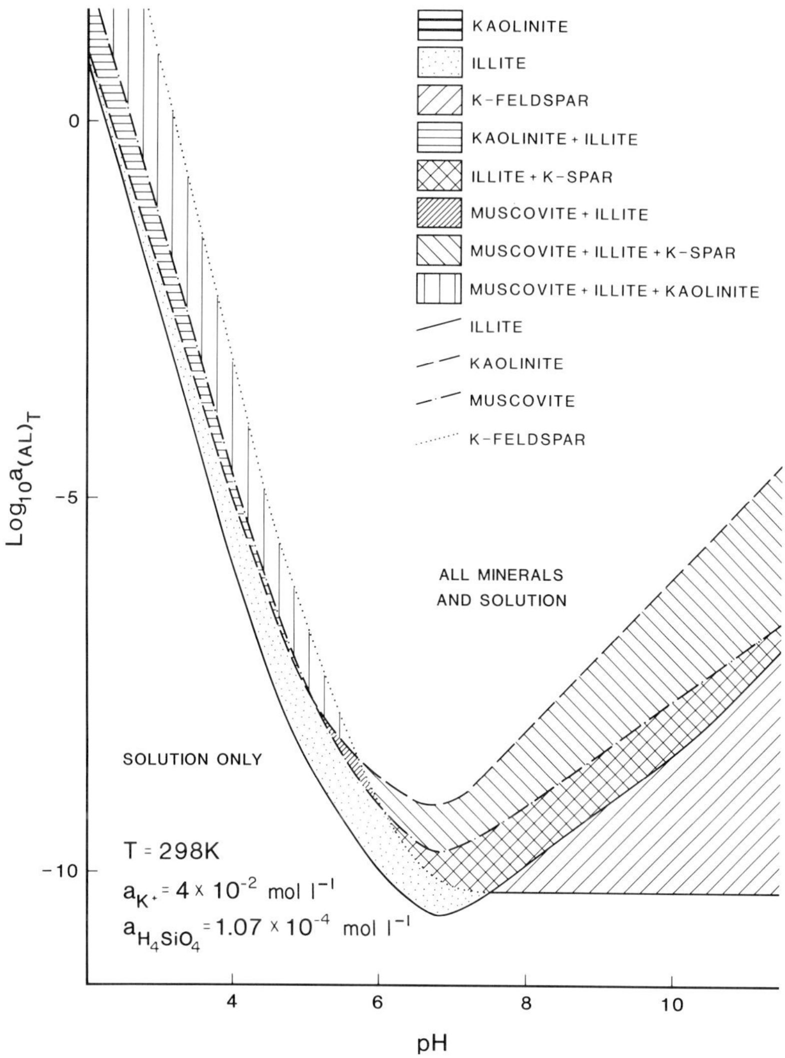

FIG. 12.

(Fig. 8). With increasing potassium ion activity in a neutral pH solution, illite becomes the most stable authigenic phase. Illite and, to a lesser extent, muscovite only show a limited sensitivity to potassium-ion activity. K-feldspar, however, is particularly sensitive; from being the most soluble phase at low activity (Fig. 9), its precipitation is favoured at high potassium ion activity (Fig. 12).

Relative stability of minerals

The various solution-mineral equilibria illustrated (Figs 8–12) indicate that illite is the least soluble, or most stable phase, over the widest range of pH and potassium ion activity considered. Kaolinite is the most stable phase at very low potassium ion activity (Fig. 8), but with increasing amounts of potassium other phases become more stable. The behaviour of muscovite is similar to that of illite, but under none of the considered conditions is muscovite more stable than illite. K-feldspar shows the most sensitivity to both pH and potassium ion activity. It appears to be very soluble at low pH and low potassium ion activity, with solubility markedly decreasing as each increases.

The application to the Bothamsall sandstones

The diagenetic history of the Bothamsall sandstones can now be investigated in terms of the solution-mineral equilibria of the phases. The preservation of point contacts and the lack of dissolution of quartz grains imply that initially a pore fluid saturated with respect to quartz entered the sandstone. Quartz overgrowths followed, indicating availability of supersaturated solutions.

Two different precipitation sequences are observed at Bothamsall: one in which illite post-dates kaolinite (case 1, Figs 5, 6 and 7), the other where kaolinite post-dates earlier illite (case 2, Fig. 3). Both kaolinite and illite precipitation post-date the onset of quartz cementation. This indicates that the initial solution was probably undersaturated with respect to both kaolinite and illite, perhaps because of an initial low aluminium and potassium ion activity.

Where illite post-dates kaolinite (case 1), aluminium activity may have increased from its initial value, so that the kaolinite solution boundary was crossed (see Fig. 8). Alternatively, pH could have increased in a solution of moderate aluminium content, but low potassium ion activity, again entering the kaolinite field (Fig. 8). Illite saturation may then have been reached in either situation by an increase in K^+ activity in the solution.

Considering case 2, potassium ion activity must have been sufficiently high ($>10^{-6}$ mol l^{-1}) for illite to precipitate before kaolinite. Kaolinite could then precipitate if the solution entered the stability field for kaolinite and illite (Figs 9–12). An increase in either pH (if initially low) or aluminium ion activity would cause this to occur.

The relative timing and magnitude of changes in pH, aluminium and potassium ion activity could thus have a dramatic effect on the authigenic assemblage and account for the variations observed (Fig. 1). Conversely, a particular observed assemblage could result from a number of different pore-fluid evolution pathways.

Muscovite alteration

The alteration of detrital muscovite grains to intergrowths of kaolinite and illite was frequently observed. The pore fluid in contact with the muscovite must have been undersaturated with respect to muscovite, but saturated in both kaolinite and illite. Such solutions are restricted in pH and aluminium and potassium ion space (Figs 9–12). Aluminium activity must have been sufficiently high ($>10^{-9}$ mol l^{-1}) for both to precipitate, otherwise illite would be the only stable phase. Potassium ion activity in excess of 10^{-6} mol l^{-1} must also be assumed for illite to precipitate. The presence in many samples of incomplete alteration suggests, moreover, that the fluid may have become muscovite saturated. Thus the pore fluid may have approached equilibrium with muscovite.

K-feldspar

Detrital K-feldspars showed varying degrees of dissolution and no evidence was seen of K-feldspar overgrowths. This would indicate that such grains were in contact with a solution undersaturated with respect to feldspar (barring kinetic factors). Feldspar, as already discussed, is very sensitive to potassium ion activity relative to the other phases. Low potassium ion activity could cause feldspar dissolution (Fig. 9). Alternatively, low pH solutions would also be undersaturated irrespective of potassium activity, and neutral pH solutions likewise if aluminium activity was low ($<10^{-7}$ mol l^{-1}). High pH solutions, unless potassium activity was low, would be supersaturated, and feldspar dissolution would not occur (Figs 11 and 12).

Summary

The precipitation of quartz, kaolinite and illite would be most favourable, in the simple example considered, in solutions of low to neutral pH, with increasing or fixed aluminium and potassium activity. Feldspar dissolution could occur in low pH or low potassium activity. Muscovite alteration to illite and kaolinite could only occur in a restricted range of solution compositions.

The possible consideration of the mineral-solution equilibria has restricted the possible pore-fluid compositions from which the diagenetic sequences evolved. However, it is necessary to consider whether the pore-fluid chemistry itself changed or whether the diagenetic sequence could have resulted from a pore fluid of more or less constant chemical composition.

Open system

In an open system the chemical composition of the pore fluid is controlled by a source external to the pore considered. The reactants are assumed to be unlimited in supply, and products removed from the system, so that the resultant fluid composition remains unchanged. The petrographic evidence of muscovite alteration and feldspar dissolution indicates that the accompanying fluid lay outside the solution-saturation curves for these phases but within those of kaolinite and illite, the precipitated phases. Such compositions lie within a restricted region of low pH and potassium ion activity for the aluminium activities commonly found in solutions (10^{-10} to 10^{-8} mol l^{-1}) (Figs 9–12, 13).

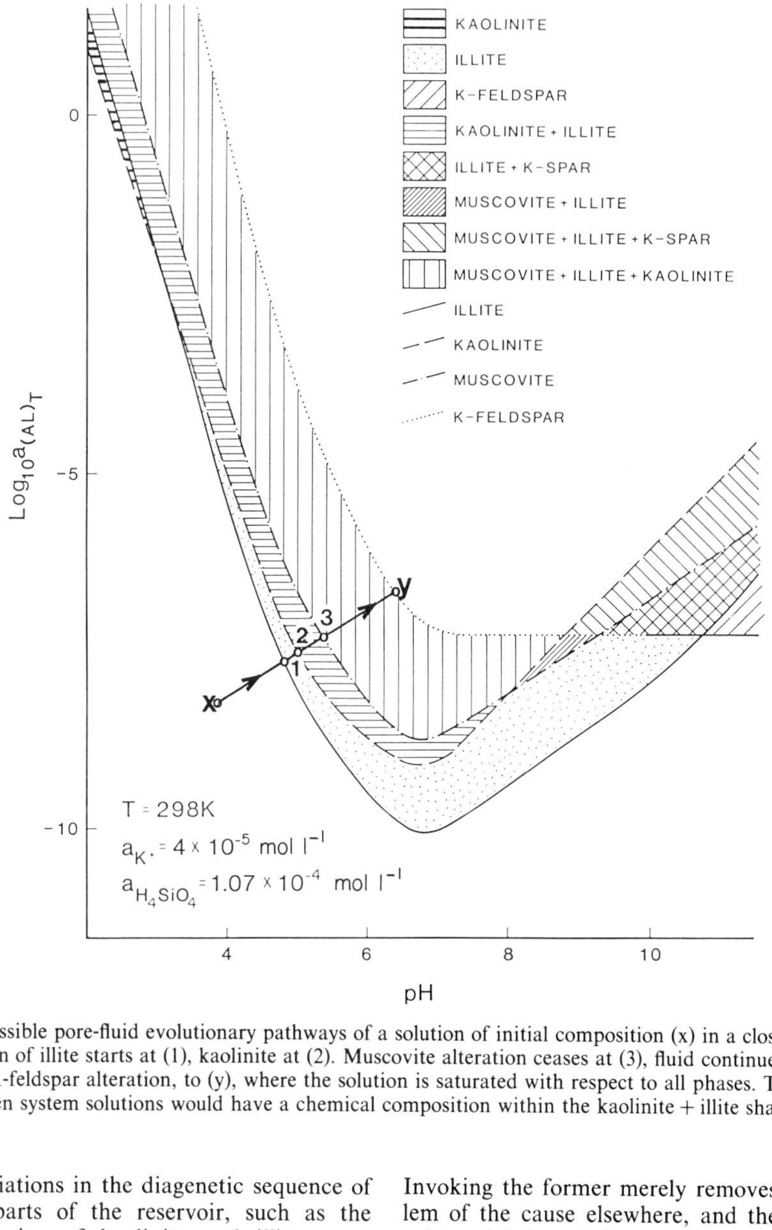

FIG. 13. Possible pore-fluid evolutionary pathways of a solution of initial composition (x) in a closed system. Precipitation of illite starts at (1), kaolinite at (2). Muscovite alteration ceases at (3), fluid continues to evolve, driven by K-feldspar alteration, to (y), where the solution is saturated with respect to all phases. The only possible open system solutions would have a chemical composition within the kaolinite + illite shaded area.

The variations in the diagenetic sequence of different parts of the reservoir, such as the relative timing of kaolinite and illite, must indicate that pore-fluid chemistry was not homogeneous within the reservoir. This could be explained by the existence of several small cells behaving as isolated open systems within the reservoir. Several problems exist, however, in the consideration of open systems. First, the constraints on the composition of the solution require a source of hydrogen and potassium ions. The breakdown of muscovite and K-feldspar are possible sources. Second, progressive sequences could only be produced via a change in fluid composition, or by a governing kinetic process. Invoking the former merely removes the problem of the cause elsewhere, and there is little substantial knowledge of the processes and kinetics of clay mineral precipitation. The diagenetic features observed at Bothamsall could have developed in an open system, but only from a very restricted range of fluid compositions.

Closed system

A closed system assumes no gain or loss of material from the volume considered (it is isochemical). This means that the fluid composition must evolve in response to the formation or destruction of the solid phases. A fluid of variable chemical composition could thus cross several mineral solubility boundaries as it responds to the ions produced from an unstable phase. A possible pore fluid evolution applicable to Bothamsall is considered in Fig. 13. A solution of low pH and low potassium ion activity (x) would cause feldspar and muscovite to dissolve:

$$KAl_3Si_3O_{10}(OH)_2 + 10H^+ = K^+ + 3Al^{3+} + 3H_4SiO_4$$

$$KAlSi_3O_8 + 4H_2O + 4H^+ = K^+ + Al^{3+} + 3H_4SiO_4.$$

Potassium, aluminium and silica would be released to the fluid and hydrogen ions consumed. Thus pH, aluminium, potassium and silica activities would increase. The fluid composition would therefore evolve until saturated with both muscovite and feldspar. When silica activity reaches quartz saturation, precipitation of quartz should occur. The increase in aluminium and potassium ions in solution would cause aluminosilicates to precipitate. In the situation considered in Fig. 13, illite precipitates first (1) due to the potassium ion activity (as in case 2 previously considered). Kaolinite would precipitate when the fluid composition evolves to the kaolinite saturation curve (2). Precipitation of both kaolinite and illite could then occur. At (3) dissolution of muscovite would cease. Only K-feldspar would dissolve until the solution becomes saturated (y). Alternatively, an increase in aluminium activity, while potassium ion activity remained low, would cause kaolinite to precipitate before illite (case 1), the other diagenetic sequence observed. No further change in pore fluid would occur as equilibrium has been reached.

Alteration and dissolution of the grains would continue in a closed system during precipitation of other phases, because precipitation would consume the ions produced from dissolution and so buffer the solution composition. Dissolution would stop only if precipitation of more stable phases was hindered in some way, kinetic factors (nucleation, growth) perhaps. Alternatively a physical factor, volume, may be important. If the pore space is effectively filled by a phase, kaolinite for example, no more can precipitate. Feldspar or muscovite grains associated with pores entirely filled by authigenic cement might thus suffer only partial replacement, whilst those in pores that remained open after cementation might suffer total obliteration. However, this treatment clearly overlooks the problem of mass balance. This is particularly pertinent to closed systems where chemicals are remobilized rather than introduced or removed in an open system.

A closed system could account for the cement sequences observed and some of the spatial and textural evidence. The closed system argument has the advantage that a range of pore-fluid chemistries could evolve to give the diagenetic features described. Initially these fluids would have low pH, aluminium and potassium ion activities; evolution would be driven by feldspar and muscovite alteration which would cease on saturation.

Conclusions

Many common aluminosilicate minerals in siliciclastic rocks show marked sensitivity not only to pH and potassium ion activity, but also to aluminium. This is due to the amphoteric behaviour of aluminium in solution. The consideration of aluminium mobility, together with all other species, is therefore important when modelling the chemical evolution of pore fluids in diagenesis. Graphical plots of aluminium versus pH and potassium ion activity, for various silicates, have been presented for 298 K and illustrate the sensitivity these minerals have to all these components. Whilst constructions at elevated temperatures might be more appropriate, these diagrams do demonstrate the importance of aluminium to mineral stability.

A simple case example from the Bothamsall oilfield has been taken to illustrate the application of the diagrams to diagenetic situations. The precipitation of the authigenic minerals (quartz, kaolinite and illite) and feldspar dissolution could result from a wide range of fluid compositions. The alteration of muscovite mica to kaolinite and illite is predicted to occur in chemically restricted fluid compositions. The possible evolutionary pathways do not necessarily preclude either an open or a closed system. Only a restricted range of pore-fluid compositions could have resulted, however, in the observed sequence from large-scale fluid movement in an open system. Chemical evolution of a silica-saturated, low aluminium and potassium-ion pore water in a closed system controlled by feldspar and muscovite dissolution, would not only result in the progressive precipitation of the quartz, kaolinite and illite, but also approach equilibrium with the unstable phases. Thus muscovite and feldspar dissolution would cease, leaving some grains only partially altered. Only subtle variations in initial pore-fluid composition are needed to generate differences in the relative timing of precipitates in different areas of the reservoir. Thus evolution of a meteoric water in a closed system driven by feldspar and, to a lesser extent, muscovite alteration, seems to be the more likely mechanism for generating the early diagenetic modification of the sandstones.

The relative importance of temperature, pressure and kinetics on the diagenetic assemblage remain to be evaluated. However, the inclusion of aluminium mobility in solution-mineral modelling does appear to be significant in the interpretation of diagenetic histories in terms of pore-fluid chemistry.

ACKNOWLEDGMENTS: Support for this research has been generously provided by a British Petroleum Scholarship. My supervisors, Charles D. Curtis (Sheffield) and Jenny Huggett (BP) are thanked for many hours of fruitful discussion and, together with Tim Young, for reading through the manuscript. My thanks are due to Anton Kearsley and Oxford Polytechnic for the use of the Back-scatter Electron Microscope facility, and to Mike Cooper for drafting the diagrams.

References

ATKINS, P. W. 1978. *Physical Chemistry*. Oxford University Press.

CURTIS, C. D. 1983. Link between aluminium mobility and destruction of secondary porosity. *Bulletin of the American Association of Petroleum Geologists*, **67**, 380–95.

DEER, W. A., HOWIE, R. A. & ZUSSMAN, J. 1966. *An Introduction to the Rock-forming Minerals*. Longman, London.

DOWNING, R. A. & HOWITT, F. 1969. Saline groundwaters in the Carboniferous rocks of the English East Midlands in relation to the geology. *Quarterly Journal of Engineering Geology*, **1**, 241–69.

GARRELS, R. M. & CHRIST, C. L. 1965. *Solutions, Minerals and Equilibria*. Harper & Row, New York.

GUION, P. D. 1971. A sedimentological study of the Crawshaw sandstone (Westphalian A) in the East Midlands coalfield area. *Unpublished MSc Thesis*, University of Keele.

HAWKINS, P. J. 1972. Carboniferous sandstone oil reservoir, East Midlands, England. *Unpublished PhD Thesis*, University of London.

—— 1978. Relationship between diagenesis, porosity reduction, and oil emplacement in late Carboniferous sandstone reservoirs, Bothamsall Oilfield, East Midlands. *Journal of the Geological Society*, **135**, 7–24.

HOVEY, J. K. & TREMAINE, P. R. 1986. Thermodynamics of aqueous aluminium; Standard partial molar heat capacities of Al^{3+} from 10 to 55°C. *Geochimica et Cosmochimica Acta*, **50**, 453–9.

HUGGETT, J. M. 1984. An S.E.M. study of phyllosilicates in a Westphalian coal measures sandstone using back-scattered electron imaging and wavelength dispersive spectral analysis. *Sedimentary Geology*, **40**, 233–47.

KAISER, W. R. 1984. Predicting reservoir quality and diagenetic history in the Frio Formation (Oligocene) of Texas. *In*: MCDONALD, D. A. & SURDAM, R. C. (eds). *Clastic Diagenesis*. American Association of Petroleum Geologists Memoir No. 37, 195–216.

KANTOROWICZ, J. D. 1985. The origin of authigenic ankerite from the Ninian Field, U.K. North Sea. *Nature*, **315**, 214–6.

MAY, H. M., KINNIBURGH, D. G., HELMKE, P. A. & JACKSON, M. L. 1986. Aqueous dissolution, solubilities and thermodynamic stabilities of common aluminosilicate clay minerals: kaolinite and smectites. *Geochimica et Cosmochimica Acta*, **50**, 1667–77.

PARKHURST, D. L., THORSTENSON, D. C. & PLUMMER, L. N. 1980. PHREEQE—a computer program for geochemical calculations. *U.S. Geological Survey, Water Resources Invest. 80–96*, NTIS Technical Report PB81-167801, Springfield, VA.

ROBIE, R. A., HEMINGWAY, B. S. & FISHER, J. R. 1978. Thermodynamic properties of minerals and related substances at 298.15 K and 1 bar (10^5 pascals) pressure and at higher temperatures. *U.S. Geological Survey Bulletin No. 1452*.

STUMM, W. & MORGAN, J. J. 1981. *Aquatic Chemistry: an Introduction Emphasizing Chemical Equilibria in Natural Waters* (2nd. ed.). Wiley, London.

SURDAM, R. C., BOESE, S. W. & CROSSEY, L. J. 1984. The chemistry of secondary porosity. *In*: MCDONALD, D. A. & SURDAM, R. C. (eds). *Clastic Diagenesis*. American Association of Petroleum Geologists Memoir No. 37, 127–49.

TARDY, Y. & GARRELS, R. M. 1974. A method for estimating the Gibbs energies of formation of layer silicates. *Geochimica et Cosmochimica Acta*, **38**, 1101–16.

WESTALL, J. C., ZACHARY, J. L. & MOREL, F. M. M. 1976. *MINEQL: a computer program for the calculatin of chemical equilibrium composition of aqueous systems*. Water Quality Laboratory, Massachusetts Institute of technology, Technical Note No. 18.

E. A. WARREN, Department of Geology, Mappin Street, University of Sheffield, Sheffield S3 7HF, UK.

Early diagenesis

Petrology (including fluorescence microscopy) of cherts from the Portlandian of Wiltshire, UK—evidence of an episode of meteoric water circulation

T. R. Astin

SUMMARY: Chert nodules from the Wockley Member of the Portland Stone (Upper Jurassic) in Wiltshire, England are comprised of three stages of chert cementation and sediment replacement: (1) opal-CT lepisphere cement and grain replacement, (2) length fast chalcedony grain and matrix replacement and cement, and (3) quartz cement. These are readily distinguished by their fluorescence, the first and last stages being non-fluorescent, the second stage showing internal zonation changing from moderate to bright fluorescence with time.

Some carbonate grain dissolution took place prior to silica precipitation. The first stage silica cement formed mainly as pore-linings, with minor replacement of grain margins. Its botryoidal, lepisphere habit implies it was opal-CT, now inverted to quartz. Extensive grain replacement took place while chalcedony cements formed. Chalcedony cements show an increase in inclusions of brown organic particles with time, joined at a late stage by iron oxide inclusions.

Silicification and carbonate dissolution occurred during a period of meteoric water penetration, probably as a result of soil formation at the top of the Portland Stone. Silica was apparently derived as opal-A from adjacent marine limestones, redistributed by groundwater flow, and precipitated as opal-CT and chalcedony. The organic inclusions probably come from the developing soil horizon, the iron oxides marking the migration of an oxidation front away from the soil horizon into the sediment. The later quartz cement lacks inclusions and was formed during later burial after stagnant groundwater conditions were re-established.

This paper describes some cherts from the Portlandian (Upper Jurassic) of Wiltshire, England (National Grid Reference ST 963 296) (Fig. 1), which is an open to marginal marine sequence of mainly limestones and sandy limestones. The cherts come from a variably fossiliferous calcareous, micritic sandstone, the Chicksgrove Plant Bed (Fig. 2), which contains terrestrial, lagoonal and marine faunal elements (Wimbledon 1976, 1980). The Plant Bed gives evidence of the earliest known non-marine conditions in the Upper Jurassic of southern England, and the fauna includes an important mammalian assemblage. As part of a multidisciplinary study of the whole bed, including its provenance, stratigraphy and palaeontology (Wimbledon *et al.*, in preparation), the cherts were investigated for information on the depositional and early diagenetic environments in which the original sediment was deposited.

Chert textures in thin section are often difficult to interpret because of their small grain size, primary void fillings and neomorphosed grains with similar textures, and primary metastable phases such as opal-CT having inverted to quartz at a later stage.

Fluorescence microscopy brings out additional textural information in these cherts mainly through its tendency to reveal the location of organic material in detrital grains and in the matrix, and as inclusions in cements. Organic matter can preserve the fine internal structure of biogenic grains, preserve grain boundaries (e.g. when included in micrite envelopes), highlight the locations of diagenesis associated with microbial activity, and reveal cement zonation.

All these effects have been seen and used to reveal primary textures of the cherts, to distinguish between the stages of silica cementation and replacement, and to relate variations in the fluorescence of some of the silica cements to the types and distribution of their inclusions.

The textural evidence has been used to relate chert formation in the Plant Bed with the period of emergence and soil formation which occurred at the base of the overlying Purbeck Formation.

Geological setting

Figure 2 gives the Portlandian sequence at the Chicksgrove Quarry. The Chicksgrove Plant Bed is about 60 cm thick and lies at the base of the micritic Wockley Member. It rests erosively on marine glauconitic shelly sandstones and is overlain by a lagoonal micrite, followed by marine micritic limestones. The Plant Bed is about 8 m below the Great Dirt Bed, a well-known palaeosol at the top of the Portland Stone (Francis 1984).

From: MARSHALL, J. D. (ed.), 1987, *Diagenesis of Sedimentary Sequences*, Geological Society Special Publication No. 36, pp. 73–85.

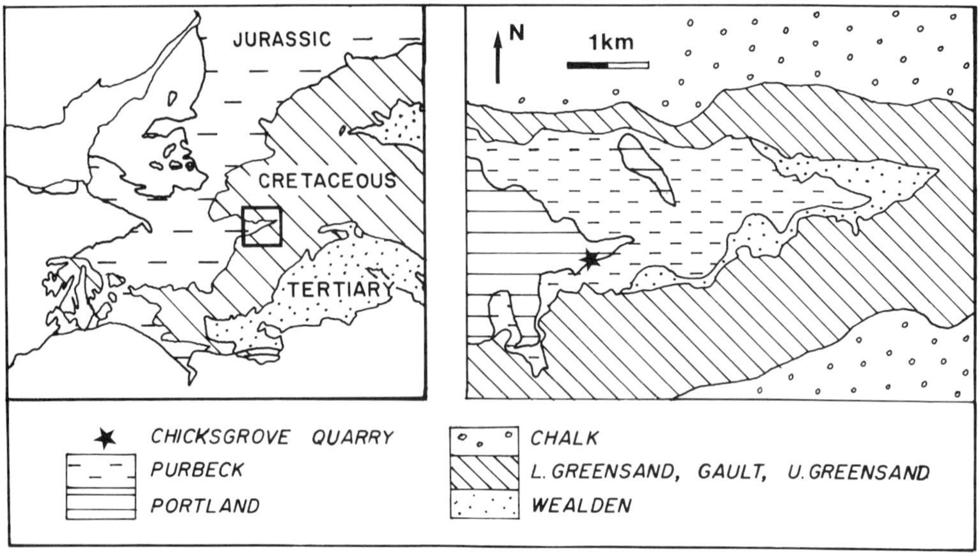

FIG. 1. Location map for Chicksgrove Quarry. The left diagram shows the broad distribution of rocks in southern England, with the boxed area expanded on the right. The map on the right is based on the British Geological Survey 1 : 50 000 series geological maps.

Most of the chert nodules occur near the bottom of the bed. They are typically 2–4 cm across and tend to be elongate along the bedding. They have compact, dark grey centres and porous, friable, pale grey outer layers. Cherts also occur in the adjacent beds. The Wockley micrites above contain scattered black, flint-like cherts which replace the limestone. In the upper part of the underlying Tisbury Member, lenticular patches and beds of crossbedded glauconitic sands are totally chert cemented. Lower down, silicification in the Tisbury Member is limited to partial replacement of shells in sandy micritic packstones.

Methodology

The chert nodules were sliced and impregnated in araldite coloured with a green dye to make porosity distinct in thin section. Twelve doubly polished thin sections were prepared and examined by optical petrographic methods and incident light excitation fluorescence.

Both UV and blue-violet light excitation were used for the fluorescence. The transmission characteristics of the UV filters used are given in Fig. 3. Dravis & Yurewicz (1985) report that for carbonate rocks blue light excitation is better than UV light excitation, because of the generally greater fluorescence intensities resulting. In the present study however, UV excitation was found to be more effective. UV light excitation results in a greater colour range in the transmitted light than blue or blue-violet excitation, and the fluorescence was sufficiently strong in this case to take advantage of this.

Two samples were analysed by semiquantitative XRD to determine the mineralogy of the cherts, and two thin sections were examined by SEM with an EDX microanalyser to investigate the composition of the inclusions.

Chert petrology

Original sediment

The initial sediment comprised a wide variety of detrital components with a mean grain size of fine sand (Wentworth scale). The largest grains included unmicritized small gastropods, wood fragments and a few large bivalve shell fragments. The fine sand consisted of angular quartz, spores and carbonate, the latter mostly micritized skeletal grains some of which are coated. The cherts included cemented and replaced grainstone, as well as replaced packstone with a matrix of carbonate silt and mud which is difficult to resolve texturally.

All the original non-silicate components (carbonate grains, organic grains, matrix) are now replaced by quartz. The replaced grains are made up of microcrystalline quartz or fibrous chalcedony.

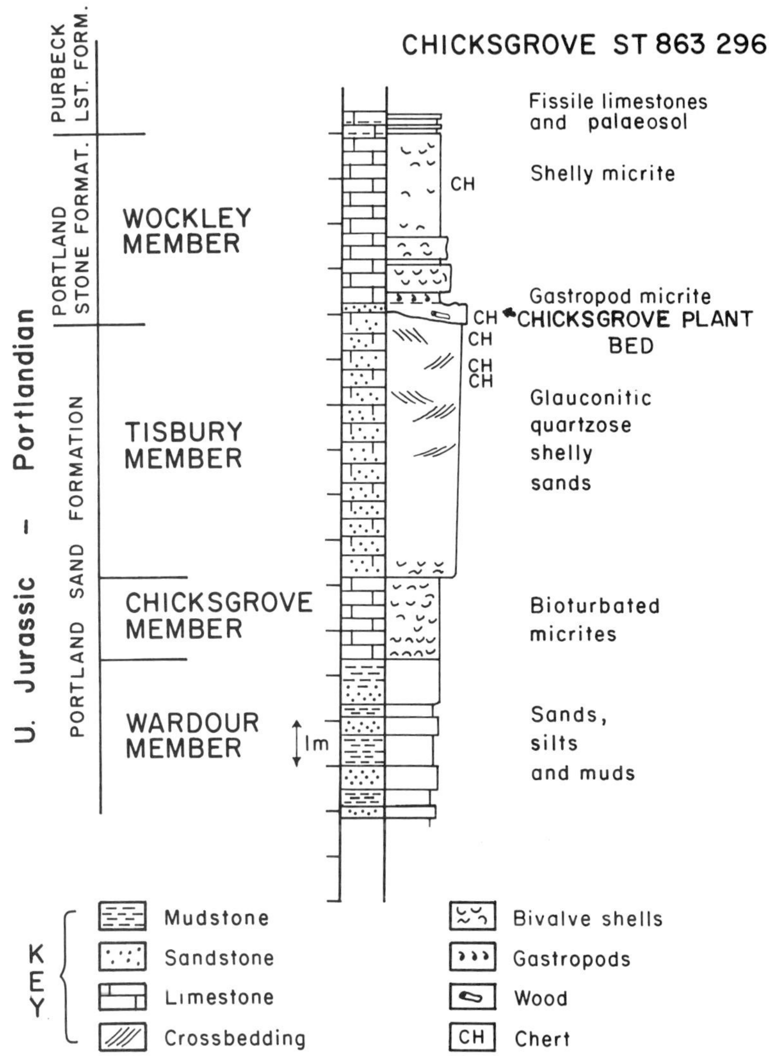

FIG. 2. The sequence present at Chicksgrove quarry (modified from Wimbledon 1976). The cherts described come from the sandstone bed at the base of the Wockley Member.

The primary sedimentary texture of all the chert is preserved unmodified by compaction. Thus delicate grains, including spores and micrite envelopes defining secondary pores, are preserved unflattened, and matrix-rich areas often preserve pellet textures (Fig. 4A, C). This contrasts with the adjacent sediment in which some compaction took place, as shown by shell breakage and a denser matrix with no surviving pelletal texture (Fig. 4K, L).

Cement stratigraphy

Three main stages of silica cementation are seen in the nodule centres, these being (1) lepispheres, (2) length fast chalcedony, and (3) quartz (Table 1). They are most clearly seen in the largest pores which originate from gastropod chambers (Fig. 4A, B, C).

The first stage consists of partial to whole lepispheres (10–13 μm diameter) lining the pores and locally coalescing into botryoidal layers. They are now made of radiating microquartz crystals giving pseudo-extinction crosses between crossed polars. Their shape suggests that they were originally opal-CT lepispheres, and this is consistent with their sizes which lie towards the upper end of those of opal-CT lepispheres reported from deep sea cherts (see, e.g. Weaver & Wise 1972, Wise & Weaver 1974,

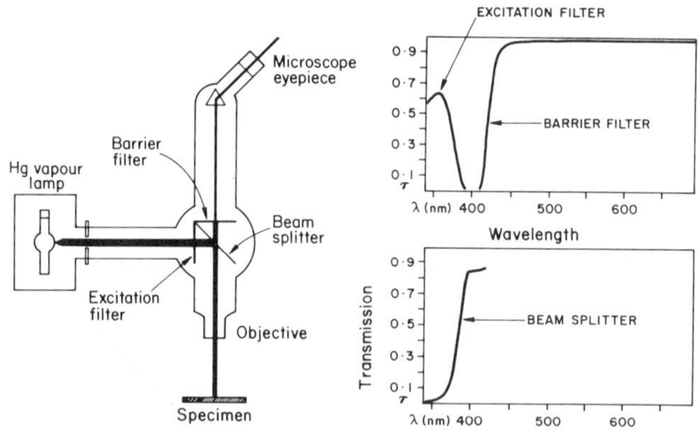

FIG. 3. An outline of the microscope arrangement for incident light fluorescence microscopy. The response characteristics of the filters used for UV fluorescence in this study are shown.

Hein *et al.* 1978, Pisciotto 1981). Opal-CT is also reported as forming botryoidal layers (Robertson & Hudson 1974). Neomorphosed opal-CT has been previously recognized in cherts for similar reasons (Meyers 1977, Keene 1983).

The second stage cements are length fast, radial fibrous quartz (chalcedony), present as pore-linings and up to 250 μm thick in the largest pores. Small pores are totally filled. In thin section the earlier chalcedony is transparent and pale green, becoming increasingly opaque towards the younger pore-centres where it is full of brown to black inclusions. There has been uncertainty about whether similar length-fast

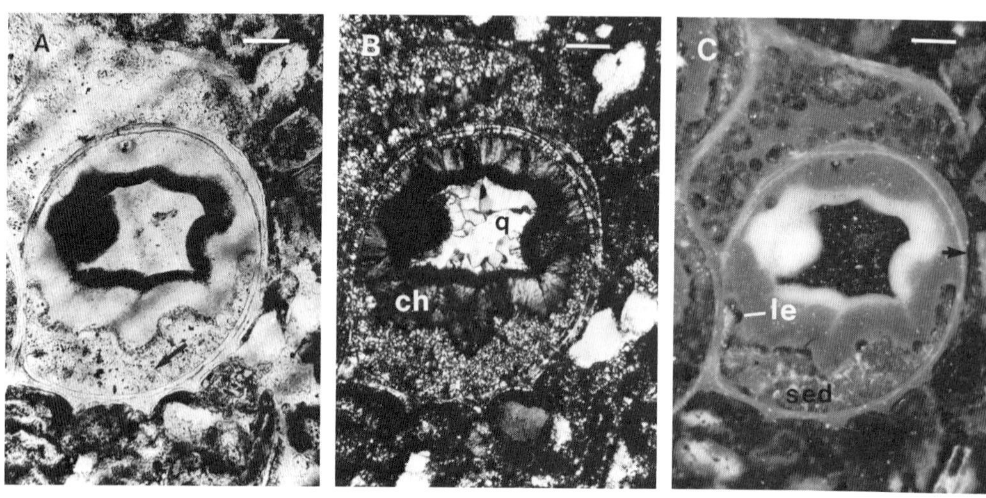

FIG. 4. Photographs of the chert textures seen in thin section. In all cases the scale bar is 100 μm.
(A) Plain polarized light (PPL). Cross-section of a gastropod showing an internal chamber filled with a combination of internal sediment (arrowed) and silica cement showing a prominent dark zone full of inclusions.
(B) Crossed polarized light (XPL). This shows the quartz cement up to and including the inclusion-rich zone to be fibrous (length fast) chalcedony, followed by a crystalline quartz cement.
(C) UV light fluorescence (UVF). This clarifies the distribution of internal sediment (sed), and shows the presence of early hemispheres of non-fluorescent cement prior to the chalcedony (le). The chalcedony changes from moderately to brightly fluorescent in the inclusion-rich zone. The late quartz is non-fluorescent.
(D) UVF. Another gastropod chamber cement sequence showing the presence of some brightly fluorescing nuclei to some early cement lepispheres (a), the moderately fluorescing earlier chalcedony (b) passing into the brightly fluorescing later chalcedony (c), and dark inclusions in the outer part of the bright zone of chalcedony cement (d). These fluoresce a reddish-brown colour typical of iron oxides. (E) PPL. Outer part of a chert nodule showing secondary voids (v) formed by opal sphere dissolution, now filled with green stained epoxy,

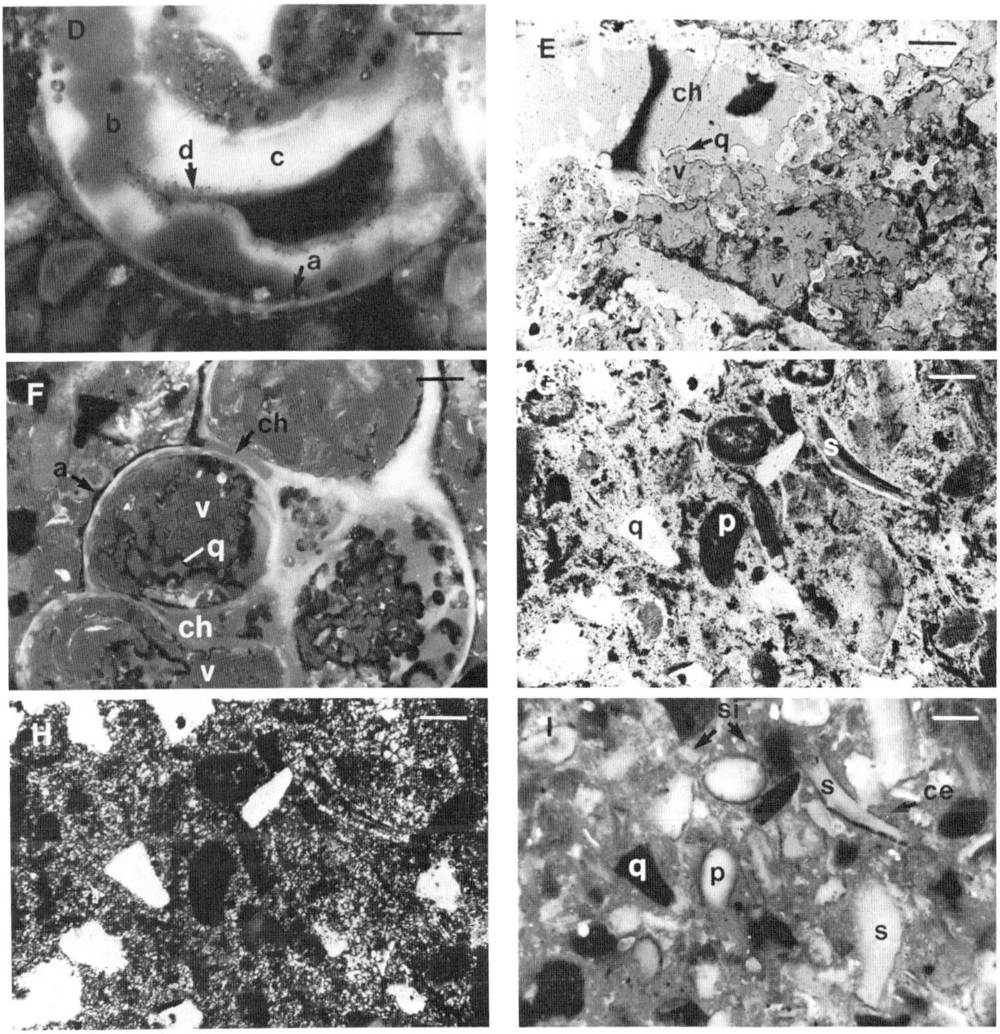

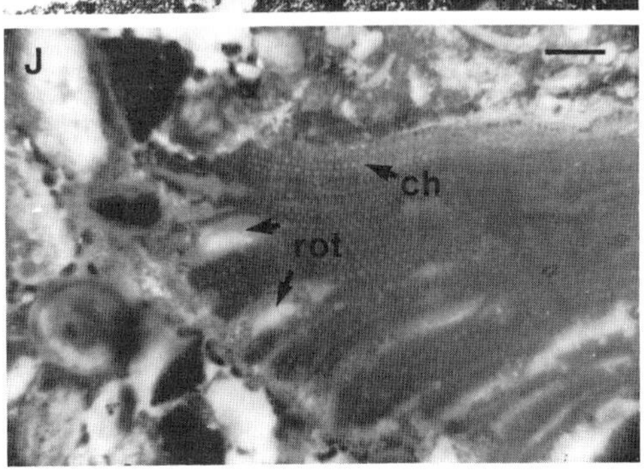

and showing up darker than the chalcedony cement (ch). The voids have a thin lining of quartz cement (q). (F) UVF. A gastropod from the margin of a nodule showing voids (v) resulting from secondarily dissolved lepispheres. The quartz lining the voids is non-fluorescent. The gastropod shell has been replaced partly by early non-fluorescent silica (a), and partly by fluorescent silica of the same generation as the chalcedony cement (ch). (G) PPL. Chert replaced sediment in which only larger quartz grains (q), shell fragments (s) and spore grains (p) are obvious. (H) XPL. This shows the microcrystalline quartz texture typical of most of the chert with few petrographic features visible. (I) UVF. This shows much more of the original sediment texture, with non-fluorescent quartz grains (q), replaced shells (s) and spores (p). Areas of pore-filling cement are revealed (ce), as are silt sized grains (si). (J) UVF. Wood showing preservation of cell walls by moderately fluorescing silica and infilled cells with later bright silica (arrowed, ch) both of the same stage as the chalcedony cements. The wood has been partly rotted prior to silicification as shown by the presence of areas infilled with chalcedony but without cell structure preserved (rot).

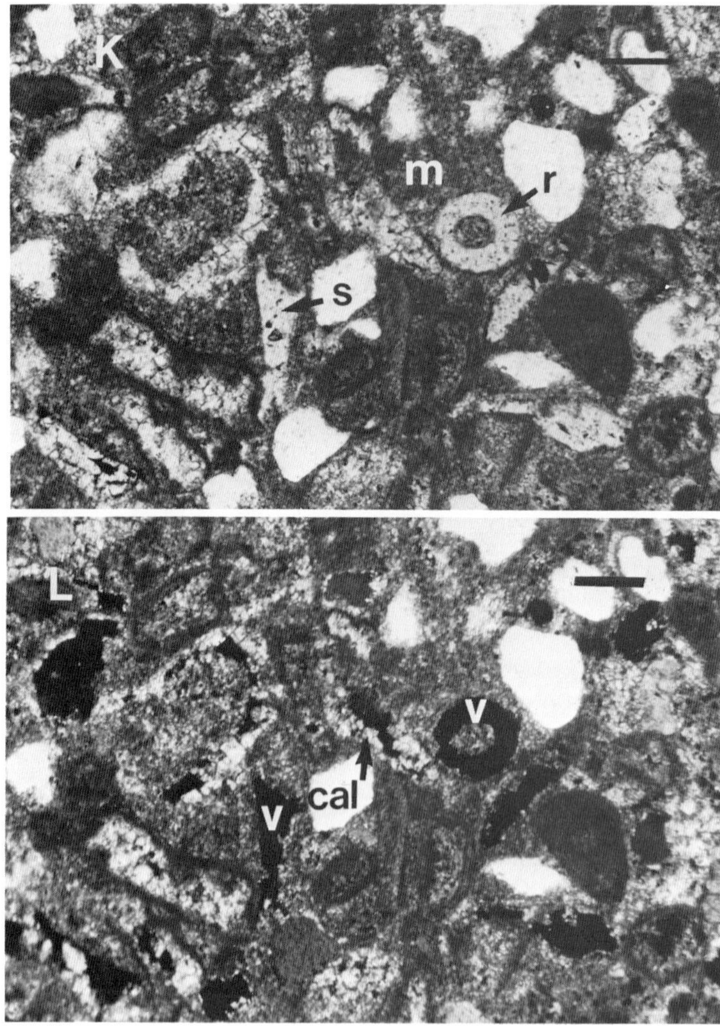

FIG. 4 (*continued*). (K) PPL. Calcareous micritic sandstone of the Chicksgrove Plant Bed away from the chert nodules. The sand includes shells (s) and a 'Raxella' type sponge spicule set in a carbonate matrix (m).
(L) XPL. Same view as (K) showing the voids (v) resulting from selective (presumed aragonitic) shell and biogenic opal spicule dissolution, and sparry calcite (cal) partly filling some of the secondary pores.

chalcedony is a primary precipitate or a replacement of opal-CT cement. Some authors prefer a replacement origin (Wise & Weaver 1974, Meyers 1977, though Meyers pointed out the uncertainty) while others favour a primary precipitate (Lancelot 1973, Keene 1983). Chalcedony is undoubtedly a primary precipitate in geodes (Mieke *et al.* 1984) and with no evidence to the contrary the Chicksgrove chalcedony is interpreted as a primary precipitate. Quartz fills the remaining space in the gastropod chambers, increasing in crystal size toward the original pore centres.

The three cement types have distinctly different fluorescence intensities. The lepispheres are weakly to non-fluorescent. They often have very bright fluorescent centres (Fig. 4C, D), seen as brown particles 8–10 μm across in ordinary light. The particles are therefore probably organic, perhaps the remains of microorganisms on which the opal-CT nucleated.

The earlier, pale green chalcedony shows moderate fluorescence, the intensity of which increases into the later chalcedony as the density of dark inclusions increases. The high intensity suggests that most of the brown inclusions are organic. The youngest, innermost parts of the chalcedony contain inclusions which fluoresce red-brown (Fig. 4D), characteristic of haematitic iron oxides. These are indistinguishable from

TABLE 1. *Summary of diagenetic textural evolution in the cherts, and the inferred controlling water chemistry at each stage*

STAGE	TEXTURAL CHANGES			U.V. FLUORESCENCE RESPONSE
	DISSOLUTIONS	REPLACEMENTS	CEMENTS	
1	Aragonite shells	Microbial alteration of wood		
2		A few shell margins replaced by Opal — CT	Opal — CT	Dead
3		Main replacement of carbonate grains and matrix by Silica	Length – fast, pale green	Dull
			Chalcedony, brown organics	Bright
			Chalcedony, brown organics + Fe oxides	Bright + Red Spots
4	Opal — CT in nodule margins		Micro – quartz	Dead

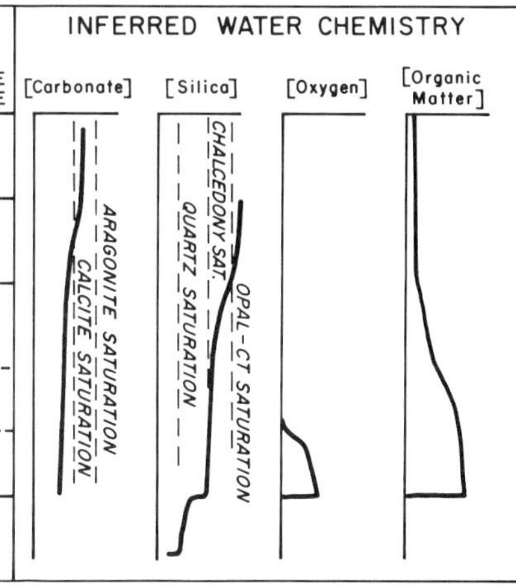

organic matter in ordinary light. Late quartz is mostly non-fluorescent. A few localized blue spots are of uncertain origin.

Two thin sections were examined by SEM with an energy dispersive X-ray (EDX) analysis system. EDX analysis of the inclusion-rich chalcedony showed no major components other than silica. Minor local iron confirmed the presence of some iron oxide inclusions and that most of the inclusions are organic and therefore undetectable by EDX.

Lepispheres and chalcedony cements fill all the primary pores between the finer grains in the nodule centres. The outer parts of the nodules, however, are porous and friable (Fig. 4E, F). Pores at the margin are partially to fully cemented by pore-lining lepispheres followed by pale green and brown chalcedony. Most of the lepispheres dissolved after the chalcedony formed, leaving secondary pores with characteristic spherical shapes. Where spheres dissolved, a thin rim of non-fluorescent microcrystalline quartz is often present in the void (Fig. 4E, F). This could be either undissolved silica from the margins of the lepispheres which subsequently recrystallized to quartz, or it could be a thin rim of stage three quartz cement deposited after lepisphere dissolution.

Matrix

The fine grained matrix is now a mass of microcrystalline quartz crystals (1 and 10 μm) within which primary grains, matrix and cements are difficult to distinguish in transmitted light. Fluoresence microscopy, however, clearly distinguishes between these. Areas of silicified primary matrix show a fawn fluorescence similar to that of the sediment inside the gastropod chambers (Fig. 4C, I). Within this matrix, many silt-sized detrital quartz and replaced carbonate grains are revealed for the first time (Fig. 4I).

The most likely precursor for the silicified matrix is lime mud, with some terrigenous silt. XRD analysis of the adjacent unsilicified sediment shows that clay minerals form only a very small part of the sediment, while micrite is abundantly present (Dr A. Parker, personal communication). Locally, the sediment within the gastropods is obviously pelleted, so much of the finer matrix may have originated as lime mud pellets.

Wood fragments often show a thin zone of dark brown matrix at their margins in ordinary light. Borings filled with the same material (diameter about 10 μm) locally penetrate the edge of the wood, truncating the primary cell structure. No bark layer seems to be preserved. The marginal brown matrix has a fluorescence colour similar to, but brighter than, that of the primary matrix (Fig. 4J).

Wood margins are interpreted as resulting from microbial degradation (probably involving fungi and bacteria) during transport and continuing into early burial. Microbial organic matter probably gives rise to the bright fluorescence. Smaller amounts of the same organic matter presumably occur in the fine sediment generating the similar fluorescence response.

Grain replacement

In ordinary light the grain margins are not always visible (Fig. 4G), but UV fluorescence shows up distinctly the distribution of grains and their internal structures (Fig. 4I).

Some shell fragments and ooids became secondary pores through carbonate dissolution prior to silicification, and were later filled with opal-CT lepispheres followed by chalcedony just as in primary pores. Along nodule margins, the secondary pores have remained partially or even wholly unfilled. Preferential leaching of aragonite grains is thought to give rise to these secondary pores. The adjacent sediment also shows selective replacement of carbonate grains with subsequent partial cementation by calcite spar (Fig. 4K).

The replaced grains often show a fluorescence colour and intensity correlatable with the cement sequence, revealing the relative timing of grain replacement. The fluorescence colours are identical with those of the cements, and some wood and carbonate grains preserve the internal organic structure, such as protein layers in bivalve shells and cellulose cell walls in wood (Fig. 4J).

Two end-member types of replacement can be recognized. In one, quartz precipitation coincided with grain removal by dissolution, no large voids forming at any stage. Original structural organic material is usually preserved, and boundaries between replacive silica from different episodes follow pre-existing internal boundaries (Fig. 4I). In the second type, a secondary pore was formed as an intermediate stage. Organic components of shells are not preserved within the replacive quartz, because organic material degradation and removal took place at the same time as carbonate dissolution and before silica replacement. Pore filling fabrics are present within the second types of grain, boundaries between different cement generations showing no relation with former internal structures.

An intermediate type also occurs, some grains showing partial loss of structure through partial dissolution prior to silicification. This has taken place, for example, in wood fragments, where the original wood structure has been locally lost, suggesting some rotting of the wood before silicification (Fig. 4J).

Timing of grain replacement

Some carbonate grains have been partially replaced by non-fluorescent silica, typically at shell margins (Fig. 4C, F, I). The silica was probably opal-CT. Most of the carbonate grains and wood are represented by combinations of weakly and brightly fluorescent silica, the fluorescence being the same as in the chalcedony cements. The cellulose cell walls of the woods were replaced first, typically by weakly fluorescent silica, and the cells subsequently infilled with brightly fluorescent silica (Fig. 4J). This is the typical sequence of events during the silicification of wood (Scurfield & Segnit 1984). Most of the spores seem to have been replaced by the later more fluorescent chalcedony (Fig. 4I). All silica replacement took place before the quartz cement formed.

Discussion

Silica source

In order to get the initial opal-CT precipitation in the cherts, a high silica concentration in the pore water is needed (Williams *et al.* 1985). This must necessarily be higher than would arise from pore waters in simple equilibrium with quartz. In principle there are two ways of achieving this while still involving quartz dissolution.

First, quartz is much more soluble in strongly alkaline ($pH \geqslant 10$) water than under lower pH conditions (Volosov *et al.* 1972). The high silica concentrations required for opal-CT precipitation at neutral to low pH could have arisen through quartz dissolving in alkaline solutions and then the pH changing. But strongly alkaline waters are rare in nature and cannot have arisen in the Chicksgrove situation because the abundant carbonate would have buffered the pore water to pH 9.9 or below (Garrels & Christ 1965).

Second, a high silica concentration could have been achieved through evaporation. Evaporative concentration is effective only at or close to the land surface, where it can give rise to soil-profile silcretes (Summerfield 1982). Such silcretes have characteristic textures, including quartz grain etching and replacement, grain fracturing, expansive fabrics, complex vug fills, and glaebule formation (Summerfield 1982) which are absent here. The Chicksgrove textures, on the other hand, show that the cherts are a type of groundwater silcrete, formed in phreatic conditions where evaporation could not play a direct role in concentration.

This argument suggests that quartz was not the primary source of the silica, and that a biogenic opal-A source is likely, as indeed it is for the large majority of cherts (see Wise & Weaver 1974, Meyers 1977, Hein *et al.* 1978, Knauth 1979, Williams *et al.* 1985 for just a few examples). As opal-A dissolves, opal-CT, being less soluble than opal-A, can precipitate and is kinetically preferred to quartz precipitation from pore waters having moderately high silica concentrations (Williams & Crerer 1985).

A few sponge spicules are indeed found in the Chicksgrove Plant Bed, possibly derived, with other marine biogenic grains, from slightly older rocks. The spicules are preserved as microquartz within the cherts, and the presence of others is recorded by voids in the upper micritic part of the Plant Bed (Fig. 4K, L). However, they were too few to have acted as the sole source of opal-A.

One untested possibility is that sponge spicules were dissolved from the overlying marine micritic limestones, perhaps with a minor contribution from the marine glauconitic sandy limestones underneath. This is suggested by the overall distribution and types of Portlandian chert at Chicksgrove. The black flint-like cherts in the Wockley Limestone suggest a local silica source in this unit, and silicification decreases downwards below the Plant Bed through the Tisbury Member, implying a source from above rather than below, and transport of dissolved silica by downward moving ground water.

Inferred water chemistry during diagenesis

Silica concentration

The change in cement type with time is from opal-CT via length-fast chalcedony to quartz. The same sequence is found in Mississippian limestones (Meyers 1977). In both, therefore, the concentration of dissolved silica in the pore water was initially high, being saturated with respect to opal-CT, and then decreased, finally falling below that leading to quartz precipitation (Table 1).

The amounts and distribution of the three cement types show that the rate of change of silica concentration varied. After initial opal-CT precipitation the silica concentration fell, but remained high enough for some time to cause the growth of fibrous chalcedony, this being the main type of silica in the nodules. To maintain a high concentration, dissolved opal-A must have been transported into the bed.

Two likely mechanisms for the fall in silica during opal-CT and chalcedony precipitation are (1) progressive dilution of silica-rich pore waters by silica-poor water, and (2) progressive depletion of the opal-A available for dissolution in the source rock (probably the overlying limestones). The first effect will have taken place following exposure as meteoric water mixed with and replaced the original marine pore water. the second effect occurred during prolonged meteoric water circulation (Fig. 5B, C).

The end of chalcedony precipitation could have resulted either from opal A depletion in the source, or from marine transgression of the exposed surface which stopped the circulation of meteoric water.

After precipitation stopped, opal-CT was dissolved at the nodule margins and quartz precipitated. The small amount of microcrystalline quartz cement which may have formed in the nodule margins at this time, being less than the opal-A dissolved, suggests that ground water was no longer contributing silica, the pore water outside the nodules becoming undersaturated in silica with respect to quartz during opal-CT dissolution.

However, the central pores became completely cemented with quartz, implying that higher silica concentrations were maintained inside the nodules than outside. This must have been set up by silica diffusion from sites of opal-CT dissolution in the nodule margins.

Opal-CT dissolution took place before opal-CT inversion to quartz, but both this and the associated quartz cementation could have taken place later than the formation of chalcedony.

Carbonate concentration

As shown by the shell replacement textures and the presence of micrite envelope 'eggshell' grains in the nodule margins, some dissolution of carbonate must have taken place to form secondary pores before silicification (Table 1). Aragonitic grains were probably lost preferentially to form these secondary voids, so there is no need to infer carbonate undersaturation with respect to calcite prior to silicification.

The opal-CT is nearly all void filling cement, with only a little grain replacement at this stage. This implies approximate carbonate saturation of the pore-waters at this stage. All the remaining carbonate in the nodules was dissolved synchronously with the formation of chalcedony when the pore water became undersaturated with respect to carbonate. This trend towards increasing carbonate dissolution, coincident with silicification, demands the introduction of quite large volumes of meteoric water (Fig. 5C).

Undersaturation is known to occur in at least two stages of a meteoric water flow path through carbonate rocks. Meteoric water, initially undersaturated with carbonate on entering the sediment, tends towards equilibrium with carbonate along the flow-path. Knauth (1979) pointed out that undersaturation recurs in the mixing zone with marine pore-water if suitable differences of P_{CO_2} exist.

Significance of inclusions in the chalcedony

The chalcedony shows a systematic increase in the abundance of included brown particles from early to late formed chalcedony. The intense fluorescence of the inclusion-rich chalcedony proves that they are largely organic.

Organic matter in the pore water during early diagenesis is likely to have come partly from

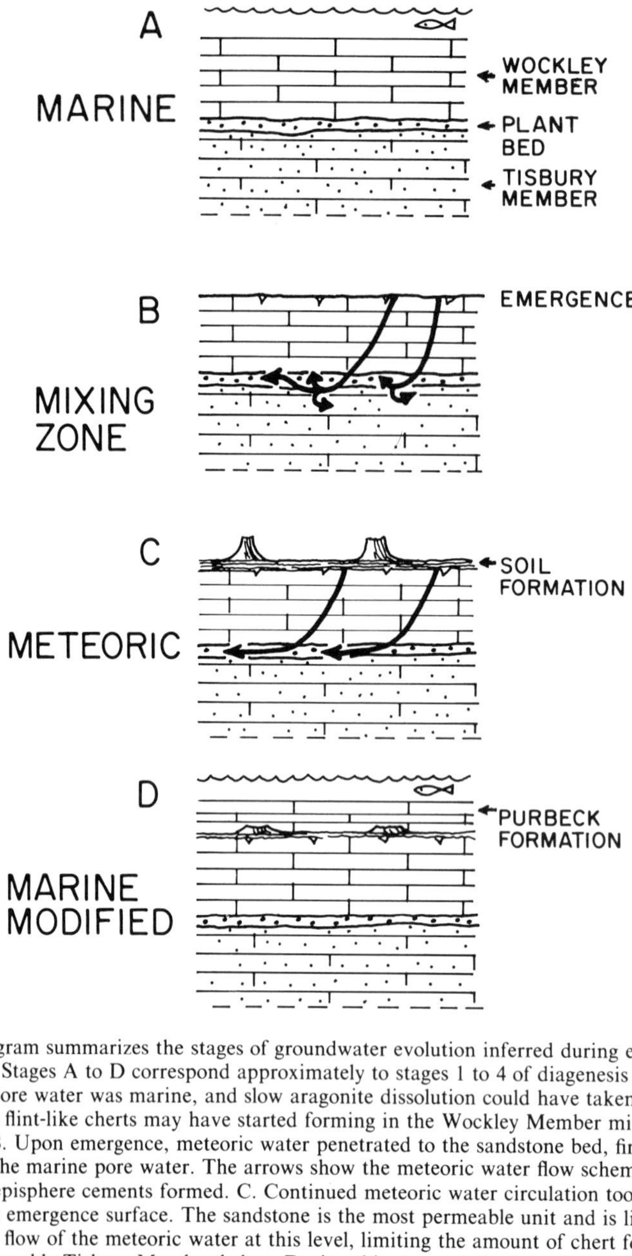

FIG. 5. This diagram summarizes the stages of groundwater evolution inferred during early diagenesis and chert formation. Stages A to D correspond approximately to stages 1 to 4 of diagenesis identified in Table 1. A. Initially the pore water was marine, and slow aragonite dissolution could have taken place in the sediments. Some flint-like cherts may have started forming in the Wockley Member micrites above the sandstone bed. B. Upon emergence, meteoric water penetrated to the sandstone bed, first mixing with and then displacing the marine pore water. The arrows show the meteoric water flow schematically. During this stage, opal-CT lepisphere cements formed. C. Continued meteoric water circulation took place during soil formation on the emergence surface. The sandstone is the most permeable unit and is likely to have caused a change to lateral flow of the meteoric water at this level, limiting the amount of chert formation in the decreasingly permeable Tisbury Member below. During this stage, chalcedony formation and carbonate dissolution took place. D. Following marine transgression and the start of Purbeck Formation sedimentation, marine pore waters were re-introduced and vigorous groundwater circulation ceased, bringing an end to chert formation. At this or a later stage, some quartz cement formed in chert centres.

physical breakdown of dissolving shells and partly from microbial activity. Both processes must have operated before and during the early stages of chalcedony formation. The moderate fluorescence shown by the early chalcedony, unrelated to obvious particulate inclusions may be caused by dissolved organic compounds from these sources being trapped during growth.

The abundant organic inclusions in the later chalcedony show that a new source of organic particles became available in much greater concentrations than ordinarily occur during early diagenesis, they must have been introduced by moving ground water. The most likely source is the humic layer of a suitable soil. The palaeosol previously mentioned lies about 8 m above the cherts (Fig. 2) and presumably had a suitable humic layer (Francis 1984).

The iron oxides shown by fluorescence microscopy to be present in the latest chalcedony (Fig. 4D) suggest that the waters became oxidizing when this formed. Here the postulated circulation of meteoric water at this time is strengthened, because it would lead to progressively deeper penetration of oxygenated water along the flow path, away from the recharge zone (soil) (Fig. 5C).

The later quartz cement is clear and non-fluorescent, implying that the pore waters no longer contained significant amounts of dissolved or particulate organic matter. This contrasts markedly with the immediately preceding chalcedony, and records a quite different pore environment when marine conditions were re-established and stopped meteoric water penetration in the area (Fig. 5D). Later quartz cement could have formed significantly later.

Relation between meteoric water circulation and chert formation

The cherts show very similar textures to those described from Mississippian limestones by Meyers (1977). He relates chert formation to an opal-A source of silica combined with meteoric water circulation, basing his argument on the timing of silicification relative to calcite cement zones of proven origin from meteoric phreatic water.

Knauth (1979) suggests that the formation of *replacive* cherts in limestones requires special circumstances for the necessary undersaturation with respect to carbonate in the meteoric water. He concluded that silicification is most likely in the zone of mixing between previously carbonate-saturated meteoric and marine waters, where local undersaturation could sometimes occur as a result of differences in P_{CO_2} between the two types of water.

In the Chicksgrove case, silicification is inferred to have taken place close to the recharge zone before the meteoric water became saturated with carbonate, the recharge zone being the soil now some 8 m above. This is consistent with the uncompacted fabrics of the sediments in the cherts, compared with the compacted matrix fabrics in adjacent sediments. Chert formation therefore took place fairly early in the diagenetic history, the only changes prior to silicification being minor microbial degradation of the wood and selective dissolution of some (probably aragonitic) carbonate grains (Table 1).

Knauth (1979) assumed that meteoric water would reach saturation with carbonate sooner than with opal-A at the start of the flow path. But this need not necessarily be the case; for example where meteoric water influx is episodic, as in a seasonal climate. In this situation, silica and carbonate concentrations move towards saturation in stagnant groundwater during the dry season. Influx of meteoric water during the wet season dilutes the groundwater, which becomes undersaturated in carbonate and opal-A, but remains saturated with respect to chalcedony close to the recharge zone.

Conclusions

Early silicification resulted from meteoric water penetrating the Portlandian limestones at a time of emergence and soil formation. The soil in question was probably what is now the Portland Dirt Bed, a palaeosol 8 m above. Meteoric water, undersaturated with respect to opaline silica and carbonate, dissolved biogenic opal from adjacent limestones and transported it into the Plant Bed.

Initially, the meteoric water mixed with little modified marine water saturated with respect to carbonate and biogenic opal (Fig. 5B) and precipitated opal-CT cement. Increasing volumes of meteoric water led first to mixing with the saline water and then to complete replacement (Fig. 5C). Hence the pore water became undersaturated with respect to carbonate, resulting in dissolution of the carbonate in the nodules and allowing extensive silicification. At the same time, carbonate was leached from adjacent limestones. Silica concentrations dropped as biogenic opal was depleted in its source rock, leading to precipitation of chalcedony after opal-CT. As an organic-rich soil developed on the exposed land, the ground water became enriched in organic particles. These formed inclusions in the later chalcedony. During the same episode an oxidation front migrated into the sediment, giving rise to iron oxide inclusions in the late chalcedony cement. Quartz cements followed after the soil horizon was buried by further sediment (Fig. 5D). They may have formed much later during deeper burial, but probably before opal-CT inverted to quartz because some opal-CT dissolution probably freed silica for the quartz cement.

The study exemplifies how an early episode of emergence and meteoric water circulation can cause silica mobilization and silicification. It shows that this silicification can occur close to the recharge zone. In the Portlandian of Wiltshire this has led locally to the pre-compaction preservation of fossils with obvious palaeontological benefits.

ACKNOWLEDGMENTS: Thanks to Bill Wimbledon of the Nature Conservancy Council for asking me to investigate the cherts and explaining their geological setting. The technical staff in the Department of Geology at Reading are thanked for their help. Ian Pryde took much care in making the thin sections, Jim Watkins helped prepare the photographs for publication and Alan Cross helped prepare the diagrams.

Penny Williams helped me with some stimulating discussions about the textures of silcretes and their genesis, and the extent to which these cherts might be considered groundwater silcretes.

I wish to thank those who led me to fluorescence microscopy, whether knowingly or unknowingly, and subsequently encouraged me to explore its possibilities; M. Talbot, J. A. D. Dickson and J. Dravis.

Thanks to the reviewers of the paper, W. J. Meyers and G. Carson, who offered numerous helpful comments for improvement of the manuscript, and my understanding of the subject. Professor P. Allen instructively helped improve my English style, though any poor construction and inaccuracies remaining are my own choice.

References

BUURMAN, P. 1972. Mineralisation of fossil wood. *Scripta Geologica*, **12**, 1–43.

DRAVIS, J. J. & YUREWICZ, D. A. 1985. Enhanced carbonate petrography using fluorescence microscopy. *Journal of Sedimentary Petrology*, **55**, 795–804.

FRANCIS, J. E. 1984. The seasonal environment of the Purbeck (Upper Jurassic) fossil forests. *Palaeogeography, Palaeoclimatology and Palaeoecology*, **48**, 285–307.

GARRELS, R. M. & CHRIST, C. L. 1965. *Solutions, Minerals and Equilibria*. Harper & Row, New York.

HEIN, J. R., SCHOLL, D. W., BARRON, J. A., JONES, M. G. & MILLER, J. 1978. Diagenesis of late Cenozoic diatomaceous deposits and formation of the bottom simulating reflector in the southern Bering Sea. *Sedimentology*, **25**, 144–81.

KEENE, J. B. 1983. Chalcedonic quartz and occurrence of quartzine (length-slow chalcedony) in pelagic sediments. *Sedimentology*, **30**, 449–54.

KNAUTH, L. P. 1979. A model for the origin of chert in limestone. *Geology*, **7**, 274–7.

LANCELOT, Y. 1973. Chert and silica diagenesis in sediments from the central Pacific. *In*: WINTERER, E. L., EWING, J. I. *et al.* (eds). *Initial Reports of the Deep Sea Drilling Project*, XVII, 377–405. US Government Printing Office, Washington, DC.

MEYERS, W. J. 1977. Chertification in the Mississippian Lake Valley Formation, Sacramento Mountains, New Mexico. *Sedimentology*, **24**, 75–105.

MIEKE, G., GRAETSDCH, H. & FLORKE, O. W. 1984. Crystal structure and growth fabric of length-fast chalcedony. *Physics and Chemistry of Minerals*, **10**, 197–9.

PISCIOTTO, K. A. 1981. Diagenetic trends in the siliceous facies of the Monterey shale in the Santa Maria region, California. *Sedimentology*, **28**, 547–71.

ROBERTSON, A. H. F. & HUDSON, J. D. 1974. Pelagic sediments in the Cretaceous and Tertiary history of the Troodos Massif, Cyprus. *In*: HSÜ, K. J. & JENKYNS, H. C. (eds). *Pelagic Sediments: on Land and Under the Sea*. Special Publication of the International Association of Sedimentologists, **1**, 403–36. Blackwell Scientific Publications, Oxford.

SCURFIELD, G. & SEGNIT, E. R. 1984. Petrification of wood by silica minerals. *Sedimentary Geology*, **39**, 149–67.

SUMMERFIELD, M. A. 1982. Silcrete. *In*: GOUDIE, A. S. & PYE, K. (eds). *Chemical Sediments and Geomorphology*. Academic Press, London.

VOLOSOV, A. G., KHODAKOVSKIY, I. L. & RYZENKO, B. N. 1972. Equilibria in the system SiO_2-H_2O at elevated temperatures along the lower three-phase curve. *Geochemistry International*, **9**, 362–77.

WEAVER, F. M. & WISE, S. W. 1972. Ultramorphology of deep sea cristobalitic chert. *Nature, London*, **237**, 56–7.

WILLIAMS, L. A. & CRERAR, D. A. 1985. Silica diagenesis, II. General mechanisms. *Journal of Sedimentary Petrology*, **55**, 312–21.

WILLIAMS, L. A., PARKS, G. A. & CRERAR, D. A. 1985. Silica diagenesis, I. Solubility controls. *Journal of Sedimentary Petrology*, **55**, 301–11.

WIMBLEDON, W. A. 1976. The Portland Beds (Upper Jurassic) of Wiltshire. *Wiltshire Natural History Magazine*, **71**, 3–11.

—— 1980. Portlandian correlation chart. *In*: COPE, J. W. C., DUFF, K. L., PARSONS, C. F., TORRENS, H. S., WIMBLEDON, W. A. & WRIGHT, J. K. *A Correlation of Jurassic Rocks in the British Isles Part Two: Middle and Upper Jurassic*. Geology Society of London, Special Report No. 15.

WISE, S. W. & WEAVER, F. M. 1974. Chertification of ocean sediments. *In*: HSÜ, K. J. & JENKYNS, H. C. (eds). *Pelagic Sediments: on Land and Under the Sea*. Special Publication of the International Association of Sedimentologists, **1**, 301–26. Blackwell Scientific Publications, Oxford.

T. R. ASTIN, Department of Geology, University of Reading, Whiteknights, PO Box 227, Reading RG6 2AB, UK.

Silicification fabrics from the Cenomanian and basal Turonian of Devon, England: isotopic results

Greg A. Carson

SUMMARY: Silicification fabrics within the Cenomanian and basal Turonian of Devon may be classified by their morphology and petrological relationships. The replacement of bioclasts by silica was often a two stage process; the early stage has $\delta^{18}O$ values ranging from 30.0 to 32.4‰ SMOW. The palaeotemperatures calculated from these are close to values for Cenomanian sea water and thus the early silicification took place in marine porewaters and before significant burial. The consistently lighter values for the later replacement (ranging from 27.2 to 29.4‰) cannot be explained by an increase in burial depth. Partial recrystallization in meteoric waters is believed to be the mechanism responsible. Isotopic data suggest that the formation of siliceous nodules (chalcedony cemented sands) within the arenaceous lower members of the Cenomanian was an early process, although petrographic evidence points to possible (later?) recrystallization. The formation of late-stage drusy quartz occurred entirely in meteoric porewaters.

Two types of silicification occur in the Cenomanian of Devon: the replacement of fossils and the formation of siliceous nodules. Silicification of bioclasts is both geographically and temporally common. It has been documented in the Ordovician brachiopods of Scotland (Curry 1986), in Carboniferous corals in Belgium (Buurman & Van der Plas 1971), brachiopods in Ireland (Brunton 1966, 1968, 1984) and New Mexico (Meyers 1977), bivalves and brachiopods from the Permian of Wyoming (Schmitt & Boyd 1981), corals and bivalves in the Liassic of Morocco (Faugères 1978), bioclasts in the Upper Jurassic (Wilson 1966) and brachiopods, bivalves (Brown *et al.* 1969) and echinoids (Holdaway & Clayton 1982) in the Upper Cretaceous of southern England. In many cases cited in the literature, the silicification was considered significant only in as much as the excellent preservation was of taxonomic value to palaeontologists (e.g. Brunton 1984 and Curry 1986). In other instances, limited geochemical work was carried out (e.g. Buurman & Van der Plas 1971), while a detailed hypothesis on the formation of different replacement fabrics was established by Holdaway & Clayton (1982) on petrographical data. In this study the work of Holdaway & Clayton (1982) is taken one stage further, both in more detailed observation of replacement fabrics and in using oxygen stable isotope methods in the interpretation of the timing of replacement.

The siliceous nodules in the Cenomanian of Devon are essentially chalcedony cemented quartz sands. Similar chalcedony cementation fabrics are well documented, from small-scale cementation of lepispheres in flint (Clayton 1982) to cementation of ooids as described by Wilson (1966).

Location

The principal field locality is a small working sand pit (known as the White Hart Sand Pit) at Wilmington, south Devon (National Grid Reference SY 209 998). Of minor importance in this study are several coastal exposures situated between Branscombe and Haven Cliff (Fig. 1). The Cenomanian consists of a series of variably indurated bioclastic and arenaceous packstones which, on the coast, may be divided into four distinct members, each one separated by a prominent glauconitized and/or phosphatized hardground (Fig. 2). Inland, the basal members of the succession change from indurated limestones to a sequence of coarse, loose quartz sands with occasional patchy carbonate cementation and development of silicified nodules and cherts. The Cenomanian in Devon shows extreme variation in thickness over very short distances. The thinnest sequence is 26 cm at Haven Cliff, while the expanded succession at Wilmington, just 11 km away, is nearly 9 m. The same scale of variation is seen over a distance of less than 1.5 km—the Cenomanian is 37 cm thick on Beer Beach and over 11 m at the Beer Stone Adit in Hooken Cliff (Jarvis & Woodroof 1984). Smith (1957) has attributed this variation to intra-Cretaceous folding. However, it is more likely to be due to intra-Cretaceous faulting (for fuller details, see Jarvis & Woodroof 1984).

Depth of burial

In order to assess the effect of any burial diagenesis, it is necessary to consider the burial history of the sequence. Sediment deposition at Wilmington after the Cretaceous was minor,

From: MARSHALL, J. D. (ed.), 1987, *Diagenesis of Sedimentary Sequences*, Geological Society Special Publication No. 36, pp. 87–102.

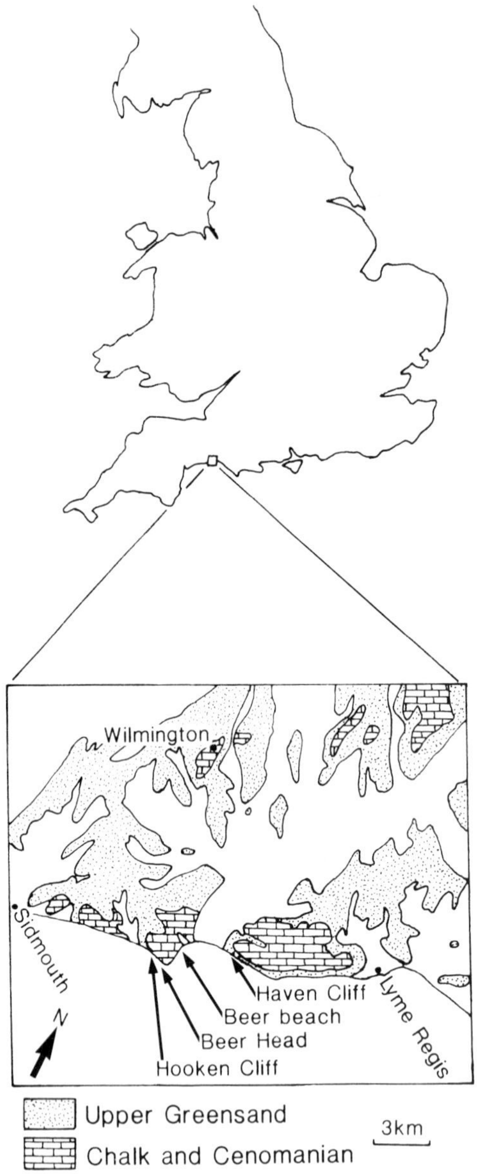

FIG. 1. Location map of field area (after Smith 1961).

Following regional uplift, SW England was subaerially exposed during the Cenozoic (see palaeographic reconstructions in Anderton *et al.* 1979). Thus an estimation of maximum burial depth for these sediments depends solely upon the amount of chalk deposition during the remainder of the Cretaceous (that is, the Turonian onwards). The maximum thickness of chalk present in Devon today is about 98 m (Rawson *et al.* 1978), extending just into the Coniacian. It is possible that deposition at Wilmington took place during the remainder of the Senonian, but the sediment was eroded during uplift. Using an average deposition rate of 15 m/10^6 yr (Black 1953), the duration of the Senonian (excluding the Maastrichtian) as 16 Ma (Van Hinte 1976) and a minimum of 10% compaction (Håkansson *et al.* 1974), a chalk pile of 216 m is produced. Including the 45 m of observed Turonian on the coast, this gives 261 m. It is, however, unlikely that 260 m of burial was attained since this part of England was in a structurally high setting during the upper Cretaceous (Hancock 1975, fig. 2).

Analytical methods

Standard reflected and transmitted light petrography were employed. Scanning electron microscopy (SEM) specimens were mounted on aluminium stubs and sputter coated with Au/Pd. They were investigated using a Philips 501 SEM with an energy dispersive X-ray (EDAX) attachment. Oxygen isotope analysis was carried out at the NERC BGS Laboratory, London. The oxygen was extracted from the manually ground silica by bromine pentafluoride treatment using a method similar to that described by Clayton & Mayeda (1963). Sample outgassing took place at 550°C overnight, pre-fluorination was at room temperature and reaction was at 450°C for 8 h. Analysis was performed on a VG Micromass 903 mass spectrometer. Results are reported in the δ notation as the ‰ deviation from SMOW. The value for the international quartz standard (NBS 28) determined by the BGS laboratory is 9.6‰, with a reproducibility of ± 0.2‰. The mean deviation for duplicate sample analyses was ± 0.45‰.

Petrography

Silicification of bioclasts

Holdaway & Clayton (1982) identify three main silicification fabrics in these fossils:

(1) A granular white crust, which they interpret as having formed when the rate of silica supply outpaced carbonate dissolution.
(2) Beekite rings, where silica supply was limited and outpaced by carbonate dissolution. They suggest that a 'liesegang' diffusion situation was achieved, causing the silica to be precipitated in concentric zones.
(3) Fine-scale replacement, which they believe was due to precipitation of silica keeping

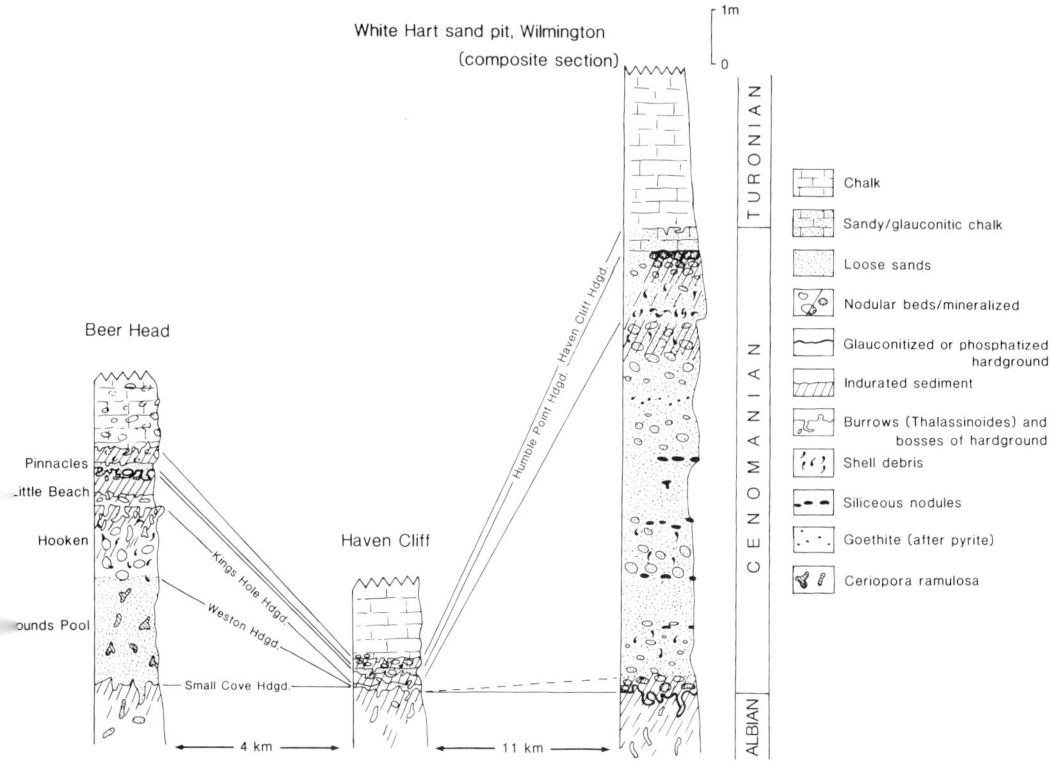

FIG. 2. Three stratigraphic profiles of the Cenomanian in Devon.

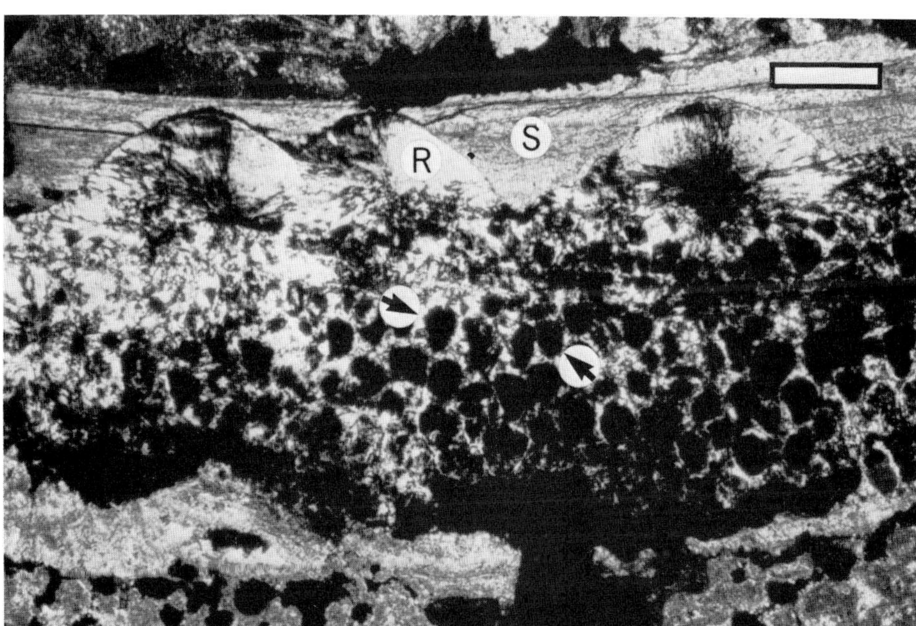

FIG. 3. Lutecite rosettes (R) replacing an oyster shell (S). Note the replacement of the vesicular shell structure by silica (arrowed). Loose, Wilmington. Crossed nicols. 0.5 mm scale bar.

pace with slow calcite dissolution, and thus accurately preserving the original crystal morphology.

They suggest that the crystal form and thus fossil type is the controlling factor in determining the type of silicification that develops. So, in echinoids, the large crystal size causes a granular crust to form (dissolution of carbonate being slow compared to the precipitation of silica); in bivalves (especially *Exogyra*), beekite forms (due to the rapid dissolution of the smaller calcite crystals), while in brachiopods, a fine-scale replacement occurs (due to the shell prisms being large enough to prevent very rapid dissolution, but not too large, preventing dissolution and causing the precipitation of a granular crust).

In the present study, the following replacement fabrics were observed in thin section:

Chalcedony 'rosettes'. These are discordant, cutting across the original calcite fabric of the shell, and occur as aggregates of chalcedonite and/or quartzine/lutecite (Folk & Pittman 1971) fibres. When they form as 'solitary' replacements, they are centred around a point, forming what must be an irregular sphere. However, most commonly they are gradational from finer grained quartz and possess hemispherical or botryoidal forms (Fig. 3). They are equivalent to cross-sections through beekite. These fabrics are most common in bivalves though they are also seen in rhynchonellids.

Fine scale replacement. This tends to be concordant with the shell fabric and occurs frequently in brachiopods and *Inoceramus*. The mode of replacement tends to vary in appearance depending upon fossil taxa. In *Inoceramus* there is clearly an inter-prism replacement, the prism boundaries acting as the principal pathways through which carbonate dissolution and silica replacement took place (Fig. 4). Where this process is most advanced, nearly all the original internal structure of the shell is lost (Fig. 5). An inter-crystalline replacement also occurs within belemnites (Fig. 6), where the silica follows the concentric and radial structure of the guard. In rhynchonellids, however, the smaller dimensions of the individual fibres commonly promote their complete silicification so that an incompletely replaced *single* fibre may be difficult to discern. Figure 7 shows a completely replaced brachial loop of a *Kingella* sp., where the silica closely follows the original fibrous structure of the shell, In a partially replaced example (Fig. 8) the silica also follows the original shell fabric, but possesses a much more irregular surface.

The present detailed investigation demonstrates that the silicification fabrics by Holdaway & Clayton (1982) are by no means mutually exclusive:

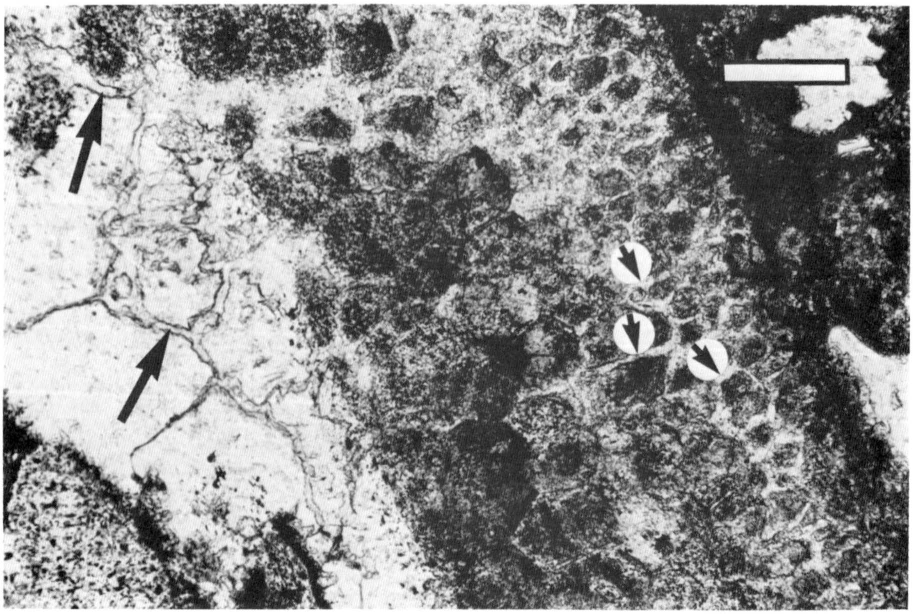

FIG. 4. Partial silicification of *Inoceramus* sp. taking place preferentially along prism boundaries (small arrows). The same phase of silicification also follows boundaries between sparry calcite cement crystals (large arrows). Base of Hooken member, loose block at Humble Point. Plane polarized light. 0.2 mm scale bar.

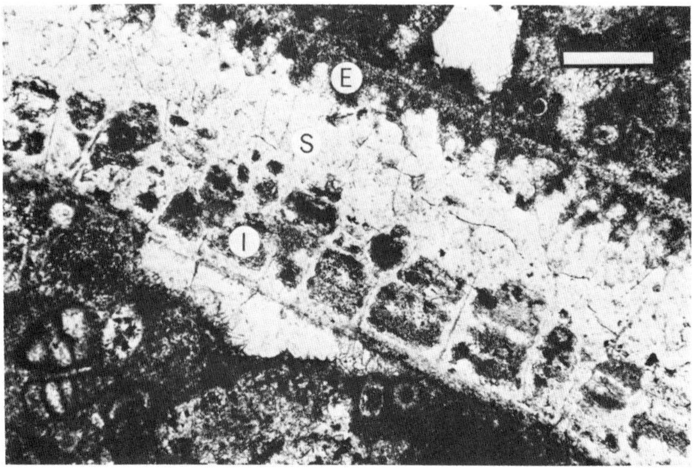

FIG. 5. Silicification of *Inoceramus* sp., more advanced than in Fig. 6. Silicification is most intense towards the centre of the fragment (S), where no remnants of original shell fabric are seen. Calcite still remains in the exterior part of the shell (E) and as 'islands' (I) isolated by silica replacement on the interior margin. Base of the Hooken member, loose block at Humble Point. Plane polarized light. 0.2 mm scale bar.

FIG. 6. Silicification of a belemnite following the concentric (C) and radial (R) structure of the guard. Wilmington sands (73 cm above the Small Cove Hardground), Wilmington. Crossed nicols. 0.2 mm scale bar.

(1) Scanning electron microscopy reveals the fine-scale replacement in brachiopods. At a macroscopic scale, however, definite concentric beekite structures are visible.
(2) Beekite is observed grading into fine-grained quartz in *Exogyra* specimens.
(3) Different styles of silicification are observed within individual specimens at different scales:

At a macroscopic scale—one valve of a pectinid (sample L24) contains at least three texturally different forms of silica (Fig. 9): a very soft (chalk-like) pure white replacement occurring as small (~ 2 mm) blebs sporadically in the shell, a white beekite replacement (from which individual zones were sampled) and a grey fine-scale replacement. Frequently, the white beekite may grade into the grey replacement (i.e. the

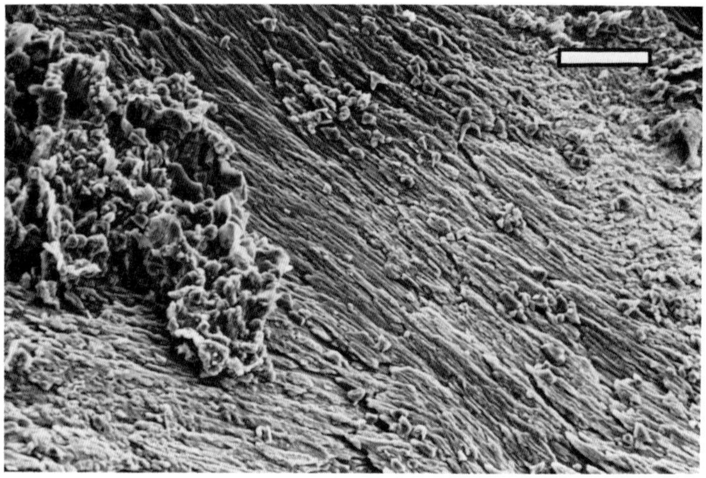

FIG. 7. SEM photograph of a completely silicified brachial loop of *Kingella* sp. Note the faithful replacement of the original shell fabric. The patch on the left is a local development of drusy quartz. Loose, Wilmington. 0.1 mm scale bar.

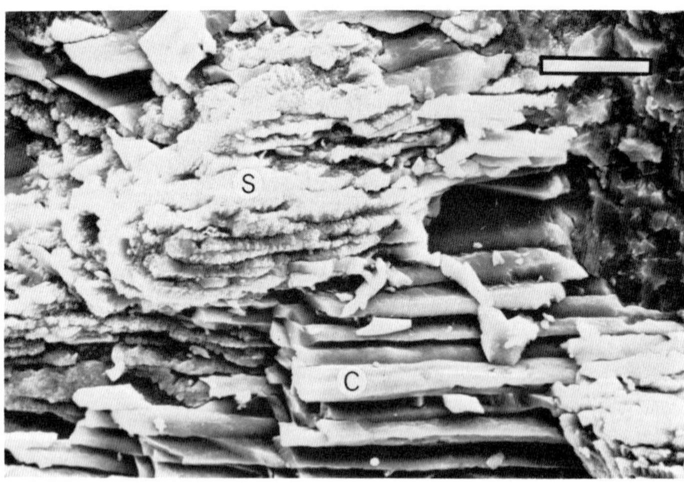

FIG. 8. SEM photograph of a partially silicified rhynchonellid. The original calcite fibres (C) possess a fresh appearance, while the silicified prisms (S) have a more irregular outline. Loose, Wilmington. 30 μm scale bar.

latter may possess concentric rings), but also the boundary may be abrupt.

At a microscopic scale—within one fragment of *Inoceramus*—replacive silica is observed in two completely different styles: replacing along prism boundaries (Fig. 10) and forming chalcedony botryoids that completely transect prism boundaries (Fig. 11).

Silicification post-dates borings and early calcite cementation (Fig. 4) and took place during early diagenesis.

A drusy fabric of mega-quartz (Folk & Pittman 1971), occluding primary porosity (Fig. 12), is occasionally associated with replaced fossils. Crystal sizes vary from less than 100 μm (near the margin) up to 800 μm (in the centre of the cavity).

Siliceous nodules

The occurrence of white siliceous nodules is restricted to the sandy facies in the White Hart

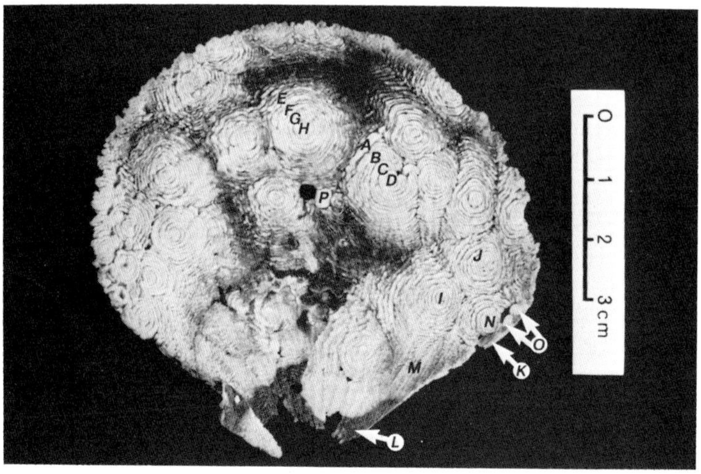

FIG. 9. Silicified pectinid (after etching in dilute HCl). Most of the shell consists of prominent beekite (A–J) and a dark fine-scale replacement (K–M) with occasional blebs of soft, white silica (N–P. N is scraped from the surface of a beekite ring). Loose, Wilmington.

FIG. 10. SEM photograph showing silica following prism boundaries in *Inoceramus* sp. Note the corroded prisms where silicification has occurred (right) compared to the unaltered shell (left). Loose, Wilmington. 0.1 mm scale bar.

Sand Pit at Wilmington. Here they occur in vague bands, the more prominent ones being ~1 m apart. Their morphology clearly reflects a relationship to burrow structures within the sediment. They vary in form from digitate to irregular masses. In many cases they may still be hollow, the interior space corresponding very well with the dimensions of *Thalassinoides* burrows found in the overlying chalk (Fig. 13).

The nodules are composed almost entirely of chalcedony cemented quartz grains (samples 436-10, 436-14, 462) and do not show any evidence of an initial lepispheric or sponge framework. The overall fabric is very similar in style and scale to those found by Wilson (1966, fig. 3) except, instead of ooids, the chalcedony cements detrital quartz. Frequently the central part of a nodule is composed of red chert,

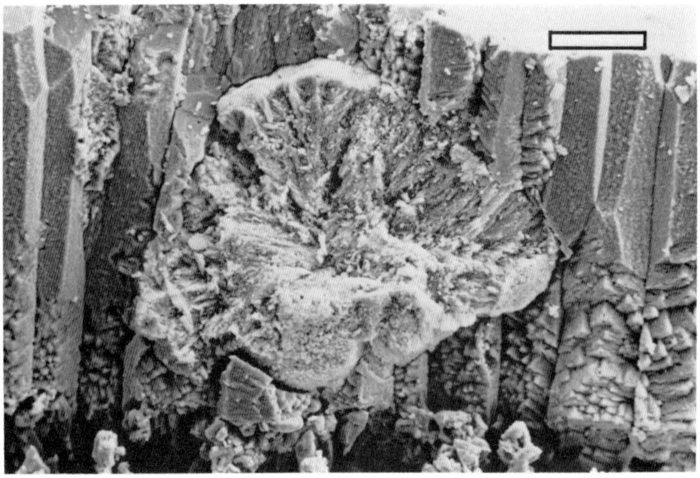

FIG. 11. From the same specimen as Fig. 10, SEM photograph of a chalcedony rosette cutting across prisms in *Inoceramus* sp. Loose, Wilmington. 0.1 mm scale bar.

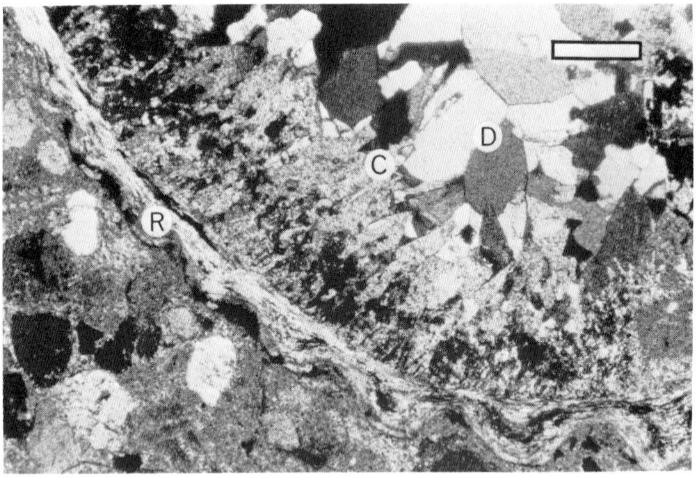

FIG. 12. Drusy quartz (D) occluding primary porosity within a silicified rhynchonellid (R). The development of 'scalenohedral' calcite (C) on the inside of the shell occurred prior to quartz precipitation. Cenomanian burrow fill, 34 cm below the Small Cove Hardground, Wilmington. 0.5 mm scale bar.

showing a conchoidal fracture even on a microscopic scale. The chert core and hollow centre appear to be mutually exclusive (although one loose specimen shows a small 'cavity' with development of quartz druse).

Investigation has also shown the following features:

(1) Hand specimen observation revealed that, despite the size of the nodule, the thickness of the exterior zone is always about the same, ~1–3 cm. Chert cores vary from ~0.5 cm wide (in digitate forms) to greater than 8 cm wide when replacing more idiomorphic burrows.

(2) In one example (Fig. 14), a patch of chalcedony, which did not contain any quartz grains, was observed immediately adjacent to the central chert, and was continuous with an area within the chert which was also devoid of quartz grains.

(3) At certain levels in the sands the nodules do not contain a chert core but possess a distinct 'pink' central area. It is suggested that the

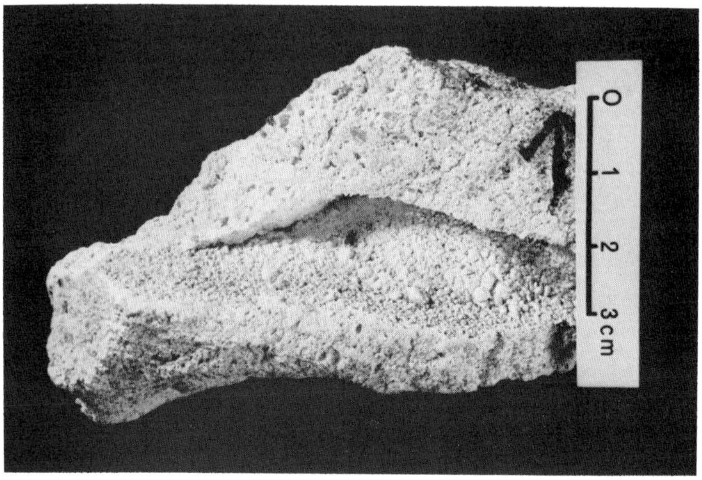

FIG. 13. Slightly compacted hollow *Thalassinoides* burrow preserved as a siliceous nodule. The chalcedony cement occurs as botryoids on the interior surface. Wilmington Sands, Wilmington.

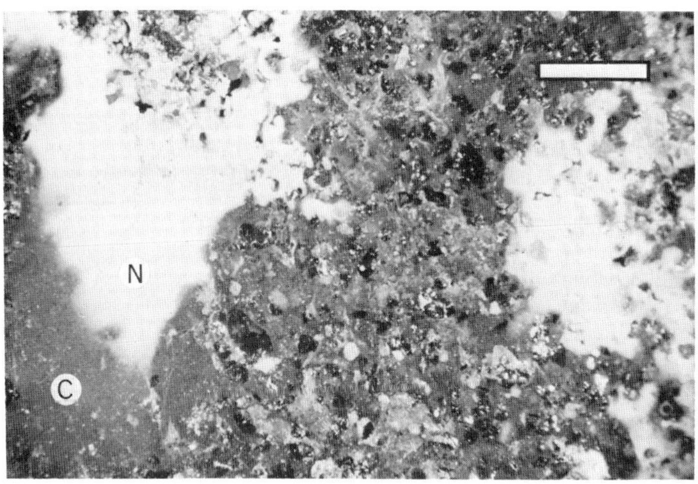

FIG. 14. Margin of chert core within a siliceous nodule. A patch of white chalcedony, free of detrital quartz (N), is adjacent to a patch of chert, also free of detrital quartz (C). This association suggests the alteration of the chert core to the fabric seen in the rest of the nodule. See text for details. Loose, Wilmington. 0.25 mm scale bar.

pink chalcedony represents a remnant of virtually complete alteration of the chert.

Scanning electron microscopy reveals the common preservation of primary porosity and occasionally moldic secondary porosity (Fig. 15), which suggests the nodule formed prior to aragonite dissolution. Silica cementation certainly took place before burrow compaction (although the burrows do have a tendency to be slightly oval in cross-section, the longer diameter being horizontal). However, the observation of chalcedony overgrowing perfectly preserved authigenic 'dog-tooth' calcite ($\sim 20\,\mu m$ long, $\sim 5\,\mu m$ wide) suggests some calcite precipitation prior to chalcedony cementation.

Thin section petrography shows that despite the macroscopic contrast between the chert core and the white chalcedonic nodule, there is no apparent difference between the two textures under crossed nicols, either in the type of cementation or in the cement/detrital quartz ratio. Under plane polarized light, however, a very slight difference in refractive index is

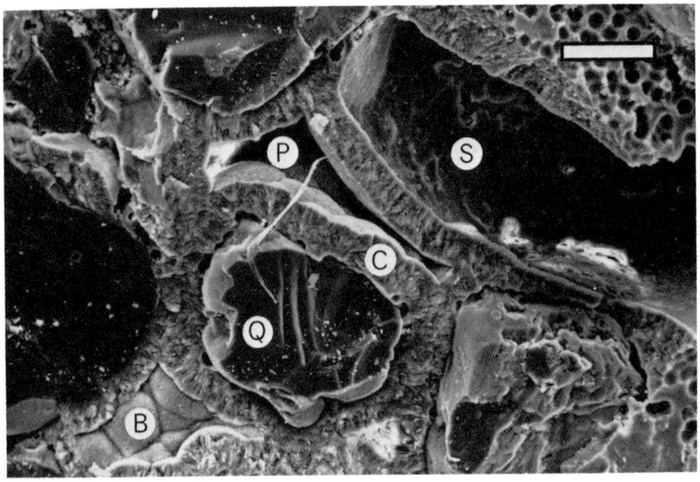

FIG. 15. SEM photograph of a siliceous nodule. Chalcedony cement (C) coats detrital quartz (Q). The surface expression of the cement is seen as botryoids (B). Note the presence of secondary porosity (S) and the preservation of primary porosity (P). Wilmington sands (300 cm below the Humble Point Hardground), Wilmington. 0.2 mm scale bar.

observed between the cement and quartz grains. This difference is greater in the nodule than in the chert, where the two phases cannot be distinguished. The most likely cause for the decreased refractive index of the chalcedony cement is the presence of water inclusions (Folk & Weaver 1952), probably due to a rapid precipitation of the cement (see below).

Another form of silicification is the replacement of carbonate within a hardground, seen only in the Small Cove Hardground. This is also found in association with *Thalassinoides* burrows, but is equally pervasive into the sediment. This occurrence will not be further dealt with here.

Isotopic results

As the quartz-water isotopic fractionation of oxygen is inversely proportional to temperature (Knauth & Epstein 1975), the $\delta^{18}O$ value of the silica reflects the relative temperature of silica precipitation and the isotopic composition of the pore waters. Here, the empirical equation quoted in Knauth & Epstein (1976) for quartz-water fractionation at low temperatures (extrapolated from the high temperature (>200°C) data of Clayton *et al.* 1972) is used to calculate formational temperatures.

$$1000 \ln \alpha_{\frac{CHERT}{WATER}} = 3.09(10^6 T^{-2}) - 3.29.$$

In this study, the difference between the $\delta^{18}O$ of the water and that of the silica is taken as being approximately the same as 1000 $\ln_\alpha$ (Friedman & O'Neil 1977). A pre-glacial sea water value of −1‰ SMOW (Shackleton & Kennett 1975) is assumed for the Cretaceous. The full list of results is given in Table 1.

Silicified fossils

The isotopic values (Table 1) are related to the type of replacement and do not appear to correlate directly with the fossil taxa. For example, brachiopod shells may be replaced by two (most often mutually exclusive) forms of silica—a very pure white variety often with a vitreous lustre and, more frequently, a patchy grey saccharoidal variety which on close inspection shows faint beekite rings. Both of these are the 'fine-scale' replacement described by Holdaway & Clayton (1982). The average values for these replacements are 32.4 ± 0.06 ($n = 2$) and 29.2‰ ± 1.28 ($n = 2$) respectively, indicating that the two varieties formed in different environments.

In *Exogyra* a similar pattern emerges, although in this case the different replacement fabrics tend to occur in the same shell: frequently, prominent pink beekite occurs near the umbo and a grey fine-scale replacement occupies the rest of the shell, which may often show beekite structures. Even though these two phases may be within one specimen, the $\delta^{18}O$ values are similar *hardground (see Fig. 2). Where no height is given, the sample was loose. SCH = Small Cove Hardground*

Sample no.	Bioclast	Quartz type	Location	Strat. height (cm)	$\delta^{18}O$
116-H	Sponge	Drusy quartz	Wilmington	+5 SCH	24.94
118-D	Brachiopod	Drusy quartz	Wilmington	−46 SCH	23.13
*118-R	Brachiopod	Fine scale replacement of shell	Wilmington	−46 SCH	28.32
118-U	Brachiopod	Unseparated quartz druse and repl. shell	Wilmington	−46 SCH	26.78
*167	Oyster	Beekite (vitreous —> white)	Haven Cliff—east		31.67
*249	Rhynchonellid	White fine scale replacement	Bovey Lane Sand Pit	+11 Humble Point	32.43
310/F	Oyster	Fine scale replacement	Mitchel Stiles Rock	on SCH	29.58
*368	Rhynchonellid	White replacement	Beer Stone Addit	−123 Haven Cliff	32.34
*431-2	*Exogyra*	Beekite	Wilmington	−292 Humble Point	30.37, 31.37, 32.78, 32.78
*431-3/F	*Exogyra*	Grey fine scale replacement	Wilmington	−292 Humble Point	29.03, 29.30
*432-2/1	*Exogyra*	Beekite	Wilmington	−292 Humble Point	31.45
*432-2/2/F	*Exogyra*	Grey fine scale replacement	Wilmington	−292 Humble Point	29.99
*432-4/2/F	*Exogyra*	Fine scale replacement	Wilmington	−292 Humble Point	28.06
436-10		White siliceous nodule	Wilmington	−300 Humble Point	24.04
436-14		White siliceous nodule	Wilmington	−300 Humble Point	22.62,23.41
436-19		Red chert core	Wilmington	−300 Humble Point	24.42
447-2/7	Pectinid	Beekite (exterior or ring)	Wilmington	−158 Humble Point	27.39
462		White siliceous nodule	Wilmington		29.01, 30.35
492-600		Detrital grains	Wilmington	−250 Kings Hole	11.18, 11.86
551/F	*Lopha*	90% quartz druse	Berry Cliff	+14 SCH	25.53
589/F	Pectinid	Lustreous beekite	Wilmington		30.80
L20-63	Oyster	Fine scale replacement	Wilmington		32.68
*L22A-1/F	*Exogyra*	Beekite	Wilmington		30.29
*L22A-2/F	*Exogyra*	Beekite	Wilmington		29.28
L22B		Siliceous nodule	Wilmington		27.37
L22B/F		Siliceous nodule	Wilmington		26.40
L23-Q/F	*Lopha*	Quartz druse (10% replacement silica)	Wilmington		24.95
*L23-R/F	*Lopha*	Replacement silica	Wilmington		29.41
*L24-1	Pectinid	Dark, late replacement	Wilmington		27.74
*L24-2	Pectinid	Beekite—three cores	Wilmington		30.00
*L24-3	Pectinid	Beekite—latest phase	Wilmington		30.52
*L24-4	Pectinid	Fine scale replacement	Wilmington		27.86
*L24-5	Pectinid	Beekite—core + 1st ring	Wilmington		26.52
*L24-6	Pectinid	Beekite—2nd ring	Wilmington		26.59
*L24-7	Pectinid	Beekite—3rd ring	Wilmington		30.15
*L24-8	Pectinid	Beekite—4/5th ring	Wilmington		29.29
*L24-9	Pectinid	Beekite—core + 1/2/(3) rings	Wilmington		30.34
*L24-10	Pectinid	Beekite—core	Wilmington		29.26
*L24-11	Pectinid	Beekite-rings 1,2	Wilmington		30.72
*L24-12	Pectinid	Beekite—rings 3,4	Wilmington		28.63
*L24-14	Pectinid	Beekite—core —> 5th ring	Wilmington		30.84
*L24-15/F	Pectinid	White	Wilmington		29.88
*L24-16/F	Pectinid	Beige	Wilmington		29.27
*L24-17/F	Pectinid	Dark brown	Wilmington		27.18
*L9-A/F	Pectinid	Light beekite	Wilmington		30.64
*L9-B/F	Pectinid	Dark beekite	Wilmington		28.73
*S1-PV/F	*Cyclothyris*	Fine scale replacement of pedicle valve	WIlmington		30.13

FIG. 16. Silicified pectinid. A grey fine-scale saccharroidal replacement (with a vague concentric structure) is truncated by prominent, white beekite. Loose, Wilmington. 2 mm scale bar.

to the equivalent phases in brachiopods, 31.6 ± 0.27 ($n = 2$) and 29.4‰ ± 0.87 ($n = 5$) respectively. Figure 16 shows a replaced pectinid (spec. L9) which is composed of only two types of silica—white porcellanous beekite rings, ~1–2 mm diameter, and a later fine-scale saccharroidal grey variety, which on close inspection also possesses concentric rings. Though only one analysis of each phase was carried out, the results are comparable with those of the *Exogyra* and brachiopods—the earlier replacement having a $\delta^{18}O$ of 30.6, the later one being 28.7‰.

TABLE 2. *Detail of isotopic results from pectinid L24. Lettered key refers to Fig. 9*

		$\delta^{18}O$	average	1σ
Individual beekite rings—from exterior to interior	A	28.6		
	B	30.7	29.7	0.96
	C	29.3		
	D	30.3		
	E	29.3		
	F	30.2	28.1	1.86
	G	26.6		
	H	26.5		
Whole beekite 'discs'	I		29.9	
	J		29.3	
Beekite average			29.3	(±0.79)
Late grey replacement	K	27.7		
	L	27.9	27.6	0.36
	M	27.2		
Soft, white beekite	N	30.8		
	O	30.0	30.5	0.42
	P	30.5		

Further analysis was carried out on a different pectinid (spec. L24—see above). In this specimen, the three distinct types of replacement were sampled (Fig. 9) and it was possible to separate individual zones within a beekite 'disc' (Holdaway & Clayton 1982). The $\delta^{18}O$ values are given in Table 2 and do not show any consistent spatial variation within individual structures. Thus the growth of the rings would have been rapid relative to the burial of the specimen. The average values of each of the two discs analysed in detail are both similar to values obtained from 'whole-rock' ring analysis. Even though petrographic evidence shows that different beekite disks have grown sequentially (Holdaway & Clayton 1982), they must all have completed their growth quite rapidly since they all possess approximately the same $\delta^{18}O$ values.

The average of all the beekite analyses from sample L24 is 29.3‰ ± 0.79 ($n = 4$) which is similar to the fine-scale grey replacement of previously mentioned specimens. The later grey replacement in sample L24 has an average value of 27.6‰ ± 0.36 ($n = 3$) and the very soft pure white replacement has an average value of 30.5‰ ± 0.42 ($n = 3$). The textural relationship shows the grey replacement post-dating the beekite, which in turn post-dates the soft, white replacement. Why the absolute values should be offset from other specimens remains uncertain.

Siliceous nodules

The isotopic results from the chert core and the exterior of a siliceous nodule are very similar (24.4, 24.0 and ~23.0‰ respectively). Both phases contain detrital quartz with a mean value of 11.5‰ ± 0.48 ($n = 2$), typical for quartz derived from acid igneous sources (Taylor 1968, fig. 8) in this case, probably the Permian granites to the north-west (although see discussion in Hancock 1969). In both the chert and the nodule, quartz grains occupy ~35% of the rock. So on adjusting the initial values, we get $\delta^{18}O$ values of 29.3 and 30.9 for the nodule, and 31.5‰ for the chert in specimen 436, and a value of 34.6 for the nodule L22B. Where chalcedony cement was isolated from a nodule (specimen 462), it gave a mean value of 29.7 ± 0.95‰ ($n = 2$).

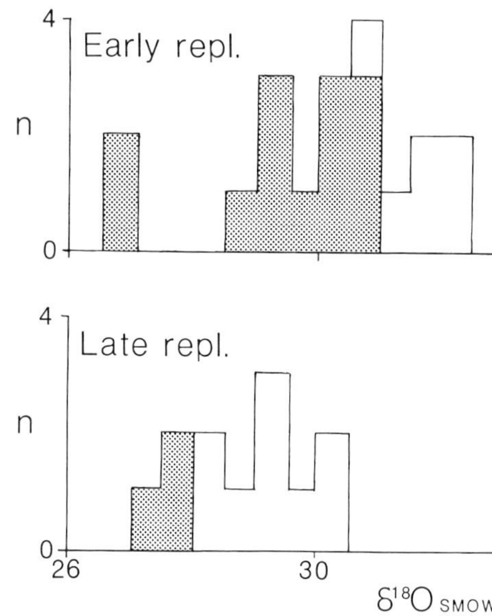

FIG. 17. Histograms of $\delta^{18}O$ values obtained for early and late replacements. Where duplicate analyses were made from one specimen, the mean value was used. Stippled area indicates values obtained from specimen L24.

Discussion

Silicified fossils

Except for specimen L24, the mean $\delta^{18}O$ values (early: from 30.6 to 32.4‰, late: from 28.7 to 29.4‰) give temperatures of 17.2–24.4°C for the earlier replacements and 30.0–32.9°C for the later ones (using a sea water $\delta^{18}O$ SMOW of −1‰). Lowenstam (1964) calculated Cenomanian sea water temperatures varying from 17 to 29°C (depending upon latitude) assuming a sea water $\delta^{18}O$ SMOW of 0 (Lowenstam & Epstein 1954). Recalculation of Lowenstam's data using a sea water value of −1‰, gives a temperature range from 13 to 25°C. These figures were calculated on data from belemnites which are nektonic organisms, and thus would reflect temperatures at some distance above the sediment surface in the water column. However, Cretaceous sea water can be assumed to have had little vertical variation in temperature (due to lack of cold bottom currents), and thus the temperature reflected by the belemnites would be comparable to that at the sediment-water interface.

The presence of a sandy facies in a predominantly carbonate succession inland at Wilmington suggests a marginal environment, although it is possible that early Cenomanian local faulting to the west contributed some of the sandy material (Jarvis & Tocher 1987). This, along with the degree of condensation of the Cenomanian on the present coast (implying a shallow water environment), would further support a vertically homogeneous water temperature. Thus the data given for the earliest silica replacement are consistent with the idea of replacement at a shallow burial depth.

However, the values for the later replacement (mean = 28.9‰ ± 1.02 ($n = 12$)) pose a problem (Fig. 17). Consistently lower values were obtained for the later replacement. The $\delta^{18}O$ difference would correspond to a temperature increase of ~10°C with a minimum increase of 8.5°C between the early and late phases in one fossil group. Four possibilities may be invoked to explain this:

(1) *Formation or recrystallization at greater depth*

From the discussion of burial depth, it seems unlikely that these sediments were buried more than ~260 m. Using an average geothermal gradient of 3.35°/100 m (measurement from Dorset—Bloomer *et al.* 1979), the temperature increase due to burial can only have been a maximum of 8.7°C. Thus the estimate of *maximum* possible overburden (~260 m) can only just account for the *minimum* calculated temperature increase, 8.5°C (equivalent to 250 m of burial).

(2) *Formation in a mixing zone*

Knauth (1979) suggested the formation of chert in an environment where the mixing of sea water and meteoric water would produce a solution undersaturated with respect to calcite but supersaturated with respect to quartz. This idea could be applied to the silicification of bioclasts and would explain the light $\delta^{18}O$ values. However, even though Wilmington represents a marginal environment, evidence for such a mode of formation (such as the association with dolomite formation) is totally lacking (cf. formation of spherulites, Elorza & Orue-Etxebarria 1985).

(3) *Late recrystallization with isotopically light meteoric waters*

Clayton (1984) illustrates that light $\delta^{18}O$ values of certain types of flint are due to a combination of inversion of opal-CT to quartz at considerable depth and recrystallization with meteoric waters during uplift. Even though they are all varieties of hydrated silica, the distinctly different morphological forms of silica replacement at Wilmington suggest that one type (the later grey replacement) may have been more susceptible to recrystallization than the others. Thus the unexpected $\delta^{18}O$ values of late replacement may not reflect its original value but a reduced value due to recrystallization. The question remains as to whether this took place at depth during burial or in a meteoric water environment.

Meyers & James (1978) noted that even though the petrography suggested that many silica cements and replacements were precipitated early, the light $\delta^{18}O$ values implied recrystallization in the presence of phreatic groundwaters. This was the case for the chalcedony, which originally would have been in the form of banded opal-CT (Meyers 1977), and the microquartz, which contain microspherules interpreted to be relics of cristobalite lepispheres (Meyers 1977). The fabrics observed in the present study show no signs of lepispheric structures, but since isotopic exchange of the silica by atomic diffusion is unlikely, as is re-equilibration of opal-CT (at less than 1000 m—Knauth & Epstein 1975) or quartz (Taylor 1968), it could be that the replacement silica has undergone some sort of recrystallization, possibly from an opal-CT precursor.

The recrystallization probably took place during uplift, after the end of the Cretaceous, when the Wilmington sands would have been subjected to (phreatic?) meteoric water input.

The geochemical data and petrographical observations do not conclusively point to any one of the three possibilities, yet a tentative suggestion is that recrystallization plays a role in explaining the light oxygen values. However, it must be noted that formation within a mixing zone was only rejected on negative evidence—further work on the associated carbonate phases might prove this to be a viable explanation.

Siliceous nodules

The $\delta^{18}O$ values for the siliceous nodules and cherts give an average temperature of 22.1°C ($n = 5$) (assuming a sea water value of $-1‰$) which corresponds very well to values given for the early fossil replacement. However, the similarity of isotopic results between the chert core and chalcedony cemented margin is perhaps surprising in view of the petrographic evidence cited earlier, which suggests a difference in water content between the interior and exterior. There are two possibilities to explain this difference.

The first is that the siliceous nodule originally crystallized as chert (as in the core) and that some later mechanism has exsolved the water on the margins of the chert. Three possible mechanisms are: recrystallization at depth, recrystallization in meteoric waters or an initial faster precipitation rate. Recrystallization at depth is unlikely since it has already been demonstrated that these cherts could not have been buried to any great depth. Clayton (1984) has suggested recrystallization of the cortex of flint may be 'seeded' by an influx of meteoric water. Yet, despite the effect of a possible meteoric phreatic phase on the fossil replacements, it is unlikely to have affected the nodules since the oxygen isotope values have not been affected in any way. Thus we are left with the hypothesis that even though the precipitation of the cement has to be fairly rapid anyway (to form hydrated silica), the exterior precipitated somewhat more rapidly than the centre of the chert. Thus the silica in the exterior was slightly more structurally disordered than that on the interior, with slightly higher concentrations of entrapped pore waters and was more prone to recrystallization. If this was so, the recrystallization must have occurred at an early stage to preserve the relatively heavy $\delta^{18}O$ values. This is contrary to the formation of flints where silicification is believed to be more intense (and thus rapid) towards the centre (Clayton 1984). However, flint formation in the chalk is initiated by lepisphere growth so a diffusion gradient would be set up with silica nucleation centres more closely spaced towards the centre of the flint (Clayton 1984). In the sands, there is no evidence for a lepispheric initiation of chalcedony precipitation and it is more likely that the initial zone of precipitation is around the circumference of a burrow than within it. This would also explain why some of the burrows remain hollow. Recrystallization during deeper burial within a closed system (Pingitore 1982), preserving the heavy $\delta^{18}O$ values, is unlikely because the sandy facies would probably have allowed free porewater circulation.

The second possibility is that later exsolution of structural water need not have occurred, the initial precipitation rate being rapid enough to trap fluid in inclusions during growth. However, this would fail to account for the constant thickness of the white exterior of the nodules, unless the rate of growth and silica supply was the same vertically and horizontally throughout the succession.

Drusy quartz in silicified fossils

All the drusy quartz analysed has a much lighter value (average = 24.2‰ ± 0.94 ($n = 3$), than the enclosing replaced shell material. Assuming formation from marine waters with a $\delta^{18}O$ of −1‰, the values for the drusy quartz give a formation temperature of 56.5°C. If they had formed from the same waters there would be a 34°C difference between the lightest fossils and the drusy quartz. This implies a difference in depth of formation of over 1000 m and we have already calculated a maximum burial depth of ~260 m. The most likely explanation for these values is that the quartz formed from isotopically light meteoric waters in a phreatic environment (Meyers & James 1978). Late recrystallization can be excluded since mega-quartz is believed to be isotopically immune to groundwater exchange (see evidence cited in Knauth & Lowe 1978, p. 214).

Conclusions

The petrographic and isotopic evidence above suggests that the replacement of bioclasts was essentially a very early diagenetic process, where the restricted microenvironment within shells or their fragments would initiate silicification (see Holdaway & Clayton 1982). However, the later replacements were more prone to a partial recrystallization which took place much later in the diagenetic history of the sediment, in a meteoric phreatic environment. The formation of drusy quartz took place in this same environment. Siliceous nodules formed penecontemporaneously with (or perhaps slightly earlier than) early fossil replacement. There is uncertainty as to whether the exterior of the nodules grew at a faster rate than the interior or recrystallized at an early stage during the burial of the sequence.

ACKNOWLEDGMENTS: This work could not have been undertaken without the tuition on the fluorination line given by Peter Greenwood and Linda Thrift (BGS, London), and their preparatory work for optimizing running conditions appropriate to these samples. Their constant guidance and many fruitful discussions were an inspiration to this study. My gratitude also goes to Jim Marshall and Chris Paul for their criticism and help in increasing the eloquence of the text and to Chris Clayton, Tim Astin and Baruch Spiro for their constructive reviews. Financial support was provided by NERC and the isotopic data are published with the permission of the Director, British Geological Survey.

References

ANDERTON, R., BRIDGES, P. H., LEEDER, M. R. & SELLWOOD, B. W. 1979. *A Dynamic Stratigraphy of the British Isles: a Study in Crustal Evolution*. Allen & Unwin, London.

BLACK, M. 1953. The constitution of the Chalk. *Proceedings of the Geological Society of London*, **1499**, 31–6.

BLOOMER, J. R., RICHARDSON, S. W. & OXBURGH, E. R. 1979. Heat flow in Britain: an assessment of the values and their reliability. *In*: CERMAK, V. & RYBACH, L. (eds) *Terrestrial Heat Flow in Europe*, pp. 293–300. Springer-Verlag, Berlin.

BROWN, G., CATT, J. A., HOLLYER, S. E. & OLLIER, C. D. 1969. Partial silicification of chalk fossils from the Chilterns. *Geological Magazine*, **106**, 583—6.

BRUNTON, C. H. C. 1966. Silicified productoids from the Viséan of County Fermanagh. *Bulletin of the British Museum (Natural History), Geology*, **12**, 173–243.

—— 1968. Silicified brachiopods from the Viséan of County Fermanagh (II). *Bulletin of the British Museum (Natural History), Geology*, **16**, 1–70.

—— 1984. Silicified brachiopods from the Viséan of County Fermanagh (III). *Bulletin of the British Museum (Natural History), Geology*, **38**, 27–130.

BUURMAN, P. & VAN DER PLAS, L. 1971. The genesis of Belgian and Dutch flints and cherts. *Geologie en Mijnbouw*, **50**, 9–28.

CLAYTON, C. J. 1982. Growth history and microstructure of flint. *International Association of Sedimentologists 3rd European Meeting, Copenhagen* (Abstract), pp. 105–7.

—— 1984. The geochemistry of chert formation in Upper Cretaceous Chalks. *PhD thesis*, University of London.

CLAYTON, R. N. & MAYEDA, T. K. 1963. The use of bromine pentafluoride in the extraction of oxygen from oxides and silicates for isotopic analysis. *Geochimica et Cosmochimica Acta*, **27**, 43–52.

——, O'NEIL, J. R. & MAYEDA, T. K. 1972. Oxygen isotope exchange between quartz and water. *Journal of Geophysical Research*, **77**, 3057–67.

CURRY, G. B. 1986. Fossils and tectonics along the Highland Boundary Fault in Scotland. *Journal of the Geological Society of London*, **143**, 193–8.

ELORZA, J. & ORUE-ETXEBARRIA, X. 1985. An example of silicification in *Gryphaea* sp. shells from Laño (south of Vitoria, Spain). *6th International Association of Sedimentologists Regional Meeting, Lleida* (Abstract), pp. 556–9.

FAUGÈRES, J.-C. 1978. Les Rides sud-rifaines. Evolution sedimentaire et structurale d'un bassin atlantico-mesogen de la marge africaine. *PhD thesis*, Part 3: Sedimentology. University of Bordeaux.

FOLK, R. L. & PITTMAN, J. S. 1971. Length slow chalcedony: A new testament for vanished evaporites. *Journal of Sedimentary Petrology*, **41**, 1045–58.

—— & WEAVER, C. E. 1952. A study on the texture and composition of chert. *American Journal of Science*, **250**, 498–510.

FRIEDMAN, I. & O'NEIL, J. R. 1977. Compilation of stable isotope fractionation factors of geochemical interest. *In*: FLEICHER, M. (ed.) *Data of Geochemistry*, Geological Survey Professional Paper 440-KK.

HÅKANSSON, E., BROMLEY, R. & PERCH-NIELSEN, K. 1974. Maastrichtian chalk of north-west Europe—a pelagic shelf sediment. *In*: HSÜ, K. J. & JENKYNS, H. C. (eds) *Pelagic Sediments: on Land and Under the Sea*. Special Publication of the International Association of Sedimentologists, **1**, 211–33. Blackwell Scientific Publications, Oxford.

HANCOCK, J. M. 1969. Transgression of the Cretaceous sea in south-west England. *Proceedings of the Ussher Society*, **2**, 61–83.

—— 1975. The petrology of the Chalk. *Proceedings of the Geologists' Association*, **86**, 499–535.

HOLDAWAY, H. K. & CLAYTON, C. J. 1982. Preservation of shell microstructure in silicified brachiopods from the Upper Cretaceous Wilmington Sands of Devon. *Geological Magazine*, **119**, 371–82.

JARVIS, I. & TOCHER, B. A. 1987. Field meeting: the Cretaceous of S.E. Devon, 14–16th March 1986. *Proceedings of the Geologists' Association*, **98**, 51–66.

—— & WOODROOF, P. B. 1984. Stratigraphy of the Cenomanian and basal Turonian (Upper Cretaceous) between Branscombe and Seaton, S.E. Devon, England. *Proceedings of the Geologists' Association*, **95**, 193–215.

KNAUTH, L. P. 1979. A model for the origin of chert in limestone. *Geology*, **7**, 274–7.

—— & EPSTEIN, S. 1975. Hydrogen and oxygen isotope ratios in silica from the JOIDES DSDP. *Earth and Planetary Science Letters*, **25**, 1–10.

—— & —— 1976. Hydrogen and oxygen isotope ratios in nodular and bedded cherts. *Geochimica et Cosmochimica Acta*, **40**, 1095–1108.

—— & LOWE, D. R. 1978. Oxygen isotope geochemistry of cherts from the Onverwacht Group (3.4 billion years), Transvaal, South Africa, with implications for secular variations in the isotopic composition of cherts. *Earth and Planetary Science Letters*, **41**, 209–22.

LOWENSTAM, H. A. 1964. Paleotemperatures of the Permian and Cretaceous periods. *In*: NAIRN, A. E. M. (ed.). *Problems in Paleoclimatology*, pp. 227–48.

—— & EPSTEIN, S. 1954. Paleotemperatures of the post-Aptian Cretaceous as determined by the oxygen isotope method. *Journal of Geology*, **62**, 207–48.

MEYERS, W. J. 1977. Chertification in the Mississippian Lake Valley Formation, Sacramento Mountains, New Mexico. *Sedimentology*, **24**, 75–105.

—— & JAMES, A. T. 1978. Stable isotopes of cherts and carbonate cements in the Lake Valley Formation (Mississippian), Sacramento Mts, New Mexico. *Sedimentology*, **25**, 105–24.

PINGITORE JR, N. E. 1982. The role of diffusion during carbonate diagenesis. *Journal of Sedimentary Petrology*, **52**, 27–39.

RAWSON, P. F., CURRY, D., DILLEY, F. C., HANCOCK, J. M., KENNEDY, W. J., NEALE, J. W., WOOD, C. J. & WORSSAM, B. C. 1978. *A correlation of Cretaceous rocks in the British Isles. Special Report of the Geological Society*, **9**.

SCHMITT, J. G. & BOYD, D. W. 1981. Patterns of silicification in Permian pelecypods and brachiopods from Wyoming. *Journal of Sedimentary Petrology*, **51**, 1297–308.

SHACKLETON, N. J. & KENNETT, J. P. 1975. Paleotemperature history of the Cenozoic and the initiation of Antarctic glaciation: oxygen and carbon isotope analysis in DSDP sites 277, 279 and 281. *Initial Reports of the Deep Sea Drilling Project*, **24**, 743–55.

SMITH, W. E. 1957. The Cenomanian Limestone of the Beer District, south Devon. *Proceedings of the Geologists' Association*, **68**, 115–35.

—— 1961. The Cenomanian deposits of south-east Devonshire. *Proceedings of the Geologists' Association*, **72**, 91–134.

TAYLOR, H. P. 1968. The oxygen isotope geochemistry of igneous rocks. *Contributions to Mineralogy and Petrology*, **19**, 1–71.

VAN HINTE, J. E. 1976. A Cretaceous time scale. *Bulletin of the American Association of Petroleum Geologists*, **60**, 498–516.

WILSON, R. C. L. 1966. Silica diagenesis in Upper Jurassic limestones of southern England. *Journal of Sedimentary Petrology*, **36**, 1036–49.

GREG A. CARSON, Department of Geological Sciences, University of Liverpool, Brownlow Street, PO Box 147, Liverpool L69 3BX, UK.

Controls on the geometry and distribution of carbonate cements in Jurassic sandstones: Bridport Sands, southern England and Viking Group, Troll Field, Norway

J. D. Kantorowicz, I. D. Bryant & J. M. Dawans

SUMMARY: The petrography and diagenesis of calcite cements in the Lower Jurassic, Bridport Sands (southern England) and Upper Jurassic, Viking Group sandstones (Troll Field, offshore Norway) have been investigated in order to assess their geometry and effect on hydrocarbon recovery.

In the Bridport Sands, sediment texture and mineralogy controlled carbonate cementation. Clay-rich fairweather sediments were weakly cemented and are now compacted. Bioclast-rich storm deposits were stabilized mechanically by early fringing cements. During burial bioclasts and fringing cements were replaced or dissolved, and pores were filled by simultaneously precipitated ferroan calcite. Thus, cemented beds are laterally continuous for several kilometres in the Bridport Sands as a consequence of the sheet-like geometry of the storm beds in which they developed.

In the Viking Group sandstones, carbonate cementation was controlled by rate of burial. Fringing cements formed locally during non-deposition or emergence. Cementation continued with non-ferroan calcite incorporating bacterially derived bicarbonate generated during prolonged residence in near surface zones of bacterial activity. Cement geometries will reflect the distribution of emergent surfaces and the longevity of residence near the surface.

These two cases demonstrate the potential for laterally extensive carbonate cements to develop in shelf sandstones. The cements in these examples have different origins but in both cases their distribution is related to the episodic nature of deposition in the shelf environment.

Carbonate minerals of diverse origin occur as cements formed during the diagenesis of shallow-marine sandstones. Cement origins include precipitation from sea water, the dissolution and reprecipitation of bioclastic aragonite and calcite, and precipitation of bicarbonate liberated during bacterial processes (see Hudson 1977). Precipitation has been recognized to occur rapidly on the seafloor (Nelson & Lawrence 1984, Høvland *et al.* 1985), during burial (Milliken *et al.* 1981), and at depths of 3 km following hydrocarbon emplacement (Kantorowicz 1985).

The effect of carbonate-cemented sandstones on hydrocarbon reservoir development depends upon cement geometry. Nodular, laterally discontinuous cemented sands are only likely to influence fluid flow locally, if at all, whilst laterally continuous cemented sandstones may present 'baffles' to fluid flow within a reservoir or divide (or 'compartmentalize') the reservoir. Whether this is beneficial or detrimental depends upon the orientation of the cemented sandstone and the nature of the proposed development programme (Fig. 1). Thus, when carbonate-cemented sandstones are encountered in the subsurface it is desirable to be able to predict their distribution. This requires an understanding of the origin of the cement and any controls on its distribution. This in turn necessitates understanding the host sediment's depositional, diagenetic and burial histories. This paper describes two case studies of carbonate cementation in shallow-marine sandstones. In the Bridport Sands carbonate cements are known to be laterally extensive, whilst in the Viking Group sandstones in the Troll Field their geometry is unknown.

Bridport Sands, onshore Dorset, UK

The Upper Liassic (Lower Jurassic) Bridport Sands outcrop in Dorset as spectacular cliffs between Bridport and Burton Bradstock. Inland the Sands have been cored in a number of boreholes, and form part of the reservoir in the Wytch Farm oil field (Fig. 2a; Colter & Havard 1981). The Bridport Sands were formed as a shallow marine bar (Davies 1967). Outcrop and borehole studies support this interpretation, and enable a more detailed model to be proposed. Throughout southern Britain the Bridport Sands comprise shallowing-upwards cycles of shelf sands, capped locally by thin limestones. Each cycle contains fine to very fine grained clay-rich sandstone beds and fine grained bioclastic sandstone beds. Clay-rich beds predominate at

From: MARSHALL, J. D. (ed.), 1987, *Diagenesis of Sedimentary Sequences*, Geological Society Special Publication No. 36, pp. 103–118.

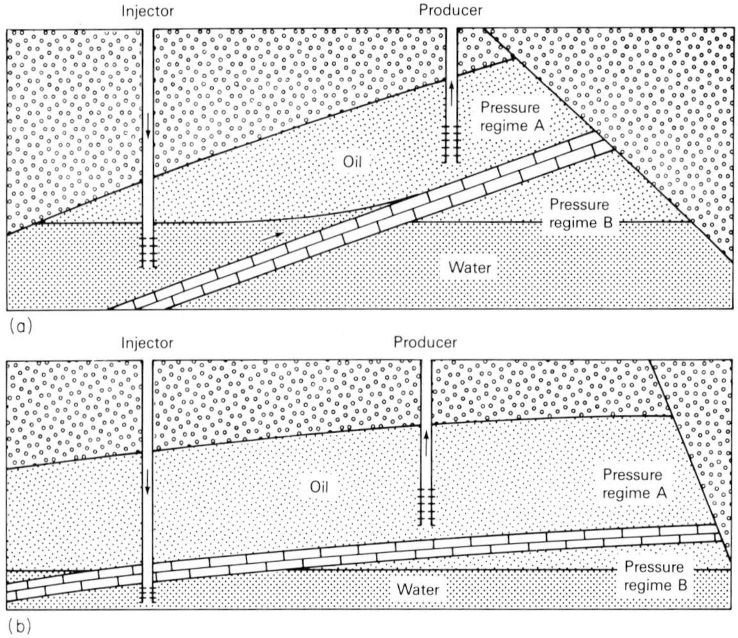

FIG. 1. (a) Cemented sand hindering development by isolating 20% of the oil zone in a separate accumulation. Accelerated water-coning is also likely as a result of the orientation of the tight streak. (b) Cemented sand aiding development by reducing possibility of bottom-water coning into the larger upper oil accumulation. However, note reduced communication between the oil- and water-saturated parts of the reservoir.

the base, enclosing occasional thick bioclastic beds. The bioclastic sands become thinner and more abundant towards the top of each cycle. Most cemented beds are continuous over 90% of the 5 km long cliff section. Based on a combination of sedimentological, palynological and ichnological data the clay-rich beds are interpreted to have been deposited on the shelf during fairweather processes and the bioclast-rich beds during storm processes.

Petrography

The Bridport Sands contain a wide variety of detrital and authigenic minerals, modal analysis of which describes a gradation between porous, clay-rich fairweather sediments and carbonate-cemented clay-poor bioclast-rich storm deposits (Table 1). In general, the fairweather sediments are weakly cemented, and the storm deposits tightly cemented. The pronounced differences between them observed at outcrop may be attributed to the effects of surface weathering (Fig. 2b).

Clay-rich sands

These are porous, very fine- to fine-grained subarkoses. They are texturally immature, containing abundant silt-sized quartz grains and up to 26% detrital clay. The framework is matrix supported. A variety of bioclasts was observed in the core but only belemnites were identified at outcrop.

The sediments have undergone various diagenetic modifications. Pyrite, often in association with organic matter, occurs as a trace percentage throughout and is the only authigenic mineral present where abundant detrital clay occurs. Where less detrital clay occurs, well-developed grain-coating berthierine has formed. Quartz overgrowths occasionally engulf berthierine, but usually are developed only where berthierine is absent (Fig. 3). Part of the remaining pore space was filled subsequently with either ferroan calcite or ferroan dolomite. Many of the bioclasts observed in thin section have undergone extensive grain interpenetration and dissolution.

Cemented sands

These are texturally mature, fine- to very fine-grained bioclast-rich sandstones with a clast-supported framework. The bioclasts present include brachiopod and bivalve shells which contain extensive iron oxide-lined algal borings,

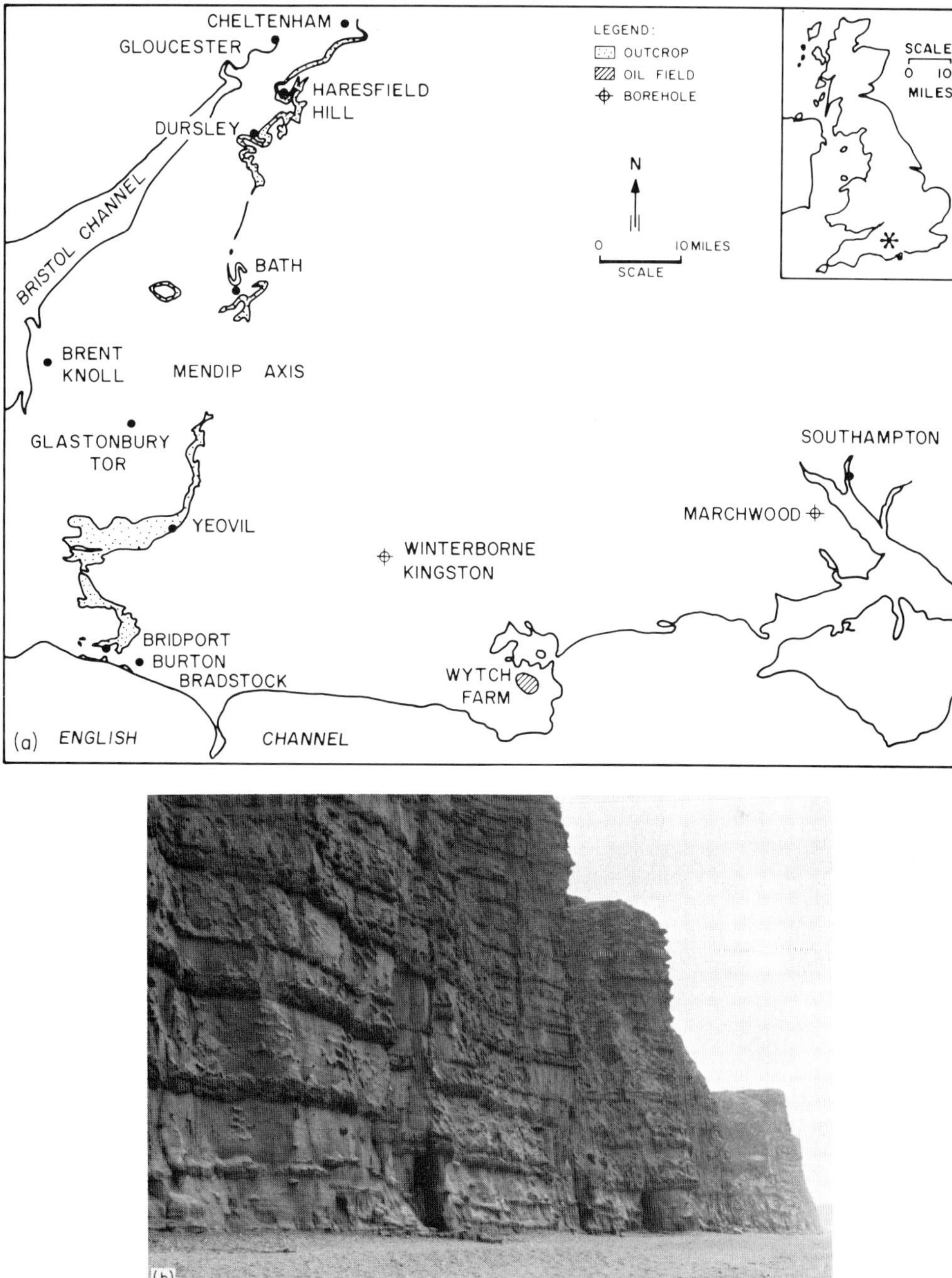

FIG. 2. (a) Location map and outcrop of the Upper Lias of Southern England. (b) Outcrop of the Bridport Sands at Bridport, Dorset. Carbonate cemented sands are hard and resistant to weathering, and so stand out relative to the more easily weathered uncemented sands.

gastropods, echinoderm plates, forams, ammonites, belemnites and bryozoa. Berthierine ooliths are common in the Marchwood Borehole but were not observed in outcrop samples.

The diagenetic history of these sediments is complex with widespread evidence of possibly concomitant carbonate precipitation and dissolution (Fig. 4). Post-depositional modifications

TABLE 1. *Petrographical characteristics of the Bridport Sand*

	Cemented	Clay-rich
Average clastic grain size	Fine	Fine
Sorting	Good	Moderate
Clastic grain to matrix ratio	3–4:1	2:1
Texture	Mature, clast supported	Immature, matrix supported
Compaction effects	None observed	Deformed micas throughout
Porosity	Mouldic and shrinkage within ooliths	Intergranular and microporosity
Typical modal mineralogy		
Quartz and feldspar	20.0	46.33
Mica	0.33	7.33
Rock fragments	0.67	1.00
Bioclasts	40.33	8.0
Berthierine ooliths	8.33	—
Detrital clay	7.67	23.33
Calcite cement	21.67	3.67
Dolomite cement	—	3.0
Quartz overgrowth	—	1.33
Pyrite	—	1.33
Porosity	1.0	4.67

FIG. 3. Scanning electron photomicrograph, Bridport Sands, Marchwood-1, 3907″ 2′. Grain-coating berthierine locally inhibits quartz overgrowth. Scale bar = 10 μm.

began with widespread algal and sponge boring and micritization (Fig. 5a). The bioclasts subsequently formed nuclei for non-ferroan calcite fringing cements and overgrowths (Fig. 5b). The remaining intergranular porosity was cemented with a mosaic of blocky and occasionally poikilotopic ferroan calcite. Where fewer bioclasts occur, poikilotopic calcite becomes more abundant. Neither fringing nor mosaic calcites luminesce nor fluoresce. The cementation history is complicated because of the uncertain timing of dissolution of aragonitic and high-Mg calcite bioclasts and early cements (Fig. 5c, d). There is widespread evidence of grain interpenetration and dissolution, whilst mouldic porosity also often occurs. These mouldic pores may be lined with euhedral calcite rhombs growing inwards from the preserved micritic envelope, or

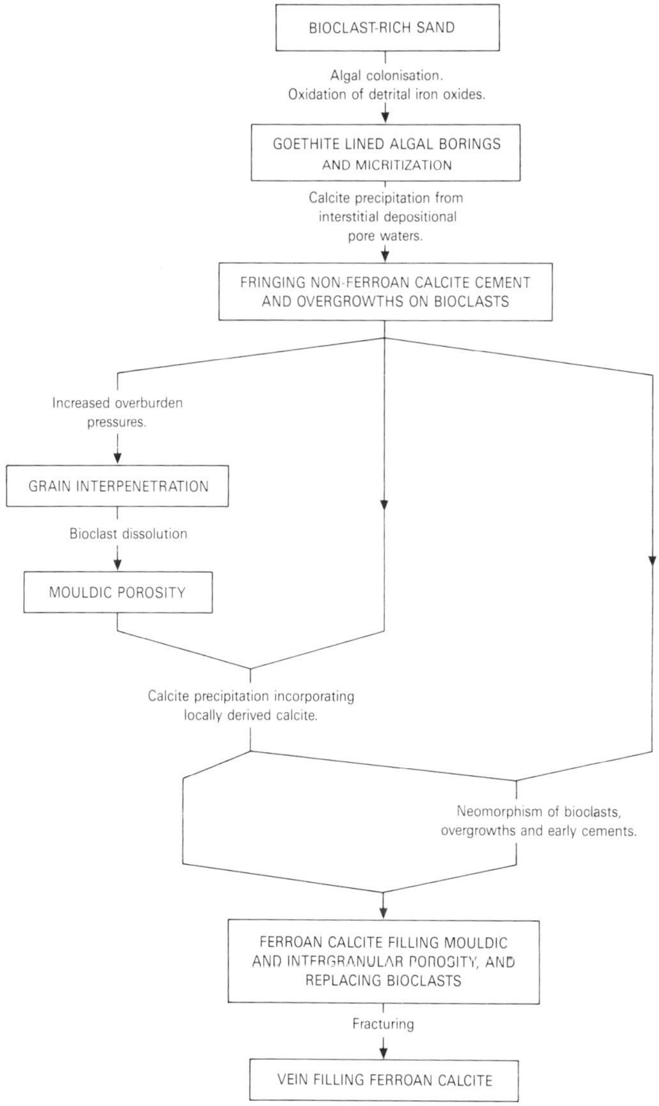

FIG. 4. Flow chart of diagenetic modifications to cemented sands in the Bridport Sand.

completely engulfed by blocky calcite (Fig. 5e). Finally, much of the aragonite or high-Mg calcite was replaced by ferroan calcite during diagenesis (Fig. 5f). Isolated ferroan dolomite rhombs occur occasionally. Ferroan calcite cemented fractures occur in many cemented beds but the calcite has not penetrated into intergranular pore space.

Carbon and oxygen stable isotope analysis

The results of stable isotope analysis are presented in Fig. 6 and Table 2.

Interpretation

The carbon isotopic composition of all the calcite and dolomite analysed from isolated cements and bulk samples is indicative of an originally marine (sea water) origin ($\delta^{13}C + 2‰$), with minor bacterial influence possibly shifting the $\delta^{13}C$ values towards $-2‰$ (see Hudson 1977). The oxygen isotopic composition, however, is indicative of precipitation at more elevated temperatures than expected near the sediment surface and presumably reflects the composition of the paragenetically later and replacive cements rather than early fringing cements. The

FIG. 5. Thin section photomicrographs, Marchwood-1 borehole. (a) Fine grained sandstone, with abundant bioclasts. The bioclast B has been bored and micritized by algae and sponges. MWD-1, 3885′6″. Scale bar = 175 μm. (b) Bioclastic sandstone with a well developed non-ferroan calcite overgrowth (stained pink) on an echinoderm plate, E, fringing cements on the other bioclasts present, and oxidized borings in places. MWD-1, 3871′6″. Scale bar = 175 μm. (c) Fine grained sandstone with non-ferroan bioclasts in various stages of diagenetic modification. Most bioclasts are fringed with calcite cement. The heavily bored bioclast B has been dissolved leaving goethite lined borings within the mouldic porosity. MWD-1, 3885′6″. Scale bar = 175 μm. (d) Fine grained bioclastic sandstone. Grain interpenetration has resulted in quartz grains penetrating through the dissolved bioclast. MWD-1 3861′11″. Scale bar = 175 μm. (e) Ferroan calcite cements this bioclast-rich sand. It occurs as an intergranular cement and within mouldic porosity in this case precipitating after dissolution of the bored bioclast B. Only the goethite filled borings and micritized envelope remain. MWD-1, 3871′6″. Scale bar = 140 μm. (f) Bioclastic-rich sandstone. The echinoderm plate E and overgrowth O have recrystallized to ferroan calcite whilst the non-ferroan brachiopod B and overgrowth have not. MWD-1, 3871′6″. Scale bar = 140 μm.

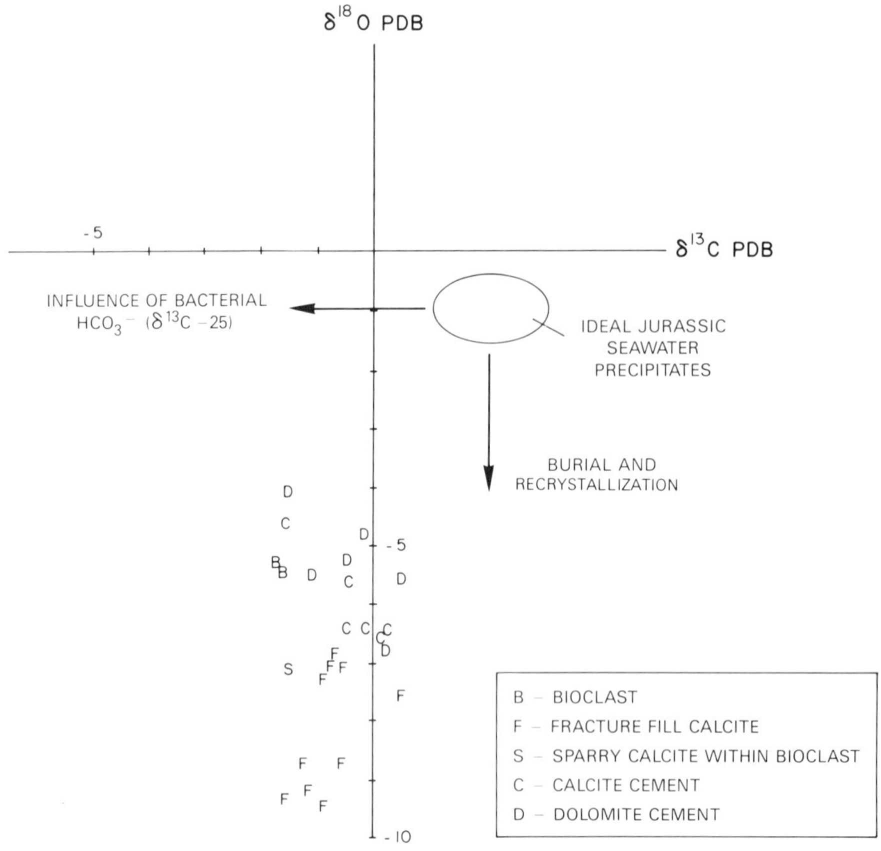

FIG. 6. Carbon and oxygen stable isotopic compositions of calcite and dolomite from Bridport Sand. The isotopic composition of calcite cement ($\delta^{13}C$ 0 to −1.5 and $\delta^{18}O$ −4.5 to −6.5) has been modified from a probable Jurassic sea water composition by two processes. Bacterially derived HCO_3^- with a $\delta^{13}C$ composition of −25 has diluted the original marine bicarbonate to more negative values. Dissolution and reprecipitation, and recrystallization during burial have led to a more negative shift in $\delta^{18}O$ values. The dolomite compositions ($\delta^{13}C$ +0.5 to −1.5 and $\delta^{18}O$ −4.0 to −6.0) possibly reflect the release of magnesium during calcite recrystallization.

shift from an assumed $\delta^{18}O$ −1.2‰ (SMOW) Jurassic sea water (Shackleton & Kennett 1975) which would have precipitated −1 to −2‰ (PDB) fringing cement, towards −4 to −6‰ (PDB) calcite may, therefore, reflect one or more of the following: (a) precipitation of the later cements during burial—the maximum burial temperatures being 80–100°C (average vitrinite reflectance $R_0 = 1.12$), (b) the influence of the low-salinity (fresh?) groundwater influx which

(a)
B
(b)
E
(c)
B
(d)
(e)
B
(f)
O
E
B

TABLE 2. *Carbon and oxygen stable isotopic data: Bridport Sands*

Sample	Details	Calcite		Dolomite	
		$\delta^{13}C$	$\delta^{18}O$	$\delta^{13}C$	$\delta^{18}O$
West Bay-2 (1)	Bioclast	−1.73	−5.14		
West Bay-2 (3)	Bioclast	−1.78	−5.26		
Burton					
Bradstock-4 (1)	Vein fill	−0.92	−7.33		
„ -4 (2)	„	−0.73	−6.89		
„ -2	„	−0.40	−7.18		
„ -1 (1)	„	−0.74	−7.25		
Hamdon Hill-1 (1)	„	−1.24	−8.67		
„ -1 (2)	„	−1.50	−9.38		
„ -1 (3)	„	−0.54	−8.76		
„ -1 (4)	„	−1.07	−9.30		
Marchwood-1					
3890′	Vein fill	+0.51	−7.59		
3890′	Cements	+0.25	−6.48	+0.50	−5.56
3863′11″	Vein fill	−0.91	−9.44		
3818′9″	Cements	−0.48	−5.62	−0.13	−4.92
3933′2″	„	−0.45	−6.53	−1.11	−5.51
3838′	„	−0.08	−6.50	−0.35	−5.62
3940′8″	„	−1.66	−4.71	−1.46	−4.08
3888′3″	„	+0.12	−6.50	+0.09	−6.64
3864′	„	−1.40	−7.07		

Edmunds *et al.* (1982) believe to have modified the originally marine pore waters, or (c) given the palaeoburial temperature and the presence of liquid hydrocarbons in the Bridport Sands at Wytch Farm (Colter & Havard 1981) the influence of thermally derived CO_2 obviously cannot be ruled out.

Controls on carbonate mineral diagenesis in the Bridport Sands

The cemented beds in the Bridport Sands contain two generations of carbonate cement: fringes or overgrowths on detrital grains and a pore-filling mosaic. The mosaic cement formed by the mobilization and redistribution of detrital and authigenic calcite and aragonite within the cemented beds themselves, and from the clay-rich beds into the cemented beds.

Berthierine

The most favourable conditions for berthierine formation are believed to be prolonged iron reduction in the suboxic zone (Coleman 1985, Curtis 1985). This allows large crystals to form. Most marine organic-rich sediments pass through oxic, suboxic and anoxic diagenetic zones during shallow burial diagenesis (Claypool & Kaplan 1974, Froelich *et al.* 1979). These are characterized by distinctive reactions between organic matter and various oxidants in the sediment or its interstitial pore waters and are often recorded by the carbon isotopic compositions of authigenic carbonates. However, the clay-rich sandstones here are dominated by the non-carbonate products of suboxic iron reduction. Iron silicates presumably formed instead of iron carbonates because produced bicarbonates were able to diffuse out of the system, preventing the build up of high carbonate alkalinity. Ferrous iron silicates have low solubility relative to calcite and precipitated rapidly in the sediment. The large crystals which formed reflect the longevity of suboxic iron reduction. The associated pyrite may have formed subsequently, when the sediments were eventually buried into the sulphate reducing zone, or when upwardly mobile H_2S rose from below. Any bicarbonate generated subsequently appears to have been swamped during cementation of the storm deposits by the abundant carbonate derived by dissolution of bioclasts. In summary, the diagenetic processes which operated in the fairweather sediments were driven by bacterial activity, however, this is recorded by the presence of iron silicates and iron sulphides rather than carbonate cements with highly negative $\delta^{13}C$.

Fringing cements

Fringing cements are not particularly diagnostic of any sedimentary environment, requiring only the presence of detrital carbonate grains as nuclei (e.g. Jørgensen 1976, Nelson & Lawrence 1984). They are most commonly found in firmgrounds and hardgrounds and hence reflect episodic non-deposition (e.g. Wilkinson *et al.* 1985). In the Bridport Sands, bioclasts were originally heterogeneously distributed, being concentrated within the storm deposits. Non-ferroan calcite or (? aragonite) fringing cements on the bioclasts mechanically stabilized the sediment framework. Bioclasts in the clay-rich sediments have occasional overgrowths, but these were not concentrated enough to stabilize the framework and prevent compaction. The distribution of fringing cements therefore reflects the texture and mineralogy of the original sedimentary deposits. There is no evidence of vadose cementation or of any other influence on fringing and overgrowth cementation.

Mosaic cements

After precipitation of the fringing cements the remaining intergranular pores in the storm sediments were filled with calcite. Petrographical evidence such as mouldic pores suggest that the cements were derived by internal redistribution of calcite. Similarly carbon stable isotope analysis only records minor bacterial influence on the original marine isotopic composition. However, it is not possible to calculate how much carbonate may have been mobilized from the fairweather beds since their initial carbonate content cannot be estimated objectively. The driving force during precipitation of the mosaic calcite was attempted equilibration of the originally detrital sedimentary assemblage with its pore waters. This assemblage included stable or metastable aragonite, high-Mg and low-Mg calcite bioclasts. The initial fringing cements were non-ferroan and may have comprised aragonite and calcite. During burial, pore waters became anoxic and temperatures increased. Ferroan calcite filled pores and replaced now unstable bioclasts and their overgrowths. Thus the carbonate mineral assemblage evolved during burial from a mixture of high and low-Mg calcite and aragonite towards ferroan calcite and dolomite. This evolution is also reflected by the gradual shift to negative $\delta^{18}O$ values. During diagenesis the clay-rich beds underwent compaction, progressively destroying intergranular microporosity. Patches of cement formed but the bioclasts were dissolved, contributing to cementation in the bioclast-rich beds. Most calcite precipitated within the pores preserved by early fringing cementation of the storm deposits. The distribution of the mosaic cements therefore reflects the texture and mineralogy of the sediments after the fringing cements precipitated.

Summary

Carbonate cementation in the Bridport Sands reflects the original sediment texture. Clay-rich fairweather sediments are weakly cemented and have undergone compaction. Interbedded clay-poor, bioclast-rich storm-deposited sediments developed early fringing cements resulting in little compaction. Pores in these uncompacted beds were filled by later cementation resulting from redistribution of bioclast-derived carbonate. Bacterial activity is only weakly reflected in the carbonate isotopic compositions of these sediments. Diagnostic criteria to predict the distribution of these cements may be found in their textural relationship to the host sediment. Mosaic calcite has formed where fringing cements were present. Fringing cements formed in the clay-free, bioclast-rich sands. This leads to the prediction that the cements will be found in the storm deposits because this is where the bioclasts were concentrated originally. These cemented beds are laterally continuous as a result of the sheet-like geometry of the storm-deposited sands.

Viking Group, Troll Field, offshore Norway

The Troll Field straddles Norwegian North Sea Blocks 31/2, 31/3, 31/5 and 31/6, and lies approximately 80 km WNW of Bergen (Fig. 7). The Troll Field reservoir comprises 400 m of Middle to Upper Jurassic Viking Group sandstones (Osborne 1985). The sedimentology of these shallow-marine shelf sands is complex, reflecting a number of transgressive and regressive cycles. Each cycle comprises coarsening upwards sands deposited during 'progradation' on to a shallow marine shelf. These sands are fine and micaceous at the base and coarsen upwards. The cycles are capped by coarse lag deposits formed by winnowing and reworking of the 'progradational' sands during subsequent transgression (Whitaker 1984, Osborne 1985). Palynofacies studies indicate that these 'progradational' sands are shelf-sand ridges or bars, rather than coastal sediments (Whitaker 1984).

The conventional microscopic, cathodoluminescence (CL) and stable isotopic characteristics of the non-ferroan calcite from two thick cemented beds (1382–1383, and 1417.5–1419 m) in the core from Well 31/2-12, and several thinner beds from the same well are described here. The texturally mature host sediments are described as bioclastic and non-bioclastic, the texturally immature sediments as micaceous. It is important to note that no specific lithofacies or

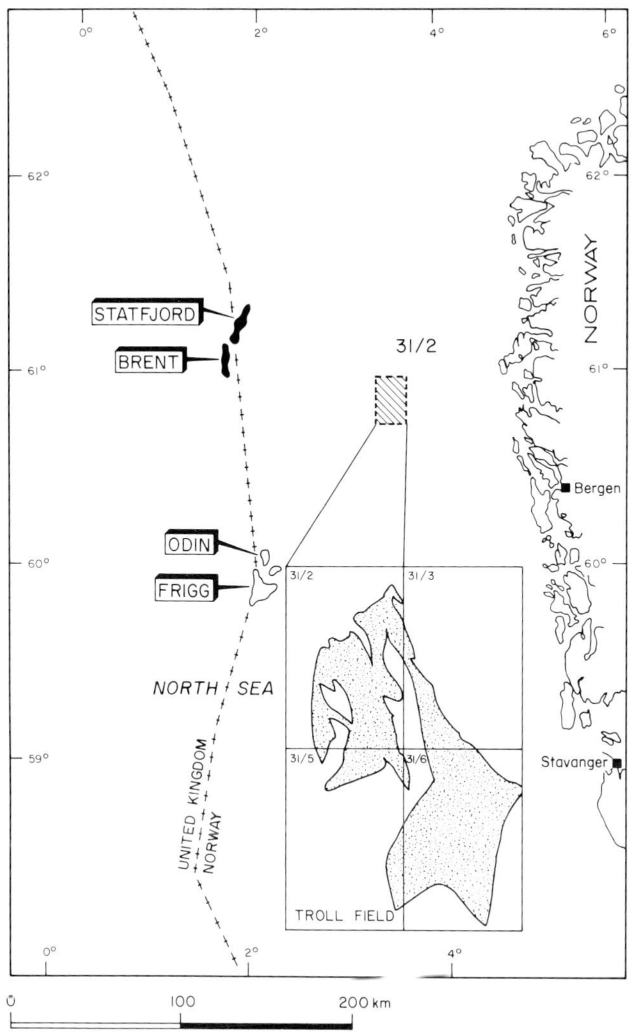

FIG. 7. Troll Field location map.

mineralogy is cemented preferentially. The sample base used in this study is extremely small and, as other cemented beds in the reservoir may have different origins, the conclusions reached here can only be extrapolated with extreme caution.

Petrography

Conventional microscopy

In the tightly cemented beds, non-ferroan calcite comprises up to 50% bulk volume. Texturally, the majority of the grains 'float' in the calcite or are only in point contact. Deformed micas were not observed in the texturally mature sediments. The picture is complicated by the amount of quartz corrosion but it may be suggested that cementation took place before any significant compaction occurred.

In the texturally mature sediments non-ferroan calcite occurs as fringing and mosaic cement. Fringing cement occurs as polygonal calcite around bioclasts and detrital clastic grains, and syntaxial overgrowths around echinoderm plate fragments. Fringing cements are succeeded and enclosed by the more volumetrically significant mosaic calcite. This mosaic calcite can be classified according to crystal size, comprising very fine or micritic calcite (15 μm in length), fine calcite (50 μm), medium to coarse blocky calcite, and poikilotopic calcite crystals (over 3 mm long).

Cathodoluminescence texture

CL microscopy revealed the fringing and mosaic cements visible in transmitted light (Fig. 8c), as well as primary calcite textures consisting of polygonal and botryoidal zoned calcite which are not visible during conventional microscopy (Fig. 8a, b and d). The rhombs occur as pendants hanging from detrital grains, or as a meniscus cement at grain contacts. Both primary textures have now been replaced. CL microscopy also reveals that microbrecciation occurred after the rhombs had formed. The mosaic cements were found to emit a variety of homogeneous or zoned luminescence patterns. However, there is no correspondence between the luminescence patterns observed in the mosaic cements and the mosaic textures observed during conventional microscopy. The homogeneously luminescing calcites display brown, bright yellow or dull-orange luminescence (Fig. 8b, c). The zoned calcites comprise from three to 10 or 12 zones, made up of either non-luminscent calcite, or brown, bright yellow and dull orange luminescent calcite (Fig. 8e). Besides containing variable numbers of zones, the zoned calcites display no systematic sequence of luminescence colour.

Zonation in carbonate cements can reflect variations in both the composition and the degree of saturation of the precipitating fluids as well as crystal growth rate. Consequently, the specific colours manifested here do not necessarily suggest that widely differing porewater conditions existed during diagenesis (Ten Have & Heijnen 1985).

In the micaceous beds CL microscopy revealed a complex fabric of grain breakage and microbrecciation (Fig. 8f).

Diagenetic sequence of non-ferroan carbonate cementation

On the basis of conventional and CL microscopy it is possible to establish a diagenetic sequence commencing with fringing cementation, followed by a variety of brown or brown and yellow zoned mosaic calcites. The sequence progresses through yellow to dull orange luminescing calcite with contemporaneous fracturing.

Stable isotope analysis

Comparing the isotopic compositions of the calcites from the Viking Group sandstones (Fig. 9 and Table 3) with the luminescence colour and transmitted light textures revealed no systematic variation. It follows that porewaters with the same stable isotopic compositions can precipitate apparently different calcites whilst apparently similar calcites can be precipitated from porewaters with widely differing isotopic compositions.

Interpretation

In both the bioclastic and non-bioclastic beds the fringing and pendant calcites have carbon isotopic compositions indicative of precipitation from sea water (see Table 3). The remaining data fall into two clusters with distinct and highly negative $\delta^{13}C$ values -21 to $-31‰$ and -41 to $-47‰$. These carbon isotopic compositions are all indicative of bacterial sources of bicarbonate (Hudson 1977, Irwin *et al.* 1977).

In the bioclastic bed, fringing calcite is post-dated by calcite with $\delta^{13}C$ values of -21 to $-31‰$ (Fig. 10a). These values are indicative of calcite incorporating bicarbonate generated by sulphate-reducing bacterial processes. The calcite is non-ferroan and encloses pyrite, confirming that sulphate reduction had occurred. If mobile H_2S fixed the ferrous iron generated during suboxic reduction, then these iron-poor carbonates formed in the sulphate reduction zone. The oxygen isotopic compositions of these samples are consistent (around $\delta^{18}O - 1‰$) and, assuming Jurassic sea water to have a $\delta^{18}O$ SMOW composition of $-1.2‰$ (Shackleton &

FIG. 8. Thin section photomicrographs, Troll Field. (a) and (b) In transmitted light only a mosaic of fine calcite cement is visible. CL reveals a zoned polygonal cement which occurs as a pendant or meniscus coating. Intergranular porosity is cemented with homogeneously luminescing cement. Fracture filling yellow calcite also occurs. 31/2-12, 1418.5 m (5), XPL and CL respectively. (c) Bioclast fringed with brown and yellow zoned calcite. Intergranular porosity is filled with homogeneous yellow or orange luminescing calcite. 31/2-12, 1417.8 m (1b), CL. (d) Brown and yellow zoned botryoidal calcite. Cement nuclei occur preferentially at the bases of detrital grains. 31/2-12, 1418.5 m (5) CL. (e) Multiple zoned calcite cementing a texturally mature sandstone. 31/2-12, 1382 m. (f) Micaceous sand cemented with brown luminescing calcite, itself fractured and infilled with yellow and then orange luminescing calcite 31/2-7, 1612 m. CL. Scale bars = 200 μm.

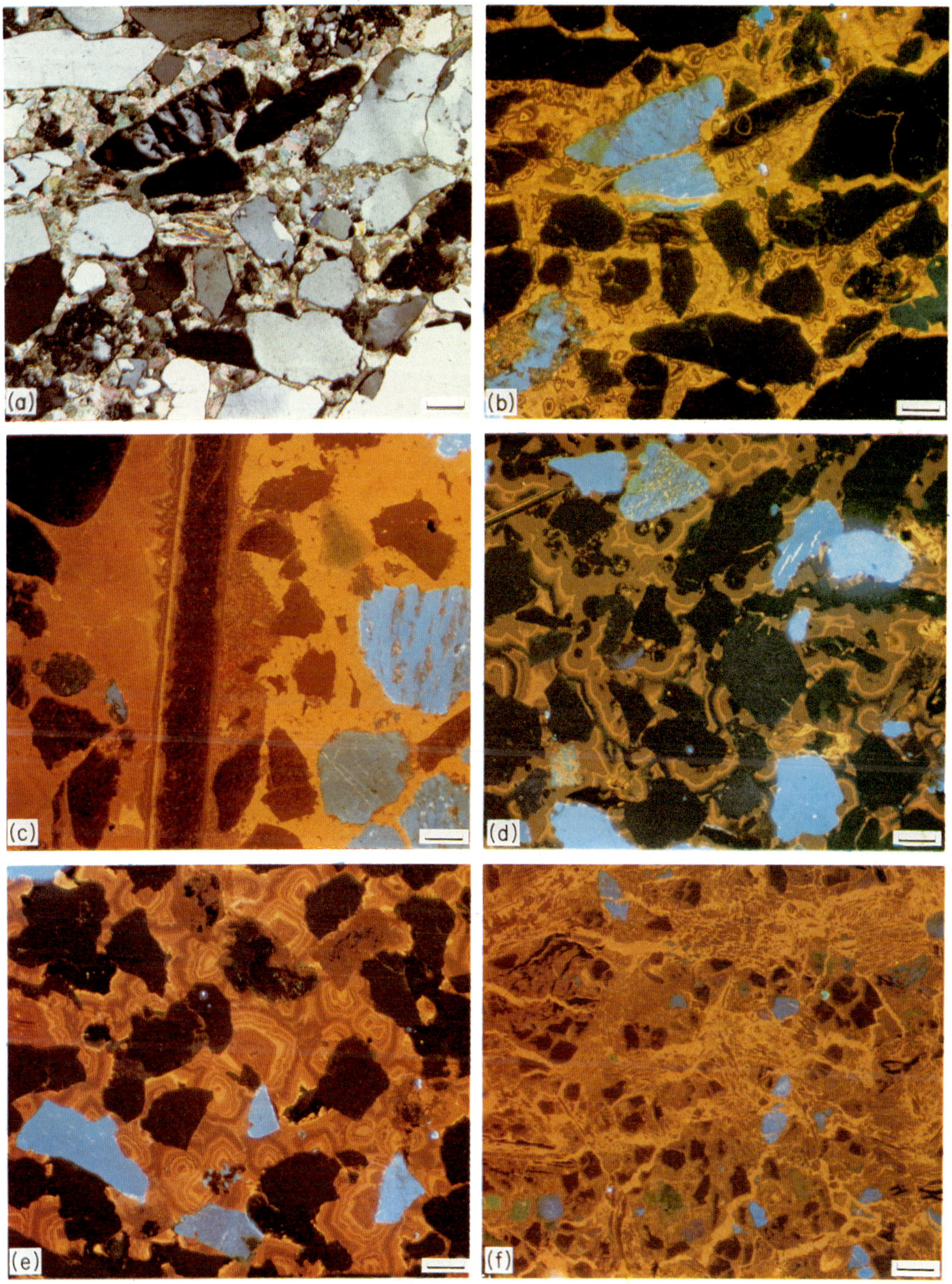
(a)
(b)
(c)
(d)
(e)
(f)

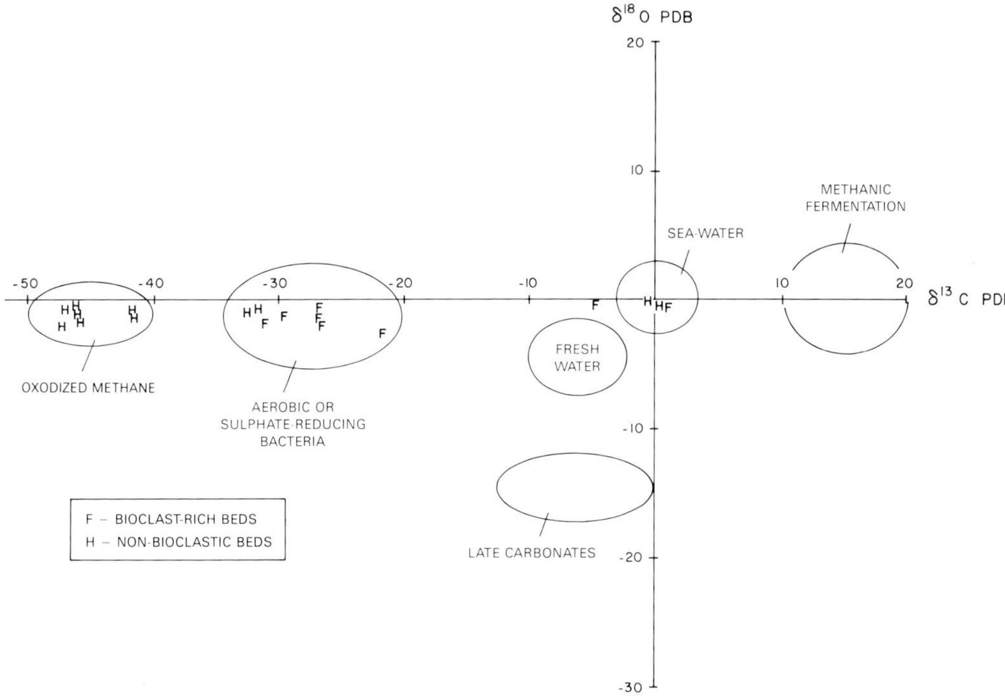

FIG. 9. Stable isotopic compositions of calcite cements, Viking Group Sandstones, Troll Field, classified according to host lithology and compared with data from Hudson (1977).

TABLE 3. *Stable isotopic compositions of calcite samples, Well 31/2-12 Troll Field, Norwegian sector, North Sea*

(a) Bioclast-rich beds

Depth (m)	Core sample	Isotope sample	$\delta^{13}C$ ‰ PDB	$\delta^{18}O$ ‰ PDB	Luminescence character	Remarks
1378.5	2	18	−26.22	−1.54	Brown	
1382.5	1	16	−21.88	−2.52	Brightly luminescing zoned yellow and dull orange	
1382.6	3	11	−26.00	−1.65	Brown	
1382.6	3	20	−3.97	−0.06	Bright yellow	Bioclast
1382.7	4	15	−29.11	1.48	Brightly luminescing zoned yellow and dull orange	
1382.8	5	4	+1.32	−0.54	Dull orange	Bioclast
1384.0	1	3	−26.52	−0.21	Brightly luminescing zoned brown and yellow	Calcite filling vein
	1	12	−31.54	−1.68	Brown	
(b) Non-bioclastic beds						
1413.0	1	10	−47.15	−1.38	Dull orange	
1417.6	1A	5	−46.16	−0.90	Dull orange	
1417.8	1B	1A	−32.59	−1.11	Bright yellow	
1417.8	1B	1B	+0.41	−0.61	Dull orange	Bioclast + early cement
1417.8	1B	1C	−41.45	−0.88	Dull orange	
1418.0	2	6	−47.13	−0.54	Dull orange	
1418.2	3	7	−41.68	−1.12	Dull orange	
1418.3	4	8	−45.86	−1.01	Dull orange	
1418.5	5A	13	−31.46	−1.93	Zoned brown and dull orange, botryoidal zonation	Early vadose cement nucleus?
1418.5	5B	14	−0.14	−0.15	Brightly luminescing zoned brown and yellow	Early vadose cement
1418.8	6	9	−46.12	−0.93	Dull orange	

Kennett 1975), formed at or near the sediment surface. There is no systematic zonation through this bed.

In the non-bioclastic bed most of the calcite which encloses pendant cement or blankets entire samples has a $\delta^{13}C$ value (−41 to −47‰) indicating that it incorporated bicarbonate generated during bacterial oxidation of methane (Fig. 10b). This occurs when biogenic and thermogenic methane serve as a substrate for continued bacterial activity. Some of the calcite in these cemented beds has a $\delta^{13}C$ value of

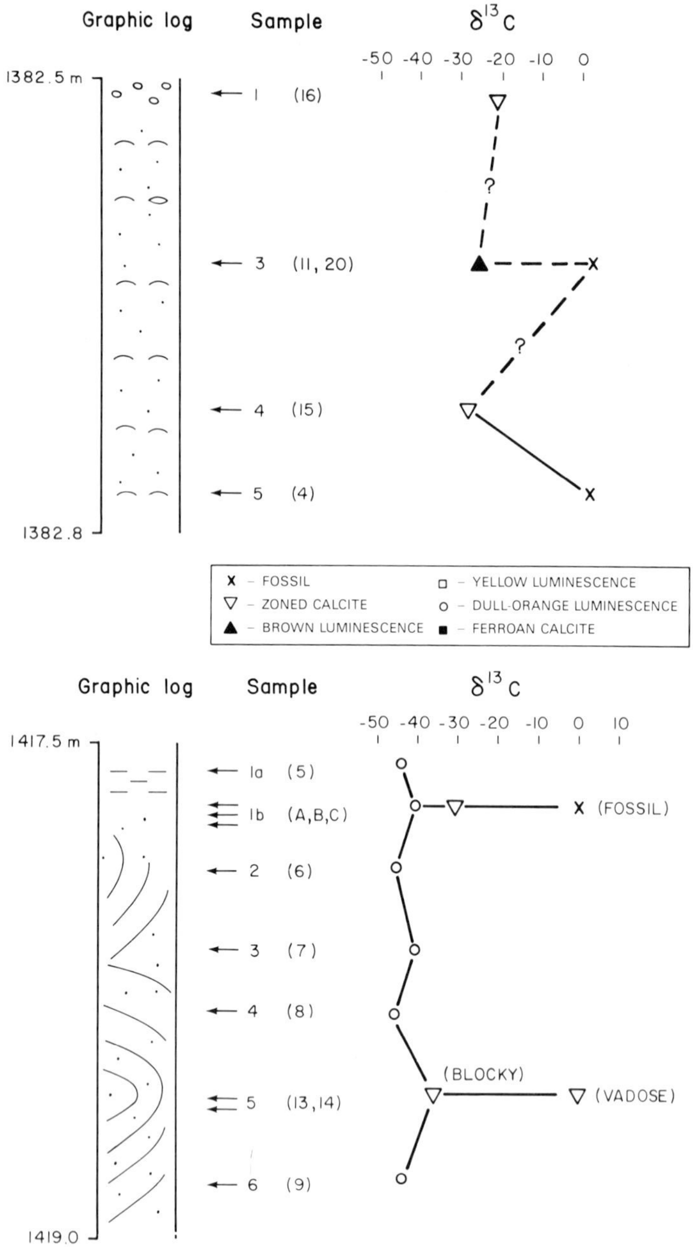

FIG. 10. (a) Stable isotopic composition of calcite cemented bed. Bioclastic bed, 31/2-12, 1382 m. (b) Stable isotopic composition of calcite cemented bed. Non-bioclastic bed, 31/2-12, 1418 m.

$-31‰$ and an origin similar to the calcite in the bioclastic bed. The samples analysed here may also have contained disproportionate mixtures of pendant calcite ($\delta^{13}C = 0‰$) and isotopically light calcite ($\delta^{13}C - 41‰$). The oxygen isotopic values are consistently around $\delta^{18}O$ $-1‰$ and again it is suggested that precipitation occurred near the seafloor.

Controls on carbonate mineral authigenesis in the Viking Group sandstones

The non-ferroan calcite in the Viking Group sandstones in Troll comprises a fringing or possibly vadose pendant cement and nucleated on these more extensive mosaic calcites that incorporated bacterially derived carbonate.

These generations of cement are separated by authigenic pyrite, kaolinite and K-feldspar.

Fringing and pendant cement

As discussed earlier, fringing cements do not appear to be particularly diagnostic of any sedimentary environment. By contrast, pendant or dripstone textures have only been described from coastal beachrock or intertidal areas, where they form in the vadose or unsaturated zone (Müller 1971). However, there is no reason why they should not form in offshore environments which undergo uplift and emergence. Einsele *et al.* (1977) discuss the effects of emergence on Holocene sediments from the continental shelf off Mauritania. They attribute carbonate lithification of sediments 100 km offshore and now 100 m below sea-level to subaerial exposure. The effects of emergence overprint marine sediments whilst subsequent transgression and reworking has removed any direct evidence of subaerial exposure. It follows that in the Viking Group sandstones one would expect to find evidence of exposure towards the top of regressive sequences or immediately beneath transgressive sands. Additional evidence for emergence is seen in the micaceous sands which occur at the base of the sequence and display fabrics reminiscent of intertidal hardground or calcrete development (Fig. 8; Assereto & Kendall 1977).

Mosaic calcite

In the Viking Group sandstones, fringing calcite is enclosed by a mosaic calcite. Thin section and oxygen stable isotopic evidence both suggest that cementation took place before significant burial had occurred. The early cements may have provided a nucleus for the mosaic or they may simply have maintained a porous framework within which the mosaic developed. Whether these cemented beds are nodular and discontinuous or continuous will depend on whether they grew around dispersed nuclei or from a finite supply of bicarbonate (Fig. 11). In the bioclastic

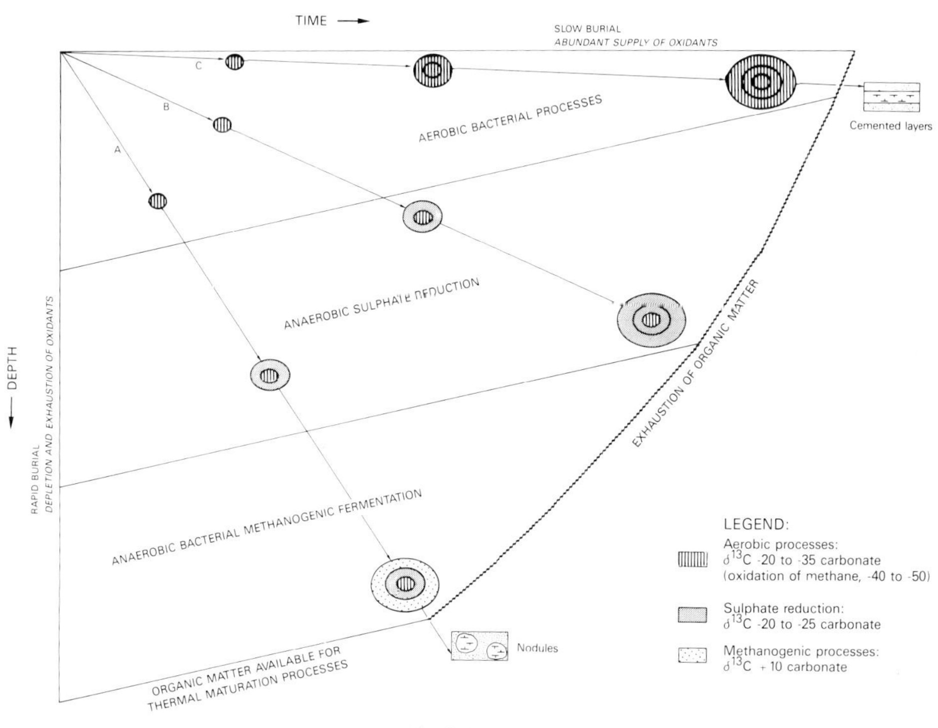

FIG. 11. Paths A, B and C illustrate the effect of burial rate on the style of cementation which may develop in marine sandstones. In path A an isotopically and mineralogically zoned nodule develops. More dispersed, patchy cementation can also occur. With faster burial than shown in path A smaller zoned nodules will form and more organic matter will be available to generate hydrocarbons. Paths B and C illustrate the isotopic and mineralogical composition of cements that develop at slower burial rates. Path C is the extreme case in which the sediment remains within the superficial bacterial zones and is completely cemented with an isotopically homogeneous non ferroan calcite. In constructing this diagram it is assumed that the boundaries between successive bacterial reaction zones are maintained at constant depths. Obviously this is not always the case.

beds there is no evidence of isotopic zonation in the mosaic cements (Fig. 10). After precipitation of the early cements these beds were buried into the sulphate-reduction zone, where they remained until tightly cemented. The thickness of the cemented bed and its uniform composition suggest that the sediments remained in this bacterial reaction zone for some time: a pathway between B and C in Fig. 11.

In the non-bioclastic cemented beds, calcite precipitated incorporating bicarbonate liberated during sulphate-reducing bacterial processes and then during oxidation of methane (Fig. 10b), giving rise to the apparent growth banding observed in the cores. In a closed system, as might exist within a shale-dominated sequence, this succession would lead to a nodule forming (Fig. 11, pathway A). However, several lines of evidence suggest that the Viking Group sandstones may not be described in this way. First, the cemented beds are interbedded with highly permeable sands. Thus methane generated elsewhere and rising through the sediment is more likely to have escaped to the surface than to have been trapped and fixed in the sulphate-reduction zone. Indeed, almost all the recorded recently-formed methanic cement in marine sandstones is found on the seafloor (e.g. Høvland *et al.* 1985, Nelson & Lawrence 1984 and references therein). Textural evidence confirms that cementation took place prior to compaction. Second, the now replaced botryoidal texture is more likely to have been aragonite formed on the seafloor rather than during burial. Third, the oxygen isotopic compositions do not show any zonation or fractionation. This is consistent with continuous precipitation from porewaters with a uniform composition, rather than in a closed system or during burial. Finally, it is not clear that anaerobic oxidation of methane is actually possible as the porewater profiles in which this process is inferred to take place are open to alternative interpretation (see, e.g. Martens & Berner 1974, Barnes & Goldberg 1976, Lovley & Klug 1986, Whiticar & Faber 1986).

In the Viking Group sandstones, mosaic calcite precipitation began when the early-formed fringing cements were buried into the sulphate-reduction zone. Continued precipitation, however, involved aerobic oxidation of methane. As the resulting cemented bed is over 1 m thick, prolonged bacterial activity must have occurred. Thus it is likely that a temporary hiatus occurred maintaining the sandstones near the seafloor for some time. It follows that three controls on the location of these cemented beds may be identified: (a) the presence of a nucleus of early fringing or pendant calcite; (b) residence near the seafloor during precipitation; (c) the permeability of the sediment enabling methane to rise into the bacterial zones above.

In the case of the non-bioclastic bed it is not clear whether the growth banding seen in the non-bioclastic bed defines the size of the cemented area, or the continued accretion of cement as further methane was supplied. If this process occurred over a large area of the seafloor then the nodular textures observed in the core could reflect coalescence of cements around discrete pendant cement nuclei (a pathway between B and C in Fig. 11). In both the bioclastic and non-bioclastic beds it is not possible to establish from core data whether or not these areas of cement will have coalesced to form continuous layers. This will depend on the longevity of the bacterial processes involved.

Summary

The cemented beds in the Viking Group sandstones in Troll contain two distinct calcite cements. First, a thin fringing or pendant cement precipitated from sea water (during regression or even temporary emergence) and, second, a mosaic calcite incorporating bacterially derived bicarbonate. To facilitate extensive cementation the sediments must have remained near the sediment surface for some time whilst bacterial processes operated. There is no direct evidence that these cemented beds are nodular, the product of closed system diagenesis. Their diagenesis reflects two independent processes operating near the sediment surface. Thus, similar cementation may occur in sediments maintained near the sediment surface, first, during periods of non-deposition, and possibly even emergence and, second, within zones of intense bacterial activity.

Discussion

These case studies demonstrate the potential for reservoir compartmentalization in shelf sandstones. Although the diagenetic histories differ a number of controls on cement geometries have been identified. The geometry of the cemented beds in the Bridport Sands reflects the sandbody geometry: they are laterally extensive. In the Viking Group sandstones in Troll the location of the cements reflects two conditions and their geometry will depend on the extent and timing of emergence of the original deposits. The question of whether or not these controls on sandstone cementation can be used for predictive purposes elsewhere is addressed below.

In the Bridport Sands early fringing cements developed around storm-deposited bioclasts. Fewer bioclasts were deposited in the clay-rich fairweather sediments with the result that less fringing cement developed. Thus it was the fast-slow nature of storm versus fairweather deposition as well as the bioclast concentration in the storm deposits which predetermined the cement location. This cementation model may be widely applicable in storm or other sediments in which potential cement nuclei are heterogeneously distributed. In the Viking Group the fringing and pendant cements owe their origin to non-deposition, if not emergence. Thus the location of cementation reflects the stop-start nature of sedimentation and burial of these sands. This style of cementation is analogous to firmground and hardground development and this model may apply to almost any shelf sediments.

Whilst the location of the cemented beds reflects large-scale sedimentological controls, the origin of the pore-filling mosaic calcite reflects smaller-scale controls. First, in the Bridport Sands the early framework cements maintained open pores which were cemented during burial by the remobilization of detrital and early authigenic calcite. Gradual equilibration of the sedimentary minerals occurred moving from aragonite, high-Mg and low-Mg calcite to ferroan calcite and dolomite. Bacterial activity influenced diagenesis as evidenced by the abundant berthierine but the bicarbonate liberated during bacterial activity has not significantly influenced the isotopic composition of the cements. Interestingly, redistribution of detrital bioclasts produces spherical nodules in more homogeneous sediments. Hence, only with an understanding of the relationship to host sediment texture is it possible to distinguish between the two in a core. In the Viking Group sandstones the fringing cement framework was also filled by mosaic calcite, but the bicarbonate incorporated here was derived from bacterial degradation of organic matter. The potential for bacterially derived cements to develop obviously exists in any organic-rich sediments subject to slow and episodic burial. Second, the Bridport Sands' cement appears to have evolved through time, leading to a progressive depletion in oxygen isotopic composition. In the Viking Group the cements have homogeneous isotopic values: during precipitation porewater compositions were relatively consistent. This style of cementation can, therefore, be distinguished from burial cements around similar fringing cements in which isotopic zonation would be expected.

There are clearly many processes involved in generating laterally extensive carbonate cemented layers. These examples show that some of the controls on the location of cements can be related to the episodic nature of shelf sedimentation, and to the textural and mineralogical variations which result. Laterally extensive cements develop quite simply because the specific sediments in which they form, or the processes responsible for their formation, occur over particularly large areas. Similar cements will not develop in more heterogeneous systems where such controls are not laterally persistent. However, laterally extensive cements can develop for a variety of other reasons. For example, Gautier & Rice (1981) describe cementation at a water table in the Cretaceous Eagle Sandstone, a coastal deposit in Montana. This cement cross-cuts depositional sedimentary structures. The potential also exists for late cementation to form laterally persistent barriers, for example at hydrocarbon-water contacts.

Laterally extensive cementation requires an abundant supply of cement. Here too there are no unique solutions, the Bridport Sands' cement, for example, forming from internally redistributed calcite, the Viking Group's cement forming as a result of bacterial activity. As can be shown from the Viking Group data, the potential for extensive cementation exists when distinct processes operate for prolonged periods. In the Bridport Sands, processes have continued for long enough to create continuous layers of totally cemented sand. In the case of the Viking Group sandstone in Troll it is not possible to establish from this data set the precise geometry of these extensively cemented beds and hence to predict whether or not laterally continuous layers have also developed here.

These case studies demonstrate the potential for laterally extensive cementation to develop in shelf sandstones. However, cementation can occur for more than one reason and thorough petrographical investigation is necessary to distinguish between various possible controls. Only after establishing the nature of controls on cementation is it possible to make accurate predictions about the geometry of cemented sands and their likely influence on hydrocarbon recovery efficiency.

ACKNOWLEDGMENTS: We thank Shell Research BV, Shell International Petroleum Maatschappij, Norske Shell and partners in the Troll Field, and the Director of the British Geological Survey (NERC), for permission to publish this paper. We thank Norske Shell, Poroperm Laboratories, Professor W. G. Mook (Groningen), and Professor J. Thorez (Liege) for providing some of the data discussed. We thank our colleagues in Rijswijk for stimulating discussions. Errors of fact and interpretation, however, are our own.

References

ASSERETO, R. L. M. & KENDALL, C. G. ST. C. 1977. Nature, origin and classification of peritidal tepee structures and related breccias. *Sedimentology*, **24**, 153–210.

BARNES, R. O. & GOLDBERG, E. D. 1976. Methane production and consumption in anoxic marine sediments. *Geology*, **4**, 297–300.

CLAYPOOL, G. E. & KAPLAN, I. R. 1974. The origin and distribution of methane in marine sediments. *In*: KAPLAN, I. R. (ed.). *Natural Gases in Marine Sediments*, pp. 99–139. Plenum Press, New York.

COLEMAN, M. L. 1985. Geochemistry of diagenetic non-silicate minerals: kinetic considerations. *Philosophical Transactions of the Royal Society of London*, **A35**, 39–56.

COLTER, L. V. & HAVARD, D. I. 1981. The Wytch Farm oilfield Dorset. *In*: ILLING, V. C. & HOBSON, G. D. (eds). *Petroleum Geology of North-West Europe*, pp. 494–503. Applied Science, London.

CURTIS, C. D. 1985. Clay mineral precipitation and transformation during burial diagenesis. *Philosophical Transactions of the Royal Society*, **A315**, 91–105.

DAVIES, D. K. 1967. Origin of friable sandstone rhythms in the Lias of England. *Journal of Sedimentary Petrology*, **37**, 1179–88.

EDMUNDS, W. M., BATH, A. H. & MILES, D. L. 1982. Pore fluid geochemistry of the Bridport Sands (Lower Jurassic) and the Sherwood Sandstone (Triassic) intervals of the Winterborne Kingston borehole, Dorset. *In*: RHYS, G. H., LOTT, G. K. & CALVER, M. A. (eds). *The Winterbourne Kingston borehole, Dorset, England*, pp. 149–63. Report Institute of Geological Sciences, 81/3.

EINSELE, G., ELOUARD, P., HERM, D., KOGLER, F. C. & SCHWARZ, H. U. 1977. Source and biofacies of Late Quaternary sediments in relation to sea level on the shelf off Mauritania, West Africa. *'Meteor' Forschungsergebnisse Reiche*, **C26**, 1–43.

FROELICH, P. N., KLINKHAMMER, G. P., BENDER, M. L. LUEDTKE, N. A., HEATH, G. R., CULLEN, D., DAUPHIN, P., HAMMOND, D., HARTMAN, B. & MAYNARD, V. 1979. Early oxidation of organic matter in pelagic sediments of the eastern equatorial Atlantic: suboxic diagenesis. *Geochimica et Cosmochimica Acta*, **43**, 1075–90.

GAUTIER, D. L. & RICE, D. D. 1981. Comparison of conventional and low-permeability reservoirs of shallow gas in the northern Great Plains. *Society of Petroleum Engineers Paper 9846.*

HØVLAND, M., TALBOT, M., OLAUSSON, S. & AASBERG, L. 1985. Recently formed methane-derived carbonates from the North Sea floor. *In*: *Petroleum Geochemistry in Exploration of the Norwegian shelf*, pp. 263–6. Norwegian Petroleum Society.

HUDSON, J. D. 1977. Stable isotopes and limestone lithification. *Journal of the Geological Society of London*, **133**, 637–60.

IRWIN, H., CURTIS, C. D. & COLEMAN, M. L. 1977. Isotopic evidence for source of diagenetic carbonates formed during burial of organic-rich sediments. *Nature, London*, **269**, 209–13.

JØRGENSEN, N. O. 1976. Recent high magnesian calcite/aragonite cementation of beach and submarine sediments from Denmark. *Journal of Sedimentary Petrology*, **46**, 940–51.

LOVLEY, D. R. & KLUG, M. J. 1986. Model for the distribution of sulphate reduction and methanogenesis in freshwater sediments. *Geochimica et Cosmochimica Acta*, **50**, 11–8.

KANTOROWICZ, J. D. 1985. The origin of authigenic ankerite from the Ninian Field, UK North Sea. *Nature, London*, **315**, 214–6.

MARTENS, C. S. & BERNER, R. A. 1974. Methane production in the interstitial waters of sulphate-depleted marine sediments. *Science*, **185**, 1167–9.

MILLIKEN, K. L., LAND, L. S. & LOUCKS, R. G. 1981. History of burial diagenesis determined from isotopic geochemistry. Frio Formation, Brazos County, Texas. *American Association of Petroleum Geologists Bulletin*, **65**, 1397–413.

MÜLLER, G. 1971. 'Gravitational' cement: an indicator for the vadose zone of the subaerial diagenetic environment. *In*: BRICKER, O. P. (ed.). *Carbonate Cements*, pp. 301–20. Johns Hopkins University Press, Baltimore.

NELSON, C. S. & LAWRENCE, M. F. 1984. Methane derived high-Mg calcite submarine cement in Holocene nodules from the Fraser delta, British Colombia, Canada. *Sedimentology*, **31**, 645–54.

OSBORNE, P. 1985. The Troll Field. *Proceedings, Norwegian Institute of Technology Seminar on North Sea Oil and Gas Reservoirs, Trondheim.*

SHACKLETON, N. J. & KENNETT, J. P. 1975. Palaeotemperature history of the Cenozoic and the initiation of the Antarctic glaciation: oxygen and carbon isotope analysis in DSDP sites 277, 279, and 281. *In: Initial Reports of the Deep Sea Drilling Project*, **29**, 743–55. Government Printing Officer, Washington DC.

TEN HAVE, A. H. M. & HEIJNEN, W. 1985. Cathodoluminescence activation and zonation in carbonate rocks: an experimental approach. *Geologie en Mijnbouw*, **64**, 297–310.

WHITAKER, M. F. 1984. The usage of palynology in definition of Troll Field geology. *6th Offshore Northern Seas Conference and Exhibition, Stavanger.*

WHITICAR, M. J. & FABER, E. 1986. Methane oxidation in sediment and water column environment—isotope evidence. *Organic Geochemistry*, in press.

WILKINSON, B. H., SMITH, A. L. & LOHMANN, K. C. 1985. Sparry calcite marine cement in Upper Jurassic limestones of southeastern Wyoming. *Society of Economic Paleontologists and Mineralogists, Tulsa. Special Publication*, **36**, 169–84.

J. D. KANTOROWICZ, I. D. BRYANT and J. M. DAWANS, Koninklijke/Shell Exploratie en Produktie Laboratorium, Postbus 60, 2280 AB Rijswijk ZH, The Netherlands.

Magnesite formation in recent playa lakes, Los Monegros, Spain

J. J. Pueyo Mur & M. Inglés Urpinell

SUMMARY: Early diagenetic magnesite is at present forming at a depth of some 20 cm in recent playa lakes in NE Spain. Magnesite formation can be considered as a result of two factors: (a) an increase in CO_2 activity caused by the decay of organic matter, and (b) the presence of post-halitic brines, strongly concentrated in magnesium, that form in summer. These two factors are mainly observed in the ephemeral salt-pan zone where the highest accumulation of magnesite is found.

The Recent lakes studied are located in Los Monegros in the central sector of the Ebro Basin, NE Spain (Fig. 1,A). They occur on evaporite bearing Miocene and Oligocene sediments. These lakes are playa lakes and dry salt lakes; they are small and have ephemeral ponds with brines.

The lakes are found between 300 and 400 m above sea-level and develop on depressions which originated through dissolution of Miocene evaporitic levels, accompanied by surface deflation. A great many of these lakes develop asymmetrical shapes conditioned by the NW dominant wind. Insolation and wind are the main factors affecting the evaporation of the waters. Rainfall in the area is low: about 350 mm yr^{-1} (Quirantes 1965).

Composition and evolution of brines

Surficial brines are of Cl^--SO_4^{2-}-Na^+-(Mg^{2+}) type (Pueyo 1978–79) and undergo strong seasonal oscillations in concentration because of evaporation and progressive precipitation of mineral phases. In most lakes there is no surficial water in the greater part of the year and even during the entire year in dry periods.

Interstitial brine accumulates in the eastern-central area of the lakes (ephemeral salt-pan) and in the western area where expandable clay minerals, with high water retention capacity, are more abundant. The interstitial salinity increases towards the lake surface as is usual in a playa lake regime, and is controlled by evaporative pumping in the vadose zone during dessication periods. An increase in Cl^-, Na^+ and Mg^{2+} concentration has also been reported from the leeside (SE) to the windward side (NW) of the playa lakes.

Brine generation depends not only on climatic conditions (aridness) but also on the influence of the evaporite-bearing substratum. Substrate evaporites developed in intermontane conditions which affected the main Tertiary depressions of the Iberian Peninsula during the Miocene. It is the recycling of these saline materials, occurring selectively according to their solubility, which gives a high salinity to the waters flowing into the lakes.

Evaporite precipitation

Different kinds of evaporitic minerals have been found. They originated through direct precipitation from the superficial waters as well as through interstitial or efflorescent precipitation from the interstitial brines.

Crystallization from the surficial waters of the lakes develops as a response to their seasonal evolution (Quirantes 1965, Pueyo 1978–79).

(a) Spring–summer precipitation: Crystallization occurs in a concentric zoned distribution according to the carbonates–gypsum–halite sequence.

(b) Winter precipitation: Crystallization of sodium sulphate in the form of mirabilite. This mineral changes easily to thenardite by dehydration induced by wind and insolation.

Efflorescent crystal-precipitates develop from interstitial brines carried up to the surface by capillarity, with bloedite, thenardite and halite being the minerals most commonly represented.

Evaporitic sediments are reworked by wind, waves, rain and organisms living in the water or at the dry lake bottom (cyanobacterial mats, small crustaceans, coleoptera, etc.).

Mineralogy of the Quaternary lacustrine deposits

To study the Quaternary sediments, sampling work in 10 playa lakes has been carried out (Pueyo & Inglés 1986). One of these lakes (the

From: MARSHALL, J. D. (ed.), 1987, *Diagenesis of Sedimentary Sequences*, Geological Society Special Publication No. 36, pp. 119–122.

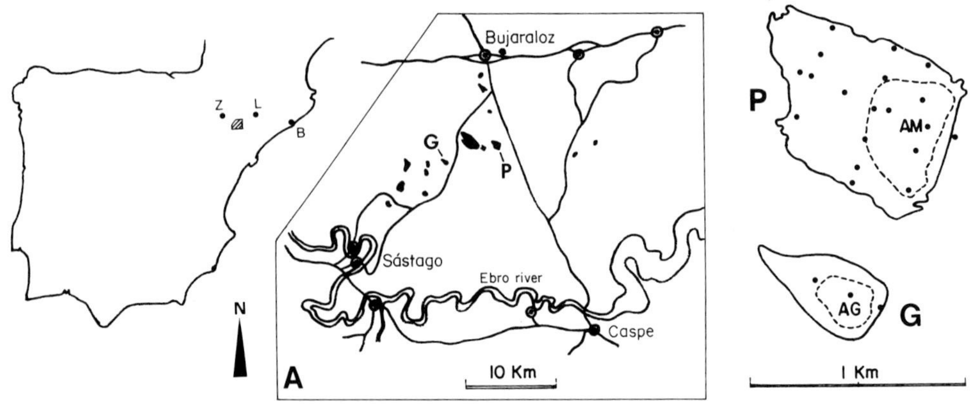

FIG. 1. Geographical situation of the studied playa lakes. P and G correspond to Pito and Guallar playa lakes, mainly reported in this paper. P. Pito playa lake and core sampling positions. G. Guallar playa lake and core sampling positions. AM sample (of Pito playa lake) and AG sample (of Guallar playa lake) are located in the ephemeral salt-pan of the lakes (dotted line).

Pito playa; Fig. 1,P) has been studied exhaustively and taken as a model which can be applied to the other ones.

The Quaternary deposits that infill the lacustrine depressions are made up of detrital and evaporitic sediments and are up to 2 m thick. Gypsum, halite, thernardite, bloedite and mirabilite all form by evaporation in the lakes, but gypsum is the only one which is preserved at any depth in the sediment. It develops as lenticular crystals and concentrates mainly in the salt-pan zone. Sometimes gypsum nodules several centimetres in size, formed by clusters of lenticular crystals, can be found. The other evaporitic minerals are dissolved during rainy seasons.

Gypsiferous and detrital sediments show opposite distribution patterns. Detrital deposits originate from erosion taking place in areas adjoining the playa lakes (up to a few kilometres). They are concentrated towards the edges and the windward side of the playas and are mainly represented by quartz, clay minerals and carbonates. Illite, chlorite, smectite and kaolinite form the clay fraction present in the Quaternary sediments and the relative abundances depend on the composition of the Tertiary substrate.

Carbonates present in the lake sediments are calcite, dolomite and magnesite. Calcite is dominant on the surface: it has a detrital origin. At depth (some 20 cm), the dolomite content tends to be higher and magnesite may appear (Fig. 2a), both minerals showing evidence of being formed during early diagenesis. The highest magnesite concentration is found below the ephemeral salt-pan area where sometimes it may be the dominant carbonate. Magnesite shows a similar horizontal distribution pattern to that of gypsum, except for the top 20 cm of the sediment, where magnesite is usually not found. Exceptions to this are the occurrence of minor amounts of surficial magnesite in the lakes where the sediments are frequently brine-filled. Minor quantities of fibrous clays (sepiolite?), not detected in the Tertiary substrate have also been found in the salt-pan area. Carbonate and clay distributions reflect the early diagenetic interactions between interstitial brines and detrital sediments.

Magnesite formation—some remarks

Although magnesite has been found in all the lakes studied, Guallar playa lake (Fig. 1,G) has been taken as a model for the study of magnesite crystals due to the great abundance of this mineral found. In the ephemeral salt pan of this lake, magnesite reaches 50% of the water-insoluble fraction (carbonates + quartz + clays), which means approximately 15% of the sediment.

Magnesite seems to have originated through interstitial precipitation from phreatic brines. The reasons for suggesting this are the almost total absence of magnesium carbonate at the surface and the subhedral shape of the magnesite crystals (Fig. 2c) which grow displacively in dominantly siliciclastic mud with some carbonate content (30% on average). Detrital components are dominant at the surface but magnesite rhombs are progressively more abundant with depth (this has been established for the upper 40–

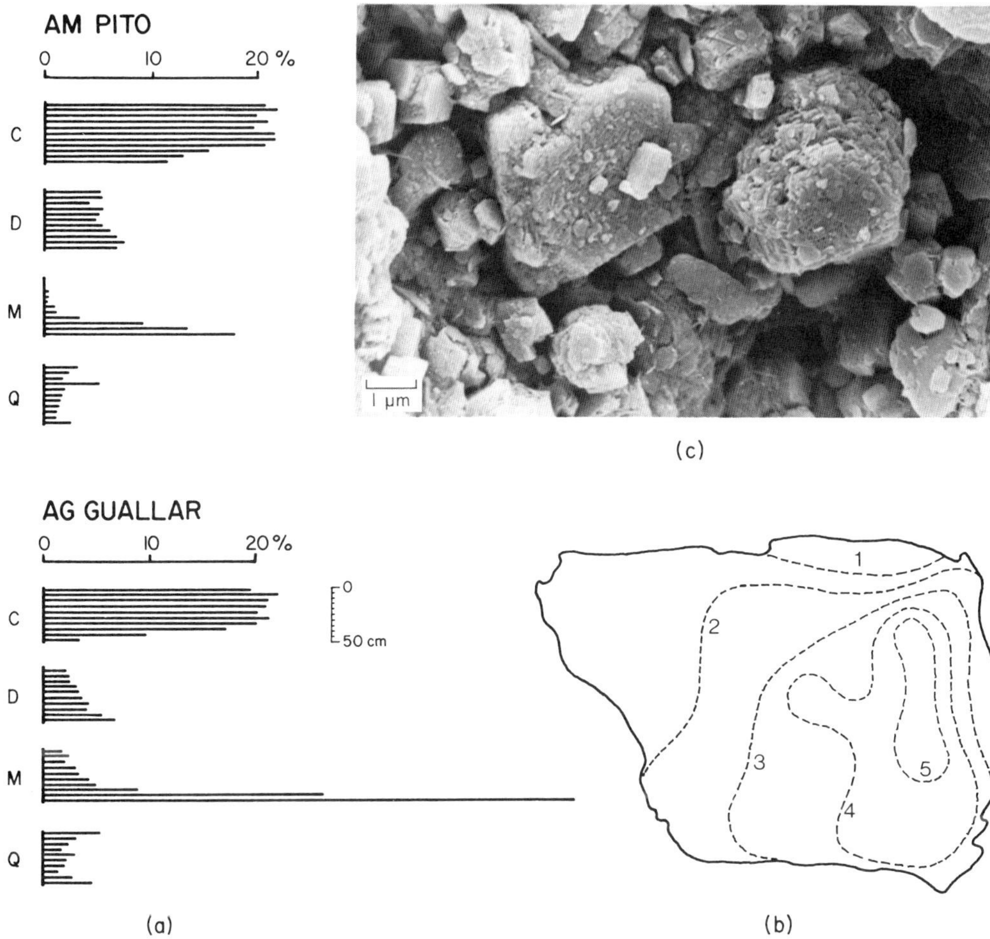

FIG. 2. (a) Distribution of minerals (carbonates and quartz) with depth at AM (Pito) and AG (Guallar) sampling points. Bars represent weight percentage of different minerals in the water-insoluble residue (C: calcite, D: dolomite, M: magnesite and Q: quartz). Samples have been taken at 5 cm intervals. (b) Map of magnesite concentration (as per cent) in Pito playa lake, at a depth of 40–45 cm. The contours were calculated using a microcomputer starting from an initial grid of 19 data points. (c) Scanning micrograph of magnesite crystals. Ultrasonic pre-treatment and sieving have been used to remove the clay fraction. Guallar playa lake, AG point at a depth of 40–45 cm.

50 cm). It is suggested that the magnesite could have precipitated interstitially as a response to the increase in the CO_2 and CO_3^{2-} activity due to bacterial decay of algal and cyanobacterial matter in a similar way to that described by Perthuisot (1974) in El Melah sebkha (Tunis), in the presence of strongly magnesic brines. Precipitation occurs in the ephemeral salt-pan zones (Fig. 2b) as they are the lowest areas of the playa lakes. These zones collect a greater amount of rain water. This causes cyanobacterial mats to develop frequently, leading to high organic (sapropelic) accumulation. During dry seasons these areas have surficial brines for a longer period subject to direct evaporation. This enables the brines to reach the post-halitic stage with a strong magnesian concentration at the end of summer.

In the Guallar playa lake these conditions are more obvious. For the last 10 years at least, a significant growth of cyanobacterial mats, coexisting in summer with magnesian brines, has been recorded (Pueyo 1978–79). Density and viscosity are high in these brines and water activity is low; such conditions favour magnesite formation and prevent nucleation of metastable hydrated magnesium carbonates, as suggested by Langmuir (1965) and Christ & Hostetler (1970).

References

CHRIST, C. L. & HOSTETLER, P. B. 1970. Studies in the system MgO-SiO_2-CO_2-H_2O. II: The activity product constant of magnesite. *American Journal of Science*, **268**, 439–53.

LANGMUIR, D. 1965. Stability of carbonates in the system MgO-CO_2-H_2O. *Journal of Geology*, **73**, 730–54.

PERTHUISOT, J. P. 1974. Les dépôts salins de la sebkha El Melah de Zarzis: Conditions et modalités de la sédimentation évaporitique. *Revue de Géographie Physique et de Géologie Dynamique*, (2) **16**, f. 2, 177–88.

PUEYO MUR, J. J. 1978–79. La precipitación evaporítica actual en las lagunas saladas del área: Bujaraloz, Sástago, Caspe, Alcañiz y Calanda (provincias de Zaragoza y Teruel). *Revista Instituto Investigaciones Geológicas. Diputación Provincial Barcelona*, **33**, 5–56.

—— & INGLÉS URPINELL, M. 1986. Substrate mineralogy, pore brine composition, and diagenetic processes in the playa lakes of Los Monegros and Bajo Aragón (Spain). *I International Symposium on Geochemistry of the Earth Surface. Granada (Spain), 16–22 March 1986*, in press.

QUIRANTES, J. 1965. Nota sobre las lagunas de Bujaraloz—Sástago. *Geographyca*, **12**, 30–4.

J. J. PUEYO MUR and M. INGLÉS URPINELL, Dep. de Geoquímica, Petrología y Prospección Geológica, Gran Via, 585, 08007 Barcelona, Spain.

Mixed-water dolomitization in a transgressive beach-ridge system, Eocene Catalan Basin, NE Spain

C. Taberner & C. Santisteban

SUMMARY: Dolomitization occurs at the top of alluvial fan deposits in the SE margin of the Eocene Catalan Basin, NE Spain. The dolomitized rocks occur just below erosion terraces which show a step-like disposition on the continental sediments and on which shallow marine facies (beach-ridge systems and reefs) are developed. Petrography shows that diagenesis in the continental sediments mainly originated under meteoric water influence, while in beach-ridge systems and reef carbonates there is evidence of early diagenesis in a marine environment. All the nearshore sediments (from alluvial fan to reefs) record a distinct geochemical pattern where a progressively more marine influence parallels the regional trend in facies across the ancient shoreline. Petrographic and geochemical data suggest that mixing of meteoric and marine waters in the sediments representing the continental-to-marine transition caused dolomitization at the top of the alluvial fan deposits. Unlike most mixing-zone dolomites, in this case mixing took place when shallow marine sedimentation took place on erosional surfaces causing the inflow of marine waters that mixed with meteoric porewaters.

Diagenetic dolomite occurs in several facies associations in the Vic area (SE margin of the Eocene Catalan Basin, Figs 1 and 2). Some of this dolomite formed during early diagenesis and has been recorded mainly in Eocene marine sediments, in a transitional zone between the continental and marine deposits of the SE sector of the studied area (Figs 2 and 3).

In this zone dolomitized sediments are generally confined to a narrow strip at the top of the continental sediments, interpreted as alluvial fan systems (Reguant 1967, Colombo 1980, Taberner 1982). The Middle Eocene transgression affected these deposits, so that nearshore sediments developed over the alluvial fan facies associations (Fig. 3). The basal sediments in the marine series are interpreted as transgressively-developed beach-ridge systems.

Field and textural features of rocks in the nearshore facies associations where dolomite has been found may be taken as evidence of dolomite development during early diagenesis prior to compaction. Petrography and preliminary geochemical results point out the possibility of mixing-zone dolomitization. This mechanism has been widely referred to as producing dolomitization in nearshore areas (Hanshaw *et al.* 1971, Badiozamani 1973, Land 1973a, b, Folk & Land 1975, Ward & Halley 1985, among others). The cases usually discussed in the literature are those regarding dolomitization of marine nearshore sediments by mixing-zone water. In this paper we show how the special features of the transgressive beach-ridge depositional system may have given rise to the dolomitization observed at the top of the alluvial fan sediments.

Sedimentary features

The beach-ridge systems

Levels interpreted as beach-ridge systems are intercalated with, or found at the top of, alluvial fan deposits in the southern and eastern margins of the studied area (Figs 1 and 2). They are Upper Lutetian–Lower Biarritzian in age, and their outcrops are arranged in a 50 km long narrow strip bounded by alluvial fan sediments and marine platform facies associations (Fig. 2). Beach system outcrops are distributed in three well-defined areas, corresponding to troughs bounded by NNW–SSE trending fractures, whose synsedimentary action during the Palaeogene sedimentation in this area has been referred to (Taberner 1982).

The position of beach levels are evidence of the location of ancient shorelines during the successive stages of the Eocene transgression. These sediments originated after the reworking of continental supplies, and the areas where they are best recorded coincide with those of greater thicknesses of alluvial fan deposits.

Beach deposits in the area studied are arranged in tabular siliciclastic bodies. Their lengths along the ancient shoreline can be in tens of kilometres and widths are up to 2 km. Beach levels are up to 4 m thick (when successive beach units are amalgamated). Two distinct groups of beach sediments in the Vic area have been distinguished: (1) constructive depositional beaches and (2) erosional beaches originating from marine reworking of alluvial supplies.

From: MARSHALL, J. D. (ed.), 1987, *Diagenesis of Sedimentary Sequences*, Geological Society Special Publication No. 36, pp. 123–139.

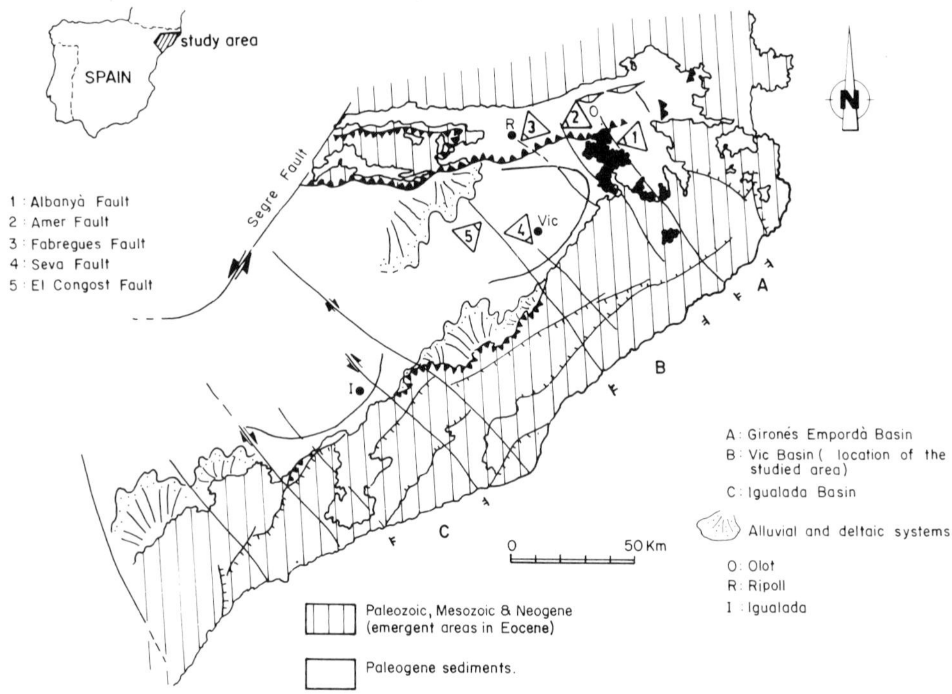

FIG. 1. Geological features of the Eocene Catalan Basin. The NNW–SSE fault system controlled the location of minor basins during the Palaeogene as well as the distribution of deposits along the southeastern sector of the Ebro Depression. The Vic sedimentation area (B) was bounded by El Congost and Amer faults. The activity of these fractures during sedimentation brought about a more complete record of the Middle Eocene transgressive stage in the southeastern margin of the Basin. The thickest sequences of beach sediments were developed in trough areas bounded by NNW–SSE trending faults.

Most beach deposits in the study area correspond to the constructive depositional type. Their external shapes and internal structures may be best compared to the beach-ridge systems described by Carter (1986) in Northern Ireland. Depositional beaches are closely related to reef carbonates, corresponding to slightly deeper sedimentation zones. In some cases, beach levels prograded over coral-reefs, but reefs are also present on depositional beach systems as a result of the transgression of the depositional system.

Erosional beaches are only represented by up to 50 cm thick lag deposits composed of cobbles and boulders, within a coarse-grained sandstone matrix. These deposits formed over up to 20 m long erosional terraces on top of the alluvial fan continental sediments.

Both kinds of beaches characterize the transition between continental and marine deposits during the Eocene (Lutetian–Biarritzian) marine transgression of the Catalan Basin. Beach units occur in a step-like arrangement, each developing, whether erosional or depositional, in the proximal areas on a horizontal to slightly seaward-dipping terrace (Figs 3 and 4), above the continental deposits. The manner in which these erosional surfaces were formed indicates dispersion of alluvial fan deposits and coincides with the development of erosional beaches. Thus, erosional terraces stemmed from the reworking of continental deposits while erosional beaches represent the lag products of reworking.

The step-like arrangement of different beach levels indicates the combined effect of transgression with tectonic pulses, causing deepening in the sedimentation area (Fig. 4). This process would proceed in the following manner (Taberner 1982):

(1) Rapid sinking of the sedimentation area and submersion of the continental deposits, thereby developing an erosional terrace from the effects of wave action.
(2) Stabilization and progradation of a depositional beach over the erosional terrace surface.

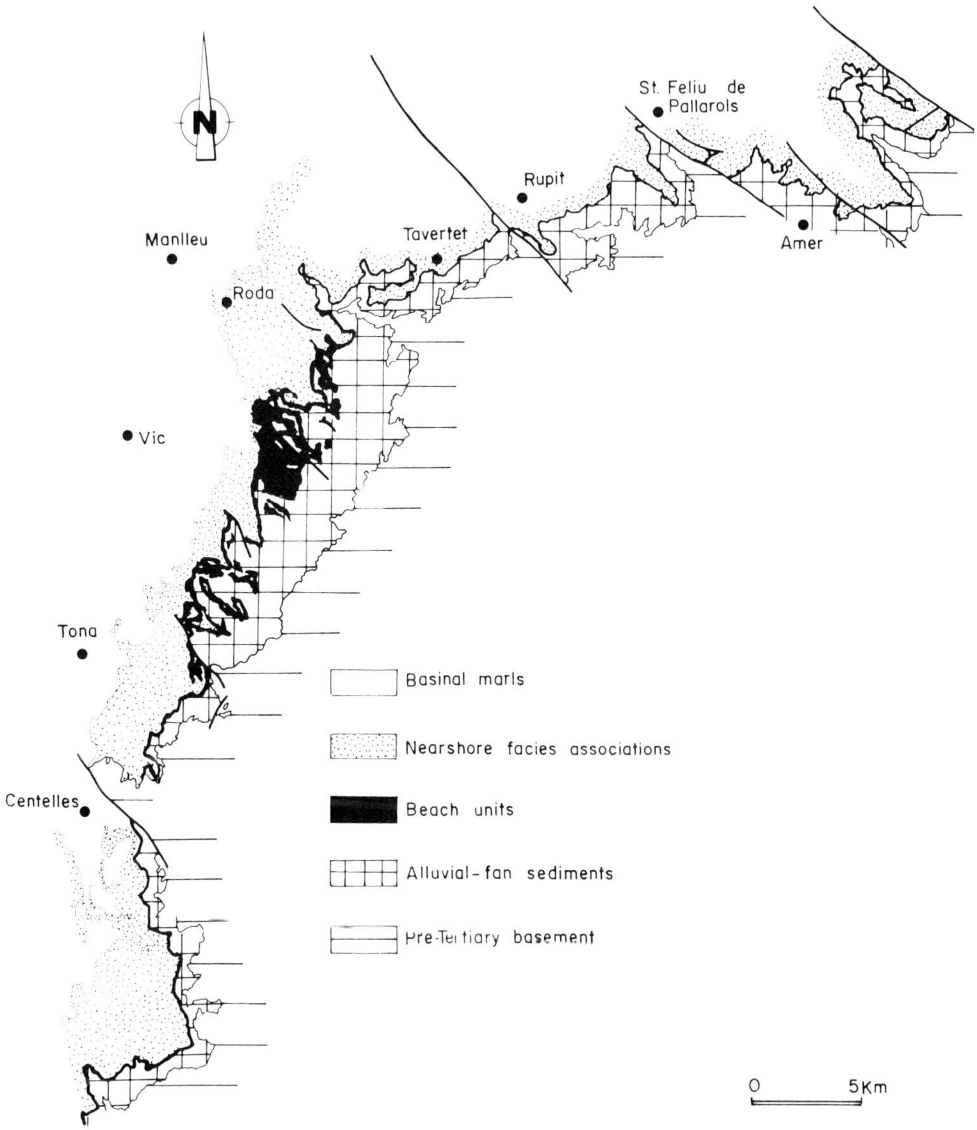

FIG. 2. Location of beach deposit outcrops in the southeastern margin of the Catalan Basin (Vic area).

Reef buildups associated with the beach systems

Beach sediments are interbedded with reef carbonates and offshore marls in the distal regions. Reefs are found overlying beach deposits: they first developed in the frontal part of beach-ridge units, and progressively overlapped them. The transgressive migration of reefs parallels that of beach-ridge systems (Fig. 3). This suggests that: (1) reefs grew on previously drowned beaches, (2) they developed in front of and sometimes simultaneously with active beach systems, (3) reefs were dynamically related to beach siliciclastic sediments, and (4) biological communities, as well as siliciclastic sedimentation systems, showed similar trends during the marine transgression. The relationships between reefs and beach levels are discussed in detail in Santisteban & Taberner (in press).

Reef buildups of the studied area consist of fringing-reefs made up of corals and coralline red algae. Taberner & Bosence (1985) found these coralline red algae indicative of shaded zones where turbid waters restricted light passage into

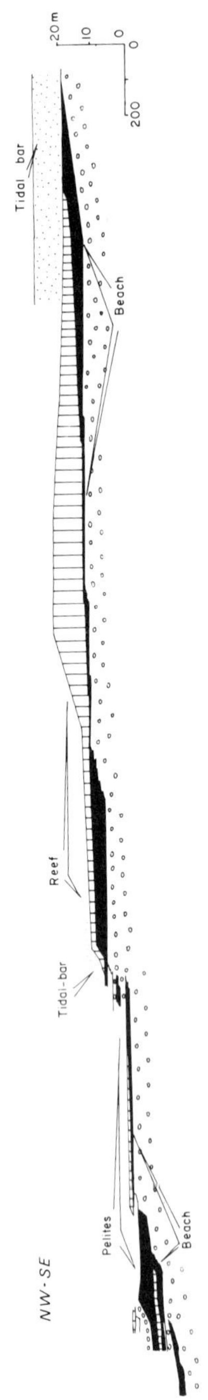

FIG. 3. Transverse cross-section illustrating the distribution of sediments related to the continental-marine transition in the eastern margin of the Vic sedimentation area (Centelles sector). Beach sediments (black) are arranged in a step-like transgressive disposition over the alluvial fan deposits (dotted) on erosion terraces. Coralgal reefs (vertical stripes) developed on previously drowned beach levels.

reef developing areas. These turbid waters were probably due to active siliciclastic sedimentation within the beach-ridge systems. Therefore, reefs would coexist with siliciclastic beach development near the shore.

Diagenetic features

The sediments we have been dealing with show evidence of meteoric as well as marine influence during early diagenesis.

This section describes the diagenetic features that characterize each of the facies associations making up the transgressive shore systems. As basic units in these systems, we have considered: (1) alluvial fan sediments, (2) transgressive beach systems, and (3) coral and red-algal reefs, developed in front of and on the beach ridges.

Alluvial fan sediments

A haematitic matrix and haematitic coatings surrounding siliciclastic grains are characteristic features of the alluvial fan sediments of the Vic area. Iron oxides were probably derived from iron-containing minerals; probably biotites and amphiboles, which are the most abundant mafic components in the likely source area (granitic and metamorphic rocks). The transformation of biotites into iron oxides is common in the studied continental rocks.

The iron oxides are likely to have formed once sediments were stabilized and buried in a phreatic environment with an oscillating water-table, in a similar way to that suggested by Walker & Waugh (1973), Walker (1976) and Glennie *et al.* (1978). However, iron coatings can also develop in meteoric phreatic conditions (Kessler 1978) or in vadose conditions (Turner 1980). We have not found any evidence to favour any individual origin for these coatings.

Primary porosity in these sediments was probably negligible: the haematitic matrix filling nearly all interparticle spaces. All residual pores are now filled in by sparry calcite cement, probably originating from precipitation in meteoric water conditions.

Alluvial fan sediments are grey at the top. These grey sediments show sedimentary structures similar to those in the red ones. The upper contact of the grey sediments corresponds to the alluvial fan top, whereas their lower contact is irregular and cross-cuts the sedimentary structures of the red alluvial fan deposits. In some cases there are patches of red sediments surrounded by grey ones at the top of the alluvial fan.

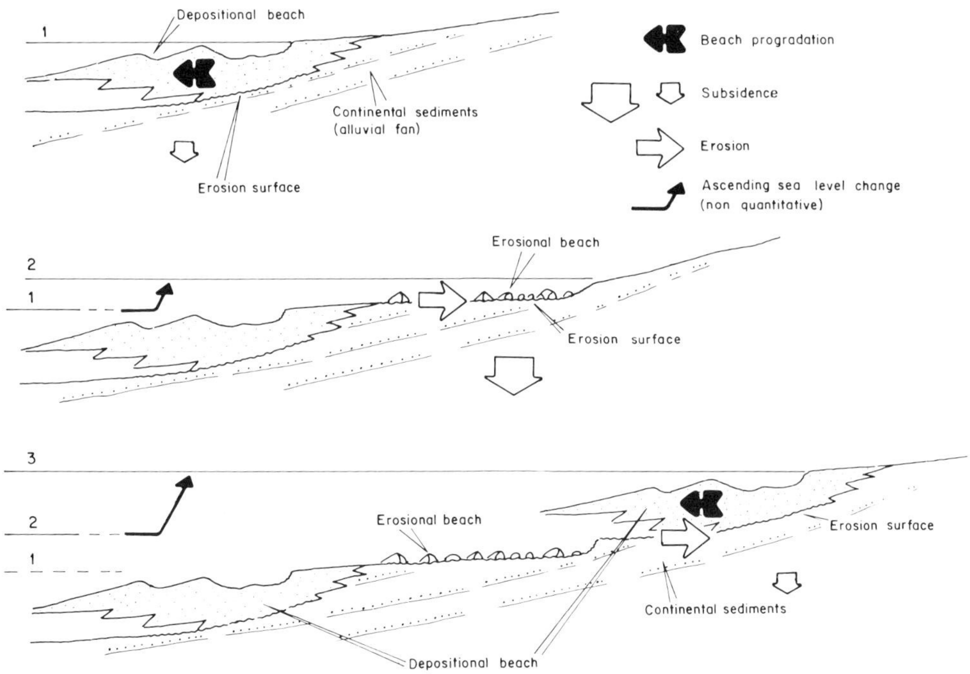

FIG. 4. Diagrams showing the development and relationships between depositional and erosional beaches as a result of differences in subsidence rate during the Lutetian–Biarritzian transgressive stage. The step-like disposition of these beach systems suggests that besides transgression, tectonic activity controlled their occurrence.

Carbonaceous remains are commonly preserved in the grey sediments. This fact, together with the grey colours themselves, may be evidence of the influence of a reducing environment. Iron oxides originating in meteoric phreatic conditions would have been reduced in this zone. The mean thickness of the reduced material ranges from 0.5 to 1 m.

The matrix in the reduced section has less terrigenous components than in the oxidized equivalents. This may be related to a leaching effect. Dolomite rhombohedra are abundant in the reduced material, increasing in abundance from the lower to the upper part of the section, but never exceeding 10% of the total rock volume. At the top, dolomite rhombohedra are the main component of the matrix between the siliciclastic grains (Fig. 5a).

The dolomite crystals are from 1 to 325 μm in size. The largest ones (> 125 μm) show the most perfect shapes and are usually zoned (Fig. 5b, c): They have dusty nuclei and cleaner well-zoned overgrowths, and usually include traces of the original clay matrix (Fig. 5e). This suggests poikilitic growth in an unlithified matrix. The largest crystals are generally found isolated in lime and clay mud surrounding the siliciclastic components.

Up to 125 μm, dolomite rhombohedra are limpid and euhedral, except for the smallest crystals, where a form is difficult to assess. Very small (up to 10 μm) crystals are arranged in homogeneous aggregates of sucrosic aspect (Fig. 5a) surrounding the larger ones.

Dolomite rhombohedra have been dissolved in the interpenetration contacts with siliciclastic components (Fig. 5b, c). Dolomites also accumulate along pressure-solution surfaces which, in turn, affected them (Fig. 5d). These facts suggest that dolomitization occurred before the compaction of sediment during early burial.

The transgressive beach systems

Deposits of the transgressive beach systems show evidence of cementation through the influence of marine waters during early diagenetic stages. However, they also have textures that may be interpreted as having originated under the influence of meteoric waters, also during early diagenesis.

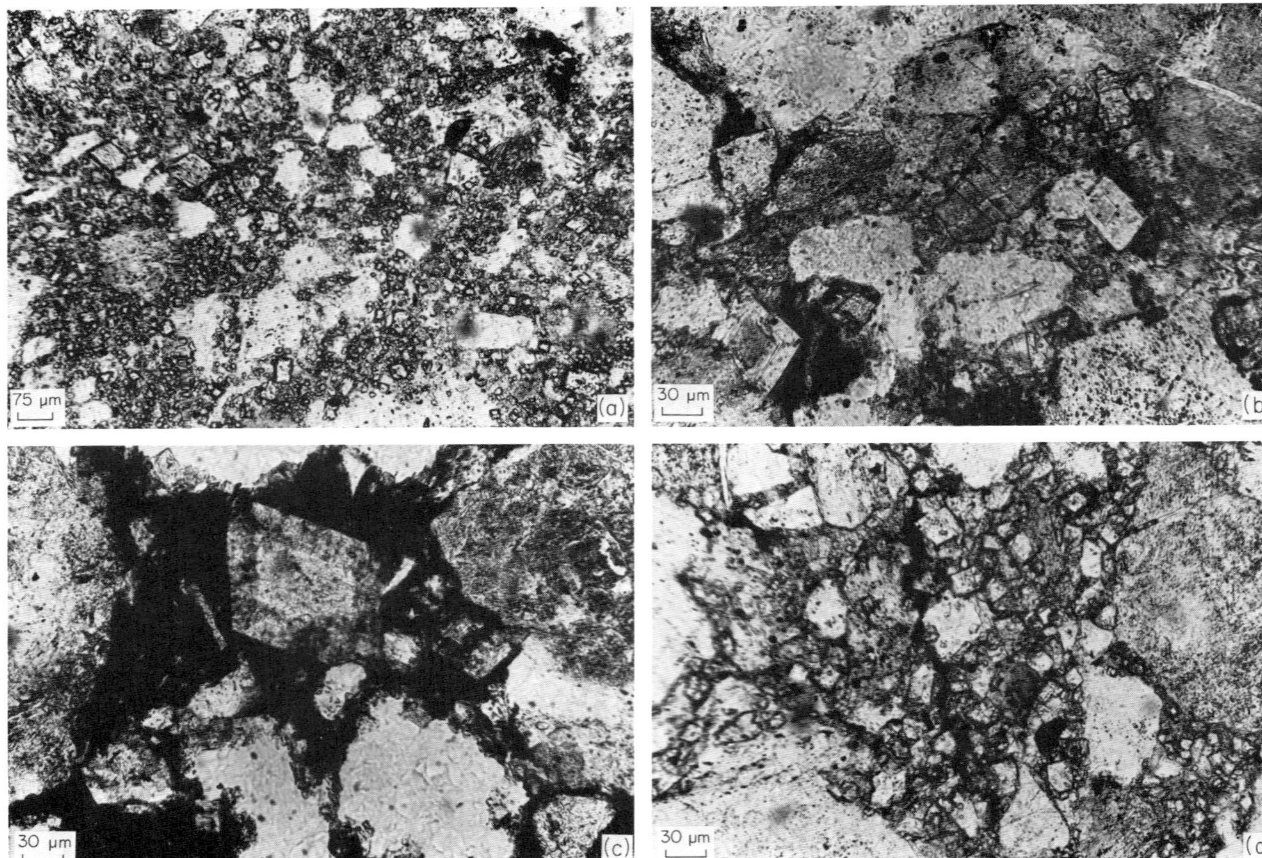

FIG. 5. (a) Dolomitized matrix in the decoloured levels at the top of alluvial-fan systems. Euhedral shapes of dolomite-rhombohedra are best displayed in the largest crystals. (b) Euhedral dolomite rhombs showing dusty nuclei and overgrowths of more limpid dolomite. Note the interpenetration contact with a siliciclastic grain in the dolomite crystal at lower left. Decoloured level. (c) Overgrowth in a large dolomite crystal. This crystal grew within a haematitic matrix and shows an interpenetration contact with the siliciclastic grain below. Both facts show its development during early diagenetic stages. Decoloured level. (d) Accumulation of dolomite rhombs in later diagenetic pressure-solution surfaces. Decoloured level. (e) Small dolomite rhombohedron partially enclosing clay minerals.

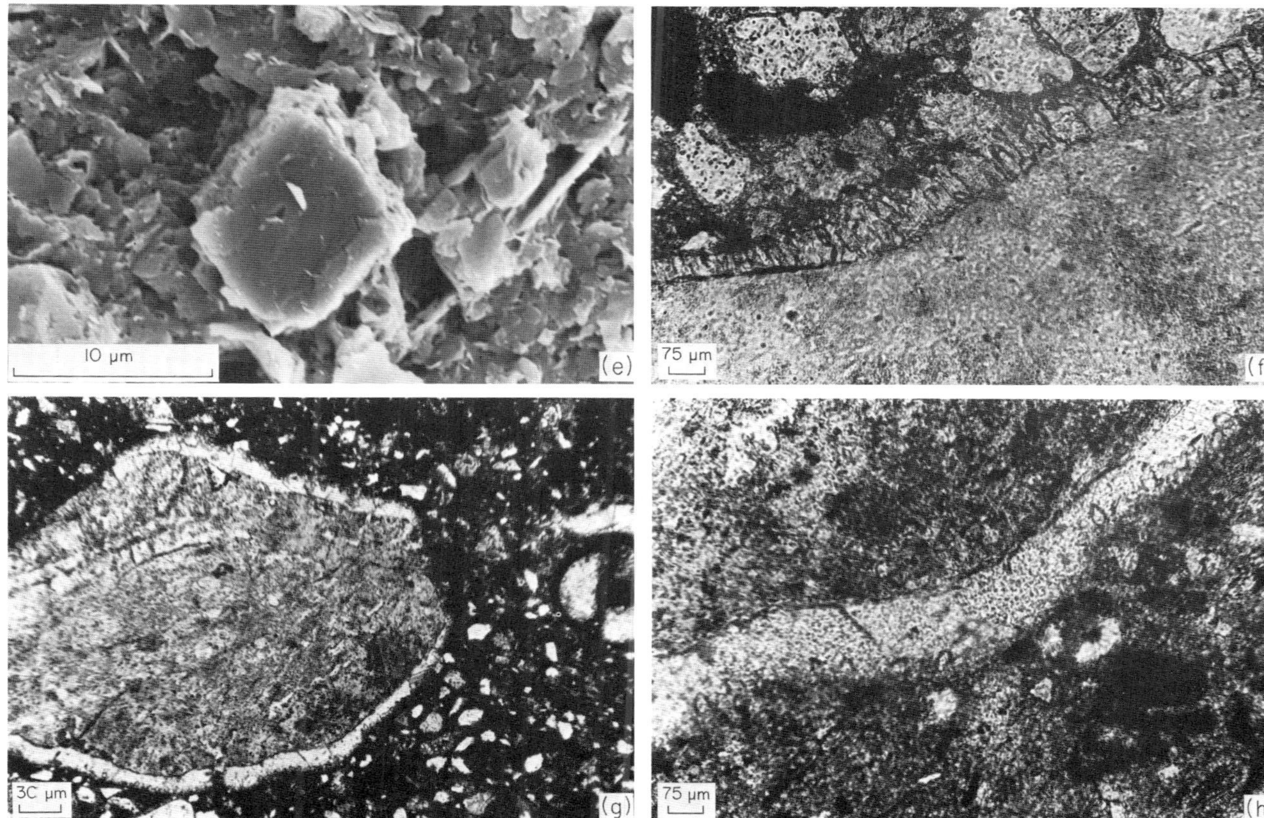

Clay minerals are represented by the plates that may be seen at the crystal surface. Decoloured level. (f) Stubby cement fringe with blunt crystal terminations. It might correspond to a HMC submarine cement precursor. Beach deposits. (g) Cavity between a potassium feldspar grain and the matrix in a beach rock. The cavity surrounds the grain and is outlined by a micrite envelope. Two generations of cements developed in the cavity. The preservation of its shape points to an infilling during early diagenesis before compaction. The existence of this cavity may be taken as evidence of dissolution of a HMC or aragonite cement rim. (h) Detail of the previous photograph. The dissolution cavity is filled in by two generations of cement, the first having a drusy aspect. This view corresponds to an area at the lower right part of the cavity in the previous picture.

Palisade fringes of low-magnesium calcite crystals developed around siliciclastic grains. These crystals are 35 μm wide and up to 75 μm long. They are prismatic in shape (Fig. 5f); and show blunted crystal terminations. Micritic envelopes, surround these cements and probably contributed to their preservation.

The above-described cements were probably not originally low-magnesium calcite. This may be inferred from the fact that they show evidence of having been dissolved during early diagenesis. This is more obvious in cases where a micritic envelope formed around the fringe of cement. The space previously occupied by the fringe is now represented by a microsparitic cement, displaying two infilling generations, the first with a drusy aspect (Fig. 5g, h).

The palisade cements possibly were composed originally of aragonite or high-magnesium calcite and therefore they may have formed in a submarine environment. Their textural features, dimensions and similarity to stubby marine cements described in recent environments by Schroeder (1972), James *et al.* (1976) and James & Ginsburg (1979), lead us to assume that they would have originally been high-magnesium calcites. Their dissolution or neomorphism to low-magnesium calcite is most likely to have occurred under the influence of meteoric waters.

Scattered dolomite rhombohedra have also been found growing in the matrix between grains. Their textural features and dimensions are similar to those described above in the decoloured sediments at the top of the alluvial fan deposits. However, they are much less abundant.

Coral and red algal buildups

Reef deposits show evidence of having been early-cemented under the influence of marine waters; but again during early diagenesis these cements would have been influenced by the movement of meteoric waters.

Two kinds of cements of likely marine origin are common in the reef carbonates: (a) micrite cements, which were possibly originally high-magnesium calcite, and (b) isopachous fibrous cements and botryoidal cements, both of which were probably originally of aragonitic composition.

The reef sediments are now composed of low magnesium calcite, but textural features suggest that the micrite matrix was cemented by an early high magnesium calcite cement: (1) borings affecting the micrite matrix as well as crusts of red algae (Taberner & Bosence 1985) are common (Fig. 6b); (2) the matrix shows different stages of boring, infilling and cementation (Taberner & Bosence 1985) seen in the heterogeneity of colours and components in the different generations of borings. Usually, the first matrix generations may be distinguished by darker colours and greater percentages of siliciclastic components, while the younger ones are lighter and purer. From this we can deduce that cement and/or internal sediment increase from the first to the later generations of borings. This leads us to the hypothesis that the cement responsible for cementation of the micrite matrix of these carbonates had high-magnesium calcite mineralogy, analogous to submarine micrite cements from recent carbonate platforms (Bathurst 1971 and James & Ginsburg 1979).

Continuous isopachous fringes of fibrous crystals are occasionally found infilling interparticle cavities (i.e. forams, bivalves, etc.). Crystals vary between 10 and 100 μm in length and are up to 1 μm in width. They are usually included in larger low-magnesium calcite sparite crystals (Fig. 6a) and occasionally their shapes are outlined by micrite relics in the sparry calcite. Neomorphic processes enabled the shapes to be preserved in the sparite cements. The occurrence of these processes lead us to assume that the previous mineralogies might have been high-magnesium calcite or aragonite. They particularly resemble recent aragonite cements described by Macintyre *et al.* (1968), Shinn (1969) and James *et al.* (1976). The fringes are affected by borings, and small fringes of the same kind of cement may develop in boring cavities. From this we can assume they are of marine origin.

Former acicular cements are also present as radial aggregates of crystals nucleated at one point (Fig. 6c). They are particularly common in the interlaminar porosities of red algal crusts (Taberner & Bosence 1985), where coalescent aggregates are found (Fig. 6d).

Dimensions of acicular crystals composing the fan-like aggregates vary from 100 to 600 μm. They are up to 7 μm wide, and usually the largest crystals are found in the most central part of the fan-like structure. The greatest variations in length of the acicular crystals are found in cases where the fan-like structures are coalescent. Isolated aggregates growing in free conditions tend to have a semicircular external shape and the crystals tend to be the same size. A hemispheric shape may be deduced for these aggregates from the different forms they show in thin-section (Fig. 6c, d).

There is evidence that this kind of cement developed in submarine conditions: it is bored together with hard skeletal components (coralline red algal crusts) and is found in cavities that have also been filled in with lime mud, cemented in the marine environment. This type of cement

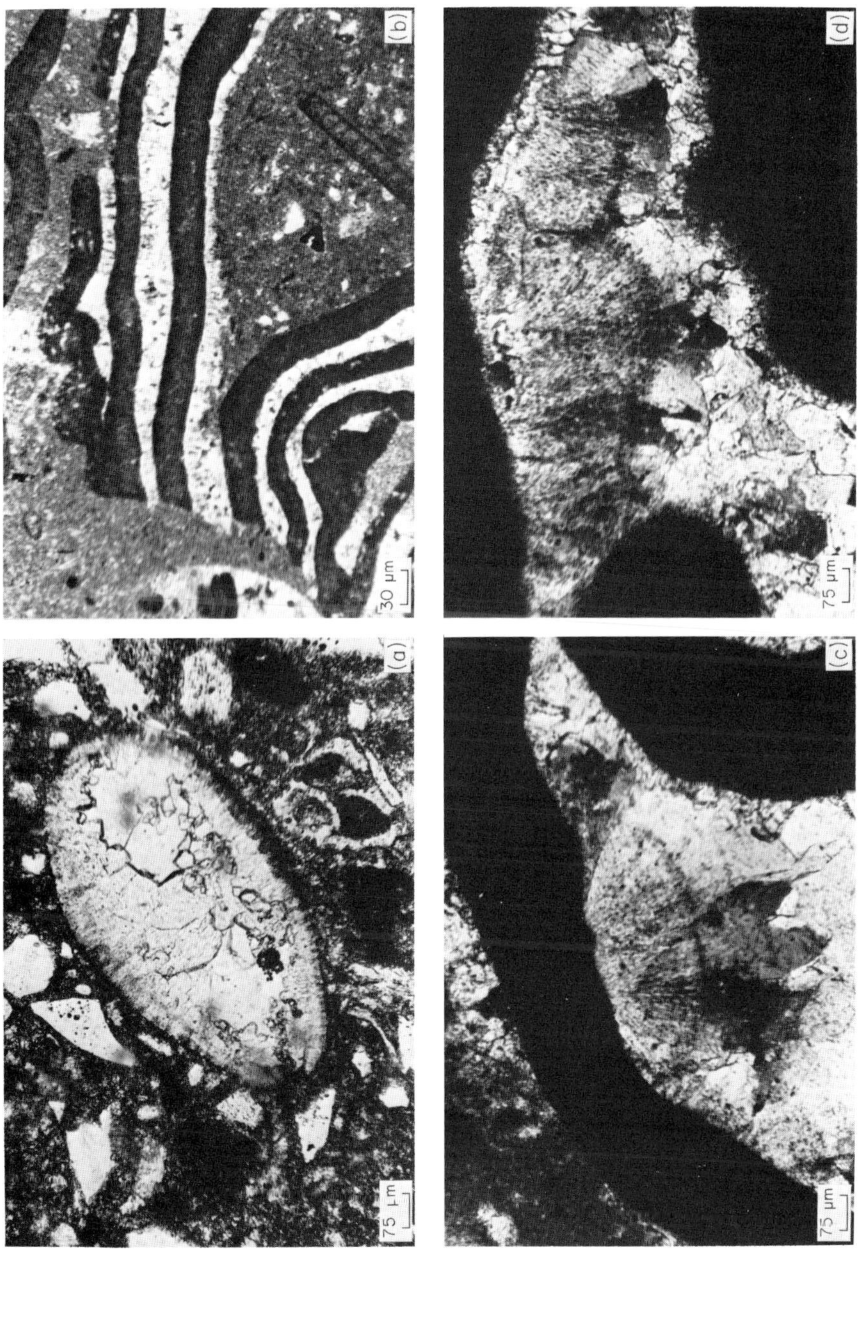

FIG. 6. (a) Fibrous crystal fringe infilling skeletal intraparticle porosity in a reef buildup. These crystals have been neomorphosed into LMC sparite. The fibrous crystals outlined by micrite relics are interpreted as evidence of a fibrous aragonite submarine cement precursor. (b) Coralline algal crusts in a reef buildup. Borings (left) affecting the matrix as well as the crusts may be taken as evidence of early cementation by HMC micrite cement in the submarine environment. (c) Isolated fan-like structure of acicular-radial crystals, neomorphosed into LMC sparite cements, developed in the primary porosity between crusts of coralline algae. The acicular crystals are outlined by micrite relics. These cements are interpreted as representing submarine botryoidal aragonite precursors. (d) Coalescent fans of acicular crystals, interpreted as submarine botryoidal aragonite cements, neomorphosed into LMC sparry calcite. Algal crust in a reef buildup.

is common among marine carbonate deposits from the studied area, while it has never been found in the continental sediments.

The shapes and arrangements of the cements described here are similar to those of recent botryoidal aragonitic cements described by Schroeder (1973), James *et al.* (1975) and Ginsburg & James (1976). However, in this case original crystals forming the radial aggregates were neomorphosed to low-magnesium sparry calcite which inherited the radial arrangement, and undulose extinction of the precursor (Fig. 6c, d). The undulose extinction might reflect the retention of subcrystalline domains of possible primary origin as proposed by Sandberg (1985). In our case the subcrystalline domains correspond to the original radial-acicular crystal fabric, observed through the presence of micrite relics. Acicular crystals arranged in a fan-like disposition would represent former submarine botryoidal aragonite cements. Neomorphism into low-magnesium sparry calcite might be related to later influence of meteoric waters.

Isotopic composition and trace element distribution

The diagenetic evolution of these nearshore deposits will be better constrained on the basis of geochemical results, our preliminary results are presented here.

Carbon and oxygen isotopic composition and trace element concentrations have been determined. The most useful trace element data are strontium, manganese, zinc and iron (Fe^{2+}). Fe^{2+} levels have been determined using colorimetric (1,10-Phenanthroline) methods, and the other elements through atomic absorption spectrophotometry of the acid (cold 10% HCl) soluble fraction. A Link System energy dispersive X-ray microanalysis system and JEOL JSM-840 SEM have been used to detect trace elements in individual crystals.

Oxygen and carbon isotopic composition has been determined using a VG SIRA-9 mass spectrometer. Extraction of CO_2 from carbonates was carried out according to the method described by McCrea (1950). A $\delta^{18}O$ correction factor of $-0.8‰$ has been applied to compare dolomite with calcite results. The dolomite/calcite ratio has been considered when correcting for the possible differences in phosphoric acid fractionation of dolomite and calcite in the studied samples. Results are expressed with respect to PDB standard.

Individual petrographic phases were generally too small to be separated, so in general analyses have been made of bulk samples. Microsampling was possible for sparitic cements. Separation of dolomite crystals has not been possible in the dolomitized rocks due to their small size, proportion and scattered distribution in the matrix. However, samples have been enriched in dolomite by means of selective dissolution in cold 1 M acetic acid for half an hour. Original dolomite/calcite ratios are about 5% (on average) and after treatment, about 60%.

For comparison sampling has also been carried out in a younger (Priabonian) barrier-reef sequence, where 60 m thick reef carbonates represent the record of almost continuous marine sedimentation. Petrographic and textural features in these deposits point out that early diagenesis occurred in a marine environment. Results from the barrier-reef deposits have been taken to represent the geochemical trends in deposits mainly influenced by marine diagenesis. Samples from this reef have been studied using the same methods as for rocks from the nearshore systems.

Trace elements

The trace element values for beach-ridge systems and associated sediments (BS), as well as in barrier-reef carbonates (R) are shown in Figs 7 and 8. These figures show the behaviour of strontium, manganese and iron (Fe^{2+}) contents. These elements may give some idea of the influence of meteoric waters (high Fe^{2+} and manganese contents) or a marine influence (high strontium values) (Veizer 1983).

Manganese and iron show good correlation in the samples of the depositional beach systems, where the highest concentrations of both elements have been found. The lowest manganese and iron contents correspond to samples from dolomitized decoloured levels and to dolomite-enriched samples. An increasing general trend from decoloured levels towards coralgal reefs can be seen for both Mn and Fe^{2+} (Fig. 7c, d).

Strontium enables discrimination of two well-defined fields representing the beach system facies associations and barrier-reef samples (Fig. 7a, b).

It is also interesting to note that the highest values in zinc correspond to the nearshore sediments (100–170 ppm), while in the barrier-reef rocks this element is usually not detected.

Sodium values in bulk samples of nearshore facies associations range from 284 to 340 ppm. However, sodium presence has been easily detected in dolomite crystals through energy dispersive microanalysis. This means that sodium contents in dolomite crystals may be much higher.

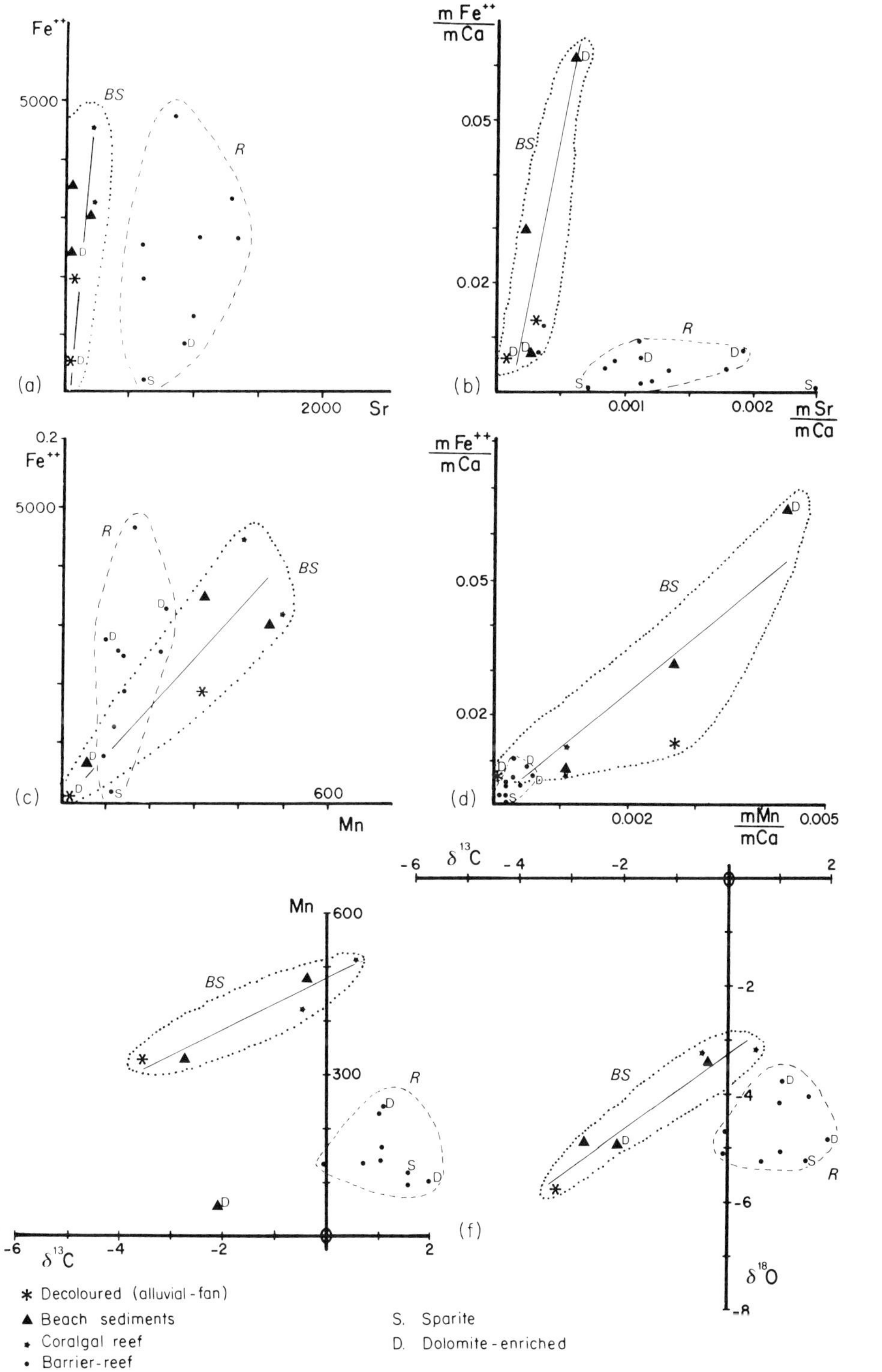

FIG. 7. Binary diagrams showing the geochemical pattern in the samples from the Vic area. BS: Beach-ridge systems; R: Barrier-reef. Values in (a) and (c) are in ppm. In (b) and (d) Fe^{2+}, Sr and Mn values are expressed in relation to Ca. As analyses were performed on bulk samples these ratios may give a clearer idea of the real variation in the carbonate fraction.

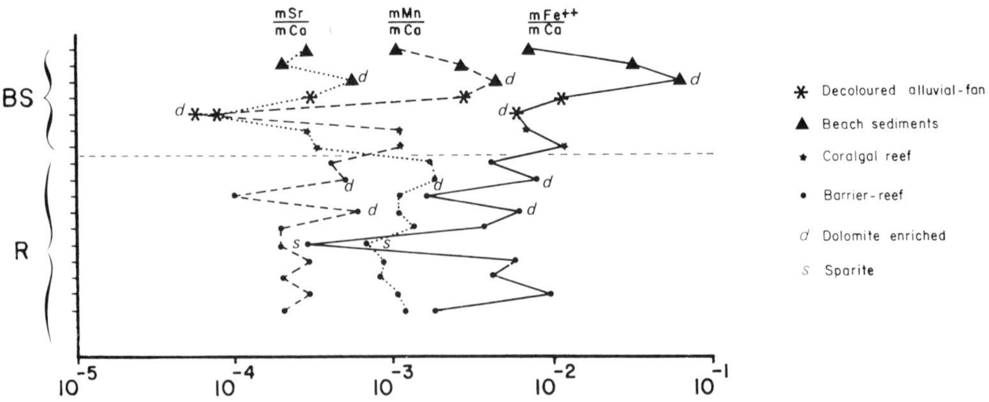

FIG. 8. Distribution of the molar ratio values of Sr, Mn and Fe^{2+} with respect to Ca in the samples analysed in the Vic area. BS: Beach-ridge systems; R: Barrier-reef. Variations in beach system molar ratios differ from those in the barrier-reef samples. Sr/Ca values are much lower in the beach rocks, while Mn/Ca and Fe^{2+}/Ca are higher. These relationships point to a meteoric influence in beach sediments and a marine signature in barrier-reef deposits.

Our trace element data may be interpreted in terms of variation in marine or meteoric influence using the approach outlined by Veizer (1983). The high manganese, iron and zinc levels and low strontium contents in the beach-system rocks may be taken as evidence of meteoric water influence. The sodium concentrations in dolomite crystals may point to a marine influence in the dolomitization process, however, Busenberg & Plummer (1985) question the use of sodium as a marker for marine conditions. The trace element distributions (Fig. 7) clearly demonstrate greater marine influence in the barrier-reef samples, where strontium values may be up to 2100 ppm.

Isotopic ratios

Isotopic ratios in samples from the beach sediments (BS), as well as from the barrier-reef carbonates (R) are shown in Fig. 7(f). $\delta^{13}C$‰PDB against manganese contents have also been plotted (Fig. 7e). These variables have been chosen as they can both be markers of meteoric influence ($\delta^{13}C$‰PDB negative values, Land *et al.* 1975, and Anderson & Arthur 1983, and high Mn contents, Veizer 1983). In our samples, the diagram in Fig. 7(e) enables a good discrimination between sample sets.

Isotopic values are in general negative. Positive $\delta^{13}C$ values have been found only in samples from the barrier-reef and in reefs related to the beach systems (Fig. 7e, f). The samples from the barrier-reef plot as a cluster (Fig. 7f) whilst those from the beach systems show a linear relationship: $\delta^{13}C$‰PDB values range from −3.5‰ to +0.6‰ as $\delta^{18}O$‰PDB ranges from −5.8‰ to −3.2‰. The lowest values in oxygen and carbon correspond to the decoloured facies, and there is a progressive shifting towards heavier values in more marine samples (reefs). The trend in the isotopic values (Fig. 7f) corresponds to the facies succession from continental to marine sediments in the beach system. This is also reflected in the diagram of Fig. 7(e), where manganese and $\delta^{13}C$ ratios show a good correlation in the beach system deposits.

The heavy $\delta^{13}C$ values recorded from the reefs associated with beach levels and in reefs at the top of the marine series may be taken as being a primary marine signature. Lighter $\delta^{18}O$ and negative $\delta^{13}C$ values may represent the influence of meteoric water (Land *et al.* 1975 and Anderson & Arthur 1983). The oxygen isotopic values might seem to be too light to be primary marine values. It is interesting to note that Renard *et al.* (1986) give $\delta^{18}O$ values between −2‰ and −2.5‰ for Eocene pelagic carbonates in North Atlantic sites. General values for pelagic carbonates proposed by Hudson (1977) are from 0‰ to +2‰. Thus it would seem that isotopic values in marine zones might have been lower during the Eocene in areas connected to the Atlantic such as in the case of the Eocene Catalan Basin.

Mechanism of dolomitization

Dolomitization is a widely reported process. One may find compilations in Land (1980, 1983) and in Machel & Mountjoy (1986). One of the frequently suggested mechanisms is for dolomitization in zones of mixed marine and meteoric water. We suggest that this is the mechanism for dolomitization in our rocks based on the whole of the petrologic, diagenetic and geochemical features observed in the sediments from the continental to marine transition. These features are:

(1) The dolomitized sediments are laterally and vertically restricted to sediments representing the continental to marine boundary.
(2) The dolomitized zone coincides with the decoloured zone at the top of the alluvial-fan sediments. This zone is likely to have been decoloured through the influence of a reducing environment.
(3) Evidence of cementation in a marine diagenetic environment in the overlying sediments (beach and reef levels) implies that there may have been a transition zone with mixed marine and continental influences in the upper part of the continental succession.
(4) Dolomite developed during early diagenetic stages before compaction. Dolomite crystals show interpenetration contacts with siliciclastic grains, and are affected by pressure-solution surfaces (Fig. 5b, c, and d).
(5) Dolomite rhombohedra are usually limpid although occasionally they have a nucleus with matrix-inclusions. This kind of dolomite, with early stages being inclusion rich and later stages of zoned growth, has been found in other dolomites thought to have formed in mixing-zones (Folk & Land 1975, Land *et al.* 1975, Land 1983, Ward & Halley 1985 and Machel & Mountjoy 1986).
(6) Trace element contents (high Mn, Zn and Fe^{2+} and low Sr contents) in the sediments of the beach system might be evidence of meteoric influence in all the facies. According to Dunham & Olson (1980) and Veizer (1983), low strontium contents are related to solutions whose Sr^{2+}/Ca^{2+} ratios are lower than that of sea water. Land (1980) assumes that the low values in strontium in ancient dolomites may be related to schizohaline environments. High values in the other elements might show the influence of meteoric waters (Veizer 1983).
(7) Isotopic compositions for $\delta^{18}O‰PDB$ and $\delta^{13}C‰PDB$ are similar to the values quoted by various authors for mixing-zone dolomites (Land *et al.* 1975, Dunham & Olson 1980 and Zhao & Fairchild, *this volume*). However, it seems that isotopic values may not be conclusive as Mattes & Mountjoy (1980) have found values in the literature ranging in $\delta^{13}C‰PDB$ from −9‰ to +4‰ (about −2.5‰ on average) and ranging in $\delta^{18}O‰PDB$ from −5‰ to +3.5‰ (−2‰ to −2.5‰ on average).

There is no doubt that these dolomites were formed during early diagenetic stages. This can be seen from their textural features and also from the fact that there is a clear sedimentological control in their distribution.

Taking into account the isotopic values and trace element contents, we do not think that they could have developed by burial-related mechanisms or be related to hypersaline conditions. If these dolomites had originated from burial, where brines would be related to formation waters or come from basinal shales, there would be a greater development of dolomites in the basinal areas, (Dunham & Olson 1980); in that case, the calcite-dolomite distribution would be the opposite to the present one. If these dolomites were related to a hypersaline environment, trace element distribution would be different, isotopic ratios would be heavier and there would also be evidence of simultaneous precipitation of evaporite minerals.

We suggest that these dolomites originated in a mixed water environment in nearshore areas where marine waters invaded continental sediments with existing meteoric porewaters. The necessary Mg^{2+} supply would come from marine waters, and dolomite formation would be favoured by the dilution of marine waters by the meteoric ones (Badiozamani 1973, Land 1973a, b and Folk & Land 1975). In this way the amount of dissolved salts in marine waters would have diminished, and, according to Baker & Kastner (1981) made it possible to overcome the inhibiting effect of seawater sulphate.

Dolomite isotopic compositions are lighter than those found in the mixing zone dolomites studied by Zhao & Fairchild (*this volume*). However the shift towards heavier values in $\delta^{13}C$ and $\delta^{18}O$ from the decoloured sediments to the reefs associated with beach-levels gives the idea of a seaward-decreasing influence of meteoric waters both in lateral and vertical directions. The progressive increase towards heavier isotopic values is considered to reflect flux gradients in meteoric aquifers flowing in marine-influenced environments (Land *et al.* 1975 and Meyers & Lohmann 1985). It can be assumed that there was a flux of fresh ground waters from the subaerially exposed areas of the alluvial fan towards the marine areas (Fig. 9). However, the inflow of marine waters in the Lutetian Biarritzian beach-ridge systems would be downwards because of the transgressive disposition of the marine facies associations over the continental ones (Figs 4 and 9). The marine water flux direction would be opposite to that of the meteoric waters (Figs 9 and 10).

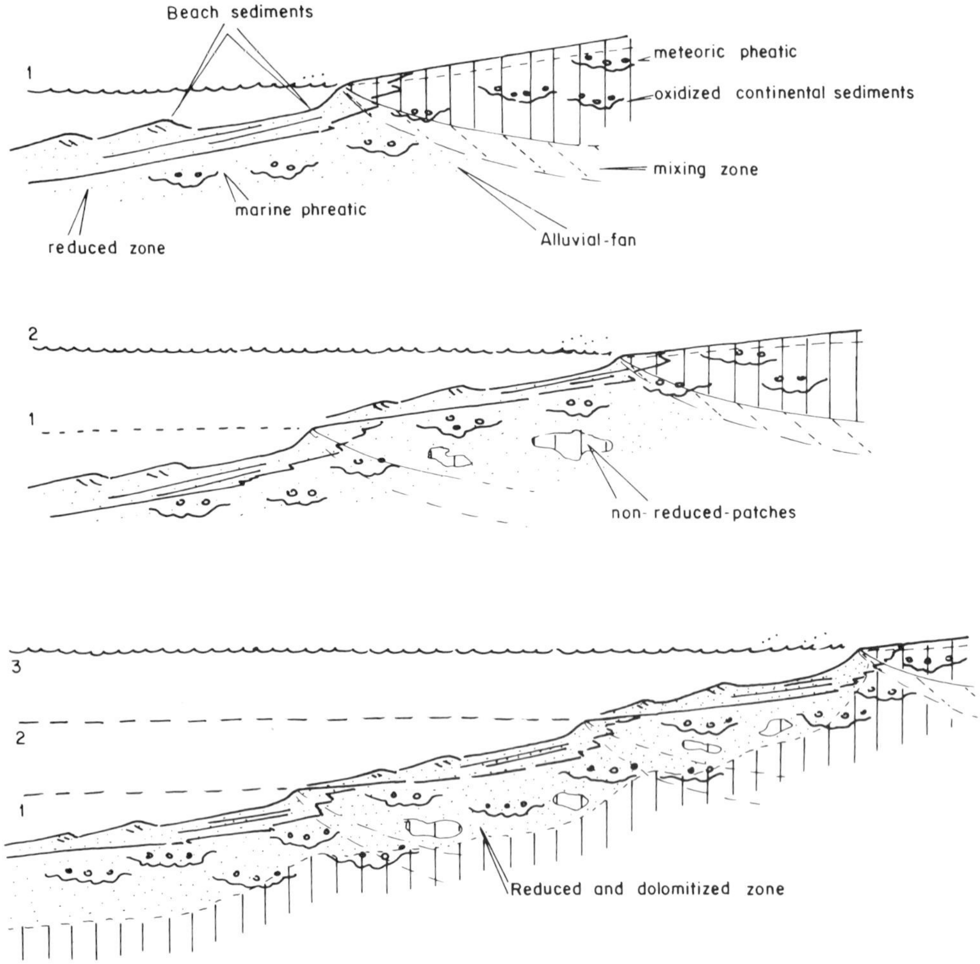

FIG. 9. Schemes showing progressive migration stages of sedimentary bodies in beach-ridge systems. This migration caused chemical changes in porewaters. Previously oxidized continental sediments are reduced in their uppermost part, just below beach levels. The reduced area also shows evidence of early dolomitization.

Some final comments

The facies associations of the Middle Eocene nearshore sediments in the SE margin of the Catalan Basin are closely interrelated in time and space. This should have had significant effects on the diagenesis. Both meteoric-influenced and marine-influenced diagenetic environments should have changed in time and space. The mixing zone should also reflect these changes (Fig. 9), but generally it should have been located at the top of the alluvial fan deposits, just in the zone representing the vertical transition from continental to marine sediments (Figs 9 and 10). There is a coincidence in the distribution of dolomitized zones and beach-ridge systems which suggests that dolomitizing fluids flowed in from the upper marine sediments. Therefore, there was both a sedimentary and palaeogeographic control in the formation and distribution of dolomites at the top of the alluvial fan sediments.

In such sediments it may be difficult to separate the marine diagenetic signatures from the continental ones, because of oscillations and migrations of the diagenetic environments. We can find evidence of meteoric diagenesis as well as of marine and mixed-water diagenesis. In this sense, Land (1980) has already explained the difficulties in interpreting the isotopic values in rocks subjected to mixing-zone processes. In our case, we are dealing with a diagenetic situation related to the deposition of marine facies associations on continental deposits. This situation would have brought about the modification of meteoric porewaters by the inflowing marine waters (Figs 9 and 10). Most cases referred to in

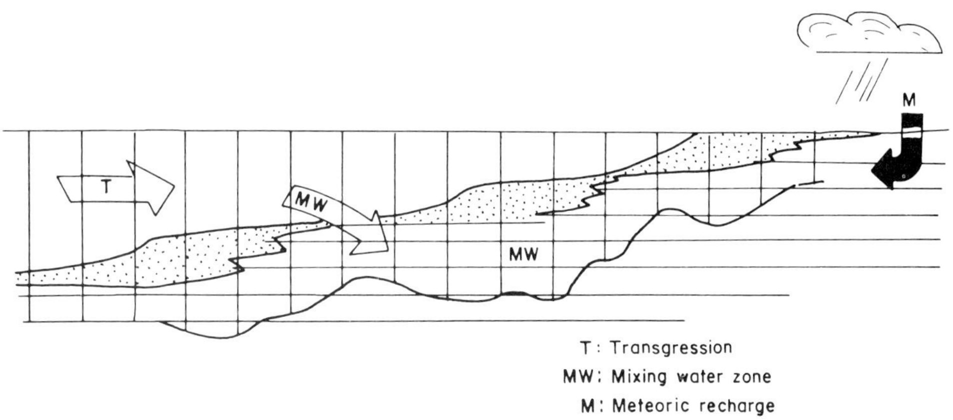

FIG. 10. Diagenetic model of dolomitization proposed for nearshore complexes at the SE margin of the Catalan Basin. Dolomitization at the top of the alluvial fan sediments developed during early diagenetic stages as a consequence of the transgression which allowed for the inflow of marine waters into continental sediments where existing porewaters had a meteoric origin. Recharge of meteoric waters would have been through subaerially exposed continental deposits. Marine waters were mixed with the meteoric ones in the zone corresponding to the decoloured levels.

the literature are the opposite to this and occur where marine waters are modified by incoming meteoric ones. Land (1980) also raises the question of the problems involved in the extrapolation of models with different dimensions and sedimentation rates. In this case, as we have already mentioned, there is a pulsating variation in the sedimentation rates. The terraces represent zones, erosional for a long time, which changed suddenly into sedimentation areas with general marine features (Fig. 4). These depositional zones are narrow and usually arranged over the continental deposits. The pulsating dynamic pattern in the sedimentary environments might have influenced the hydrology of porewaters and, consequently, the diagenetic products.

The collection of more data will perhaps enable us to detect variations in the meteoric and marine water-tables during the Middle Eocene transgressive stage in the Catalan Basin. Some aspects that will also have to be considered to achieve this are: (1) the variations in sedimentation rates throughout the different areas; (2) the area influenced by each successive mixing-zone; (3) the relative velocities of transgression-regression and (4) to decide in each case if nearshore dynamics could have enabled the development of the mixing zone by the inflow of marine waters in meteoric aquifers, establishing whether these different situations are responsible for variations in the isotopic composition or in the trace element distribution, or whether there is any evidence here for more conventional mixing where marine fluids are displaced by meteoric ones.

ACKNOWLEDGMENTS: We are indebted to Dr C. Pierre (Laboratoire de Géologie Dynamique de l'Universitié Pierre et Marie Curie, Paris VI), who performed analysis of the isotopic composition in the studied samples.

We very much appreciated the help of R. Fontarnau, C. Carulla and A. Dominguez for instruction in the use of the electron microscope and analysing system and the 'Servei de Microscopia Electrònica de la Universitat de Barcelona' for use of equipment and laboratory facilities. We also are grateful to the 'Servei d'Analisi Quimica de la Facultat de Geologia' and especially to R. Marimón, J. Sánchez and E. Balart for their help with trace elements. The help of J. Ros with the drawings was also of great value.

We wish to convey our deep appreciation to Dr Marshall and Dr Fairchild for their constant support and continuing effort in reviewing this paper. Their comments were extremely valuable. We also wish to acknowledge Dr Nichols for his kind help on sedimentological aspects.

References

ANDERSON, T. F. & ARTHUR, M. A. 1983. Stable isotopes of oxygen and carbon and their application to sedimentologic and paleoenvironmental problems. *In*: *Stable Isotopes in Sedimentary Geology*. Society of Economic Paleontologists and Mineralogists. Short Course, **10**, 1.1–151.

BADIOZAMANI, K. 1973. Dorag dolomitization model, Ordovician. Wisconsin. *Journal of Sedimentary Petrology*, **43**, 965–84.

BAKER, P. A. & KASTNER, M. 1981. Constraints on the formation of sedimentary dolomite. *Science*, **213**, 214–6.

BATHURST, R. G. C. 1971. *Carbonate Sediments and their Diagenesis*, Elsevier, Amsterdam.

BUSENBERG, E. & PLUMMER, L. N. 1985. Kinetic and thermodynamic factors controlling the distribution of SO_4^{2-} and Na^+ in calcites and selected aragonites. *Geochimica et Cosmochimica Acta*, **49**, 713–25.

CARTER, R. W. G. 1986. The morphodynamics of beach ridge formation; Magilligan, Northern Ireland. *Marine Geology*, **73** (3/4), 191–214.

COLOMBO, F. 1980. Estratigrafía y sedimentología del terciario inferior continental de los Catalánides. *Unpublished PhD Thesis*, University of Barcelona.

DUNHAM, J. B. & OLSON, E. R. 1980. Shallow subsurface dolomitization of subtidally deposited carbonate sediments in the Hanson Creek Formation (Ordovician–Silurian) of Central Nevada. *In*: ZENGER, D. H., DUNHAM, J. B. & ETHINGTON, R. L. (eds). *Concepts and Models of Dolomitization*, pp. 139–61. Society of Economic Paleontologists and Mineralogists. Special Publication, **28**.

FOLK, R. L. & LAND, L. S. 1975. Mg/Ca ratio and salinity: two controls over crystallization of dolomite. *American Association of Petroleum Geologists Bulletin*, **59**, 60–8.

GINSBURG, R. N. & JAMES, N. P. 1976. Submarine botryoidal aragonite in Holocene reef limestone. Belize. *Geology*, **4**(7), 431–6.

GLENNIE, K. W., MUDD, G. C. & NAGTEGAAL, P. J. C. 1978. Depositional environment and diagenesis of Permian Rotliegendes sandstones in Leman Bank and Sole Pit areas of the U.K. Southern North Sea. *Journal of the Geological Society of London*, **135**, 25–34.

HANSHAW, B. B., BACK, W. & DEIKE, R. G. 1971. A geochemical hypothesis for dolomitization by ground water. *Economic Geology*, **66**, 710–24.

HUDSON, J. D. 1977. Stable isotopes and limestone lithification. *Journal of the Geological Society of London*, **133**, 637–60.

JAMES, N. P. & GINSBURG, R. N. 1979. Morphology, sedimentology, organism distribution and Quarternary history. *In*: *The Seaward Margin of Belize Barrier and Atoll Reefs*. International Association of Sedimentologists. Special Publication, **3**. Blackwell Scientific Publications, Oxford.

KESSLER II, L. G. 1978. Diagenetic sequence in ancient sandstones under desert climatic conditions. *Journal of the Geological Society of London* **135**, 41–9.

LAND, L. S. 1973a. Contemporaneous dolomitization of Middle Pleistocene reefs by meteoric water, North Jamaica. *Marine Science Bulletin*, **23**, 64–92.

—— 1973b. Holocene meteoric dolomitization of Pleistocene limestones, North Jamaica. *Sedimentology*, **70**, 411–24.

—— 1980. The isotopic and trace element geochemistry of dolomite: the state of art. *In*: ZENGER, D. H., DUNHAM, J. B. & ETHINGTON, R. L. (eds). *Concepts and Models of Dolomitization*, pp. 87–110. Society of Economic Paleontologists and Mineralogists. Special Publication, **28**.

—— 1983. The application of stable isotopes to studies of the origin of dolomite and to problems of diagenesis of clastic sediments. *In*: *Stable Isotopes in Sedimentary Geology*, pp. 4.1–22. Society of Economic Paleontologists and Mineralogists. Short Course, **10**.

——, SALEM, M. R. I. & MORROW, D. W. 1975. Paleohydrology of ancient dolomites. Geochemical evidence. *American Association of Petroleum Geologists Bulletin*, **59**(9), 1602—25.

MACHEL, H. G. & MOUNTJOY, E. W. 1986. Chemistry and environments of dolomitization—a reappraisal. *Earth Science Reviews*, **23**(3), 175–222.

MACINTYRE, J. G., MOUNTJOY, E. W. & D'ANGLEJAN, B. F. 1968. An occurrence of submarine cementation of carbonate sediments off the west coast of Barbados W.I. *Journal of Sedimentary Petrology*, **38**, 660–3.

MATTES, B. W. & MOUNTJOY, E. W. 1980. Burial dolomitization of the Upper Devonian Miette Buildup, Jasper National Park, Alberta. *In*: ZENGER, D. H., DUNHAM, J. B. & ETHINGTON, R. L. (eds). *Concepts and Models of Dolomitization*, pp. 259–97. Society of Economic Paleontologists and Mineralogists. Special Publication, **28**.

MCCREA, J. M. 1950. On the isotopic chemistry of carbonates and a paleotemperature scale. *Journal of Chemistry and Physics*, **18**, 849–57.

MEYERS, W. J. & LOHMANN, K. C. 1985. Isotope geochemistry of regionally extensive calcite cement zones and marine components in Mississippian limestones, New Mexico. *In*: SCHNEIDERMANN, N. & HARRIS, P. M. (eds). *Carbonate Cements*, pp. 223–39. Society of Economic Paleontologists and Mineralogists. Special Publication, **36**.

REGUANT, S. 1967. El Eocene marino de Vic. *Memorias del Instituto Geológico y Minero de España*, **LXVII**, 1–330.

RENARD, M., BERTHENET, F., CLAUSER, S. & RIGHEBOIS, G. 1986. Geochemical events (trace elements and stable isotopes) recorded on bulk carbonates near the Eocene–Oligocene boundary. Application to the Contessa Section (Gubbio, Umbria, Italia). *In*: POMEROL, C. & PREMOLI-SILVA, I. (eds). *Terminal Eocene Events*, pp. 331–48.

SANDBERG, P. 1985. Aragonite cements and their occurrence in ancient limestones. *In*: SCHNEIDERMANN, N. & HARRIS, P. M. (eds). *Carbonate Cements*, pp. 33–57. Society of Economic Paleontologists and Mineralogists. Special Publication, **36**.

SANTISTEBAN, C. & TABERNER, C. (in press). Sedimentary models of siliciclastic deposits and coral reef interrelation. *In*: *Carbonate-Clastic Transitions*. Developments in Sedimentology. Elsevier, Amsterdam.

SCHROEDER, J. H. 1972. Fabrics and sequences of submarine carbonate cements in Holocene Bermuda cup reefs. *Geologische Rundschau*, **61**, 708–30.

—— 1973. Submaine and vadose cements in Pleistocene Bermuda Reef-Rock. *Sedimentary Geology*, **10**, 179–204.

SHINN, E. 1969. Submarine lithification of Holocene carbonate sediments in the Persian Gulf. *Sedimentology*, **12**, 109–44.

TABERNER, C. 1982 (1983). Evolución ambiental y diagenética de los depósitos del Terciario inferior

(Paleoceno y Eoceno) de la cuenca de Vic. *PhD Thesis*. Universidad de Barcelona.

—— & BOSENCE, D. W. J. 1985. Ecological sucession from corals to Coralline algae in Eocene patch reefs of N. Spain. *In*: TOOMEY & NITECKI (eds). *Paleoalgology: Contemporary Research and Application*, pp. 226–36. Springer-Verlag, Berlin.

TURNER, P. 1980. Continental red beds. *In*: *Developments in Sedimentology*, **29**. Elsevier, Amsterdam.

VEIZER, J. 1983. Chemical diagenesis of carbonates: theory and application of trace element technique. *In*: *Stable Isotopes in Sedimentary Geology*, pp. 3.1–100. Society of Economic Paleontologists and Mineralogists Short Course, **10**.

WALKER, T. R. 1976. Diagenetic origin of continental red beds. *In*: FALKE, H. (ed.). *The Continental Permian in Central West and South Europe*, pp. 240–82. NATO Advanced Study Institute Series: Series C. Mathematical and Physical Sciences. Reidel, Dordrecht, Holland.

—— & WAUGH, B. 1973. Intrastratal alteration of silicate minerals in late Tertiary fluvial arkose. Baja California. Mexico. *Geological Society of America Bulletin*. Abstracts with Program, **7**, 853–4.

WARD, W. C. & HALLEY, R. B. 1985. Dolomitization in a mixing zone of near-seawater composition. Late Pleistocene, northeastern Yucatan Peninsula. *Journal of Sedimentary Petrology*, **55**, 407–20.

ZHAO, X. & FAIRCHILD, I. 1987. Mixing zone dolomitization of Devonian carbonates, Guangxi, South China. *This volume*.

C. TABERNER, Departament de Geoquímica, Petrologia i Prospecció Geològica, Fctat. de Geologia, Universitat de Barcelona, Gran Via 585, 08007 Barcelona, Spain.

C. SANTISTEBAN, Departament de Geologia, Fctat. de Biologia, Universitat de Valencia, C/Dr Moliner 50, Burjassot, Valencia, Spain.

Early diagenetic phosphate cements in a turbidite basin

R. D. A. Smith

SUMMARY: Early diagenetic phosphate cements of late Llandovery age developed in fine-grained turbidites deposited in the Welsh Basin. Bottom water conditions fluctuated between dysaerobic and anaerobic. Phosphate enrichment of layers close to the tops of fine-grained turbidites occurred very rapidly after turbidite deposition. Early phosphate cement precipitation caused lithification of these layers prior to significant sediment compaction. The cements consist of apatite crystallites (0.5–5 μm in diameter) which filled pore spaces and also grew displacively in the cleavages of detrital phyllosilicates. Two possible mechanisms for the localization of cements close to turbidite tops are (a) bacterial concentration of phosphate in near-surface layers of transition between oxic and sulphidic environments, and (b) adsorption of upward-diffusing dissolved phosphate from underlying anoxic porewaters by ferric and manganese oxyhydroxides present above levels of oxygen and nitrate exhaustion. Phosphate cements are absent from adjacent Lower Silurian shelf facies deposited in fully oxic bottom water environments, and are also absent from turbidites deposited in totally anoxic bottom waters. In the former case oxidation of labile organic matter close to the sediment-water interface prevented a source of dissolved phosphate from being buried, whereas in the latter case phosphate would have been free to diffuse out into bottom waters.

Early diagenetic phosphate cements are a common feature of Lower Silurian fine-grained turbidites in Wales (Cave 1979), but they have not previously been studied in any detail. This study focuses on phosphate cements developed in the deposits of a late Llandovery turbidite system of *griestoniensis* Zone age (Figs 1 and 2) and forms part of a sedimentological basin analysis of the Welsh Basin for this zone (Smith 1987a and in preparation).

The main aims of this paper are to attempt to answer the following questions: (1) What was the development history of the cements relative to individual turbidite depositional events? (2) What were the phosphate-concentrating processes and which factors controlled the localization of phosphate cements within individual turbidites? (3) Which factors controlled the large-scale (basin-wide) distribution of phosphate cements?

The *griestoniensis* Zone sequence is ideal for such a study in that (a) a good biostratigraphic framework for correlation exists utilizing graptolite zones in basinal facies and brachiopod zones in shelf facies (Wood 1906, Cocks *et al.* 1984); (b) the late Llandovery palaeogeography is well-known (Ziegler *et al.* 1968, Bridges 1975, Smith 1987a) and (c) the sequence contains numerous reference levels, in the form of turbidite event deposits, against which early diagenetic cement development can be considered. Since the cements developed in a deep-water basin they escaped the winnowing and concentration processes normally suffered by shallow water phosphate cements (e.g. Mullins & Rasch 1985, Cook & Shergold 1986) hence preserving their spatial relationships to their enclosing event deposits.

In this paper new fabric and mineralogical data are presented and integrated with palaeogeographical data to develop a qualitative model for the controls on marine apatite cement precipitation in the early Silurian Welsh Basin.

Methods

Samples collected from various mid-Wales exposures of *griestoniensis* Zone turbidites were slabbed and polished in order to study the relationships between phosphate cements and individual turbidite and hemipelagic units. Colours were determined with reference to a Geological Society of America Rock Colour Chart using the Munsell system.

Epoxy resin-mounted thin sections of over 20 samples were prepared for study using optical microscopy, backscattered scanning electron microscopy (BSEM) and energy-dispersive X-ray microanalysis, (EDXRA), electron microprobe analysis (EMP) and cathodoluminescence (CL). The sections were cut perpendicular to bedding and were polished using Buehler TEXTET laps and 1 μm diamond paste abrasive.

For BSEM and EDXRA the specimens were coated with 100 Å of carbon, mounted on aluminium stubs and examined in a Phillips 501B SEM equipped with a four element KE backscatter detector and a Link AN10000 X-ray microanalyser. A number of specimens were also studied in a Jeol JSM-840 SEM equipped with a Jeol dipole solid state backscatter detector and a

From: MARSHALL, J. D. (ed.), 1987, *Diagenesis of Sedimentary Sequences*, Geological Society Special Publication No. 36, pp. 141–156.

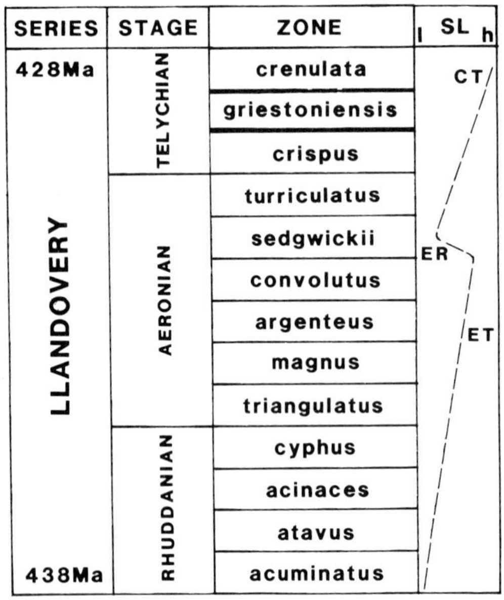

FIG. 1. Biostratigraphy of the Llandovery Series and sea-level curve generalized from McKerrow (1979). CT, continental trangression; ET, eustatic transgression; ER, eustatic regression. Absolute ages are from the Harland *et al.* (1982) time-scale.

TRACOR Northern TN2000 X-ray microanalyser. BSEM produces images in which brightness is proportional to mean atomic number. It enables petrographic examination of fine-grained rocks with far higher resolution than is available using optical microscopy (Pye & Krinsley 1984).

A cold-cathode luminoscope with an approximately 25 kV generator built by W. Wilson and Dr J. A. D. Dickson (see Fairchild 1980) was used for CL examination of the samples.

Bulk-rock mineralogies were determined by X-ray powder diffraction in a Phillips powder diffractometer using Cu K_α radiation. Minerals were identified by reference to the ASTM Powder Diffraction File and data in Brindley & Brown (1980). The CO_2 contents of the apatites were obtained from the positions of the (004) and (410) peaks using the method of Gulbrandsen (1970).

Eight specimens were prepared for autoradiographic study with the aim of detecting any enrichments of uranium in phosphatic layers. CR39 plastic was attached to polished surfaces of slabbed samples and exposed for five weeks. The density of alpha tracks developed in the plastic is proportional to the concentration of alpha

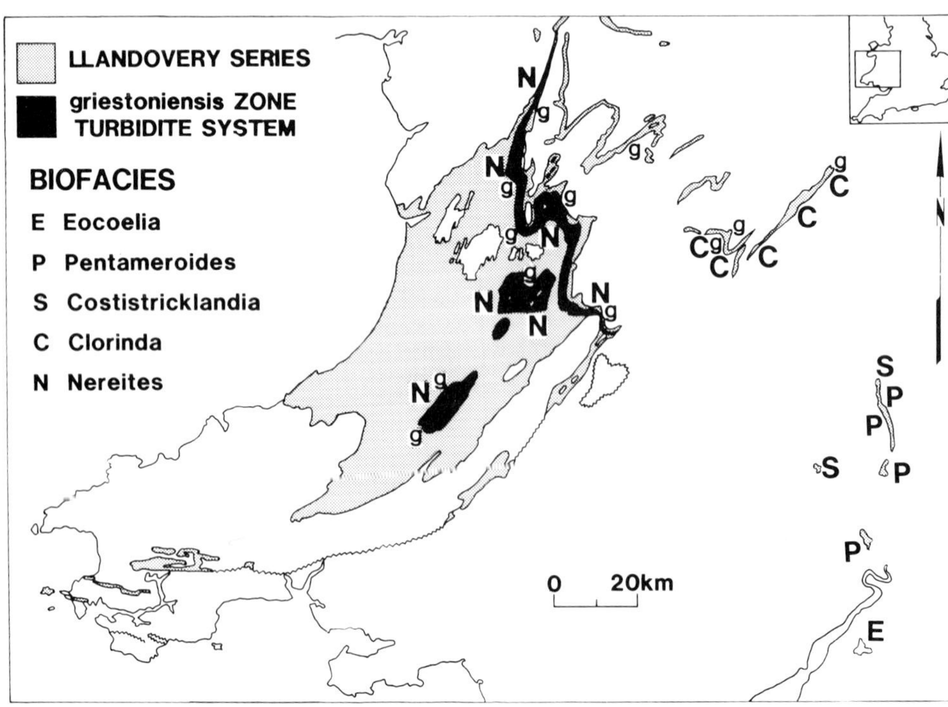

FIG. 2. Location map showing outcrop patterns of the Llandovery Series and deposits of the *griestoniensis* turbidite system, and the distribution of benthic biofacies (see Fig. 3) of *griestoniensis* Zone age. g denotes the presence of pelagic faunas of graptolites. The distribution of shelly 'communities' of the shelf is after Ziegler *et al.* (1968).

particle-emitting elements (mainly uranium and thorium). The plastic was then etched in 6 M NaOH at 80°C for 18 hr to enlarge the alpha tracks and viewed down a petrological microscope.

Palaeogeographical setting

The early Palaeozoic Welsh Basin was situated on continental crust close to the leading edge of a northward-drifting microcontinental fragment, 'Eastern Avalonia' of Soper *et al.* (1987). The basin geometry was strongly controlled by an array of long-lived lineaments dominantly trending between NE–SW and N–S (Kokelaar *et al.* 1984, Woodcock 1984, Smith 1987a). Petrochemical studies of Welsh volcanic rocks indicate that the basin occupied a back-arc setting from early to late Ordovician (Arenig to Caradoc) times (Kokelaar *et al.* 1984). The leading edge of Eastern Avalonia may have become a transform margin after the main, late Ordovician, volcanic shutdown (Leggett *et al.* 1983). However, documented strike-slip displacements on Welsh Lineaments affecting Lower Palaeozoic rocks are all small (e.g. Fitches & Campbell 1987) and were probably mainly generated during the late Caledonian (Acadian) and Variscan deformation maxima. During the Silurian the remnant back-arc basin began to be modified by a phase of sinistral transpression, associated with Avalonia–Laurentia collision which culminated in early to mid-Devonian time (Smith 1987a, b, Soper *et al.* in press). Palaeomagnetic evidence indicates a low latitude position in the southern hemisphere for the early Silurian Welsh Basin (Piper 1985).

The preserved deposits of the late Llandovery *griestoniensis* Zone turbidite system comprise remnants of the distal part of a large (minimum length of 100 km) elongate turbidite system. The sequence consists of a thin-bedded fine-grained (silt-mud) turbidite background punctuated by packets of thicker-bedded sandstones (sandstone lobes). Palaeo-water depths of the order of 500–1000 m have been estimated for the late Llandovery Welsh Basin (Smith 1987a).

Biofacies distributions indicate that the early Silurian basinal water column was stratified. Dissolved oxygen levels of bottom water in the basin were too low to support skeletonized benthos and at times even excluded soft-bodied benthos which enabled the preservation of fine laminations in hemipelagic facies (Smith 1987a, c). At present laminated sediments devoid of macrobenthos are preserved beneath bottom waters containing less than 0.1 ml l^{-1} dissolved oxygen (Savrda *et al.* 1984, Thomson *et al.* 1985). By analogy with biofacies patterns in modern oxygen-deficient basins, bottom water environments in the late Llandovery Welsh Basin fluctuated between dysaerobic and anaerobic conditions (cf. Rhoads & Morse 1971, Savrda *et al.* 1984, Thomson *et al.* 1985, Edwards 1985, Savrda & Bottjer 1986). The Llandovery was a time of high sea-level (Fig. 1) with high organic productivities giving rise to a strong tendency to oxygen-depletion in the water column (Leggett 1980). Figures 2 and 3 show the distribution of

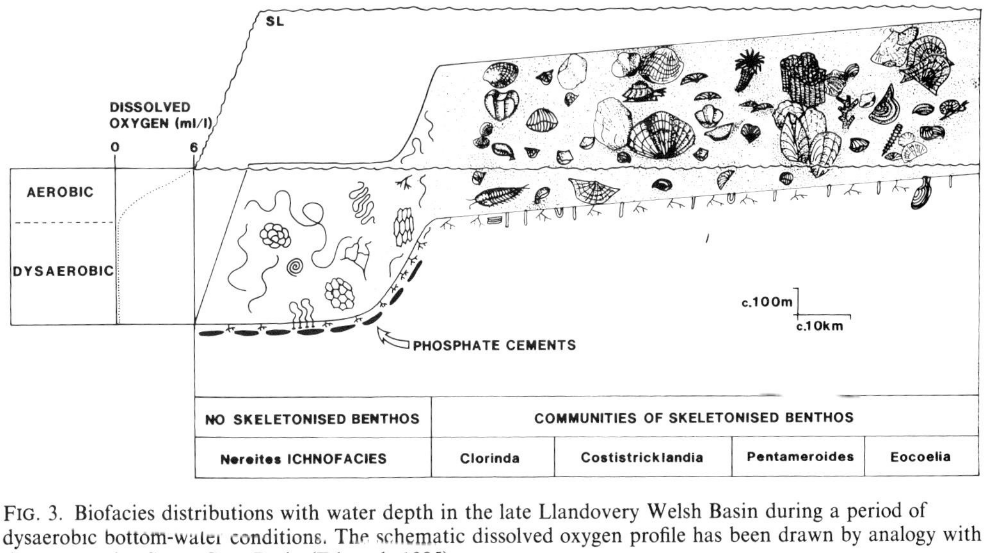

FIG. 3. Biofacies distributions with water depth in the late Llandovery Welsh Basin during a period of dysaerobic bottom-water conditions. The schematic dissolved oxygen profile has been drawn by analogy with the present-day Santa Cruz Basin (Edwards 1985).

biofacies for the *griestoniensis* Zone sediments and their interpreted significance in terms of a lateral gradient in palaeo-oxygenation.

Morphology and occurrence

The phosphate cements, when not reworked, occur close to the tops of fine-grained (silt-mud) turbidites (Fig. 4), or occasionally in the basal coarse silt divisions. Turbidites containing phosphatic layers range in thickness from about 1.5 to 10 cm. Bouma (1962) T_{c-e}, T_{d-e} or simply T_e divisions are represented. Many of the finer-grained turbidites are very similar to examples described by Stow & Shanmugam (1980). Inter-turbidite hemipelagic facies are either poor in organic carbon and bioturbated or richer in organic carbon and finely laminated. The phosphate cements occur in bedding-parallel thin sheets, layers of strongly oblate concretions, and more rarely in isolated concretions. Sheet and concretion thicknesses range from about 1 to 10 mm. Turbidites commonly have paler coloured top layers, which range in thickness from 1 mm to 2 cm and are usually bioturbated. Bioturbation in these top layers may be either in the form of pervasive microbioturbation associated with burrows up to 1 mm in diameter, or consist of isolated *Chondrites* burrow systems with burrow diameters of about 0.1–0.3 mm. Phosphate sheets or layers of phosphatic concretions frequently occur at, or a small distance above, colour changes, which may be very subtle. However, they may also occur without any obvious associated colour change. The phosphate cements are black when fresh, but weather white. Turbidite colours range from olive-black (Munsell colour 5Y 2/1) to light olive-grey (Munsell colour 5Y 5/2). Phosphatic layers often show evidence of physical disruption by bioturbating organisms, and wet-sediment faulting and

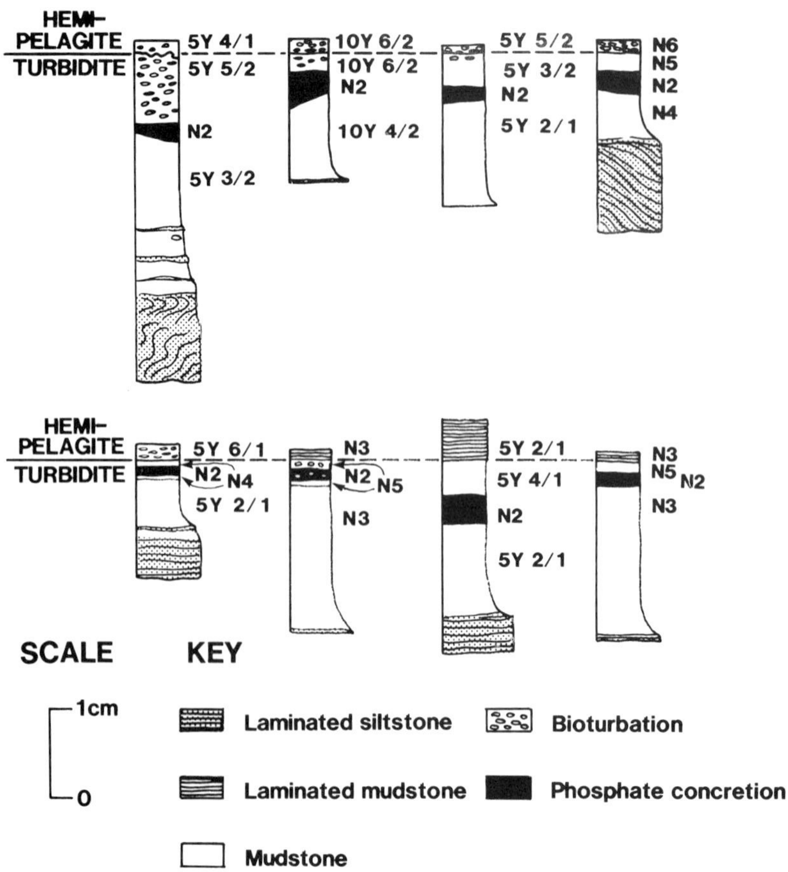

FIG. 4. Representative silt-mud turbidite-hemipelagite couplets with phosphatic layers. Colour designations use the Munsell system notation.

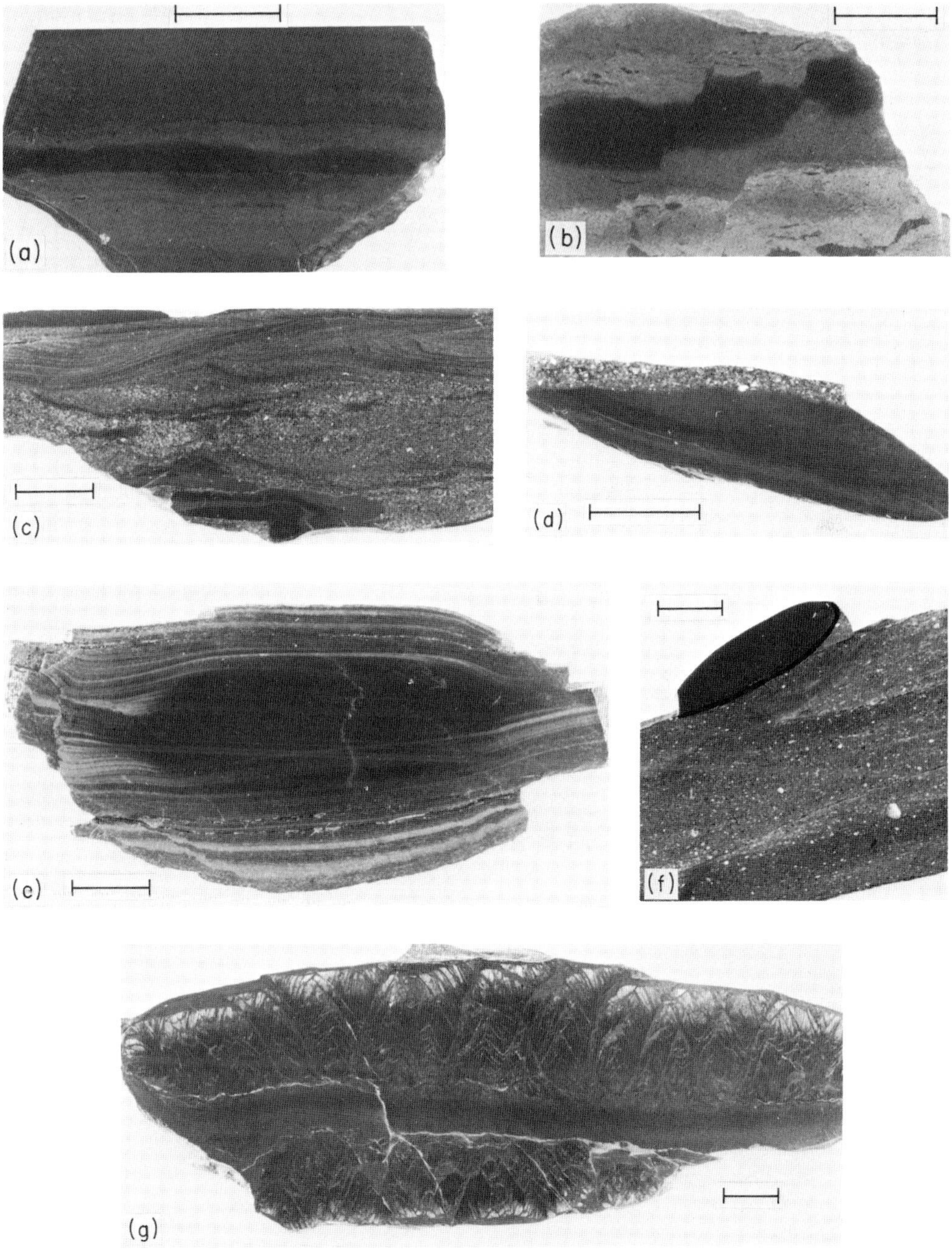

FIG. 5. Evidence for early unconsolidated stages of phosphatic layer development (a–d) and for early lithification (e–g). (a) Disruption of a phosphatic layer (dark) by bioturbation. Apatite crystallites have been mixed into the underlying part of the enclosing turbidite. (b) Wet-sediment microfaulting affecting a phosphatic layer. Note the contrast in intensity of bioturbation above and below the phosphatic layer. (c) Plastic deformation of a phosphatic layer within a mudstone rip-up clast present near the top of a sandstone turbidite. (d) Truncation of a phosphatic layer at the edge of a mudstone rip-up clast. (e) Curving of laminae in a siltstone turbidite around a phosphatic concretion. (f) An undeformed ripped-up phosphate concretion present in a matrix-rich sandstone turbidite. (g) Silicified cone-in-cone concretion with a phosphatic sheet core. All scale bars represent 1 cm.

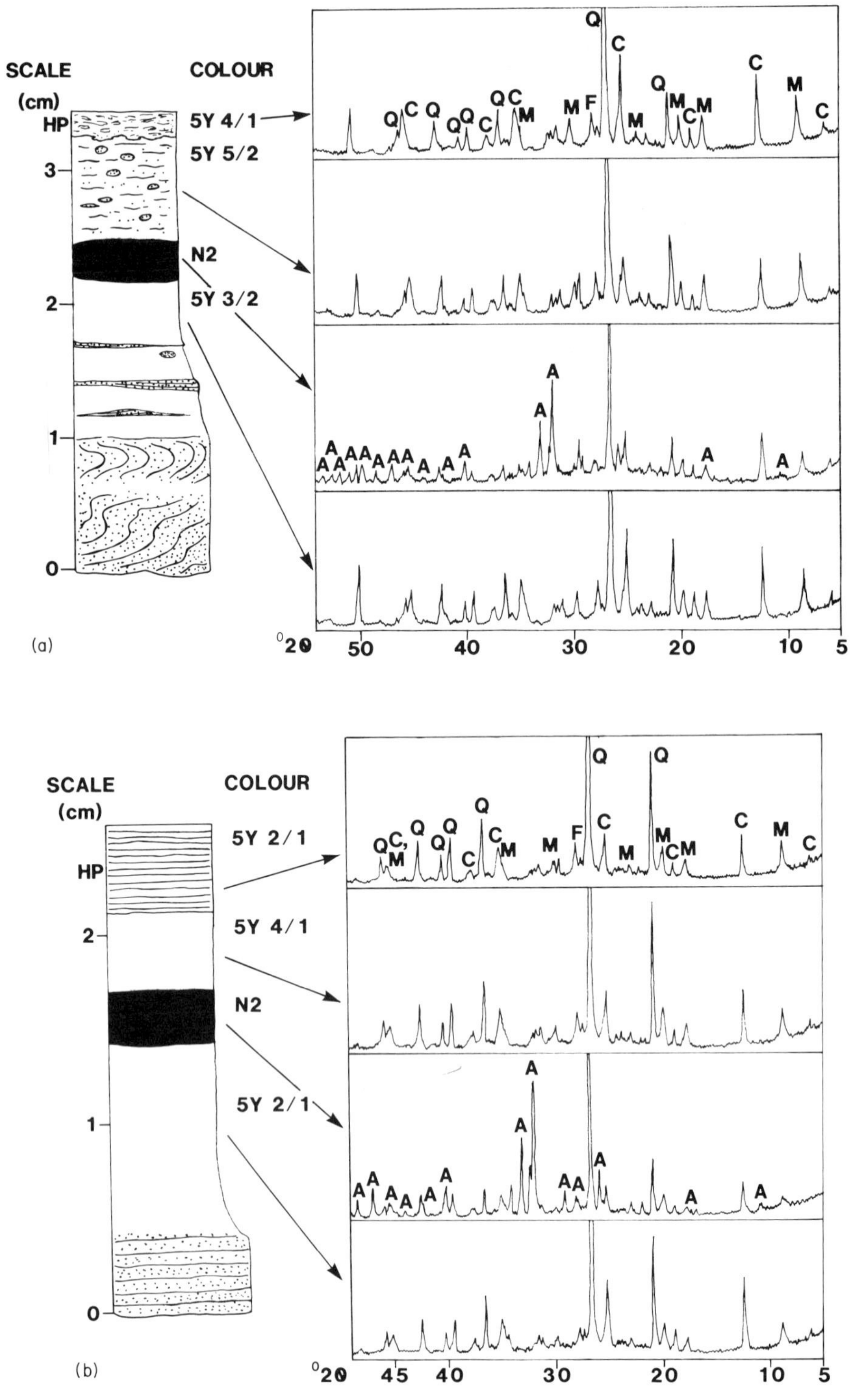

FIG. 6. Turbidite-hemipelagite mineralogies shown by XRD traces (Cu K_α radiation). Colour designations use the Munsell system notation. HP, hemipelagite; A, apatite; C, chlorite; F, feldspar; M, white mica; Q, quartz. Note the increased abundances of chlorite beneath phosphatic layers.

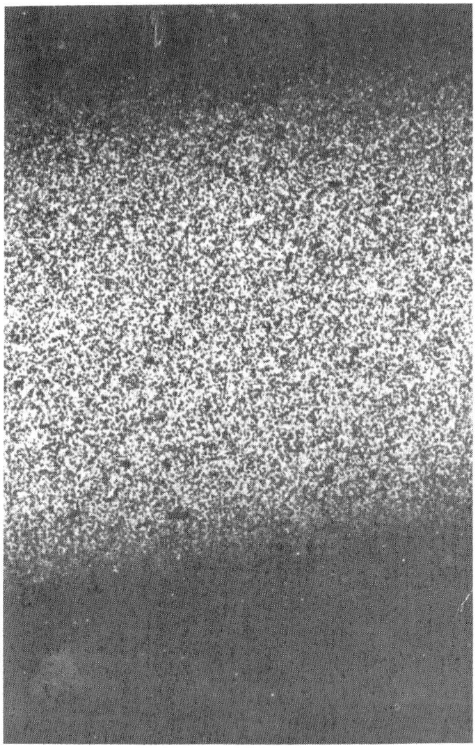

FIG. 7. Cathodoluminescence image of a 5 mm thick phosphatic layer showing the distribution of luminescent apatite cement.

folding (*sensu* Pickering 1987), either *in situ* or in rip-up clasts present within thicker turbidite sandstones (Fig. 5a, b, c). Phosphate sheets often form cores to later diagenetic cone-in-cone concretions (Fig. 5g).

Phosphate cements have not been recorded from Lower Silurian shelf facies of the Midland Platform (Bridges 1975 and author's observations).

Microfabrics and mineralogy

Turbidite mineralogies, determined using XRD, EDXRA and EMP, are dominated by an assemblage of quartz, chlorite and white mica (often intergrown in chlorite-mica stacks ranging up to 150 μm maximum diameter), feldspar and pyrite (Fig. 6). Accessories include titanium dioxides, florencite, xenotime and zircon. Chlorites have iron-rich (ripidolite) compositions. Pyrite is more abundant in darker coloured turbidites associated with laminated hemipelagites, in which it occurs in the form of framboidal aggregates as well as euhedra. Framboidal pyrite aggregates are occasionally intergrown with chalcopyrite and arsenopyrite.

Phosphatic layers contain up to 15% yellow-luminescing apatite (Fig. 7) in the form of 0.5–5 μm crystallites with hexagonal sections (Figs 8a and 9a). The apatite crystallites formed an inter-particle porosity-filling cement (Fig. 7), but also grew displacively in detrital phyllosilicates as shown by the splitting apart of white mica lamellae in early diagenetic chlorite mica stacks (Fig. 9b). Dimberline (1986) argues that at least some of these stacks result from early diagenetic modification of detrital biotites. Detrital smectite is another probable precursor phase (Craig *et al.* 1982). The apatites have negligible carbonate content (using the peak pair method of Gulbrandsen 1970), probably the result of carbonate loss from an original francolite (carbonate-fluorapatite) composition during burial diagenesis and incipient metamorphism since the turbidites are presently dominantly of lower to upper anchizone grade (cf. McClellan 1980, McArthur 1985, Smith 1987b). No uranium enrichments in phosphatic layers were detected using autoradiography.

Below colour changes, near the tops of turbidites, there is frequently a marked increase in small chlorite-dominated chlorite-mica intergrowths of the order of 5–30 μm in diameter (Figs 6 and 9c). Such colour changes thus appear to be dominantly controlled by chlorite abundances. Pyrite also increases in abundance below colour changes in the darker, more organic carbon-rich turbidites. Below colour changes in such turbidites, it is present as framboidal aggregates (up to 150 μm diameter) as well as euhedra ranging from about 2 to 15 μm diameter (Fig. 8d). Above colour changes pyrite occurs as scattered euhedra 2–5 μm in diameter.

The interfaces between bioturbated hemipelagites and underlying turbidites are often highly disrupted by bioturbation. Bioturbated hemipelagites have a very similar mineralogical composition to the turbidites, but contain fewer chlorite-mica stacks. The laminated hemipelagites, however, also contain dolomite rhombs (up to 60 μm diameter) which are usually ferroan, but sometimes preserve non-ferroan dolomite cores (Fig. 8). In addition, they contain more abundant pyrite in the form of framboidal aggregates (up to 60 μm diameter) and euhedra (2–40 μm diameter). Dolomite rhombs have sharp crystal faces and often have pyrite and quartz inclusions. These features suggest that they probably formed within the sediment as authigenic precipitates.

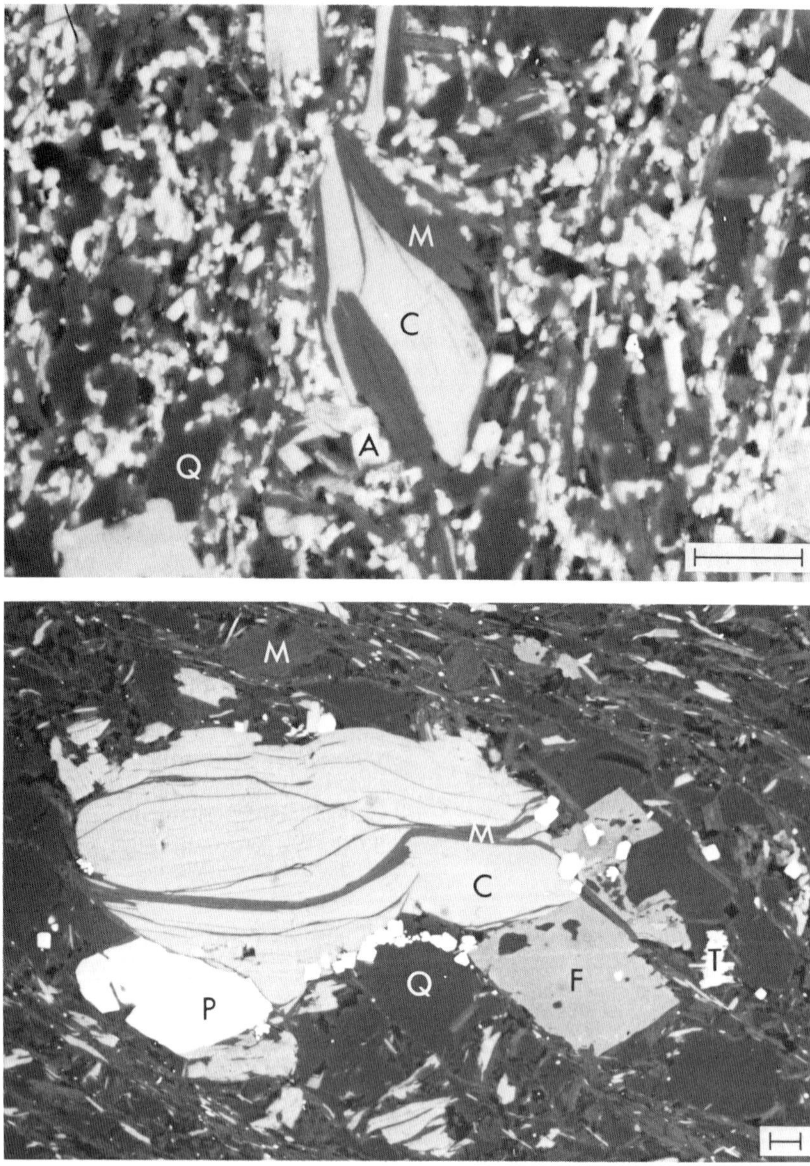

FIG. 8. (a) BSEM image of microfabric and mineralogy of a phosphatic layer. (b) BSEM image of microfabric and mineralogy of a laminated hemipelagite. A, apatite; C, chlorite; F, ferroan carbonate; M, white mica; P, pyrite; Q, quartz; T, titanium dioxide. Scale bars represent 10 μm.

Discussion

Early unconsolidated stages of phosphatic layer development

An early unconsolidated stage in the development of the phosphatic layers is indicated by a number of features:

(1) Evidence for disruption of phosphatic layers by burrowing animals (Fig. 5a). This indicates that phosphate enrichment initiated very soon after turbidite deposition since bioturbation is limited to the top few millimetres or centimetres and turbidite repeat times were probably short. An order of magnitude estimate for the mean repeat time between turbidite deposition events of

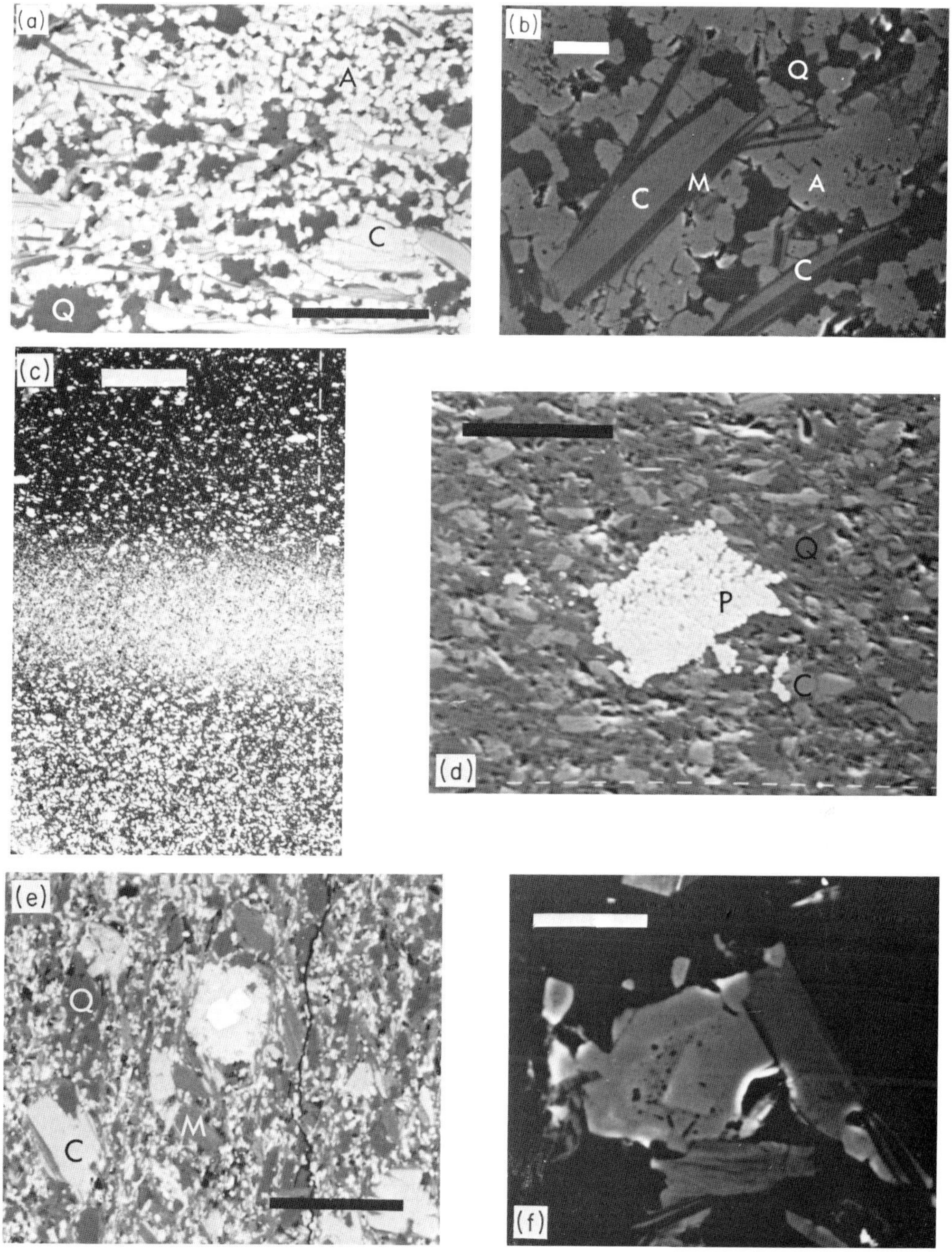

FIG. 9. BSEM images illustrating turbidite microfabrics and mineralogies. (a) Apatite crystallites in a siltstone turbidite. Note pressure solution embayments in detrital quartz grains. Scale bar = 50 μm. (b) Apatite crystallites which have grown in a detrital phyllosilicate grain. Note the splitting apart of white mica lamellae. Scale bar = 10 μm. (c) High contrast image showing bright apatite-rich band and a marked enrichment of bright chlorites below this band. Scale bar = 500 μm. (d) Framboidal pyrite aggregate present in an organic-carbon rich turbidite beneath the phosphatic layer. Scale bar = 100 μm. (e) Clustering of apatite crystallites around pyrite euhedra in a phosphatic layer. Scale bar = 50 μm. (f) Organic inclusions present in an unusually large apatite crystallite. The surrounding quartz and mica has been blacked out using contrast and brightness controls. Scale bar = 10 μm. A, apatite; C, chlorite; M, white mica; P, pyrite; Q, quartz.

about 180 years has been calculated for the most proximal preserved parts of the turbidite system using the following assumptions: (a) the 13 graptolite biozones of the Llandovery Series (Cocks *et al.* 1984) represent equal periods of time; (b) The duration of the Llandovery Series was 10 Ma (using the Harland *et al.* 1982 time-scale); (c) the mean number of turbidites present per 10 m of section is 105, based on a 45 m measured section near Blaen Twrch, NE Dyfed (SN67304893) and (d) The succession of *griestoniensis* Zone age near Blaen Twrch is about 400 m thick (Smith 1987b). Since assumption (c) is based on all turbidites in the Blaen Twrch section, including thick sandstone turbidites, it will tend to lead to an overestimate of the repeat times of the thin-bedded fine-grained turbidites which contain *in situ* phosphatic layers.

(2) Phosphatic layers are affected by wet-sediment faulting (Fig. 5b).

(3) Plastic deformation of phosphatic layers occurred in some ripped-up clasts of unlithified turbidite mudstone (Fig. 5c). Plastic deformation of phosphate layers associated with loading of overlying sandstone turbidites has also been observed.

(4) Phosphatic layers may be truncated at the edges of rip-up clasts (Fig. 5d).

Many authors have reported similar early unconsolidated stages of development in recent phosphatic sediments (e.g. Burnett 1977, 1980, Baturin & Bezrukov 1979, Burnett *et al.* 1983, O'Brien & Veeh 1980). The features described above may record the presence of either an amorphous calcium phosphate precursor phase, or the presence of numerous isolated apatite crystallites which would provide nuclei for further growth on subsequent supply of phosphate.

Early lithification

Early lithification of the phosphatic layers, either through crystallization of an apatite mineral from an amorphous precursor, or through precipitation of more apatite crystallites, is indicated by the following features:

(1) In cases of cement development within laminated turbidite siltstones laminae clearly curve around concretions (Fig. 5e). This indicates that lithification took place prior to significant sediment compaction.

(2) Rip-up clasts of isolated undeformed phosphate concretions with thin mudstone veneers are common in thick turbidite sandstones of the *griestoniensis* Zone succession (Fig. 5f). In these cases the phosphate concretions had clearly become lithified prior to the enclosing turbidite mudstone.

(3) Phosphatic sheets commonly form the cores to later diagenetic concretions with cone-in-cone structure (Fig. 5g). These are at present composed of microgranular quartz, but are interpreted as having been originally composed of calcite which grew displacively into still unlithified enclosing mudstones (Woodland 1964, Marshall 1982). Marshall's stable isotopic data indicated that his Jurassic study examples of cone-in-cone structure were formed at depths of tens to hundreds of metres in overpressured sediments. In the case of the *griestoniensis* Zone sequence, however, since ripped-up concretions with phosphate cores have been found in sandstone turbidites, at least some of these concretions may have formed at shallower depths. In one example, a 35 cm × 12 cm concretion with a core consisting of a phosphatic sheet (5 mm thick) occurs within a coarse to medium-grained sandstone turbidite (40 cm thick) near Ty-Poeth, Powys (SO 866 072). Turbidity currents depositing such sand turbidites excavated scours up to 10 m or more deep in the channel-lobe transition zone (Mutti & Normark 1987) of the *griestoniensis* Zone turbidite system (Smith 1986, 1987a). Hence this area is a possible source for such ripped-up cone-in-cone concretions.

Localization of cements near turbidite tops

There is still considerable uncertainty about many details of the mechanisms of phosphate enrichment and precipitation in sediments. However, there is general agreement on some points. The source of phosphate for marine phosphate cements is widely thought to be a by-product of the microbial degradation of organic matter (e.g. Berner 1974, 1980, Burnett, 1977, Baturin & Bezrukov 1979, Baturin 1982, De Lange 1986). In anoxic conditions porewaters become greatly enriched in phosphate relative to phosphate concentrations in sea water (e.g. Nissenbaum *et al.* 1972, Sholkovitz 1973). Phosphate may become enriched in sediments through a variety of processes, including the following: (a) adsorption on to iron and manganese oxyhydroxides (e.g. Berner 1973, Krom & Berner 1981, Borgaard 1983, Watanabe & Tsunogai 1984, De Lange 1986, Schaffer 1986) (b) adsorption on to illite-mica (Bremner 1980) (c) replacement of carbonate grains by apatite (e.g. Manheim *et al.* 1975, de Kanel & Morse

1978, Bentor 1980, Berner 1980, Prévot & Lucàs 1986); precipitation of phosphate minerals from porewaters (e.g. Berner 1974, 1980, Bentor 1980, Burnett 1977) and (d) *in situ* concentration of phosphate in bacterial cells (O'Brien & Veeh 1982, Williams & Reimers 1983). Apatite precipitation may be favoured by relatively low pH (Nathan & Sass 1981), low magnesium concentrations in porewaters (Martens & Harris 1970), and the presence of catalysts such as the enzyme carbonic anhydrase (McConnell *et al.* 1961). The following discussion attempts to constrain the major controls on phosphate concentration and apatite cement precipitation in the Lower Silurian turbidites of the Welsh Basin.

The origin of thin paler-coloured bioturbated top layers of some of the *griestoniensis* Zone turbidites is attributed to a downward migrating oxidation front process analogous to the phenomenon recently documented and inferred from late Pleistocene to Recent turbidites of the Madeira Abyssal Plain (Colley *et al.* 1984, Colley & Thomson 1985, Wilson *et al.* 1985, 1986, Jarvis & Higgs 1987, Thomson *et al.* 1987). Wilson *et al.* (1985) suggest that the well-known vertical sequence of diagenetic zones present in 'steady state' hemipelagic regimes (e.g. Froelich *et al.* 1979) may still be present beneath downward-burning oxidation fronts active in turbidites, but in a highly compressed form. Their sample spacing, however, was not close enough to be able to demonstrate this. Oxidation fronts in the Welsh Basin turbidites migrated far shorter distances (rarely greater than 1 cm) than the Madeira Abyssal Plain examples probably primarily due to the low levels of dissolved oxygen in the bottom waters. In addition, short turbidite repeat times would prevent large distances of downward migration. The short migration distances account for the failure to observe in the Welsh examples the marked uranium enrichments recorded at relict oxidation fronts in the Madeira Abyssal Plain turbidites.

One possible explanation for the localization of phosphate cements at, or a short distance above, the interface between the paler top layers and the lower parts of turbidites is that phosphate enrichment occurred at these levels due to the adsorption of dissolved phosphate, diffusing upwards along a concentration gradient from underlying anoxic porewaters, by iron and manganese oxyhydroxides present above the relict oxidation fronts. The phosphate was probably a by-product of the microbial degradation of organic matter in the underlying anoxic interstitial waters. The turbiditic sediments may have been already enriched in phosphate in shallow marine environments prior to resedimentation in the basin (cf. Krom & Berner 1981).

Ferric and manganese oxyhydroxides would still be present in thin 'post-oxic' (Berner 1981) or 'suboxic' (Froelich *et al.* 1979) layers present above levels of nitrate exhaustion. Hence, phosphate cements not associated with bioturbated top layers may have been localized close to levels of nitrate exhaustion at times when bottom waters contained insufficient dissolved oxygen for downward burning oxidation fronts to develop. Since nitrate reduction accounts for very small proportions of the total oxidation of organic matter in marine sediments (e.g. Müller & Mangini 1980, Reeburgh 1983, Jones 1985) no strong colour change due to organic matter depletion would occur across such a level. Subtle colour changes appear to be due to the enrichment of small chlorite-mica intergrowths, together with pyrite, in the lower parts of turbidites (Figs 6 and 9c). Such enrichments suggest that ferrous iron was being immobilized at an early stage in the sulphidic diagenetic environment (Berner 1981) present below the thin weakly oxic or post-oxic surface layer.

In cases where turbidites with phosphate cements are overlain by unbioturbated laminated hemipelagites oxic or post-oxic conditions at the sediment surface were clearly only short-lived since the absence of bioturbation in the laminated hemipelagites indicates that a sulphidic diagenetic environment extended up to the sediment surface during their accumulation (cf. Maynard 1982). The ferroan dolomite rhombs present in the laminated hemipelagites possibly grew at the top of the methanogenesis zone, with their non-ferroan cores precipitating in the overlying sulphate reduction zone. Within the sulphate reduction zone iron would be preferentially incorporated in sulphides (cf. Pye 1985, Burns & Baker 1987). The presence of minute *Chondrites* burrow systems in some turbidite top layers beneath laminated hemipelagites indicates seafloor dissolved oxygen concentrations only slightly higher than critical levels required by macrobenthos for survival (Bromley & Ekdale 1984, Savrda & Bottjer 1986). Under such conditions it is difficult to envisage sufficient inorganic concentration of phosphate in the temporarily weakly oxic or post-oxic surface layers significantly to influence phosphate cement localization. Another possibly important factor in localizing phosphate cements close to turbidite tops in these cases is phosphate enrichment associated with high microbial growth rates resulting from nutrient cycling across the oxic-anoxic transition layer (e.g. nitrate-nitrite cycling: Anderson 1982, Anderson *et al.* 1982). Williams & Reimers (1983) suggested that phosphate cement development in the Miocene Monterey Formation of California may have been controlled by the presence of non-photosynthetic filamentous bacteria. Such bacteria, which obtain energy by oxidizing reduced sulphur species, have been recently reported from sediments off Peru and Chile (Gallardo 1977) and California (Soutar & Crill 1977, Williams & Reimers 1983). These forms are adapted to microaerophilic conditions present in thin zones of transition between overlying oxic and underlying sulphidic environments, exactly analogous to those inferred to have existed close to the tops of many of the Welsh turbidites.

On subsequent burial of turbidites with phosphate-enriched top layers, ferric iron in the iron oxyhydroxides would be reduced and the adsorbed phosphate released. In addition, bacteria active at oxic-anoxic interfaces would die in the sulphidic environment created during burial beneath succeeding turbidites and their remains would become available for degradation leading to regeneration of phosphate. Apatite might precipitate either directly, if porewaters were depleted in magnesium (Martens & Harris 1970), or via the ageing of an amorphous calcium phosphate precursor (cf. Jahnke *et al.* 1982, Gulbrandsen *et al.* 1984). Clustering of apatite crystallites around authigenic pyrite euhedra occasionally seen in BSEM images (e.g. Fig. 9e) can be interpreted as the product of phosphate release on reduction of ferric oxyhydroxides in the sulphate reduction zone with synchronous precipitation of pyrite. It is also possible that enzymes such as carbonic anhydrase, a common constituent of bacteria (Veitch & Blakenship 1963), may have catalysed apatite precipitation since this enzyme is known to stimulate the

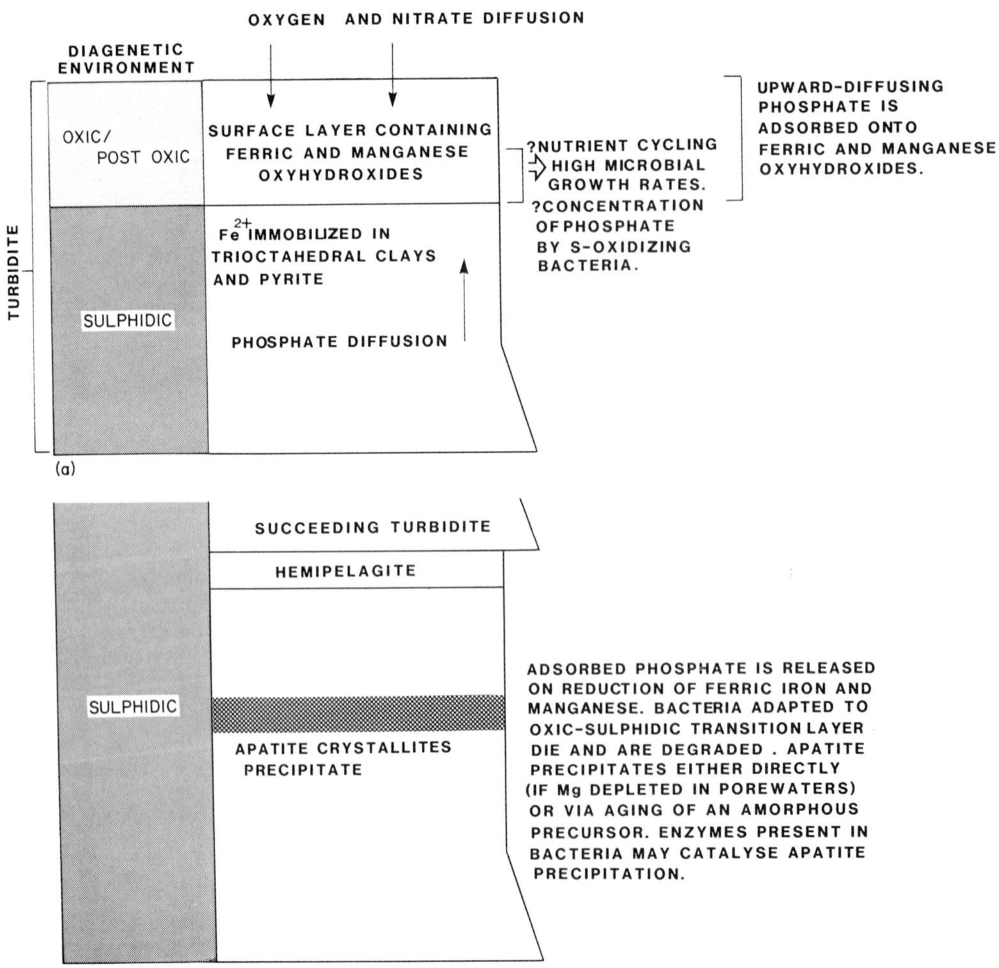

FIG. 10. Schematic model for possible processes involved in the localization (a) and precipitation (b) of early diagenetic phosphate cements in Lower Silurian Welsh Basin turbidites. Diagenetic environments are designated using the terminology of Berner (1981).

precipitation of carbonate-hydroxyapatite (McConnell *et al.* 1961). O'Brien & Veeh (1982) speculate that the operation of an *in vivo* mechanism in bacteria may prevent mineralization of bacterially 'fixed' phosphate until cell death when such an inhibitor would cease to be active, enabling mineralization to occur. Possible support for such a scenario in the Welsh Basin cements comes from the presence of organic inclusions, sometimes seen to be elongated (with approximate dimensions of 2×0.3 μm), in some of the larger apatite crystallites (e.g. Fig. 9f). The absence of carbonates in turbidite mudstones suggests that the pH was low enough to prevent their precipitation. A relatively low pH increases the stability of apatites relative to that of calcium carbonates (Nathan & Sass 1981). Figure 10 diagrammatically summarizes the inferred, probably dominant, controls on early diagenetic apatite cement development in the *griestoniensis* Zone turbidites.

Controls on basin-wide distribution of phosphate cements

In Upper Llandovery sequences in Wales and the Welsh Borderland phosphate cements appear to be restricted to turbiditic deposits of the oxygen-deficient Welsh Basin. The apparent absence of phosphate cements in the shelf facies is simply explained as the result of oxidation of labile organic matter close to the sediment-water interface, preventing the burial of a phosphate source. This would have been enhanced by lower sedimentation rates on the shelf.

Phosphate cements are also absent from the overlying Wenlock turbidite sequence which was deposited in totally anoxic (euxinic) bottom waters (Dimberline & Woodcock 1987). This probably results from the diffusive loss of dissolved phosphate out into the bottom waters (cf. Watanabe & Tsunogai 1984).

Conclusions

(1) Phosphate enrichment in thin layers close to the tops of fine-grained turbidites occurred soon after deposition of turbidites in dysaerobic bottom water environments of the Welsh Basin.

(2) Phosphatic layers became lithified early, prior to significant sediment compaction.

(3) The localization of phosphate cements close to the tops of turbidites is thought to be due to early enrichment of phosphate at these levels. Two possible mechanisms are: (a) adsorption of dissolved phosphate diffusing upwards from underlying anoxic interstitial water by ferric and manganese oxyhydroxides, and (b) concentration by bacteria adapted to the oxic-sulphidic transition layer. The initiation of cement precipitation may have been catalysed by enzymes present in bacterial cells.

(4) The basin-wide control on the distribution of phosphate cements appears to have been primarily bottom water oxicity. This is because bottom water oxygen levels (in addition to sedimentation rate) control: (a) the flux of labile organic matter, the major source of phosphate, through the sediment-water interface and (b) the presence or absence of iron and manganese oxyhydroxides, and certain environmentally specific bacteria in layers close to the sediment surface. These limit the escape of dissolved phosphate into bottom waters.

Similar phosphate cements might be expected in other successions deposited in dysaerobic turbidite basins.

ACKNOWLEDGMENTS: This work has greatly benefited from numerous discussions with the following people: Max Coleman, Sue Colley, Peter Cook, Charles Curtis, Tony Dickson, Andrew Dimberline, Harry Elderfield, Cathy Goss, Ian Jarvis, Anton Kearsley, Ken Pye and John Thomson. Dick Cave first drew my attention to the presence of the phosphate cements. Peter Cook, Andrew Dimberline, and Jim Marshall suggested useful improvements to the manuscript. I would like to thank Dave Newling and Anton Kearsley for their assistance with the SEM work and Tony Abrahams for assistance with the XRD work. Tony Dickson and Alison Searle helped with the CL work. Dave Martell helped with the autoradiography work. I acknowledge a NERC research studentship. Cambridge Earth Science Series No. 977.

References

ANDERSON, J. J. 1982. The nitrite-oxygen interface at the top of the oxygen minimum zone in the eastern tropical North Pacific. *Deep-Sea Research*, **29**, 1193–201.

——, OKUBO, A., ROBBINS, A. S. & RICHARDS, F. A. 1982. A model for nitrite and nitrate redistributions in oceanic oxygen minimum zones. *Deep-Sea Research*, **29**, 1113–40.

BATURIN, G. N. 1982. *Phosphorites on the Sea Floor.* Developments in Sedimentology. Elsevier, Amsterdam.

—— & BEZRUKOV, P. L. 1979. Phosphorites on the sea floor and their origin. *Marine Geology*, **31**, 317–32.

BENTOR, Y. K. 1980. Phosphorites—the unsolved problems. *In*: BENTOR, Y. K. (ed.). *Marine Phosphorites—Geochemistry, Occurrence, Genesis*, pp. 3–18. Society of Economic Paleontologists and Mineralogists Special Publication, **29**.

BERNER, R. A. 1973. Phosphate removal from seawater by adsorption on volcanogenic ferric oxides. *Earth and Planetary Science Letters*, **18**, 77–86.

—— 1974. Kinetic models for the early diagenesis of nitrogen, sulfur, phosphorus and silicon in anoxic marine sediments. *In*: GOLDBERG, E. D. (ed.). *The Sea*, **5**, 427–50. Wiley, New York.

—— 1980. *Early Diagenesis. A Theoretical Approach.* Princeton Series in Geochemistry. Princeton University Press.

—— 1981. A new geochemical classification of sedimentary environments. *Journal of Sedimentary Petrology*, **51**, 359–65.

BORGAARD, O. K. 1983. Effect of surface area and mineralogy of iron oxides on their surface charge and anion-adsorption properties. *Clays and Clay Minerals*, **31**, 230–32.

BOUMA, A. H. 1962. *The Sedimentology of some Flysch Deposits.* Elsevier, Amsterdam.

BREMNER, J. M. 1980. Concretionary phosphorite from SW Africa. *Journal of the Geological Society of London*, **137**, 773–86.

BRIDGES, P. E. 1975. The transgression of a hard substance shelf: the Llandovery (Lower Silurian) of the Welsh Borderland. *Journal of Sedimentary Petrology*, **45**, 79–94.

BRINDLEY, G. W. & BROWN, G. (eds) 1980. *Crystal Structures of Clay Minerals and their X-ray Identification.* The Mineralogical Society, London.

BROMLEY, R. G. & EKDALE, A. A. 1984. Chondrites: a trace fossil indicator of anoxia in sediments. *Science*, **224**, 874–972.

BURNETT, W. C. 1977. Geochemistry and origin of phosphorite deposits from off Peru and Chile. *Bulletin of the Geological Society of America*, **88**, 813–23.

—— 1980. Apatite-glauconite associations off Peru and Chile: palaeo-oceanographic implications. *Journal of the Geological Society of London*, **137**, 757–64.

——, ROE, K. K. & PIPER, D. Z. 1983. Upwelling and phosphorite formation in the ocean. *In*: SUESS, E. & THIEDE, J. (eds). *Coastal Upwelling and its Sediment Record, Part A*, pp. 377–96. Plenum Press, New York.

BURNS, S. J. & BAKER, P. A. 1987. A geochemical study of dolomite in the Monterey Formation, California. *Journal of Sedimentary Petrology*, **57**, 128–39.

CAVE, R. 1979. Sedimentary environments of the basinal Llandovery of mid-Wales. *In*: HARRIS, A. L., HOLLAND, C. H. & LEAKE, B. E. (eds). *The Caledonides of the British Isles—reviewed*, pp. 517–26. Special Publication of the Geological Society of London No. 8.

COCKS, L. R. M., WOODCOCK, N. H., RICKARDS, R. B., TEMPLE, J. T. & LANE, P. D. 1984. The Llandovery Series of the type area. *Bulletin of the British Museum (Natural History)*, **38**, 131–82.

COLLEY, S. & THOMSONS, J. 1985. Recurrent uranium relocations in distal turbidites emplaced in pelagic conditions. *Geochimica et Cosmochimica Acta*, **49**, 2339–48.

——, ——, WILSON, T. R. S. & HIGGS, N. C. 1984. Post-depositional migration of elements during diagenesis in brown clay and turbidite sequences in the North East Atlantic. *Geochimica et Cosmochimica Acta*, **48**, 1223–35.

COOK, P. J. & SHERGOLD, J. H. 1986. Proterozoic and Cambrian phosphorites—nature and origin. *In*: COOK, P. J. & SHERGOLD, J. H. (eds). *Phosphate Deposits of the World, Volume 1*, pp. 369–86. Cambridge University Press.

CRAIG, J., FITCHES, W. R. & MALTMAN, A. J. 1982. Chlorite-mica stacks in low-strain rocks from central Wales. *Geological Magazine*, **119**, 243–56.

DE KANEL, J. & MORSE, J. W. 1978. The chemistry of orthophosphate uptake from seawater onto calcite and aragonite. *Geochimica et Cosmochimica Acta*, **42**, 1335–40.

DE LANGE, G. J. 1986. Early diagenetic reactions in interbedded pelagic and turbiditic sediments in the Nares Abyssal Plain (western north Atlantic): Consequences for the composition of sediment and interstitial water. *Geochimica et Cosmochimica Acta*, **50**, 2543–61.

DIMBERLINE, A. J. 1986. Electron microscope and microprobe analysis of chlorite-mica stacks in the Wenlock turbidites, mid Wales, U.K. *Geological Magazine*, **123**, 299–306.

—— & WOODCOCK, N. H. 1987. The southeast margin of the Wenlock turbidite system. *Geological Journal*, **22**, 61–71.

EDWARDS, B. D. 1985. Bioturbation in a dysaerobic bathyal basin: California Borderland. *In*: CURRAN, H. A. (ed.). *Biogenic Structures: their Use in Interpreting Depositional Environments*, pp. 309–31 Society of Economic Paleontologists and Mineralogists Special Publication, 35.

FAIRCHILD, I. J. 1980. Stages in Precambrian dolomitization, Scotland: cementing versus replacemen textures. *Sedimentology*, **27**, 631–50.

FITCHES, W. R. & CAMPBELL, S. D. G. 1987. Tectoni evolution of the Bala Lineament in the Wels Basin. *Geological Journal*, **22**, 131–53.

FROELICH, P. N., KLINKHAMMER, G. P., BENDER G. R., LUEDTKE, N. A., HEATH, G. R. CULLEN D., DAUPHIN, P., HAMMOND, D. & HARTMAN, F 1979. Early oxidation of organic matter in pelagi sediments of the eastern equatorial Atlantic suboxic diagenesis. *Geochimica et Cosmochimic Acta*, **43**, 1075–90.

GALLARDO, V. A. 1977. Large benthic microbi communities in sulfide biota under Peru–Chi subsurface counter current. *Nature*, **268**, 331—

GULBRANDSEN, R. A. 1970. Relation of carbon dioxic content of apatite of the Phosphoria Formation regional facies. *United States Geological Surv Professional Paper, 700B*, 9–13.

—— , ROBERSON, C. E. & NEIL, S. T. 1984. Time and crystallisation of apatite in sea water. *Geochimica et Cosmochimica Acta*, **48**, 213–8.

HARLAND, W. B., COX, A. V., LLEWELYN, P. G., PICKTON, C. A. G., SMITH, A. G. & WALTERS, R. 1982. *A Geological Time Scale*. Cambridge University Press.

JARVIS, I. & HIGGS, N. 1987. Trace element mobility during early diagenesis in distal turbidites: Late Quarternary of the Madeira Abyssal Plain, N Atlantic. *In*: WEAVER, P. P. E. & THOMSON, J. (eds). *Geology and Geochemistry of Abyssal Plains*, pp. 179–213. Geological Society Special Publication, 31. Blackwell Scientific Publications, Oxford.

JAHNKE, R. A., EMERSON, S. R., ROE, K. K. & BURNETT, W. C. 1982. The present day formation of apatite in Mexican continental margin sediments. *Geochimica et Cosmochimica Acta*, **47**, 259–66.

JONES, J. G. 1985. Microbes and microbial processes in sediments. *Philosophical Transactions of the Royal Society of London*, **A315**, 3–17.

KOKELAAR, B. P., HOWELLS, M. F., BEVINS, R. E., ROACH, R. A. & DUNKLEY, P. N. 1984. The Ordovician marginal basin of Wales. *In*: KOKELAAR, B. P. & HOWELLS, M. F. (eds). *Marginal Basin Geology*, pp. 245–71. Special Publication of the Geological Society of London, 16. Blackwell Scientific Publications, Oxford.

KROM, M. D. & BERNER, R. A. 1981. The diagenesis of phosphorus in a nearshore marine sediment. *Geochimica et Cosmochimica Acta*, **45**, 207–16.

LEGGET, J. K. 1980. British Lower Palaeozoic black shales and their palaeo-oceanographic significance. *Journal of the Geological Society of London*, **137**, 139–56.

—— , MCKERROW, W. S. & SOPER, N. J. 1983. A model for the crustal evolution of Southern Scotland. *Tectonics*, **2**, 187–210.

MANHEIM, F., ROWE, G. T. & JIPA, D. 1975. Marine phosphorite formation off Peru. *Journal of Sedimentary Petrology*, **45**, 243–51.

MARSHALL, J. D. 1982. Isotopic composition of displacive fibrous calcite veins: reversals in porewater composition during burial diagenesis. *Journal of Sedimentary Petrology*, **52**, 615–30.

MARTENS, C. S. & HARRIS, R. C. 1970. Inhibition of apatite precipitation in the environment by magnesium ions. *Geochimica et Cosmochimica Acta*, **34**, 621–5.

MAYNARD, J. B. 1982. Extension of Berner's 'New Geochemical classification of sedimentary environments' to ancient sediments. *Journal of Sedimentary Petrology*, **52**, 1325–31.

MCARTHUR, J. M. 1985. Francolite geochemistry—compositional controls during formation, diagenesis, metamorphism and weathering. *Geochimica et Cosmochimica Acta*, **49**, 23–35.

MCCLELLAN, G. H. 1980. Mineralogy of carbonate-fluorapatite. *Journal of the Geological Society of London*, **137**, 675–81.

MCCONNELL, D., FRAJOLA, W. J. & DEANER, D. W. 1961. Relation between inorganic chemistry and biochemistry of bone mineralization, *Science*, **133**, 281–2.

MCKERROW, W. S. 1979. Ordovician and Silurian changes in sea level. *Journal of the Geological Society of London*, **136**, 137–45.

MÜLLER, P. J. & MANGINI, A. 1980. Organic carbon decomposition rates in sediments of the Pacific Manganese Nodule Belt dated by ^{230}Th and ^{231}Pa. *Earth and Planetary Science Letters*, **51**, 94–114.

MULLINS, H. T. & RASCH, R. F. 1985. Sea-floor phosphorites along the Central California continental margin. *Economic Geology*, **80**, 696–715.

MUTTI, E. & NORMARK, W. R. 1987. Comparing examples of modern and ancient turbidite systems: problems and concepts. *In*: LEGGETT, J. K. & ZUFFA, G. G. (eds). *Deep Marine Sedimentology: Models and Case Histories*. Graham & Trotman, London.

NATHAN, Y. & SASS, E. 1981. Stability relations of apatites and calcium carbonates. *Chemical Geology*, **34**, 103–11.

NISSENBAUM, A., PRESLEY, B. J. & KAPLAN, I. R. 1972. Early diagenesis in a reducing fjord, Saanich Inlet, British Columbia, I, Chemical and isotopic changes in major components of interstitial water. *Geochimica et Cosmochimica Acta*, **36**, 1007–27.

O'BRIEN, G. W. & VEEH, H. H. 1980. Holocene phosphorite on the East Australian continental margin. *Nature*, **288**, 690–2.

—— & —— 1982. Are phosphorites reliable indicators of upwelling? *In*: SUESS, E. & THIEDE, J. (eds). *Coastal Upwelling, its Sediment record*, pp. 399–419. Plenum Press, New York.

PICKERING, K. T. 1987. Wet-sediment deformation in the Upper Ordovician Point Leamington Formation: an active thrust-imbricate system during sedimentation, Notre Dame Bay, north-central Newfoundland. *In*: JONES, M. E. & PRESTON, R. M. F. (eds). *Deformation of Sediments and Sedimentary Rocks*, pp. 213–39. Geological Society Special Publication, 29. Blackwell Scientific Publications, Oxford.

PIPER, J. D. A. 1985. Palaeomagnetism in the Caledonian-Appalachian orogen: a review. *In*: GEE, D. G. & STURT, B. A. (eds). *The Caledonide Orogen: Scandinavia and Related Areas*, pp. 35–49. Wiley, London.

PRÉVOT, L. & LUCÀS, J. 1986. Microstructure of apatite replacing carbonate in synthesised and natural examples. *Journal of Sedimentary Petrology*, **56**, 153–9.

PYE, K. 1985. Electron microscope analysis of zoned dolomite rhombs in the Jet Rock Formation (Lower Toarcian) of the Whitby area, U.K. *Geological Magazine*, **122**, 279–86.

—— & KRINSLEY, D. H. 1984. Petrographic examination of sedimentary rocks in the SEM using backscattered electron detectors. *Journal of Sedimentary Petrology*, **54**, 877–88.

REEBURGH, W. S. 1983. Rates of biogeochemical processes in anoxic sediments. *Annual Reviews of Earth and Planetary Sciences*, **11**, 269–98.

RHOADS, D. C. & MORSE, J. W. 1971. Evolutionary and ecologic significance of oxygen-deficient marine basins. *Lethaia*, **4**, 413–28.

SAVRDA, C. E. & BOTTJER, D. J. 1986. Trace fossil model for the reconstruction of palaeo-oxygenation in bottom waters. *Geology*, **14**, 3–6.

——, ——, & GORSLINE, D. S. 1984. Development of a comprehensive oxygen-deficient marine biofacies model: Evidence from Santa Monica, San Pedro, and Santa Barbara Basins, California continental borderland. *American Association of Petroleum Geologists Bulletin*, **68**, 1179–92.

SCHAFFER, G. 1986. Phosphate pumps and shuttles in the Black Sea. *Nature*, **321**, 515–7.

SHOLKOVITZ, E. R. 1973. Interstitial water chemistry of the Santa Barbara Basin sediments. *Geochimica et Cosmochimica Acta*, **37**, 2043–73.

SMITH, R. D. A. 1986. Giant erosional structures in the channel-lobe transition zone of a Welsh Basin turbidite system. *Abstracts of the annual meeting of the British Sedimentological Research Group, Nottingham.*

—— 1987a. The *griestoniensis* zone turbidite system, Welsh Basin. *In*: LEGGETT, J. K. & ZUFFA, G. G. (eds). *Deep Marine Clastic Sedimentology: Models and Case Histories*, pp. 89–107. Graham & Trotman, London.

—— 1987b. Structure and deformation history of the Central Wales Synclinorium, NE Dyfed: evidence for a long-lived basement structure. *Geological Journal*, **22**, 183–98.

—— 1987c. Palaeo-oxygenation in a turbidite basin: reconstruction of lateral gradients and variations through time using trace fossil evidence. *Abstracts of the 8th European Meeting of the International Association of Sedimentologists, Tunis.*

SOPER, N. J., WEBB, B. C. & WOODCOCK, N. H. 1987. Late Caledonian (Acadian) transpression in North West England: timing, geometry and geotectonic significance. *Journal of the Geological Society of London*, in press.

SOUTAR, A. & CRILL, P. A. 1977. Sedimentation and climatic patterns in the Santa Barbara basin during the 19th and 20th centuries. *Bulletin of the Geological Society of America*, **88**, 1161–72.

STOW, D. A. V. & SHANMUGAM, G. 1980. Sequence of structures in fine-grained turbidites; comparison of recent deep sea and ancient flysch sediments. *Sedimentary Geology*, **25**, 23–42.

THOMSON, J., COLLEY, S. HIGGS, N. C., HYDES, D. J., WILSON, T. R. S. & SORENSEN, J. 1987. Geochemical oxidation fronts in NE Atlantic distal turbidites and their effects in the sedimentary record. *In*: WEAVER, P. P. E. & THOMSON, J. (eds). *Geology and Geochemistry of Abyssal Plains*, pp. 167–77. Special Publication of the Geological Society of London, 31. Blackwell Scientific Publications, Oxford.

THOMSON, J. B., MULLINS, H. T., NEWTON, C. R. & VERCOUTERE, T. L. 1985. Alternative biofacies model for dysaerobic communities. *Lethaia*, **18**, 167–79.

VEITCH, F. P. & BLAKENSHIP, L. C. 1963. Carbonic anhydrase in bacteria. *Nature*, **197**, 76–7.

WATANABE, Y. & TSUNOGAI, S. 1984. Adsorption-desorption control of phosphate in anoxic sediment of a coastal sea, Funka Bay, Japan. *Marine Chemistry*, **15**, 71–83.

WILLIAMS, L. A. & REIMERS, C. 1983. Role of bacterial mats in oxygen-deficient marine basins and coastal upwelling regimes: preliminary report. *Geology*, **11**, 267–9.

WILSON, T. R. S., THOMSON, J., COLLEY, S., HYDES, D. & HIGGS, N. C. 1985. Early organic diagenesis: the significance of progressive subsurface oxidation fronts in pelagic sediments. *Geochimica et Cosmochimica Acta*, **49**, 811–22.

——, ——, HYDES, D. J., COLLEY, S., CULKIN, F. & SORENSEN, J. 1986. Oxidation fronts in pelagic sediments: diagenetic formation of metal-rich layers. *Science*, **232**, 972–5.

WOOD, E. M. R. 1906. The Tarannon Series of Tarannon. *Quarterly Journal of the Geological Society of London*, **62**, 644–701.

WOODCOCK, N. H. 1984. Early Palaeozoic sedimentation and tectonics in Wales. *Proceedings of the Geologists' Association of London*, **95**, 323–35.

WOODLAND, B. G. 1964. The nature and origin of cone-in-cone structure. *Fieldiana: Geology*, **13**, 185–305.

ZIEGLER, A. M., COCKS, L. R. M. & MCKERROW, W. S. 1968. The Llandovery transgression of the Welsh Borderland. *Palaeontology*, **11**, 736–82.

R. D. A. SMITH, Department of Earth Sciences, University of Cambridge, Downing Street, Cambridge CB2 3EQ, UK.

Mixing zone dolomitization of Devonian carbonates, Guangxi, South China

Zhao Xun & Ian J. Fairchild

SUMMARY: Regional facies relationships indicate that dolomitization around the Lower–Middle Devonian boundary in Guangxi region is largely restricted to platform margin facies. In the Liujin area dolostones are represented by the Tangding and succeeding Yingtang Formations. The latter comprises alternations of laminated fine-grained dolostones (type A dolomite) and massive, more coarsely-crystalline dolostones (type B dolomite). Type B is richer in marine fossils and other allochems, and increases in importance upwards. The Formation is bounded above by a disconformity marked by a haematitic marl, interpreted as a subaerial erosion surface, and is succeeded by limestones lacking dolomite.

Petrographic observations indicate that the two dolomite types represent the dolomitization of originally different sedimentary facies by a single dolomitization process. The dolomite is dominantly replacive, but also partly cementing. Variably negative $\delta^{13}C$ and $\delta^{18}O$ values (interpreted as original) and covariation of these parameters with Sr and Na points to dolomitization by mixed meteoric-marine fluids. Stratigraphic variation in geochemical data is inconsistent with a single episode of subaerial exposure and therefore points to repeated emergence.

Dominantly replacive dolostones can form by the mixing-zone mechanism given sufficiently stable conditions.

Land (1980) stated that the essence of the 'dolomite problem' was the dolomitization of platform carbonates. Of these, dolostones that contain a normal marine fauna are particularly intriguing. Following the documentation of Holocene and Pleistocene dolomite formation in meteoric-marine mixing zones (e.g. Land 1973a, b), this setting has been used by many authors to explain examples of ancient platform dolomite. Recently, Machel & Mountjoy (1986) have vigorously disputed the effectiveness of mixing-zone dolomitization and pointed out the equivocal origin of many alleged examples. They contrast the massive, replacive nature of many ancient occurrences of platform dolostones with the abundance of dolomite cement in undisputed mixing-zone dolomitic rocks (e.g. Kaldi & Gidman 1982, Ward & Halley 1985). Indeed, where geochemical evidence has been available for alleged ancient mixing-zone dolostones, it has often been at best consistent with the proposed model rather than strongly supporting it (e.g. the rather heavy carbon and oxygen isotopes in a Carboniferous example described by Choquette & Steinen 1980). In contrast, the majority of ancient platform dolostones have rather negative oxygen isotope values, and Land (1980) concluded that crystallization or recrystallization of dolomite in contact with meteoric and/or hot waters must have occurred. In this paper we describe an example of a dolostone unit where the field data suggesting dolomitization associated with emergence are backed by geochemical evidence, particularly a sympathetic variation of carbon and oxygen isotopes with values somewhat lighter than expected for marine waters. Machel & Mountjoy's (1986) arguments are re-examined in the light of this evidence.

The strata concerned are located in the Guangxi region, South China (Fig. 1) where platform carbonates largely of Middle to Upper Devonian age succeed Lower Devonian non-marine siliciclastic deposits. Much research effort has been expended on these rocks over several decades because of their associated mineralization and their petroleum potential, although this information is largely in unpublished reports of the Guangxi Geological Bureau (but see also Hou 1978, Liao *et al.* 1979, Bai *et al.* 1982, pp. 22–6, Kuang 1982, Dineley 1984). This article synthesizes information produced by the Guangxi Geological Bureau in collaboration with the Chinese Academy of Sciences concerning field relationships and palaeontology, and introduces new laboratory data obtained in the UK.

Methods

Having observed various samples in stained thin section, selected samples were studied by cathodoluminescence (CL) in polished section, X-ray diffraction (XRD), particularly for characterizing dolomite stoichiometry; the geochemistry of

From: MARSHALL, J. D. (ed.), 1987, *Diagenesis of Sedimentary Sequences*, Geological Society Special Publication No. 36, pp. 157–170.

the acid-soluble fraction was analysed by inductively-coupled plasma emission spectrometry (ICP); most of these samples were also analysed for carbon and oxygen isotopes and some were studied in the Scanning Electron Microscope (SEM), including a little quantitative analysis.

Samples for XRD and ICP were drilled from specimens and ground to 10–50 μm. XRD was carried out at 40 kV and 20 mA using Cu Kα radiation. ICP analysis was carried out on solutes from cold 10% HCl dissolution of samples at dilutions of 1:200 (50 mg sample in 10 ml final solution) and using matrix-matched standards. Accuracy and precision are only around 10 relative % because of initial problems in establishing standardization procedures. A larger (0.1–3 g) portion of each powder was also dissolved in HCl for determination of insoluble residue (IR) content. Comparison of XRD and ICP results indicate negligible calcite impurity in the dolomite.

For oxygen and carbon stable isotope analysis, 10 mg of each sample powder were reacted with 100% phosphoric acid at 25° for 72 hr and analysed along with a laboratory standard dolomite on a VG 903 mass spectrometer. A phosphoric acid fractionation factor of 1.01109 (Sharma & Clayton 1965) was used in calculating results, which are expressed with respect to the PDB standard and are reproducible to 0.05‰.

Field relationships

Studies of numerous cross-sections of Devonian carbonates in the Guangxi region indicate that several types of dolomite are present. One type occurs beneath exposure surfaces and so a mixing-zone origin was proposed. The dolomitized strata are either reefs or bioclastic bank accumulations that have topographic relief, making them more likely to have suffered subaerial exposure than surrounding strata. Three examples are given below.

First, in the Beishan section and associated with a base-metal deposit is a stromatoporoid-algal reef within the Middle to Upper Devonian Dongganglin and Guilin Formations. The reef is

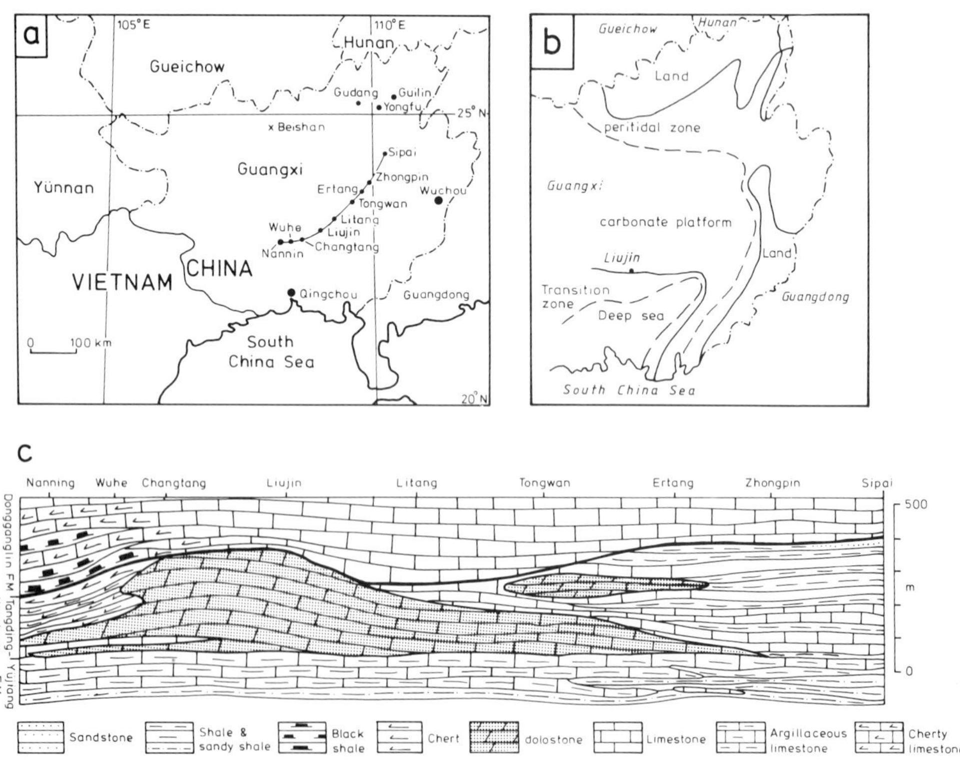

FIG. 1. (a) Geography of the Guangxi region including location of line of section in (c). (b) Palaeogeography of the Guangxi region around the boundary of Lower and Middle Devonian times. (c) Reconstruction of lithofacies relationships of Lower and Middle Devonian shelf sediments in the Guangxi region. The Liujin section provided the specimens used in the present study. An approximate time-line corresponding to the top of the Yingtang Formation is shown as a heavy line.

100 m thick of which the upper 50 m is dolomitized completely and the lower half partially.

Second, in the Gudang section, the 270 m thick Dongganglin Formation is dolomitic throughout, and contains 40–75% transported bioclasts (stromatoporoids, echinoderms, ostracodes, calcispheres, corals and brachiopods) and some *in situ* stromatoporoids and crinoids in a lime-mud matrix. It passes laterally shorewards into laminated dolostones with few fossils (calcispheres and tentaculitids). The top is marked by a scoured surface and conglomerate with iron oxide-stained pebbles. Pores in the dolomite contain bitumen and there is much pyrite and base-metal mineralization which also appears to post-date dolomitization.

The third example is from the regional standard cross-section at Liujin where dolostones of the Tangding and Yingtang Formations straddle the Lower–Middle Devonian boundary. Palaeontological evidence indicates that the Tangding Formation corresponds with the Upper Zlickovian stage of the European Lower Devonian. The Yingtang Formation, for which laboratory data are given in this paper, contains fossils indicative of the Dalejian to Eifelian stages of the Middle Devonian, and the succeeding Dongganglin Formation is of Givetian age (Dineley 1984).

Figure 1(c) summarizes the stratigraphic relationships. The dolostones at Liujin were located at the margin of a carbonate platform (Fig. 1b) separating restricted shelf carbonate facies (eventually passing landwards into terrigenous clastics) from slope-related cherty limestones.

The Tangding Formation is mostly composed of dark grey fine-grained dolostones with millimetre-scale lamination, birdseye structure and

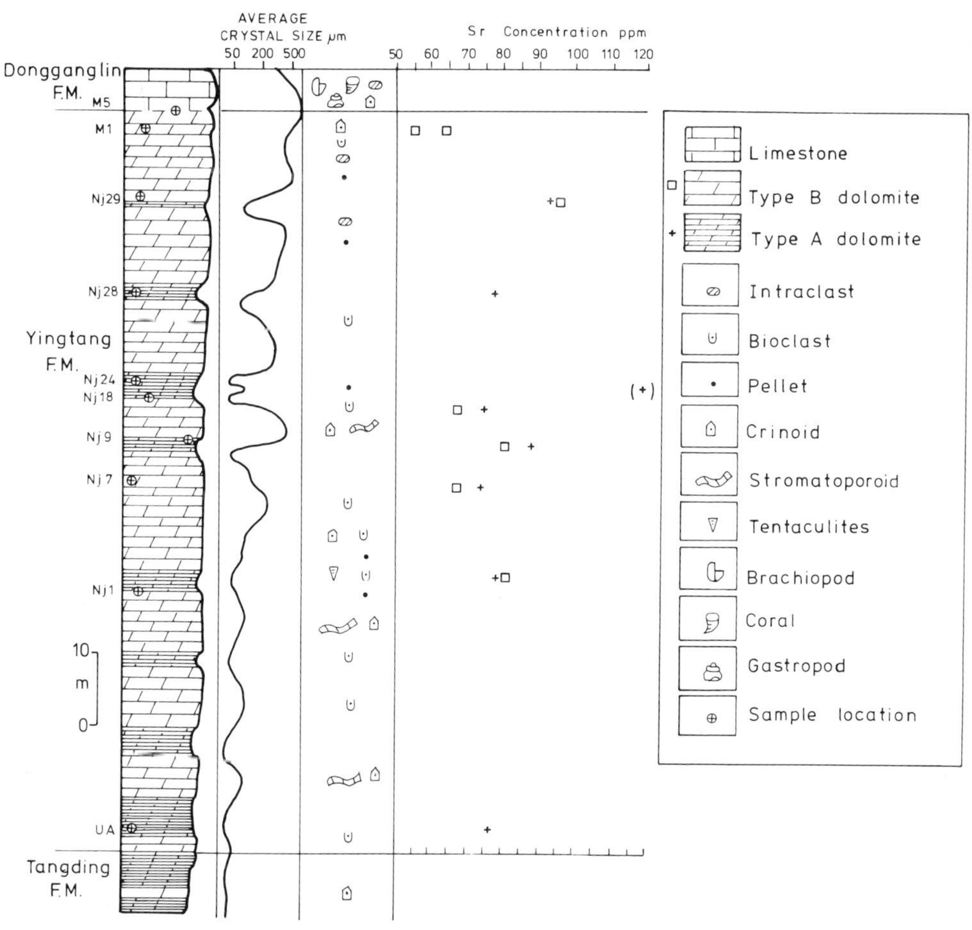

FIG. 2. Stratigraphy of the Yingtang Formation in the Liujin cross-section together with the stratigraphic distribution of Sr (key as in Table 1).

local silicification including length-slow chalcedony and dolomitized calcarenites. The sediments contain a fauna dominated by nektonic/planktonic tentaculitids, ostracodes and ammonoids, with the addition of echinoderms and gastropods higher up.

The Yingtang Formation (Fig 2) likewise contains some laminated dolostones, here referred to as *type A dolomite*. These dolostones contain fossil moulds and show synsedimentary brecciation in places. However, increasingly upwards, the sediments consist of more massive, porous and coarsely-crystalline dolostones, referred to as *type B dolomite*. Although the fabric-destructive nature of the dolomitization makes recognition of primary characteristics difficult, bioclasts can be recognized: dominantly crinoids and stromatoporoids, with subordinate bryozoans, gastropods, brachiopods and corals. There are also intraclasts and peloids (possibly originally ooids). Some sinusoidal lamination and normal and reverse grading suggestive of intense wave or current action are recognizable in places. Overall, the Yingtang Formation appears to represent a bank or shoal with intercalations of finer sediments representing more sheltered environments. It is notable that base-metal deposits co-exist with the dolomite, especially on the upper part of the onshore side of the bank.

The top of the Yingtang Formation is marked by a discontinuity surface with a haematitic marl layer (Kuang 1982) followed upwards by a conglomerate of carbonate pebbles and then dolomite-free limestones with a diverse open-marine fauna (Fig. 2). This time-horizon was also marked by uplift in the continental sediments in the northern basin margin (northern Guangxi and southern Gueichow, Bai *et al.* 1982) where a prominent disconformity is found.

The conclusion drawn from the field data was that the type B dolostones were formed by mixing-zone dolomitization associated with exposure at the top of the Yingtang Formation. The type A dolostones were thought to have been dolomitized prior to emergence. The laboratory data discussed below point to a mixing-zone origin for both types of dolomite, but with repeated exposure during deposition of the Yingtang Formation in addition to the final episode of emergence.

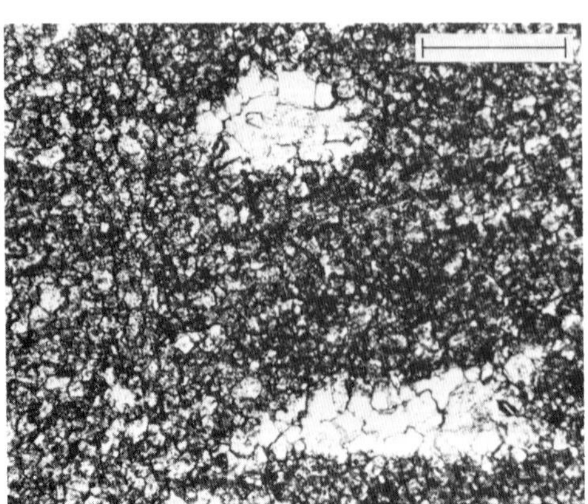

FIG. 3. Typical inclusion-rich type A dolomite as seen in plane polarized light (sample UA). Moulds of probable tentaculitids are present. Scale bar 100 μm.

Petrography

The type A dolostones are completely dolomitized carbonate mudstones or wackestones, with some local dolomite-cemented fossil moulds

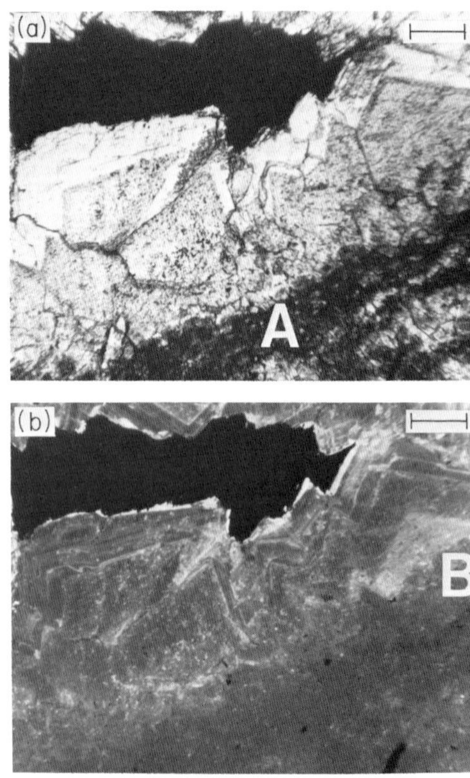

FIG. 4. Sample Nj-1 seen in plane polarized light (a) and under CL (b). Scale bar 100 μm. Dark area, labelled A in (a) is type A dolomite seen in (b) to have identical luminescence characteristics to the earlier growth stages (labelled B) of the zoned type B dolomite. The latter terminates against a vug at the top of the photographs (ragged edges may be an artefact).

(Fig. 3) and lamination. The matrix has crystal sizes typically of a few tens of microns, and contains a high density of inclusions of fluid, calcite, iron ?oxide, pyrite, clay (identified by qualitative analysis on SEM) and very finely-disseminated organic carbon. Such dolomite also makes up the matrix of a sedimentary breccia in one sample.

The type B dolostones (Figs 4 to 6) are composed of more coarsely-crystalline dolomite, zoned by density of inclusions (mostly of calcite and fluid, Fig. 5a). Type B is also interlaminated with type A in some samples, indicating that the two dolomite types represent original sedimentary layers of different composition. Fossil-rich layers (presumed original grainstones or packstones) are always preserved as type B dolomite which is fabric-destructive (Fig. 5b). Conversely, type A dolomite seems to represent replacement of micritic or dolomicritic layers rich in organic carbon.

A combination of observations in transmitted light and by cathodoluminescence reveals the growth history of the dolomite. In a given sample

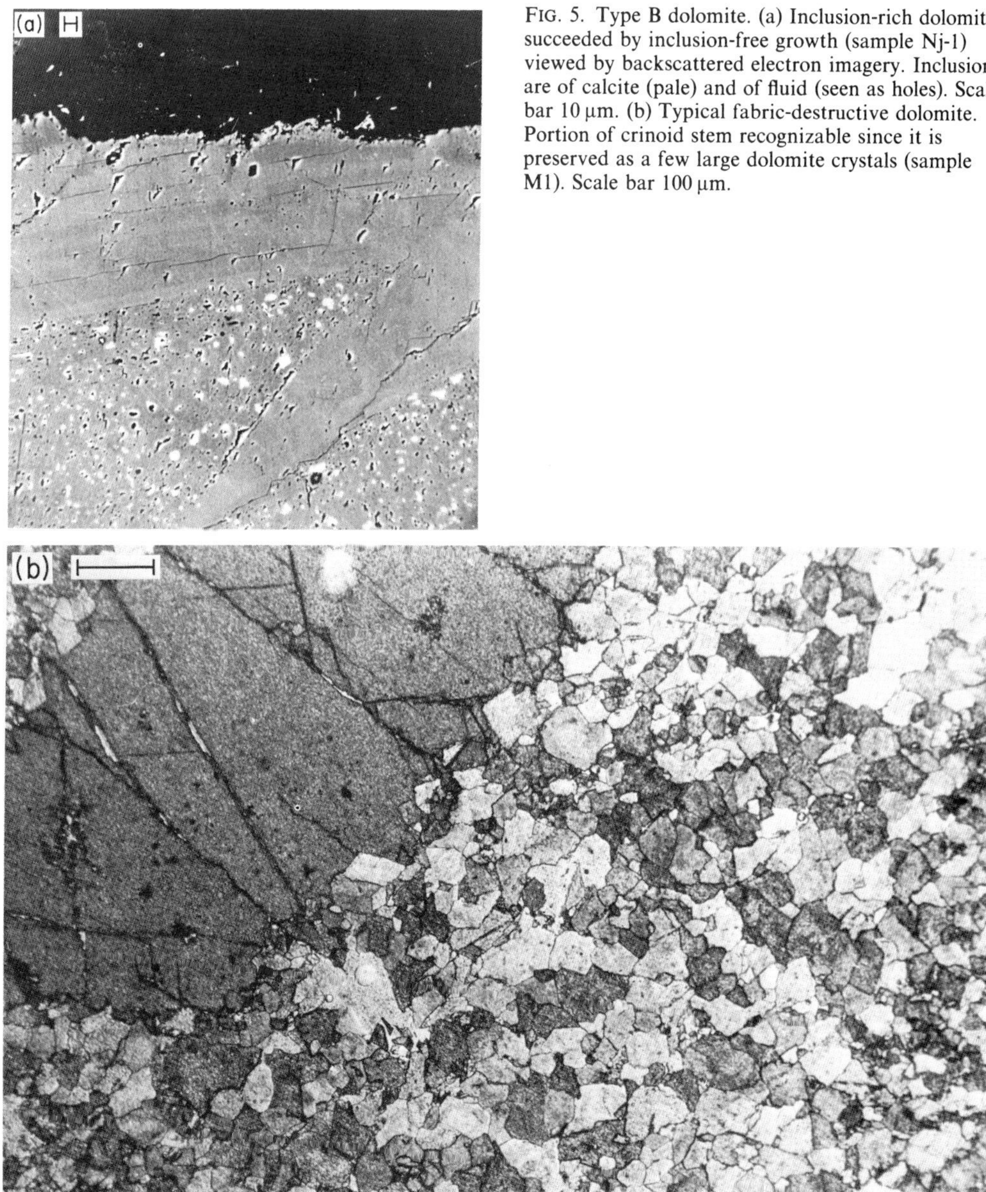

FIG. 5. Type B dolomite. (a) Inclusion-rich dolomite succeeded by inclusion-free growth (sample Nj-1) viewed by backscattered electron imagery. Inclusions are of calcite (pale) and of fluid (seen as holes). Scale bar 10 μm. (b) Typical fabric-destructive dolomite. Portion of crinoid stem recognizable since it is preserved as a few large dolomite crystals (sample M1). Scale bar 100 μm.

type A has the same luminescence characteristics as the earlier growth stages of type B dolomite (e.g. Fig. 4). Typically, CL is very dull red, becoming brighter near crystal boundaries, although a more brightly-luminescing core is shown in specimens Nj-24 and M1 (see Fig. 2 for stratigraphic location). Half to 95% of each sample is represented by early growth stages of dolomite (types A, B or both) as just described.

Later stages of growth of type B dolomite generally show a reduction in the proportions of inclusions, locally with a sharply-defined and narrow inclusion-free zone followed sharply by more inclusion-bearing dolomite and then finishing with inclusion-free growth. Since some of the inclusions are calcite crystals (Fig. 5a) far larger in size than calcites known to have formed by exsolution of dolomite (Wenk & Zhang 1985), it can be deduced that the calcite inclusions are relics of the original sediment, now replaced. At some stage the inclusion-free growth became cementing rather than replacive since the dolomite crystals border cavities in several samples.

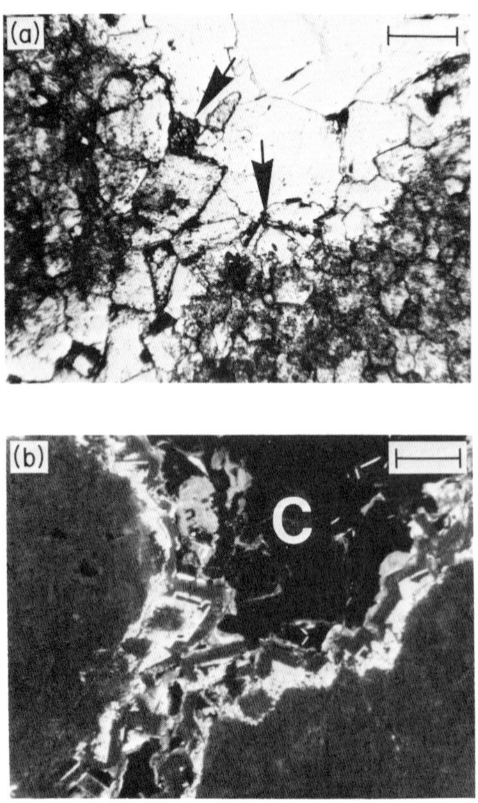

FIG. 6. Sample Nj-29 seen in plane polarized light (a) and under CL (b). Scale bar 100 μm. Dolomite crystals projecting into vugs are coated by haematite (arrowed in a) and succeeded by calcite cement (c).

CL zonation is invariably shown by the dolomites, although commonly only in the later stages of growth. Nj-28 and Nj-29 show an apparently identical zonation of later growth stages, but other samples differ from each other in varying degrees and show no correlation if inclusion density as well as luminescence intensity are considered (e.g. compare Figs 4 and 6).

Nj-29 displays a haematitic rind (?with clay) on the margins of dolomite crystals which grew into former vugs (Fig. 6) and the final event in this and samples Nj-7 and Nj-24 was a calcite filling of the vugs.

Sample M5, just below the disconformity at the top of the Formation, displays small interstitial areas of undolomitized limestone, as well as later vug-fills of calcite.

In summary, the dolomite of the Yingtang Formation is dominantly replacive. In a given sample dolomite crystals started to grow simultaneously in both carbonate mudstones and wackestones rich in organic carbon and in other lithologies. Crystal nuclei were more closely spaced in the carbonate mudstones and wackestones so that growth was concluded earlier to yield type A dolomite with its relatively fine crystal size. In other lithologies type B dolomite continued to grow and eventually became cementing. Although some vugs are primary cavities in fossils, most are not obviously related to the primary rock fabric and so are probably secondary. In general the zonation of the crystals is not identical in different samples, therefore growth of the dolomite was not necessarily synchronous throughout the Formation. Following dolomitization, haematite rinds and calcite cements formed locally.

Geochemistry

Table 1 shows the data and Figs 7 to 10 illustrate the most important geochemical relationships. Subsamples of types A and B are distinguished on the plots. In some cases it was possible to subsample areas of dolomite dominated by relatively earlier (type A or early zones of type B) or later (type B) growth stages. Wherever subsamples differing in relative age of crystal growth are available they are connected by an arrow on the plots to show the geochemical trends with time for each sample.

One clear feature of the diagrams is the complete overlap of the chemical data for types A and B dolomite and the conflicting trends of variation from early to late stages of growth in different samples. This indicates that type A and all stages of type B dolomite can potentially be

TABLE 1.

Sample	Type A dolomite	Type B dolomite	Crystal size, μm	$\delta^{13}C$ ‰	$\delta^{18}O$ ‰	Sr ppm	Na ppm	Fe ppm	Mn ppm	mole% $CaCO_3$ in dolomite	%IR
UA	+		40	−3.26	−4.24	76	215	972	128	55.7	5.29
Nj1-1	+		52	−2.41	−2.62	78	215	198	53	55.7	1.31
Nj1-2		Early and late	131	−2.81	−3.70	79	210	445	65	53.2	4.43
Nj7-1		Early	86	−2.42	−4.57	74	195	196	59	54.2	0.32
Nj7-2		Late	242	−3.58	−5.50	67	179	196	70	56.9	0.25
Nj9-1	+		63	−2.32	−4.23	88	196	437	80	52.8	2.72
Nj9-2		Early and late	157	−2.32	−4.51	81	164	394	56	52.6	0.71
Nj18-1	+		66			75	188	287	69	52.8	2.02
Nj18-2		Early	156			68	227	704	73	52.6	0.90
Nj24	(+)		99	−1.17	−4.93	120	215	408	91	54.2	1.18
Nj28	+		68			78	167	512	130	52.9	5.32
Nj29-1	+		69	−0.84	−2.34	93	255	433	80	51.7	5.90
Nj29-2		Early and late	131	−0.81	−1.53	96	255	289	81	52.4	8.82
M1-1		Early	86			56	175	323	59	52.3	0.29
M1-2		Late	202			65	167	175	56	52.4	0.18

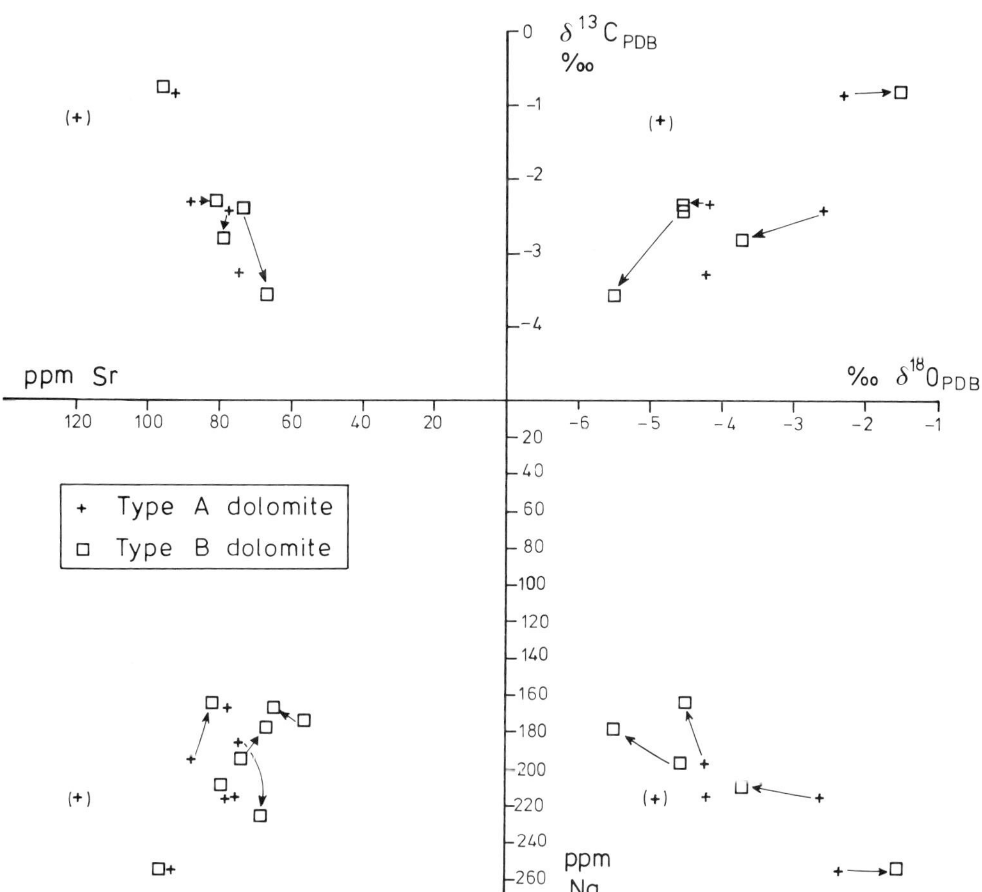

FIG. 7. Plot of $\delta^{13}C$, $\delta^{18}O$, Sr and Na to illustrate their covariation. Key as Table 1; significance of correlations listed in Table 2.

TABLE 2. *Correlations between parameters expressed as the Pearson correlation coefficient (calculated by the methods of Lemaitre 1982), with significance level in parentheses (only correlations significant at minimum 90% level included)*

Covariant parameters	Significance level	
	without Nj-24	with Nj-24
$\delta^{13}C$-Sr	+0.92(99)	+0.81(97)
$\delta^{18}O$-Na	+0.88(98)	+0.81(97)
Sr-IR	+0.75(99)	+0.40(<50)
Fe-Mn	+0.71(98)	+0.70(98)
$\delta^{13}C$-%$CaCO_3$	−0.82(96)	−0.73(92)
$\delta^{13}C$-$\delta^{18}O$	+0.81(96)	+0.60(66)
Na-IR	+0.64(95)	+0.61(94)
$\delta^{18}O$-Sr	+0.80(95)	+0.19(<50)
Sr-Na	+0.62(93)	+0.51(78)
IR-Mn	+0.61(92)	+0.57(89)
$\delta^{18}O$-IR	+0.75(90)	+0.76(94)

explained by a single process with reversible geochemical characteristics.

The data show clear covariations between several variables (Table 2). One sample (Nj-24) does not fit these trends as far as $\delta^{18}O$ is concerned. This sample contains some clear dolomite veins with bright luminescence, similar to veins (separated from a younger Devonian sample in another cross-section) which yielded $\delta^{18}O$ of −12. Although such veins do not occupy more than 10% of the sample, as seen in thin section, it is possible that a higher density of veins was met with when drilling out the sample powder. The oxygen isotope composition of Nj-24 is thus set aside in the following discussion.

The data show a clear positive covariation of $\delta^{13}C$, Sr, $\delta^{18}O$ and Na (Fig. 7), inverse correlation of $\delta^{13}C$ and Ca content of dolomite (Fig. 8), positive correlation of Fe and Mn (Fig. 9) and positive correlation of IR with Mn (Fig. 10), Sr, Na, and $\delta^{18}O$ (Table 2). It is notable that crystal size shows no correlation with geochemical parameters. This is in agreement with the lack of overall geochemical trends with crystal growth, as already noted.

The correlation of IR with Na could be explained by some leaching of these elements from insoluble residue (Veizer *et al.* 1977), although the correlations with Sr, Mn and $\delta^{18}O$ are unexpected. However, considerable geochemical variation occurs in samples with low content of IR (e.g. Fig. 10). Also, the IR of Nj-29 is clearly post-depositional haematite, different from the IR of other samples. Removal of this data point removes the statistically significant correlation of IR with Na and $\delta^{18}O$.

Isotopes

The assumption is made in the following discussion that the dolomites preserve their original isotope values. This is warranted by their preserved elemental zones. In contrast, many ancient finely crystalline dolomites seem to have undergone some depletion in ^{18}O, presumably due to recrystallization (Land 1980). Hence we do not compare the Yingtang dolomites with Palaeozoic fine-grained dolomites, but with well-preserved Devonian calcitic components which allow the isotopic composition of ancient sea water to be inferred. Recent work on Devonian brachiopods (Popp *et al.* 1986, Veizer *et al.* 1986) suggests that if one assumes no temperature shift

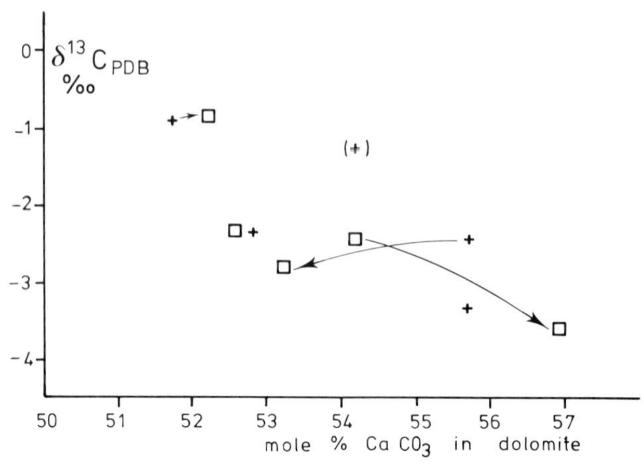

FIG. 8. Plot of $\delta^{13}C$ versus Ca content of dolomite (see also Tables 1 and 2).

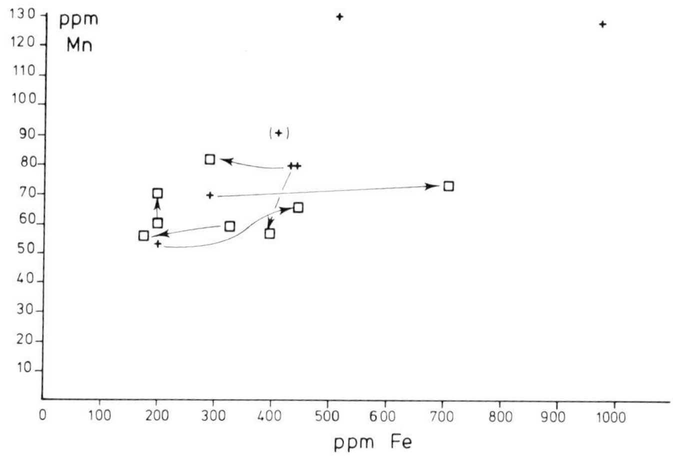

FIG. 9. Plot of Mn versus Fe (see also Tables 1 and 2).

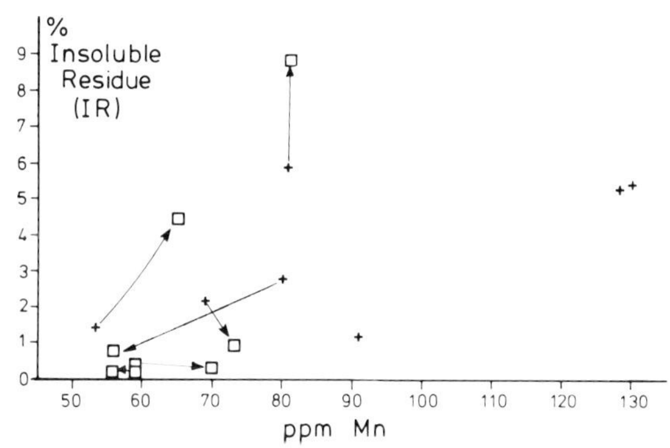

FIG. 10. Plot of IR versus Mn (see also Tables 1 and 2).

in low-latitude seas, the hydrosphere would have been depleted in ^{18}O by about 3‰ compared with today: hence sea water $\delta^{18}O$ would have been of the order of −4. Since there is no evidence that evaporites ever formed in or directly above the Yingtang Formation, hypersaline fluids need not be considered here. Using the dolomite-water fractionation equation advocated by Land (1983) and removing the Sharma & Clayton (1965) correction to the data, if sea water were the dolomitizing agent, palaeotemperatures would fall in the range 22–37°C. One can also consider the isotopic range of fluids corresponding to the dolomites' isotopes under isothermal conditions. If the temperature was 20°C, the fluids would have ranged from around −4 to −8 SMOW. These data seem most consistent with near surface dolomitization by mixed meteoric-marine waters, with a dominant marine component.

Marine Devonian carbonates have carbon isotopic composition around zero to +1 (Holser 1984). Since the Yingtang dolomites are isotopically negative they must have incorporated excess ^{12}C from organic or atmospheric sources. Given the conclusions of the last paragraph, a freshwater source of excess ^{12}C would have been available if the carbonate bank became emergent. A meteoric lens would have developed with isotopically negative carbon arising from the dissolution of carbonates by rainwater charged with atmospheric CO_2, and perhaps by CO_2^- incorporation into groundwater from decay of land plants colonizing the emergent surface. Two other possible mechanisms for ^{13}C-depletion are introduction of isotopically light carbon to porewater by bacterially mediated reactions such as sulphate reduction, and introduction of light CO_2 by thermal breakdown of organic matter. These are discussed below.

A positive covariation of oxygen and carbon isotopes can arise in different circumstances. The mechanism we prefer is that of mixing of meteoric and marine waters (e.g. Allan & Matthews 1982, Ward & Halley 1985). The slope of the covariation will vary in different circumstances, depending on both the isotopic composition of the groundwater and the degree of inheritance of carbon isotopes from precursor carbonate, the latter perhaps being relatively small in this case. An alternative mechanism arises during burial diagenesis where increasing temperatures, leading to progressively more negative oxygen isotopes in cements, are accompanied by increasing input of isotopically light carbon to pore waters by decarboxylation (e.g. Dickson & Coleman 1980, Tucker 1982). This is thought inapplicable to the Yingtang dolomites for the following reasons: (1) there is no relation between growth stage and isotopic values: (2) oxygen isotope values are too heavy; (3) the sharp top to the dolomitized strata suggests near-surface dolomitization; (4) variable amounts of burial diagenetic dolomite are present in higher parts of the Devonian sections, they differ in being more coarsely crystalline (average around 200 μm) and more negative oxygen isotope values (12 analyses average −8‰).

The existence of the covariation is also an argument against incorporation of much carbon via sulphate reduction since sulphate availability will be least in the least saline fluids. The limited range of isotopic variation contrasts with dolomites in mixed carbonate-siliciclastic organic-rich sediments (e.g. Baker & Burns 1985, Behrens & Land 1972) where bacterial reactions are important in supplying carbon for carbonate. Nevertheless, it is recognized that the depletion in carbon is more than might be expected for a mid-Palaeozoic mixing zone, and a subordinate source from bacterial reactions (probably including Fe- and Mn-reduction) may be required.

Elemental chemistry

The Sr and Na values are relatively low for dolostones and typical of examples interpreted as non-evaporitic or mixing-zone in origin (e.g. Land 1973a, Choquette & Steinen 1980, Churnet & Misra 1981). However, the *absolute* values are not direct indicators of salinity since Sr and Na may be variably inherited from precursor carbonate and Na incorporation is now known to be related to lattice defects (Busenberg & Plummer 1985). Defect density depends on speed of growth of crystals which is a function of supersaturation of solution. Supersaturation may be a simple function of salinity in individual hydrological situations, but is an independent variable in general. Also some Na and Sr has also probably been derived from fluid inclusions and/or by leaching of IR. Although quantitative modelling is not thought appropriate in this instance, qualitatively the good correlation of both Sr and Na with $\delta^{18}O$ and $\delta^{13}C$ is best attributed mostly to salinity variations of the dolomitizing fluids.

The good correlation of Fe and Mn with each other and not with other geochemical parameters indicates a separate control on their abundance. Chemical zonation of Mn is indicated by the luminescence zonation given that Fe is well below the concentrations required to quench luminescence (Fairchild 1983). One sample was microprobed and Fe-zoning was found to be present. Complex Mn or Fe-Mn zonation is known from inferred phreatic (Meyers 1978) and mixing-zone (Choquette & Steinen 1980) carbonates and is likely to result principally from Eh variations in the source area, related to changing water-table position. Values of iron in particular are too high to have formed under oxygenated conditions, but are compatible with mixing-zone fluids.

The reasons for departures from dolomite Ca-Mg stoichiometry have been much debated, but are not resolved. Despite the expectation that dilute water 'limpid' dolomite would be stoichiometric because of ideal conditions of slow growth (Folk & Land 1975), such dolomite can be extremely calcian (Ward & Halley 1985). Evaporitic dolomite forming in modern sabkhas with demonstrably high pore water Mg/Ca ratios may also be deficient in Mg (e.g. Patterson & Kinsman 1982), although there is some evidence that within a given setting, higher Mg/Ca in porewater is associated with more stoichiometric dolomite (Sass & Katz 1982). Ancient evaporitic dolomites tend to be near-stoichiometric whilst non-evaporitic ones are more variable and more Ca-rich on average (Lumsden & Chimahusky 1980). In general, one would expect that fine-grained and poorly-crystalline calcian dolomites would tend to recrystallize and become more stoichiometric during diagenesis, while originally coarser dolomite may retain its original chemistry and preserve its original growth zones.

The Yingtang dolomites show a wide range in Ca content from 51.7 to 56.9 mole% $CaCO_3$. Sample Nj-1 analysed by microprobe showed intra-crystalline variations of up to 1.7 mole%. Since the type B dolomites preserve fine zonation, these chemical variations are probably original. Ward & Halley (1985) noted considerable variations in stoichiometry with growth in Pleistocene Yucatan mixing-zone dolomite crystals. The Yingtang dolomites show a good inverse correlation of $\delta^{13}C$ with Ca content: the more 'marine' samples are more stoichiometric. Since sea water would be expected to have a much higher Mg/Ca ratio than fresh water on an emergent carbonate shoal, this relationship is the one predicted by Sass & Katz (1982). Note, however, that the least stoichiometric dolomites described by Sass & Katz (1982) were inferred to have formed in more saline waters, but with Mg/Ca lowered by dolomitization in a diagenetically isolated system (Fig. 12). Hence, since relatively low Mg/Ca can occur in waters of both low and high salinities, so can Ca-enriched dolomite.

Discussion

The stratigraphic evidence that dolomitization was near-surface, the fact that open-marine sediments on palaeotopographic highs were preferentially dolomitized, and the isotopic values and their co-variation are the main lines of evidence used to support the hypothesis of dolomitization by mixed meteoric-marine fluids. However, the initial hypothesis that this could relate entirely to the final episode of emergence is contradicted by the stratigraphic variation of $\delta^{13}C$, $\delta^{18}O$, Sr and Na (palaeosalinity indicators in this case) which do not show uniform lightening of isotopes or decreases in minor elements upwards (Fig. 2, Table 1). This points to the necessity for a series of periods of emergence during the deposition of the Yingtang Formation. Although such emergence surfaces have not yet been found, they need never have been prominent and fieldwork to date has not been geared to looking for such features. The existence of repeated emergence removes the necessity for large-scale erosion at the top of the Yingtang Formation, but some erosion (?up to a few tens of metres) must have occurred to account for the presence of largely dolomitized strata at the surface when the succeeding limestones were deposited.

The overall change from inclusion-rich to inclusion-poor (ultimately cementing) dolomite is common throughout the Formation although, as suggested above, this must be at least partly diachronous. The lack of fine-scale mimic replacement and presence of calcite inclusions suggest that dolomitizing solutions were only weakly supersaturated for dolomite (few nucleation sites) and not greatly undersaturated for calcite. Sibley (1982) argued for dolomitization of low-magnesian calcite by solutions slightly oversaturated for calcite to yield the inclusion-rich cores common to much replacive dolomite. Later stages of growth in the Yingtang Formation were apparently associated with more rapid dissolution of calcite leading to cementation instead of replacement by dolomite. Vuggy porosity is most abundant near the top of the Formation, presumably related to the final episode of emergence. It is possible that all the leaching throughout the Formation relates to this time. The patchy nature of the later calcite cementation suggests an origin related to early phreatic processes, rather than later burial cementation.

The extent of mixing-zone dolomitization in the underlying Tangding Formation is unknown. The greater lateral persistence of this Formation (lower half of dolostone lens of Fig. 1c) and the occurrence of length-slow chalcedony suggests that penecontemporaneous dolomitization may be important in this formation.

This paper contains inferences which appear to contradict the statements of Machel & Mountjoy (1986) concerning the efficiency of mixing-zone dolomitization: they stress the incomplete nature of the dolomitization and the predominance of dolomite cement. Based on isotopic evidence one can distinguish two kinds of Neogene mixing-zone dolomites: those with highly negative isotopic values (particularly for carbon) which appear to be forming in near freshwater solutions with Mg supplied perhaps by diffusion (Land 1973b, Magaritz *et al.* 1980) and those with heavier isotopic values indicating either a high proportion of sea water in the mixed fluid, or else evaporation (Land 1973a, Sibley 1980, Ward & Halley 1985). The Yingtang dolostones, although intermediate, seem closer isotopically to the second type which also contains the examples of more complete dolomitization. It seems likely that these Neogene examples do not fully illustrate the range of potential mixing-zone dolostones. They formed during times of geologically brief sea-level fall in the chaotic eustatic pattern of the Neogene. They do include examples of replacive dolomite with crystals on the 100 μm scale (Sibley 1980, 1982), but show highly variable fabrics and degrees of dolomitization probably related to their unstable history. In the Yingtang example we require overall more stable conditions allowing dolomitization to proceed to completion. Periodic emergence caused by sedimentation leading to the production of freshwater lenses above calcite saturation and highly undersaturated for dolomite could lead at shallow depth to mixing-zone dolomitization in solutions slightly supersaturated for dolomite and slightly undersaturated or oversaturated (Sibley 1982) for calcite leading to the observed textures. It should also be borne in mind that the dolomitization may have been aided by factors other than the thermodynamic

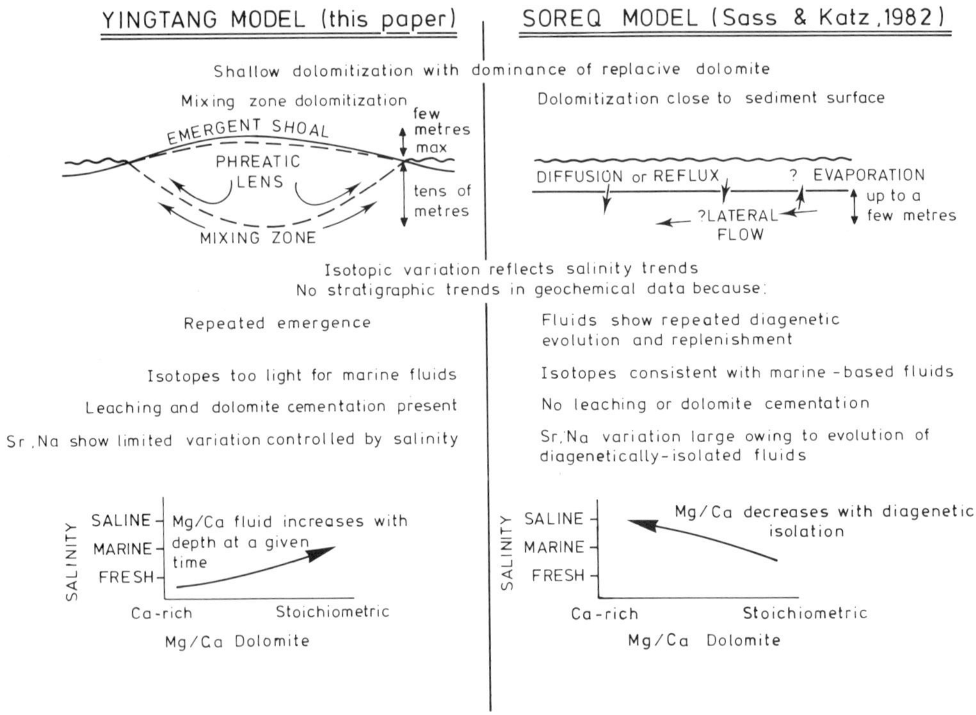

FIG. 11. Comparison of the mixing-zone model for the Yingtang Formation with the model for the Cretaceous Soreq Formation of Israel as described by Sass & Katz (1982). Note that the Na modelling of these authors is invalidated by the work of Ishikawa & Ichikuni (1984) and Busenberg & Plummer (1985).

changes resulting from mixing. For example, it could be argued that kinetic triggers such as sulphate reduction or removal or organic inhibitors were important. Unfortunately the available data do not allow us to evaluate these possibilities.

The shallow subtidal model advocated by Machel & Mountjoy (1986) may well explain many platform dolomites, but suffers even more from the lack of latter-day analogues than does the mixing-zone model since described examples of Holocene shallow marine dolomitization have variable carbon isotopic values indicating a link with early bacterial reactions, and hence a different model The lack of a stable sea-level probably also explains the restricted range of subtidal dolomite occurrences.

Figure 11 summarizes the main characteristics of the model of dolomitization of the Yingtang Formation by repeated emergence, and makes a comparison with the model of Sass & Katz (1982) based on the Cretaceous Soreq Formation of Israel. The latter is used by Machel & Mountjoy (1986) as a prime example of the shallow subtidal type of dolomitization. Although several similarities exist, clearly two different models are required to explain these two dolomitic formations. This study confirms the necessity of studying not only the numerical value of geochemical parameters but also the relationships between them in order to assess the dolomitization mechanism (Sass & Katz 1982).

ACKNOWLEDGMENTS: Zhao Xun is supported by the Chinese Government and the British Council. Isotope work was carried out at the NERC Stable Isotope Facility at the British Geological Survey, London where Charles Branch and Baruch Spiro gave valuable instruction and assistance. ICP work was carried out at Kings College, London: the help of Alison Warren and Nick Walsh is gratefully acknowledged. At Birmingham, Graham Hendry is thanked for his patient assistance with XRD and advice on analytical procedures, Paul Hands for rock-sectioning, and Carl Burness for draughting. We are grateful to Baruch Spiro, Julian Andrews, Maurice Tucker and Jim Marshall for their comments on the manuscript.

References

ALLAN, J. R. & MATTHEWS, R. K. 1982. Isotope signatures associated with early meteoric diagenesis. *Sedimentology*, **29**, 791–817.

BAI, S.-L., JIN, S.-Y. & NING, Z.-S. 1982. *The Devonian Biostratigraphy of Guangxi and Adjacent Areas*. Beijing University Press House (in Chinese).

BAKER, P. A. & BURNS, S. J. 1985. Occurrence and formation of dolomite in organic-rich continental margin sediments. *American Association of Petroleum Geologists Bulletin*, **69**, 1917–30.

BEHRENS, E. W. & LAND, L. S. 1972. Subtidal Holocene dolomite, Baffin Bay, Texas. *Journal of Sedimentary Petrology*, **42**, 155–61.

BUSENBERG, E. & PLUMMER, L. N. 1985. Kinetic and thermodynamic factors controlling the distribution of SO_4^{2-} and Na^+ in calcites and selected aragonites. *Geochimica et Cosmochimica Acta*, **49**, 713–25.

CHOQUETTE, P. W. & STEINEN, R. P. 1980. Mississippian non-supratidal dolomite, Ste, Genevieve Limestone, Illinois Basin: evidence for mixed-water dolomitization. *In*: ZENGER, D. H., DUNHAM, J. B. & ETHINGTON, R. L. (eds). *Concepts and Models of Dolomitization* pp. 163–96. Society of Economic Paleontologists and Mineralogists Special Publication, 28.

CHURNET, H. B. & MISRA, K. C. 1981. Genetic implications of the trace element distribution pattern in the Upper Knox carbonate rocks, Copper Ridge District, East Tennessee. *Sedimentary Geology*, **30**, 173–94.

DICKSON, J. A. D. & COLEMAN, M. L. 1980. Changes in carbon and oxygen isotope composition during limestone diagenesis. *Sedimentology*, **27**, 107–18.

DINELEY, D. L. 1984. *Aspects of a Stratigraphic System: the Devonian*. Macmillan, London.

FAIRCHILD, I. J. 1983. Cathodoluminescence of natural calcites and dolomites: new data and review. *Sedimentology*, **30**, 579–83.

FOLK, R. L. & LAND, L. S. 1975. Mg/Ca ratios and salinity: two controls over crystallization of dolomite. *American Association of Petroleum Geologists Bulletin*, **59**, 60–8.

HOLSER, W. T. 1984. Gradual and abrupt shifts in ocean chemistry during Phanerozoic time. *In*: HOLLAND, H. D. & TRENDALL, A. F. (eds). *Patterns of Change of Earth Evolution*, pp. 171–205. Springer-Verlag, Berlin.

HOU, H. 1978. The Devonian system in South China. *In*: CHINESE GEOLOGICAL ACADEMY (ed.). *Conference on Devonian in South China*, pp. 214–40. Geological Press House (in Chinese).

ISHIKAWA, M. & ICHIKUNI, M. 1984. Uptake of sodium and potassium by calcite. *Chemical Geology*, **42**, 137–46.

KALDI, J. G. & GIDMAN, J. 1982. Early diagenetic dolomite cements; examples from the Permian Lower Magnesian Limestone of England and the Pleistocene carbonates of the Bahamas. *Journal of Sedimentary Petrology*, **52**, 1073–86.

KUANG, G.-D. 1982. The Devonian cross-sections of Liujing and Loufu in the Guangxi region. *In*: *Devonian Cross-sections in Guangxi*, pp. 1–47. Guangxi Geological Institute (in Chinese with English abstract).

LAND, L. S. 1973a. Contemporaneous dolomitization of Middle Pleistocene reefs by meteoric water, North Jamaica. *Bulletin of Marine Science*, **23**, 64–92.

—— 1973b. Holocene dolomitization of Pleistocene limestones, North Jamaica. *Sedimentology*, **20**, 411–24.

—— 1980. The isotopic and trace element geochemistry of dolomite: the state of the art. *In*: ZENGER, D. H., DUNHAM, J. B. & ETHINGTON, R. L. (eds). *Concepts and Models of Dolomitization*, pp. 87–110. Society of Economic Paleontologists and Mineralogists Special Publication, 28.

—— 1983. The application of stable isotopes to studies of the origin of dolomite and to problems of diagenesis of clastic sediments. *In*: ARTHUR, M. A., ANDERSON, T. F., KAPLAN, I. R., VEIZER, J. & LAND, L. S. (contributors). *Stable Isotopes in Sedimentary Geology*, pp. 4-1–22. Society of Economic Paleontologists and Mineralogists Special Publication, 10.

LEMAITRE, R. W. 1982. *Numerical Petrology*. Elsevier, Amsterdam.

LIAO, W.-H., SHU, H.-G., WANG, C. Y., CAI, C. Y., RUAN, Y.-P., MU, D.-C. & LU, C.-L. 1979. *Description of some Standard Devonian Sections in South-west China*, pp. 221–9. Press House of Science (in Chinese).

LUMSDEN, D. N. & CHIMAHUSKY, J. S. 1980. Relationship between dolomite non-stoichiometry and carbonate facies parameters. *In*: ZENGER, D. H., DUNHAM, J. B. & ETHINGTON, R. L. (eds). *Concepts and Models of Dolomitization*, pp. 123–37. Society of Economic Paleontologists and Mineralogists Special Publication, 28.

MACHEL, H.-G. & MOUNTJOY, E. W. 1986. Chemistry and environments of dolomitization—a reappraisal. *Earth Science Reviews*, **23**, 175–222.

MAGARITZ, M., GOLDENBERG, L., KAFRI, U. & ARAD, A. 1980. Dolomite formation in the seawater-freshwater interface. *Nature*, **287**, 622–4.

MEYERS, W. J. 1978. Carbonate cements: their regional distribution and interpretation in Mississippian limestones of southwestern New Mexico. *Sedimentology*, **25**, 371–99.

PATTERSON, R. J. & KINSMAN, D. J. J. 1982. Formation of diagenetic dolomite in coastal sabkha along Arabian (Persian) Gulf. *American Association of Petroleum Geologists Bulletin*, **66**, 28–43..

POPP, B. N., ANDERSON, T. F. & SANDBERG, P. A. 1986. Textural, elemental and isotopic variations among constituents in Middle Devonian limestones, North America. *Journal of Sedimentary Petrology*, **56**, 715–27.

SASS, E. & KATZ, A. 1982. The origin of platform dolomites: new evidence. *American Journal of Science*, **282**, 1184–213.

SHARMA, T. & CLAYTON, R. N. 1965. Measurement of O^{18}/O^{16} ratios of total oxygen and of carbonates. *Geochimica et Cosmochimica Acta*, **29**, 1347–53.

SIBLEY, D. F. 1980. Climatic control of dolomitization, Seroe Domi Formation (Pliocene), Bonaire, N. A. *In*: ZENGER, D. H., DUNHAM, J. B. & ETHINGTON, R. L. (eds). *Concepts and Models of Dolomitization*, pp. 247–58. Society of Economic Paleontologists and Mineralogists Special Publication, 28.

—— 1982. The origin of common dolomite fabrics: clues from the Pliocene. *Journal of Sedimentary Petrology*, **52**, 1087–1100.

TUCKER, M. E. 1982. Precambrian dolomites: petrographic and isotopic evidence that they differ from Phanerozoic dolomites. *Geology*, **10**, 7–12.

VEIZER, J., FRITZ, P. & JONES, B. 1986. Geochemistry of brachiopods: oxygen and carbon isotopic records of Paleozoic oceans. *Geochimica et Cosmochimica Acta*, **50**, 1679–96.

——, LEMIEUX, J., JONES, B., GIBLING, M. R. & SAVELLE, J. 1977. Sodium: palaeosalinity indicator in ancient carbonate rocks. *Geology*, **5**, 177–9.

WARD, W. C. & HALLEY, R. B. 1985. Dolomitization in a mixing zone of near-seawater composition, Late Pleistocene, Northeastern Yucatan peninsula. *Journal of Sedimentary Petrology*, **55**, 407–20.

WENK, H.-R. & ZHANG, F. 1985. Coherent transformations in calcian dolomite. *Geology*, **13**, 457–60.

ZHAO XUN* & IAN J. FAIRCHILD, Department of Geological Sciences, University of Birmingham, PO Box 363, Birmingham B15 2TT, UK. *Permanent address: Bureau of Geology and Mineral Resources of Guangxi, 1, Jianzheng Road, Nannin City, Guangxi, China.

Regional studies and burial diagenesis

Diagenesis of carbonate cements in Permo–Triassic sandstones in the Wessex and East Yorkshire–Lincolnshire Basins, UK: a stable isotope study

A. H. Bath, A. E. Milodowski & B. Spiro

SUMMARY: The $^{13}C/^{12}C$ and $^{18}O/^{16}O$ ratios of carbonate cements in Permo–Triassic fluvio-deltaic sandstones differ between the Wessex and East Yorkshire–Lincolnshire Basins. The isotopic data indicate different conditions of deposition and early post-depositional carbonate precipitation which supports petrological and mineralogical studies of the same sandstones focussed on the development of reservoir properties. Isotopic compositions of carbonate cements vary across the Wessex Basin; they indicate early derivation of dissolved carbonate during recrystallization of detrital limestone. Within the basin, calcite cements show little variation in $\delta^{13}C$ but considerable differences in $\delta^{18}O$ between early calcrete calcites and isotopically-light later calcites; $\delta^{18}O$ decreases to values close to isotopic equilibrium with present water while $\delta^{13}C$ is not in isotopic equilibrium with bicarbonate in present-day brines. Two distinct generations of replacive dolomite, early non-ferroan and later ferroan, are found in the basin centre and have different $\delta^{18}O$ but similar $\delta^{13}C$ values. $^{34}S/^{32}S$ and $^{18}O/^{16}O$ ratios of anhydrite cements could indicate a remote Zechstein source of sulphate or a mixed marine sulphate-sulphide source. The Permo–Triassic sandstones of the East Yorkshire–Lincolnshire Basin, in contrast, contain dominantly dolomite cement. $\delta^{18}O$ and $\delta^{13}C$ values are relatively heavy and indicate early dolomitization of calcite over large areas. $\delta^{18}O$ values of minor calcites indicate that dedolomitization is continuing in the present-day groundwater regime. The isotopic differences between carbonate cements in the two basins is tentatively attributed to the differing early post-depositional conditions and to the differences in carbonate chemistry between calcite dissolution/precipitation reactions and dolomitization reactions.

The Wessex and East Yorkshire–Lincolnshire Basins are onshore structures in southern and eastern England respectively (Fig. 1). They are both filled with Permian and Mesozoic sediments which reach maximum thicknesses of over 3000 m in the Wessex Basin and over 2000 m in the East Yorkshire–Lincolnshire Basin.

The Triassic Sherwood Sandstone in the Wessex Basin (Fig. 2) was derived probably from sources to the south following early Permian subsidence which was variable and strongly controlled by faulting. These sediments, predominantly arenaceous until deposition of the late Triassic Mercia Mudstone, were laid down under fluviatile or sub-aerial conditions in an arid or semi-arid climate (Henson 1970, Lott & Strong 1982). Thicknesses of the Sherwood Sandstone exceed 100 m in much of the basin, and reach a maximum of over 300 m. Detailed study of depositional facies has shown extensive and complex lateral as well as vertical variation, suggesting transport and deposition by braided river systems (Henson 1970, Lott & Strong 1982). Triassic sedimentation culminated in the deposition of the Mercia Mudstone. This represents a low energy evaporitic environment in which silts and clays were deposited, possibly with the occasional development of playas. Subsequently, in the Jurassic, marine transgression led to accumulation of clays and limestones as further basinal subsidence occurred along growth faults. Maximum burial depth of basal Triassic sandstones is thought to have been around 2.5 km in the fault-bounded Winterborne Kingston trough near the centre of the basin; this is only slightly more than their present depth (Fig. 3).

The Permo–Triassic sandstones in the East Midlands, Lincolnshire and East Yorkshire are fluvial deposits on the eastern England Shelf (Fig. 4). The relatively stable eastern England Shelf is separated from the southern North Sea Basin, in which subsidence and deposition were more rapid, by the Dowsing Fault Zone (Fig. 4). Subsequent subsidence of the shelf has resulted in burial and cover by Triassic clays and marls; the Permo–Triassic strata dip eastwards regularly from outcrop running N–S through Nottinghamshire and Yorkshire. The depths at which the Permo–Triassic sandstones occur on the east coast of England are between 1100 and 1900 m. Here, as represented by the Cleethorpes borehole, the Triassic Sherwood Sandstone is up to 400 m thick, whereas the Basal Permian Sands are around 25 m thick (Fig. 5).

Because of the potentially favourable aquifer properties of these two sandstone sequences and their burial depth, they have been assessed for their potential as low enthalpy geothermal

From: MARSHALL, J. D. (ed.), 1987, *Diagenesis of Sedimentary Sequences*, Geological Society Special Publication No. 36, pp. 173–190.

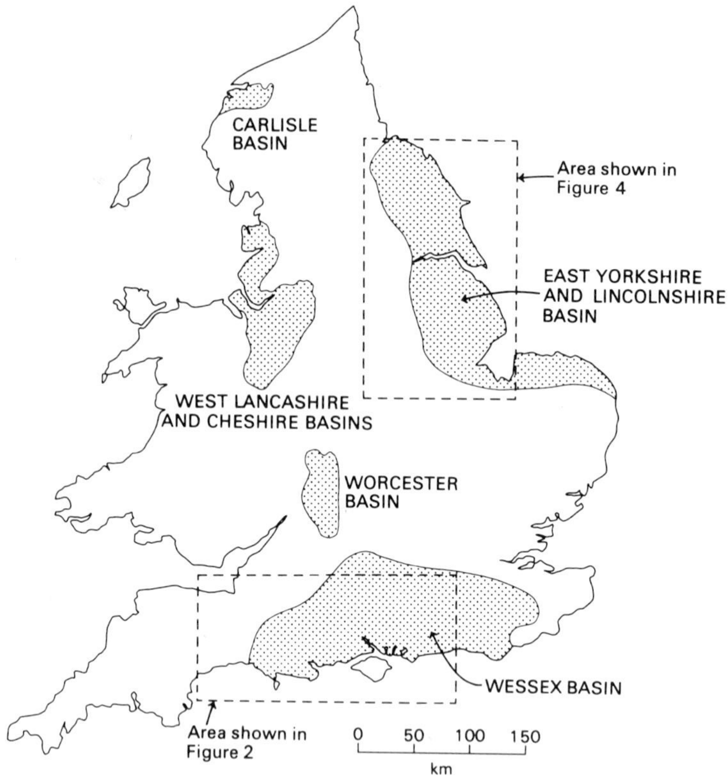

FIG. 1. Locations of the Wessex and East Yorkshire–Lincolnshire sedimentary basins; the other three Mesozoic basins in England and Wales are also shown.

energy sources (Downing & Gray 1986). This paper describes the stable isotopic evidence for the environments of formation and for sources of carbonate and sulphate cements; the study was part of a broader petrological and mineralogical study of factors determining the reservoir properties of the Permo–Triassic sandstones (Milodowski *et al.* 1986, 1987, Strong & Milodowski, *this volume*). The Permo–Triassic sequences of both the Wessex Basin and North Sea Basin are

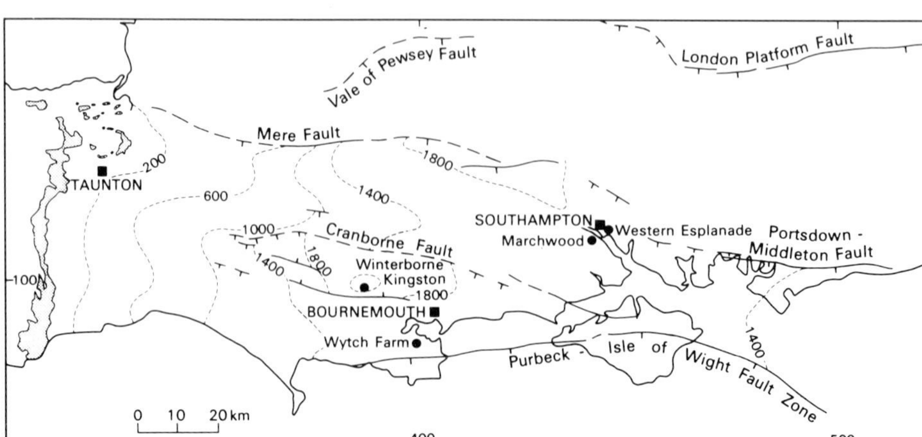

FIG. 2. Locations of drillcore samples from the Wessex Basin. Samples were also studied from the Sherwood Sandstone outcrop on the western edge (shown stippled). Contours show depth (metres below sea-level) to top of the Sherwood Sandstone Group (after Downing & Gray 1986).

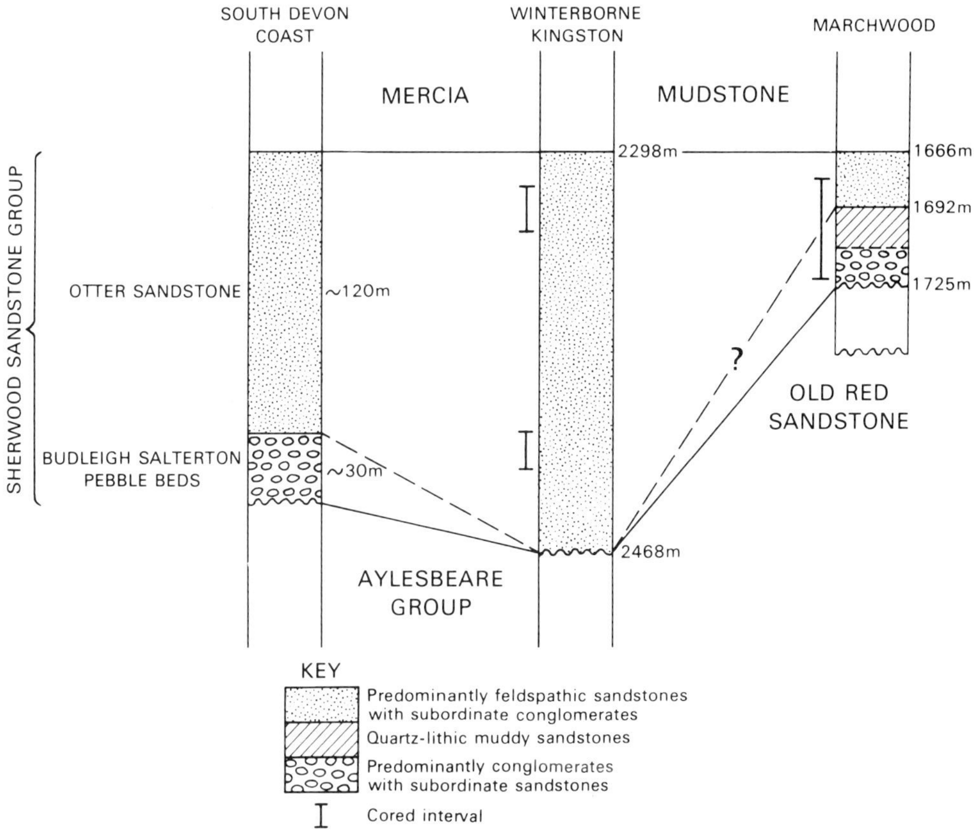

FIG. 3. Simplified stratigraphic sections showing thicknesses and lithologies of the Triassic Sherwood Sandstone Group across the Wessex Basin from the western outcrop in South Devon to the Marchwood geothermal well on the eastern side of the basin; the Winterborne Kingston sequence represents the depositional centre of the basin.

important locally as hydrocarbon reservoirs—in the onshore Wytch Farm oilfield and in offshore gasfields respectively.

Petrological background

The petrology and diagenetic history of the Sherwood Sandstone in the Wessex Basin is described in detail by Strong & Milodowski (*this volume*). In summary, the formation has a lower conglomeratic unit referred to at outcrop on the western margin of the basin as the Budleigh Salterton Pebble Beds (Fig. 3). This is overlain by a thick sequence of feldspathic sandstones, the Otter Sandstones at outcrop, exhibiting a series of fining-upwards cycles often with erosional conglomeratic bases (Edmonds & Williams 1985). Cementation by carbonate is a major feature of the sandstone throughout the basin; discrete thin bands of anhydrite (≤ 1 m) and anhydritic mudstones and siltstones are also present in wells drilled close to the basin depocentre, particularly in the lower part of the sequence. Calcite cement forms nodular expansive calcrete concretions at the tops of some cycles in the marginal parts of the basin, e.g. outcrop on the western rim and boreholes at Southampton on the eastern rim.

The major diagenetic processes identified by Strong & Milodowski (*this volume*) are: (i) early cementation by calcite and anhydrite inhibiting further compaction followed by local later-stage removal of anhydrite; (ii) variable degree of compaction depending on framework grain (quartz, feldspar) characteristics and the development of overgrowths, followed by later-stage partial dissolution of some framework grains, and (iii) subsequent invasion of this 'secondary porosity' by late stage cements (calcite, dolomite, anhydrite). Earlier studies of drillcore indicated

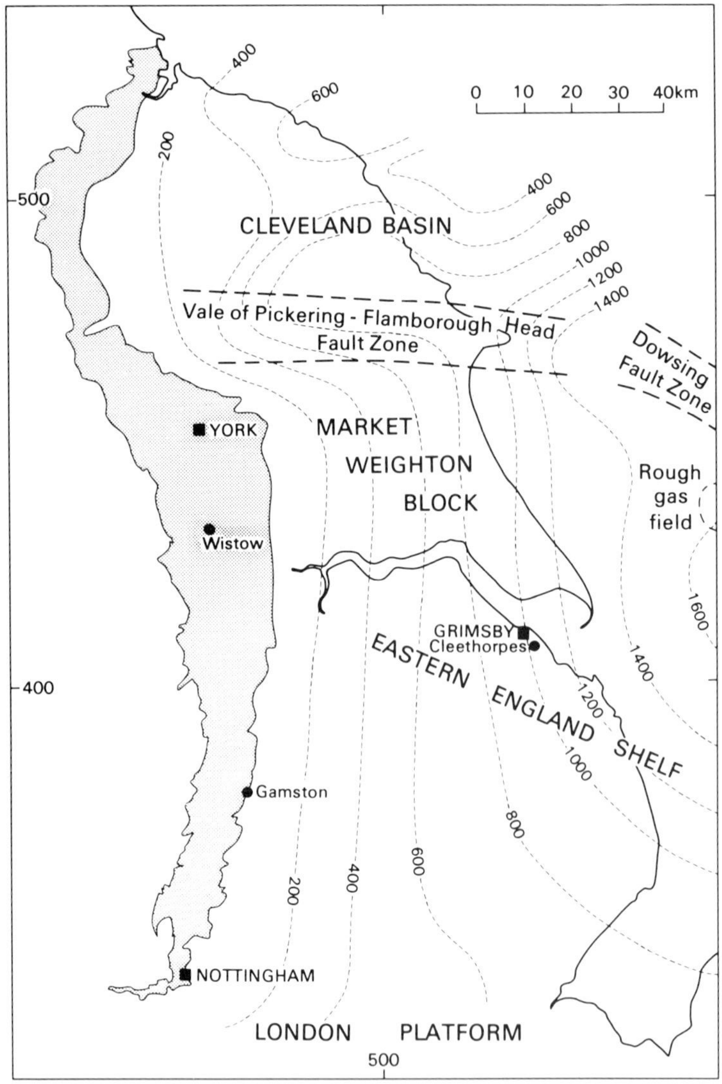

FIG. 4. Locations of drillcore samples from the East Yorkshire–Lincolnshire Basin and the Sherwood Sandstone outcrop. Contours show depth (metres below sea-level) to top of the Sherwood Sandstone Group (after Downing & Gray 1986).

that framework grain (K-feldspar) dissolution in some zones in late diagenesis has compensated for porosity reduced by compaction (Knox *et al.* 1984). On the other hand, Burley (1984) interpreted the present porosity distribution as resulting from the removal of much of a widespread early compaction-resisting carbonate cement. As a result of further basin-wide petrological studies, Milodowski *et al.* (1986) have suggested that widespread early anhydrite cementation might also account for resistance to compaction, being subsequently dissolved in many parts of the basin.

The Sherwood Sandstone in the East Yorkshire–Lincolnshire Basin is dominated by feldspathic sandstones although the proportion of K-feldspar is lower than in the Wessex Basin (Milodowski *et al.* 1987; Fig. 6). Carbonate cementation is only weakly and patchily developed, dolomite being the major cement accompanied by only minor patches of calcite. The varying cementation makes the sandstone friable, particularly at outcrop where carbonates and feldspars show evidence of extensive dissolution. Anhydrite occurs as a major cement near the top of the buried sandstone, but is absent at

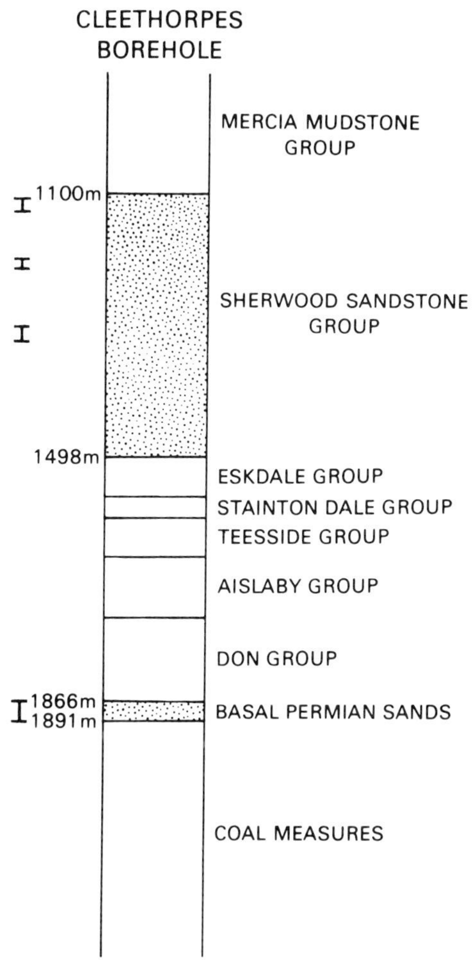

FIG. 5. Simplified stratigraphic sequence in Permo–Triassic sediments intersected by the Cleethorpes geothermal well in the East Yorkshire–Lincolnshire Basin. Samples of feldspathic sandstone with dolomite cements from the Sherwood Sandstone Group and from the Basal Permian Sands were studied. The intervening five groups are evaporitic units within the Zechstein sequence. Key as in Fig. 3.

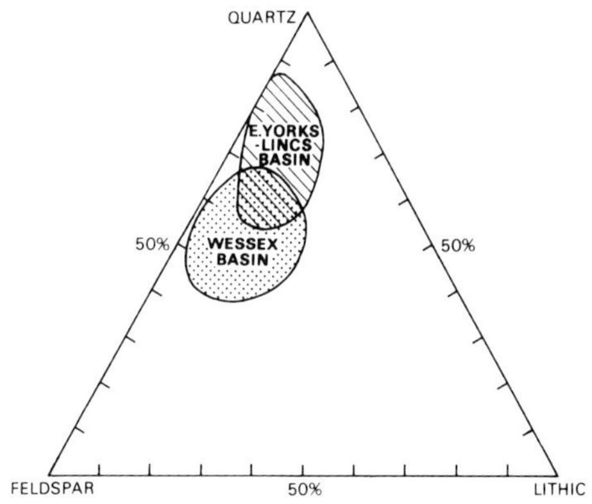

FIG. 6. Triangular diagram showing approximate proportions of quartz, feldspar and lithic grains in the Triassic Sherwood Sandstone samples in this study.

outcrop. The Basal Permian Sands at depth in the basin comprise massive poorly-sorted sandstones with minor conglomerates and interbedded siltstones. The sandstones have around 5–10% K-feldspar, and both these and the conglomerates are variably cemented by dolomite and anhydrite. Dolomite in the conglomerates is clearly replacive of pre-existing carbonate; the bulk of the dolomite throughout the sequence is non-ferroan, but late-stage ankerite overgrowths are observed rimming the dolomite.

The major diagenetic events which affect carbonate and sulphate cementation of the Sherwood Sandstone Group in the two basins are summarized in Fig. 7, which illustrates the similarities and differences in this respect between the two basins.

Samples and methods

Four suites of drillcore samples from the Wessex Basin (Fig. 2), and a drillcore from the Cleethorpes well in NE Lincolnshire (Fig. 4) were available for study. The Cleethorpes well and the Winterborne Kingston, Marchwood and Western Esplanade (near and in Southampton respectively) wells in the Wessex Basin, were all drilled specifically for the investigation of geothermal potential in the deep aquifers (Downing & Gray 1986). In addition, a drillcore from a hydrocarbon well at Wytch Farm was studied. In both basins, samples of outcropping Permo-Triassic sandstone were collected for comparison; outcrops occur only on the western rim of the Wessex Basin running N–S through Devon and Somerset (Fig. 2), whilst Sherwood Sandstone outcrops from Nottinghamshire northwards into East Yorkshire (Fig. 4). Two drillcores from or close to outcrop in eastern England were also studied: at Wistow in South Yorkshire and Gamston in Nottinghamshire.

Samples for stable carbon and oxygen isotopic analysis of carbonates were selected from zones for which full petrographic and mineralogical information was available (Milodowski *et al.* 1986, 1987, Strong & Milodowski, *this volume*). Cathodoluminescence microscopy was particularly valuable for identifying different generations of carbonate, detecting mineral zonation, as well as estimating the carbonate homogeneity.

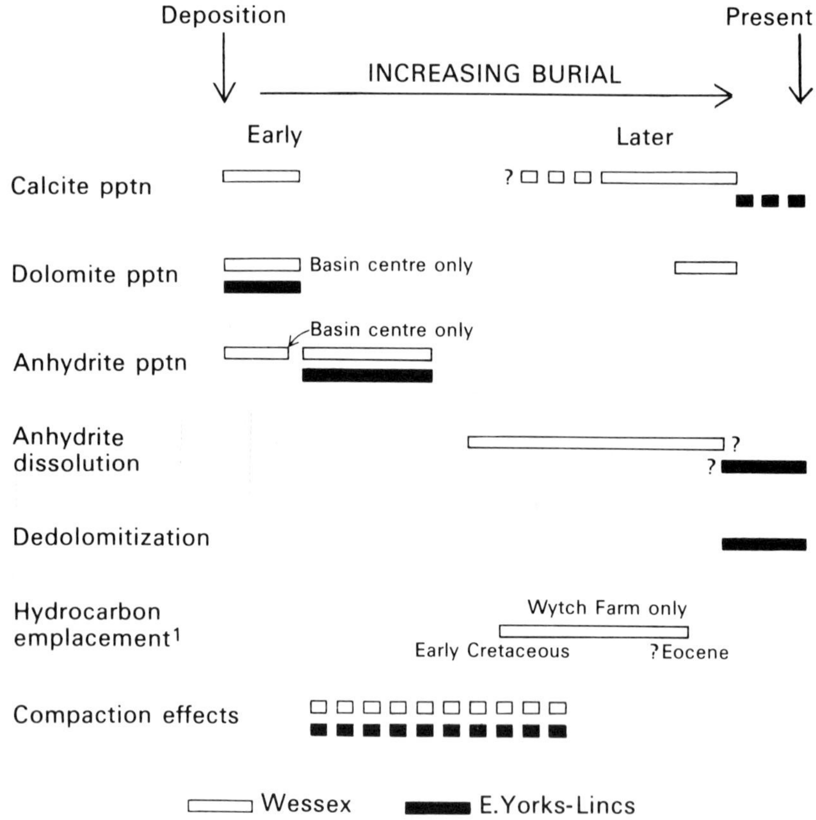

FIG. 7. Summary comparison of major diagenetic events which affect carbonate and sulphate cementation in the Sherwood Sandstone Group in the Wessex Basin and East Yorkshire–Lincolnshire Basin. ([1]Data on timing of hydrocarbon emplacement into the Sherwood reservoir at Wytch Farm from Colter & Havard 1981.)

Due to the fine grain-size it was not feasible to separate calcite from dolomite, or to obtain samples of discrete cements from complex samples. Most of the samples analysed had a dominant single generation of cement.

Early- and later-diagenetic calcites, as inferred by petrographic relationships, are clearly characterized by dull and bright orange cathodoluminescence respectively. In addition dolomite is distinctive for its dull red luminescence. Therefore these characteristics were used to select appropriate samples and to estimate the amounts of particular carbonate generations. However, detailed petrographic and scanning electron microscope studies in parallel with the cathodoluminescence observations suggested that the latter gave only a coarse classification; it is likely that the so-called early and later carbonates are themselves subdivided into precipitates formed under ranges of environments within these broad classifications. Brief descriptions of the samples according to the above observations are listed in Appendices 1 and 2 for the Wessex Basin and East Yorkshire–Lincolnshire Basin respectively.

The preparation of samples for stable isotope analysis follows that described by McCrea (1950). Small amounts of whole-rock samples were ground for 15 minutes in a ball mill, and 10–100 mg subsamples, depending on the estimated carbonate content, were taken for $^{13}C/^{12}C$ and $^{18}O/^{16}O$ analysis. Carbon dioxide was extracted from samples by reacting with excess 100% orthophosphoric acid *in vacuo* at a constant temperature of 25.2 ± 0.05°C. $\delta^{18}O$ was calculated using the constants $\alpha^{18}_{CO_2\text{-calcite}} = 1.01025$ (Friedman & O'Neil 1977) and $\alpha^{18}_{CO_2\text{-dolomite}} = 1.011$ (Sharma & Clayton 1965).

Isotopic analysis was performed on a VG-903 triple collector mass spectrometer. Results are quoted on the conventional del (δ) scale in per mil (‰) deviation from the PDB standard. The normal analytical precision of duplicate analyses is better than ± 0.1‰ for both $\delta^{13}C$ and $\delta^{18}O$.

Results and discussion

Wessex Basin

The stable isotopic compositions of total carbonate in samples from the Wessex Basin are listed in Table 1. These results are plotted as $\delta^{13}C$ versus $\delta^{18}O$ in Fig. 8, which includes data from the previous study of Marchwood drillcore by Knox *et al.* (1984). Isotopic ratios ($\delta^{2}H$, $\delta^{18}O$) of present-day brines in the deep Sherwood Sandtone are given in Table 2, which also shows the $\delta^{13}C$ of dissolved bicarbonate in brine at Marchwood.

TABLE 1. *$^{13}C/^{12}C$ and $^{18}O/^{16}O$ analyses of total carbonate content of samples from the Sherwood Sandstone in the Wessex Basin*

No.	Sample number	$\delta^{13}C$	$\delta^{18}O$
		‰ vs PDB	
Western Outcrop			
1	HW03 (calcite pebbles)	−1.26	−5.65
2	HW03 (matrix)	−2.63	−7.22
3	MV01	−7.38	−9.49
4	MV02	−2.67	−8.33
5	WS02	−1.73	−6.20
6	BL01	−8.90	−6.09
7	LB01	−8.97	−5.33
8	LB02	−2.04	−7.72
9	LB03	−11.23	−4.94
10	LBV01	−7.76	−5.78
Winterborne Kingston drillcore			
11	WK01	−6.69	−11.33
12	WK06	−7.66	−8.44
13	WK09	−6.83	−11.62
14	WK11	−7.70	−10.09
15	WK18	−7.63	−1.13
16	WK19	−7.45	−1.00
17	WK22	−8.99	−0.43
Marchwood drillcore			
18	E59535	−6.24	−10.11
19	E59536 (cornstones)	−5.16	−2.08
20	E59536 (bulk)	−4.81	−2.84
21	E59538	−7.62	−6.44
22	MW4	−7.76	−8.41
23	MW9	−8.45	−5.15
24	MW19	−7.37	−12.96
25	MW22	−8.98	−5.86
26	MW23	−8.33	−7.78
27	MW24	−8.93	−6.14
Western Esplanade (Southampton) drillcore			
28	E59553	−5.54	−9.24
29	E59557	−7.07	−7.64
30	E59559	−6.84	−7.08
31	E59562	−6.24	−9.41
32	E59564	−6.78	−7.68
33	E59565	−6.37	−9.16
Wytch Farm X14 drillcore			
34	E59540	−5.43	−4.80
35	E59542	−5.67	−4.78
36	E59544	−5.16	−6.66
37	E59546	−6.57	−5.01
38	E59551	−7.24	−6.42

Outcrop

Carbonates in the outcrop samples of the Sherwood Sandstone fall mostly into two groups according to isotopic compositions (Fig. 8). A distinctive group (five analyses: 1, 2, 4, 5, 8) have high $\delta^{13}C$ (−2.7 to −1.2‰; B in Fig. 8) which suggests provenance from Carboniferous or Devonian detrital limestone (A in Fig. 8). Most of these samples come from the Budleigh Salterton Pebble Beds and contain discrete limestone pebbles; in one case a pebble has been separated and analysed (sample 1). The data indicate recrystallization in a meteoric water environment (either the early post-depositional meteoric water which might have had $\delta^{18}O_{H_2O}$ around −4 to −6‰ vs SMOW at around 20–30°C or present meteoric water with $\delta^{18}O_{H_2O}$ around −6 to −7‰ at 10°C). The $\delta^{13}C$ of dissolved carbonate must have been dominated by detrital calcite with a possibility of exchange only with atmospheric CO_2 ($\delta^{13}C \approx -7$‰), and with negligible contribution from CO_2 of soil organic derivation. These conditions might have occurred in the highly permeable Pebble Beds due to rapid meteoric water flux. This process continues presently at outcrop where the flux and hydrochemistry of groundwater in the Pebble Beds contrasts with conditions in the Otter Sandstone (Walton 1982).

A second group of samples (6, 7, 10), from the Otter Sandstone formation, have isotopic compositions (−9 to −7.5‰ $\delta^{13}C$, −6 to −5‰ $\delta^{18}O$) similar to those found previously for calcrete carbonate in Marchwood drillcore (Knox *et al.* 1984). These widespread early calcretes formed by precipitation in fine- to medium-grained sediments in the shallow subsurface (evident from the expansive nature of the cement and lack of compaction of the sediment). Some organic activity in the soil zone, even in the semi-arid Triassic environment, would have generated CO_2 which would have had $\delta^{13}C$ between −25‰ and −13‰, depending on the plant types (Smith & Epstein 1971). The following evaporation-reaction-exchange model describes the processes and possible isotopic changes (A to D in Fig. 8): (1) $\delta^{18}O_{H_2O}$ increased in this groundwater by up to 6‰ due to evaporative fractionation. (2) Dissolved carbonate in this groundwater had variable $\delta^{13}C_{HCO_3}$ depending on the behaviour of the dissolved carbonate system as open or closed with respect to evaporative loss of CO_2 or to exchange with the reservoir of organically derived CO_2 in the soil zone. (3) Reaction of detrital carbonate with dissolved soil zone CO_2 would give $\delta^{13}C_{HCO_3}$ between −13‰ and 6‰ depending on the isotopic compositions of carbon sources (Smith

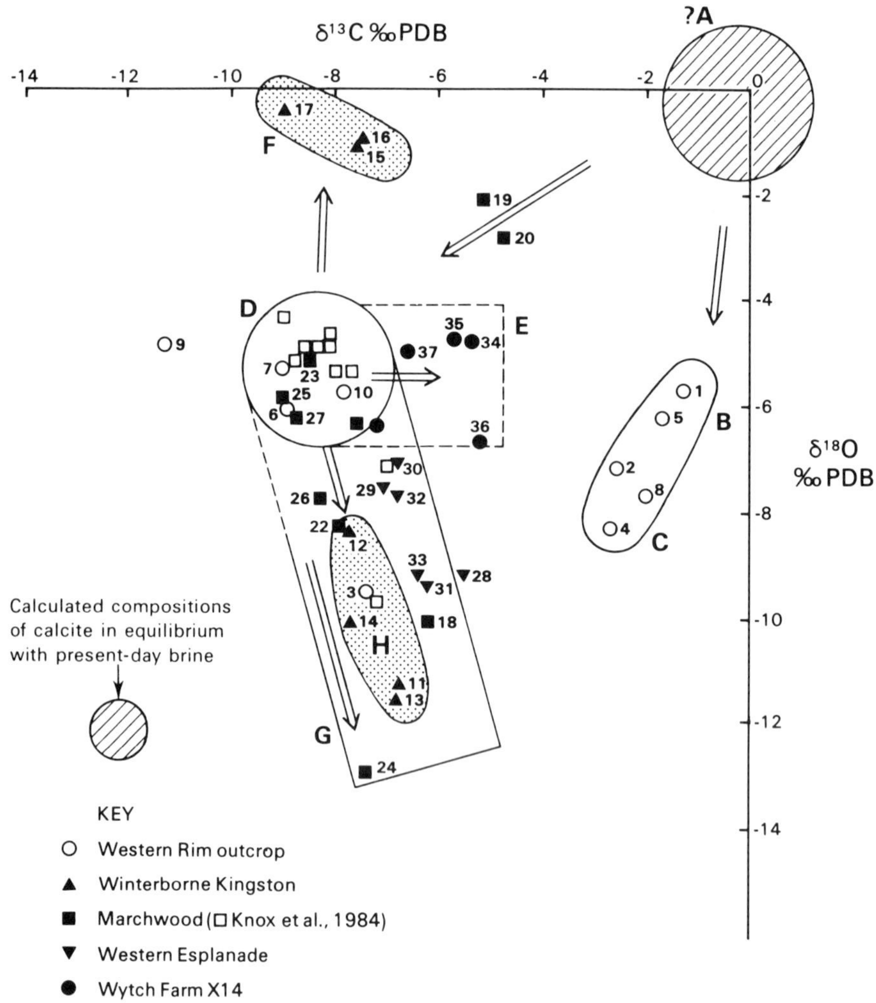

FIG. 8. Plot of $\delta^{13}C$ against $\delta^{18}O$ for samples from the Sherwood Sandstone in the Wessex Basin. Some analyses previously reported by Knox *et al.* (1984) are also shown. Cross-hatched areas denote inferred compositions of detrital calcite and of calcite in equilibrium with present-day brine. Stippled areas indicate dolomite compositions. The arrows illustrate generalized isotopic evolution pathways. A—inferred compositions of detrital ?Carboniferous/Devonian limestone; B—early recrystallized calcite at the basin rim, trending towards C possibly due to mixing with later calcites precipitated from solutions expelled towards rim during compaction; D—widespread early calcrete development in basin: E—early calcite occurring locally (Wytch Farm) with possible minor contribution of CO_2 from degradation (fermentation) of organics; F—localized dolomitization of calcrete in sabkha developed at basin centre (Winterborne Kingston); G—widespread later carbonates precipitated through maximum burial; H—localized later dolomite precipitated at basin centre (Winterborne Kingston).

& Epstein 1971, Wigley *et al.* 1978, Fontes & Garnier 1979). (4) Evaporative outgassing of CO_2 would increase $\delta^{13}C_{HCO_3}$ of remaining dissolved carbonate (Emrich *et al.* 1970).

The net result of these processes is that carbonate precipitating from these solutions will have $\delta^{13}C_{CaCO_3}$ values between $-11‰$ and $-4‰$, or higher if CO_2 loss occurs. Exchange between atmospheric CO_2 ($-7‰$ $\delta^{13}C$) and dissolved carbonate would result in precipitating carbonates with $\delta^{13}C$ around $+3‰$ ($\alpha^{13}_{CaCO_3-CO_2} \approx 1.0102$ at 20°C; Emrich *et al.* 1970). This is considerably higher than observed $\delta^{13}C$ values and suggests that exchange with atmospheric CO_2 was not a dominant process in the formation of early calcretes in this basin. This range of

TABLE 2. $^{18}O/^{16}O$, $^{2}H/^{1}H$ and $^{13}C/^{12}C$ *analyses of brines from the Sherwood Sandstone reservoir in the Wessex Basin*

Source of sample	$\delta^{18}O$ ‰ vs SMOW	$\delta^{2}H$ ‰ vs SMOW	$\delta^{13}C$ ‰ vs PDB	Salinity (mg l^{-1} Cl)
Winterborne Kingston (Triassic Drill Stem Test)	−2.7	−24		180,000
Dorset (Triassic DST)	−2.5	−30		160,000
Marchwood (Triassic pump-test)	−3.1	−33	−15.8	63.000
Western Esplanade (Triassic pump-test)	−2.2	−35	−15.5	76.000
Dorset (Triassic drillcore water)	−3.9	−34		59,000
Dorset (Triassic DST)	−1.2	−25		170,000

possible isotopic behaviours in evaporative carbonate formation is illustrated by examples of modern calcretes (Talma & Netterberg 1983, Salomons *et al.* 1978) and by active sabkha dolomitization in the Arabian Gulf (McKenzie 1981).

The one anomalous composition of outcrop material (sample 3; −7.4‰ $\delta^{13}C$ and −9.5‰ $\delta^{18}O$) is of calcite coating a fracture surface in the Pebble Beds and suggests precipitation in different groundwater conditions. Its petrological identification as a fracture-filling calcite cutting across the dominantly early calcite-cemented Pebble Beds and the isotopic similarity to calcite cements from depth in the basin suggest that this might have precipitated from groundwater expelled towards the basin rim during burial and compaction. The trend of isotopic compositions (B to C in Fig. 8) in the bulk of calcite cements at outcrop might be due to minor components with this origin.

Winterborne Kingston

This represents the thickest sequence in the basin depocentre (Fig. 3). The cementing phases here are both anhydrite and carbonate, which is predominantly replacive dolomite (Strong & Milodowski, *this volume*). Samples 15, 16 and 17, which have very high (50–80%) contents of dolomite, are from the base of the Sherwood Sandstone and have little or no anhydrite; they have distinctively heavy $\delta^{18}O$ (0–1‰) with $\delta^{13}C$ values between −9 and −7‰. The high $\delta^{18}O$ indicates that the dolomite was precipitated from freshwater-derived solutions which had been isotopically fractionated due to evaporation. It is thought unlikely that the $\delta^{18}O$ could have been due to seawater, although the presence of marine brines locally during late Triassic deposition of the overlying Mercia Mudstones cannot be definitely excluded. However the light values for $\delta^{13}C$ indicate that this was controlled by the carbon isotopic composition of pre-existing calcite—probably calcrete similar to that found elsewhere in the basin (i.e. D to F in Fig. 8). Dolomitization within a sabkha or playa environment is inferred to have taken place here. The distinct isotopic composition suggests that the aridity and groundwater discharge typical of continental sabkha conditions were developed only at the centre of this basin during the Triassic.

Those samples (11, 13, 14) containing significant anhydrite and predominantly ferroan dolomite have a dramatically different $\delta^{18}O$ (−10 to −12‰) but similar $\delta^{13}C$ (−6.7 to −6.8‰). This lighter ferroan dolomite is replacive and must be late stage, since the light $\delta^{18}O$ values indicate formation at significant burial depths near the present maximum (H in Fig. 8). The remaining sample (12) has an intermediate composition, reflecting the mixture of early calcite and later dolomite which is shown by petrography.

It is concluded from the combined petrological and isotopic evidence that zones having high original contents of calcrete-type carbonates were not penetrated by the fluids responsible for anhydrite cementation. The anhydrite was possibly derived during burial from evaporites in the overlying Mercia Mudstone or from syndepositional anhydrite near the basin depocentre—the sulphur and oxygen isotope data are presented later. Thus the early calcrete distribution is thought to have controlled subsequent diagenesis and permeability: the zones with sparse early cementation allowed sulphate solutions and Mg-rich dolomitizing solutions to penetrate subsequently.

Marchwood and Western Esplanade

The present data cover a wider range than those previously reported for Marchwood by Knox *et al.* (1984), which are included in Fig. 8. A narrow range of values for calcrete calcite $\delta^{18}O$ (of about −5‰ and $\delta^{13}C$ from −9 to −8‰) probably represents the evaporation and exchange process already described (D in Fig. 8). Higher $\delta^{18}O$ and $\delta^{13}C$ values (about −2‰ and −5‰ respectively) which are found in analyses (19 and 20) from a hand-picked calcite clast ('cornstone') and the corresponding bulk sample must indicate relatively rare early post-depositional conditions with isotopically heavier ($\delta^{18}O$ from −3 to −4‰) water. Enrichment of $\delta^{18}O$ and $\delta^{13}C$ with this trend during evaporative precipitation of calcite has been observed in some modern calcretes (Salomons *et al.* 1978).

The calcrete compositions occur as an end-member of a mixing series of isotopic compositions (i.e. D to G in Fig. 8), in which the samples might be situated according to their proportional contents of the early eodiagenetic calcrete and the later mesodiagenetic calcite (see Strong & Milodowski, *this volume*). These positions roughly correspond, in most cases, to the proportions estimated according to the contrast between the nodular calcrete and the pore-filling luminescent later calcite (Appendix 1). The samples dominated by this late calcite (samples 18, 24, 28, 31, 33) have $\delta^{18}O$ mostly around −10‰ and $\delta^{13}C$ around −6‰; sample 24 from Marchwood has the lowest $\delta^{18}O$ (−13‰). It is concluded that these calcites have precipitated at burial temperatures up to 70°C, if a basinal brine with similar isotopic composition to the present ($\delta^{18}O_{H_2O} \approx -3‰$ vs SMOW; Table 2) is assumed. This is close to the present-day temperature at this depth in the basin, which according to burial curves in Milodowski *et al.* (1986) could have been reached from the late Cretaceous to the present.

The composition of calcite which would be in isotopic equilibrium with this present-day Marchwood brine at the formation temperature of 70°C has been calculated, using the values in Table 2 ($\delta^{18}O_{H_2O} \approx -3‰$ vs SMOW and $\delta^{13}C_{HCO_3} \approx -16‰$), $\alpha^{18}_{CaCO_3-H_2O} \approx 1.021$ and $\alpha^{13}_{CaCO_3-HCO_3} \approx 1.0036$ (Emrich *et al.* 1970) at 70°C. This calculation gives $\delta^{18}_{CaCO_3} \approx -12‰$ vs PDB and $\delta^{13}_{CaCO_3} \approx -12‰$. This calculated $\delta^{13}C$ is very different to the observed calcite values, though $\delta^{18}O$ is similar to some late-stage calcite cements (Fig. 8), therefore it is tentatively concluded that the observed data show no evidence of present-day precipitation of cements in the deep Wessex Basin. There is no immediate explanation for the apparent change in $\delta^{13}C_{HCO_3}$ values between the solution which precipitated late-stage calcites and the present-day brine.

Wytch Farm

These drillcore samples contain both anhydrite and calcite cements—cathodoluminescence suggests mixtures of late and early calcite generations (Appendix 1). Three samples (34, 35, 37) have similar $\delta^{18}O$ but slightly higher $\delta^{13}C$ values (E in Fig. 8) compared with the early calcite cements in the Marchwood and Southampton samples. The difference might reflect relative differences in the contributions of different inorganic carbon components (i.e. atmospheric and soil zone CO_2, dissolved detrital carbonate) at the time of formation. It is also possible that this difference indicates a minor isotopically heavier CO_2 contribution resulting from bacterial fermentation of organic matter. The apparent absence here of carbonate cements with lower $\delta^{18}O$, and hence representing deeper burial stages, could be explained by the 'blocking' action of hydrocarbons which were generated in Jurassic strata and migrated into the Sherwood reservoir in the Wytch Farm area during the later Jurassic/early Cretaceous (Colter & Havard 1981).

Wytch Farm drillcore, like that at Winterborne Kingston, is characterized by anhydrite cement, up to 15%. The two samples (38, 36) with lowest anhydrite contents also have slightly lower $\delta^{18}O$ values (−6.4 and −6.7‰); the low abundance of early anhydrite cement in these zones might have permitted groundwater penetration and calcite cementation at a slightly later stage of burial than elsewhere. In summary, therefore, the sequence of diagenetic processes inferred in this part of the basin is (i) early post-depositional calcite cementation possibly with a minor contribution from organic breakdown in adjacent strata, followed by (ii) anhydrite cementation which limited both groundwater flows and further carbonate cementation, followed by (iii) migration of hydrocarbons into the formation which prevented the deposition of the later, deep-burial, calcite cements found elsewhere in the basin.

East Yorkshire–Lincolnshire Basin

Dolomite is the dominant carbonate cement in all samples studied from this basin (Milodowski *et al.* 1987). This contrasts with the Wessex Basin in which calcite was the dominant carbonate with the exception of samples from Winterborne Kingston. Two suites of samples were studied from outcrop (including a shallow drillcore from Wistow), and from the well at Cleethorpes (Appendix 2). The stable isotope analytical results are shown in Table 3, and are illustrated in Fig. 9.

Triassic sandstone outcrop in South Yorkshire–Nottinghamshire

The four outcrop samples and five samples from Wistow drillcore show a range of composition from −3.6 to −1.1‰ $\delta^{13}C$ and −5.9 to −2.4‰ $\delta^{18}O$. They fall close to the range found in

TABLE 3. *$^{13}C/^{12}C$ and $^{18}O/^{16}O$ analyses of total carbonate content of samples from the Permo–Triassic sandstones in the East Yorkshire–Lincolnshire Basin*

No.	Sample number	$\delta^{13}C$ (‰ vs PDB)	$\delta^{18}O$ (‰ vs PDB)
Cleethorpes drill core (Triassic Sherwood Sandstone)			
1	CL(SH)1 (bulk sample)	+0.03	−4.36
2	CL(SH)1 (calcite nodules)	+0.04	−12.79
3	CL(SH)6	+0.19	−4.22
4	CL(SH)7	+0.34	−3.38
5	CL(SH)13	+0.28	−4.15
6	CL(SH)20	−0.61	−4.06
Cleethorpes drillcore (Basal Permian Sands)			
7	CL(BP)3	−1.23	−4.43
8	CL(BP)4	+0.05	−2.58
9	CL(BP)9	−2.34	−2.07
10	CL(BP)11	−3.00	−2.20
Wistow drillcore (Sherwood Sandstone)			
11	WSS-2	−3.59	−4.49
12	WSS-3	−3.09	−4.26
13	WSS-7	−2.78	−3.85
14	WSS-9	−2.95	−4.04
15	WSS-14	−2.77	−4.45
Outcropping Permo–Triassic sandstone in South Yorkshire			
16	789-1V	−2.15	−2.35
17	178-1HX1	−2.96	−5.91
18	1207-1V	−1.08	−2.90
19	526-H121	−3.29	−5.15

previous study of Sherwood Sandstone drillcore samples from shallow burial depth at Gamston in Nottinghamshire (Edmunds *et al.* 1982; also shown for comparison in Fig. 9). The isotopic, chemical and mineralogical information on the Gamston samples and co-existing groundwater was interpreted as indicating that the carbonate comprises isotopically heavy dolomite with minor amounts of isotopically light calcite which is actively precipitating as a product of dedolomitization (Edmunds *et al.* 1982, Evans *et al.* 1984, Bath *et al.* in press). The present outcrop samples similarly have a spread of isotopic compositions representing mixtures between dolomite and subordinate calcite (positively identified in 178-1HX1 and WSS-2 and traces in others).

Petrographic observations show that the dolomite here is an early-stage cement, in some cases replacing pre-existing clasts (Appendix 2; Milodowski *et al.* 1987). Isotopic data suggest that the dolomite formed soon after deposition, probably in an environment influenced by Mg-enriched groundwater evolving from nearby Permian Magnesian Limestone. These early dolomite cements have inherited the $\delta^{13}C$ of precursor detrital (?Carboniferous) limestones (≥ 0‰ $\delta^{13}C$) more directly than the bulk of the early calcite cements in the Wessex Basin. This may be because the dolomitization takes place without the necessity of additional CO_2 influx:

$$2CaCO_3 + Mg^{2+} \longrightarrow CaMg(CO_3)_2 + Ca^{2+}$$

whereas calcite dissolution and recrystallization is driven by change in CO_2 activity (with resultant influence on $\delta^{13}C$ due to isotopic exchange with CO_2):

$$CaCO_3 + CO_2 + H_2O \longrightarrow Ca^{2+} + 2HCO_3^-.$$

The $\delta^{18}O$ values for the early post-depositional dolomite fall in a slightly higher range than those for calcrete calcites in the Wessex Basin. This difference might be due to the variation in oxygen isotope fractionations between calcite or dolomite and solution (around 2‰; Fritz & Smith 1970). It is also possible that this difference indicates palaegeographical effects or variably higher extents of evaporative enrichment of $\delta^{18}O_{H_2O}$ in water recharging the early sub-surface of the East Yorkshire–Lincolnshire Basin. The surface conditions might have been similar to those observed in modern continental sabkhas in which dolomite precipitates have high $\delta^{13}C$ and $\delta^{18}O$ values (McKenzie 1981). However it is doubtful whether this surface phenomenon alone could account for pervasive dolomitization throughout a sequence on this scale (Machel & Mountjoy 1986).

Although isotopic analyses of separated calcite have not been possible, their approximate isotopic compositions are inferred from the trend of data in Fig. 9 (B to D). The estimated $\delta^{13}C$ and $\delta^{18}O$ values correlate with values which represent isotopic equilibration with present-day groundwater which has $\delta^{18}O_{H_2O} \approx -8.5$‰ vs SMOW and $\delta^{13}C_{HCO_3} \approx -13$ to -10‰ (Table 4; Bath *et al.* 1979, Edmunds *et al.* 1982). This tends to confirm that these calcites are precipitating ('dedolomite calcite') actively in the present groundwater regime by the reaction:

$$CaMg(CO_3)_2 + CO_2 + H_2O \longrightarrow CaCO_3 + Mg^{2+} + 2HCO_3^-.$$

Cleethorpes: Sherwood Sandstone

These carbonates are recognized petrographically as predominantly early replacive dolomite, and have higher $\delta^{13}C$ than the outcrop samples (Fig. 9). The source of these higher values is not clear—possibly different extents of atmospheric equilibration, or different early diagenetic conditions affecting the carbonate system (A to C in Fig. 9). In marked contrast, a nodule of late calcite-cemented sandstone extracted from CL(SH)1 has a very light $\delta^{18}O$ value of -12.8‰ which represents isotopic equilibrium with present groundwater (-6.9‰; Table 4) at the

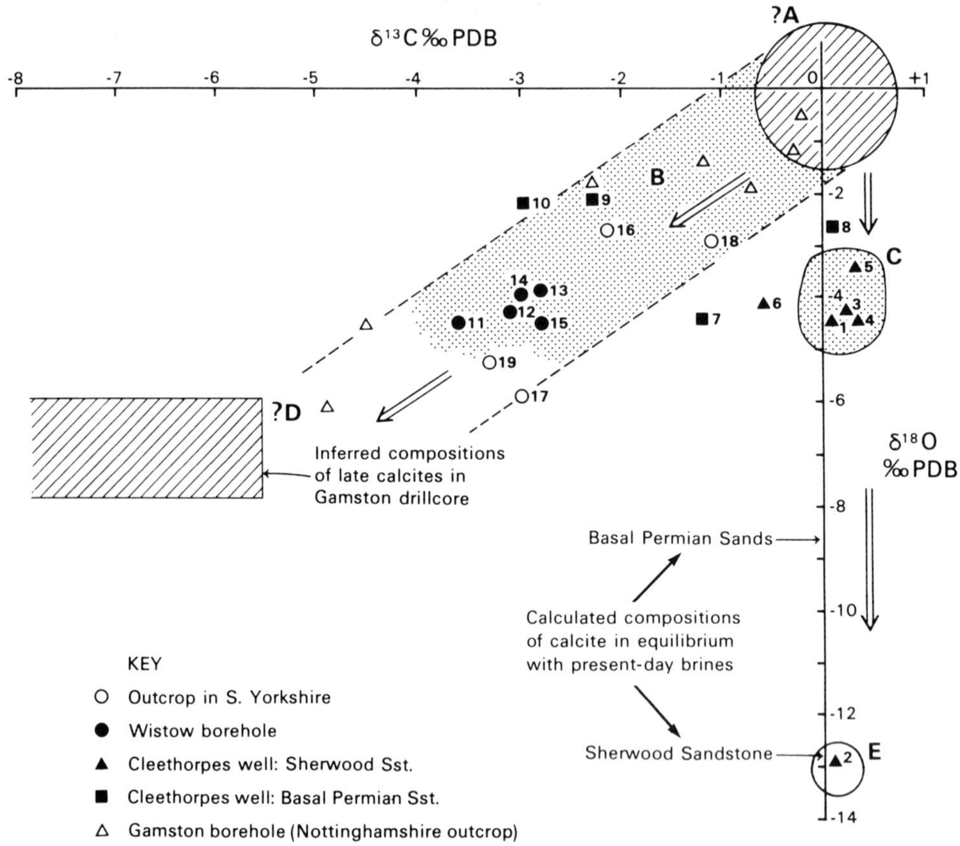

FIG. 9. Plot of $\delta^{13}C$ against $\delta^{18}O$ for samples from the Sherwood Sandstone in the East Midlands and the East Yorkshire–Lincolnshire Basin. Analyses of samples from the Basal Permian Sands in the Cleethorpes drillcore are also shown. Stippling indicates ranges of dolomite compositions. The open triangles show previously reported compositions of carbonate in drillcore samples from Gamston. Cross-hatched areas denote inferred compositions of detrital calcite and of dedolomite calcites in outcrop and shallow buried sandstone; the latter were estimated by progressive acid-leaching experiments on the Gamston samples (Edmunds *et al.* 1982, Evans *et al.* 1984). The arrows illustrate generalized isotopic evolution pathways. A—estimated composition of detrital ?Carboniferous limestone; B and C— ubiquitous early dolomitization of detrital calcite, varying compositions in west (outcrop) and east (Cleethorpes) respectively; D—inferred compositions of calcite due to late and present-day dedolomitization in shallow groundwater in west of basin; E—sparse calcite due to late dedolomitization in deeply buried sandstones (Cleethorpes).

observed temperature of 50°C; however, the $\delta^{13}C$ of dissolved carbonate is not known, and the relationship remains uncertain. It is possible that this is a late-stage or active precipitate (C to E in Fig. 9); the contrast in $\delta^{13}C$ with the inferred composition of dedolomite calcite at outcrop is thought to be a consequence of the depth and the absence of active circulation of meteoric water with isotopically-light dissolved bicarbonate. Progressively less negative $\delta^{13}C$ values for dissolved bicarbonate have been observed in groundwater down-gradient in the East Midlands Sherwood Sandstone aquifer (Bath *et al.* 1979; Edmunds *et al.* 1982).

Cleethorpes: Basal Permian Sands

The four samples analysed have scattered $\delta^{13}C$ values which fall across the field for other data from this area (Fig. 9). Two samples are dolomite-cemented conglomerates in which up to 20% of the cement is a late-stage ankeritic rim. However there is no direct correlation of ankerite proportions with isotopic compositions. Petrographic evidence suggests that these dolomites and ankerites formed at different stages in the burial history, and this could account for the range in $\delta^{13}C$. They have similar or slightly higher $\delta^{18}O$ values relative to the Sherwood

TABLE 4. *$^{18}O/^{16}O$, $^{2}H/^{1}H$ and $^{13}C/^{12}C$ analyses of groundwaters in Permo–Triassic aquifers in the East Midlands and the East Yorkshire–Lincolnshire Basin (Bath et al. 1979, Downing et al. 1985, Parker et al. 1986)*

Source of sample	$\delta^{18}O$	$\delta^{2}H$	$\delta^{13}C$ HCO_3	Salinity
	‰ vs SMOW		‰ vs PDB	(mg l^{-1} Cl)
Modern recharge to Notts Sherwood Sst aquifer	−8.6 to −7.7	−58 to −50	−13.5 to −10.3	18 to 174*
Late Pleistocene water in Notts/Lincs Sherwood Sst	−9.8 to −9.1	−66 to −57	−10.6 to −8.5	6 to 21
Modern recharge to S Yorks Sherwood Sst aquifer	−8.2 to −8.0	−56 to −51	−11.4	18 to 120*
Recent recharge in confined Sherwood Sst in S Yorks	−7.8 to −7.1	−57 to −46	−17	9 to 22
Sherwood Sandstone aquifer in Cleethorpes Well	−6.9	−51	n/a	45200
Basal Permian Sands aquifer in Cleethorpes Well	+0.4	−12	n/a	133 000

* Higher Cl values here attributed to contaminant sources affecting recharge.

TABLE 5. *$^{34}S/^{32}S$ and $^{18}O/^{16}O$ analyses of sulphate mineral phases in the Triassic sandstones in the Wessex Basin*

Sample number	$\delta^{34}S$	$\delta^{18}O$	Notes
Winterborne Kingston drillcore			
WK11	+7.85	+14.53/+14.88	*c.* 5–10% anhydrite
Wytch Farm X14 drillcore			
E59542	+9.30/+9.80	+15.43/+15.54	*c.* 10–15% anhydrite
E59546	+10.84	+15.35	*c.* 5–10% anhydrite

$\delta^{34}S$ values are in ‰ variation with respect to the Canyon Diablo meteorite troilite standard, CD. $\delta^{18}O$ values are in ‰ versus SMOW.

Sandstone dolomites, which is noteworthy in view of the sharply contrasted values for $\delta^{18}O$ in present-day brines in the two formations (Table 4: Downing *et al.* 1985). This suggests that these early dolomites precipitated in broadly similar environments, since then the two formation brines have evolved independently.

Stable isotopic composition of anhydrite cement

Three $^{34}S/^{32}S$ and $^{18}O/^{16}O$ analyses were carried out on the diagenetic anhydrite occurring in the Sherwood Sandstone at Winterborne Kingston and Wytch Farm in the Wessex Basin. The objective was to investigate the source and diagenesis of the anhydrite. The results are given in Table 5; these are in the range of +7.8 to +10.8‰ $\delta^{34}S$ and +14.5 to +15.5‰ $\delta^{18}O$. The isotopic composition of dissolved sulphate (1400 mg l^{-1}) in the brine from the Marchwood well has already been reported: +7.5 to +8.7‰ $\delta^{34}S$ and +11.6‰ $\delta^{18}O$ (Bath & Darling 1981). As expected, the sulphur isotope compositions suggest that the dissolved sulphate is derived from the solid anhydrite.

The $\delta^{34}S$ values are lower than those found in marine sulphate deposits through the stratigraphical record (Claypool *et al.* 1980). The minimum of $\delta^{34}S$ in evaporites are in Permian Zechstein sulphates, for which values are around +10‰ (Claypool *et al.* 1980, Taylor 1983). It is suggested that the low $\delta^{34}S$ values of anhydrite in Sherwood Sandstone indicate that the sulphate was originally derived by remobilization of Zechstein sulphate. However, the Wessex Basin has no Zechstein deposits in the Permo–Triassic sequence, therefore a remote source with transport by surface- and/or groundwater into the basin is inferred. Alternatively the anhydrite may have been produced from a mixture of marine sulphate and oxidized sulphides possibly from the overlying Mercia Mudstone. Taylor

(1983) has interpreted $\delta^{34}S$ data for Mercia Mudstone sulphate horizons in the East Midlands as indicating locally varying proportions of marine and continental brine inputs to the late Triassic depositional environment in that area.

The $\delta^{18}O$ values of the anhydrite (*c.* +15‰ $\delta^{18}O$) are significantly heavier than the Zechstein minimum (*c.* +10‰) in the $\delta^{18}O_{SO_4}$ time-curve (Claypool *et al.* 1980). This might also indicate that the anhydrite contains some non-Zechstein sulphate; alternatively it could indicate that there had been some oxygen-isotope exchange between water and dissolved sulphate resulting in an enrichment of ^{18}O in the sulphate. The rate at which this exchange might take place under these conditions is uncertain as it depends on pH as well as temperature. Present evidence suggests this would be very slow even at 70°C, at which the equilibrium isotopic fractionation $\alpha^{18}_{SO_4-H_2O}$ is about 1.021 (Chiba & Sakai 1985). However, isotope exchange seems unlikely since the anhydrite precipitation is attributed to an earlier stage of burial diagenesis when temperatures and therefore rates of exchange would have been lower.

Conclusions

Inevitably the models to interpret the stable isotopic evolution of the meteoric water diagenetic environments are somewhat speculative. They are inferences based on evidence from modern meteoric water aquifers and evaporite environments (calcretes and sabkhas). However the scale of the processes, both in time and in space in the Triassic rocks, and the uncertain detail of the Triassic palaeoenvironment introduce much uncertainty into these comparisons.

The Sherwood Sandstone in the Wessex Basin is cemented predominantly with calcites whose isotopic compositions suggest precipitation in a range of burial environments. This range extends from early post-depositional calcrete formation in freshwater to later deep-burial calcite recrystallization in basinal brines. The carbonate for early calcretes was probably derived by dissolution of detrital Carboniferous or Devonian limestone fragments; the $\delta^{13}C$ values also indicate a minor contribution from soil-zone organic-derived CO_2 except in calcite found in the present outcrop for which the data suggest recrystallization without CO_2 exchange. This outcrop of the Wessex Basin marginal facies has abundant recrystallized detrital carbonate which is still being dissolved in the present groundwater regime. An exception to calcite cementation is found in the basin's depocentre, where dolomite with relatively high $\delta^{18}O$ indicates an evaporitic environment.

The isotopic compositions of calcite cements in a drillcore from the Wytch Farm Sherwood Sandstone oil reservoir differ slightly from those at other sites; this is attributed to processes involving the inhibition of groundwater flux in the formation, first by early anhydrite cementation and later by hydrocarbon influx.

The restricted range of $\delta^{13}C$ values suggests that the carbonate system in the Wessex Basin has remained virtually closed over its burial history since the early formation of calcretes. The later carbonate cements have been derived by remobilization of the pre-existing cements with little change in $\delta^{13}C$.

Permo–Triassic sandstone samples from the East Yorkshire–Lincolnshire Basin are dominated by early dolomite precipitation. These dolomites are characterized by relatively heavy $\delta^{13}C$ and $\delta^{18}O$ ratios. This indicates early dolomitization of pre-existing detrital calcite in sabkha conditions, possibly associated with a widespread flux of Mg-enriched groundwater into the basin coupled with arid conditions as groundwater discharged through the low-relief surface.

The difference in $\delta^{13}C$ values between the two basins may reflect differences in the carbon chemistries of the two reactions: calcite recrystallization and dolomitization. The resultant $\delta^{13}C$ value depends both on the composition of precursor carbonate and also on the contribution of external CO_2 (atmospheric or soil zone) to the reaction.

A further difference between the two basins is found in the effective degrees and compositions of later carbonate cementation. The petrological and isotopic evidence indicates that the Wessex Basin has abundant calcite precipitated at various stages of burial up to the maximum: the isotopic compositions form a series between the early and deep-burial calcites (which, however, are not in carbon isotopic equilibrium with present formation fluid). This apparently continuous process of precipitation in a closed carbonate system might be driven simply by the pressure and temperature changes and compaction during burial.

The only isotopic evidence for carbonate reaction after early dolomite cementation in the East Yorkshire–Lincolnshire Basin is that for a progressive and still active dedolomitization due to the influx of fresh groundwater through the outcropping rim of the basin, bringing with it isotopically light carbon from superficial sources. This zone is therefore subjected at present to enhanced dedolomitization, and any calcite precipitated has a characteristically light carbon and oxygen isotopic composition. The isotopic evidence for this process in the deeper basin—only one isotopically light calcite positively identified—is limited, although there is petrological evidence for sparse late calcite. The groundwater flux into the deeper basin over its geological history would have been limited. However, the light isotopic composition of the present brine indicates a direct influence of meteoric water here, in contrast to the isotopically heavy evolved brines in the Wessex Basin.

Finally, there is no substantial evidence that the organic reactions preceding hydrocarbon migration into the Sherwood reservoir at Wytch Farm have greatly influenced isotopic compositions of carbonate cements. These carbonates have $\delta^{13}C$ values slightly higher than elsewhere, but the difference is small and its significance is not clear. However, the absence of late diagenetic calcites with isotopic compositions representing the deepest burial conditions suggests that the influx of hydrocarbon prior to the deepest burial stage somehow inhibited further calcite precipitation in both oil and water zones.

ACKNOWLEDGMENTS: This study was carried out within the UK Geothermal Energy Assessment Project, funded by Department of Energy, project manager R. A. Downing. The assistance of G. Strong and S. Holloway in the petrographical study and the structural interpretation respectively is acknowledged. C. Branch carried out C and O isotope analyses; M. E. Patton (Department of Earth Sciences, University of Waterloo, Canada) carried out S and O isotope analyses. I. George assisted with sample preparation. The paper benefitted from reviews by Charles Curtis, Greg Samways and Jim Marshall. The manuscript was typed by P. Bath and V. Jones. This paper is published with permission from the Director, British Geological Survey (NERC).

References

BATH, A. H. & DARLING, W. G. 1981. Stable isotope analyses of water, carbonate and sulphate from the pump-test of Marchwood No. 1 borehole. *Stable Isotope Technical Report No. 9*, British Geological Survey (unpublished).

——, EDMUNDS, W. M. & ANDREWS, J. N. 1979. Palaeoclimatic trends deduced from the hydrochemistry of a Triassic sandstone aquifer, United Kingdom. *In*: *Isotope Hydrology 1978*, pp. 545–68. IAEA, Vienna.

——, MILODOWSKI, A. E. & STRONG, G. E. In press. Fluid flow and diagenesis in the East Midlands Triassic sandstone aquifer. *In*: GOFF, J. C. & WILLIAMS, B. P. J. (eds). *Fluid Flow in Sedimentary Basins and Aquifers*. Special Publication of the Geological Society of London. Blackwell Scientific Publications, Oxford.

BURLEY, S. D. 1984. Patterns of diagenesis in Sherwood Sandstone Group (Triassic), United Kingdom. *Clay Minerals*, **19**, 403–40.

CHIBA, H. & SAKAI, H. 1985. Oxygen isotope exchange rate between dissolved sulfate and water at hydrothermal temperatures. *Geochimica et Cosmochimica Acta*, **49**, 993–1000.

CLAYPOOL, G. E., HOLSER, W. T., KAPLAN, I. R., SAKAI, H. & ZAK, I. 1980. The age curves of sulfur and oxygen isotopes in marine sulphate and their mutual interpretation. *Chemical Geology*, **28**, 199–260.

COLTER, V. S. & HAVARD, D. J. 1981. The Wytch Farm oilfield, Dorset. *In*: *Petroleum Geology of the Continental Shelf of North West Europe*, pp. 494–503. Institute of Petroleum, London.

DOWNING, R. A., ALLEN, D. J., BIRD, M. J., GALE, I. N., KAY, R. L. F. & SMITH, I. F. 1985. Cleethorpes No 1 Geothermal Well—a preliminary assessment of the resource. *Investigation of the Geothermal Potential of the U.K.* British Geological Survey, Keyworth.

—— & GRAY, D. A. (eds). 1986. Geothermal energy—the potential in the United Kingdom. *Report, British Geological Survey*. HMSO, London.

EDMONDS, E. A. & WILLIAMS, B. J. 1985. The geology of the country around Taunton and the Quantock Hills. *Memoirs, British Geological Survey*, Sheet 295.

EDMUNDS, W. M., BATH, A. H. & MILES, D. L. 1982. Hydrochemical evolution in the East Midlands Triassic sandstone aquifer, England. *Geochimica et Cosmochimica Acta*, **46**, 2069–82.

EMRICH, K., EHHALT, D. H. & VOGEL, J. C. 1970. Carbon isotope fractionation during the precipitation of calcium carbonate. *Earth and Planetary Science Letters*, **8**, 363–71.

EVANS, G. V., OTLET, R. L., WASSELL, L. L. & BATH, A. H. 1984. Verification of the presence of carbon-14 in secondary carbonates within a sandstone aquifer and its hydrological implications. *In*: *Isotope Hydrology 1983*, pp. 577–90. IAEA, Vienna.

FONTES, J.-C. & GARNIER, J.-M. 1979. Determination of the initial ^{14}C activity of the total dissolved carbon: a review of the existing models and a new approach. *Water Resources Research*, **15**, 399–413.

FRIEDMAN, I. & O'NEIL, J. R. 1977. Compilation of stable isotope fractionation factors of geochemical interest. *In*: FLEISCHER, M. (ed.) *Data of Geochemistry*, Chapter KK. US Geological Survey Professional Paper 440-KK.

FRITZ, P. & SMITH, D. 1970. The isotopic composition of secondary dolomites. *Geochimica et Cosmochimica Acta*, **34**, 1161–73.

HENSON, M. R. 1970. The Triassic rocks of South Devon. *Proceedings of the Ussher Society*, **2**, 172–7.

KNOX, R. W. O'B., BURGESS, W. G., WILSON, K. S. & BATH, A. H. 1984. Diagenetic influences on reservoir properties of the Sherwood Sandstone (Triassic) in the Marchwood geothermal borehole, Southampton, England. *Clay Minerals*, **19**, 441–56.

LOTT, G. K. & STRONG, G. E. 1982. The petrology and petrography of the Sherwood Sandstone (?Middle Triassic) of the Winterborne Kingston borehole, Dorset. *In*: RHYS, G. H., LOTT, G. K. & CALVER, M. A. (eds). *The Winterborne Kingston Borehole, Dorset, England*, pp. 135–42. *Report Series*, 81/3. Institute of Geological Sciences, London.

MCCREA, J. M. 1950. On the isotopic chemistry of carbonates and a paleotemperature scale. *Journal of Chemical Physics*, **18**, 849–57.

MCKENZIE, J. A. 1981. Holocene dolomitization of calcium carbonate sediments from the coastal sabkhas of Abu Dhabi, U.A.E.: a stable isotope study. *Journal of Geology*, **89**, 185–98.

MACHEL, H.-G. & MOUNTJOY, E. W. 1986. Chemistry and environments of dolomitization—a reappraisal. *Sedimentology*, **23**, 175–222.

MILODOWSKI, A. E., STRONG, G. E., WILSON, K. S., ALLEN, D. J., HOLLOWAY, S. & BATH, A. H. 1986. Diagenetic influences on the aquifer properties of the Sherwood Sandstone in the Wessex Basin. *Investigation Geothermal Potential U.K.* British Geological Survey.

——, ——, ——, HOLLOWAY, S. & BATH, A. H. 1987. Diagenetic influences on the aquifer properties of the Permo–Triassic sandstones in the East Yorkshire–Lincolnshire Basin. *Investigation Geothermal Potential U.K.* British Geological Survey.

PARKER, J. M., FOSTER, S. S. D., SHERRATT, R. & ALDRICK, J. 1985. Diffuse pollution and groundwater quality of the Triassic sandstone aquifer in southern Yorkshire. *Report Series, British Geological Survey*, Vol. 17, No. 5, HMSO, London.

SALOMONS, W., GOUDIE, A. & MOOK, W. G. 1978. Isotopic composition of calcrete deposits from Europe, Africa and India. *Earth Surface Processes*, **3**, 43–57.

SHARMA, T. & CLAYTON, R. N. 1965. Measurements of O^{18}/O^{16} ratio of total oxygen carbonates. *Geochimica et Cosmochimica Acta*, **29**, 1347–53.

SMITH, A. N. & EPSTEIN, S. 1971. Two categories of $^{13}C/^{12}C$ ratios for higher plants. *Plant Physiology*, **47**, 380–4.

STRONG, G. E. & MILODOWSKI, A. E. 1987. Aspects of the diagenesis of the Sherwood Sandstone of the Wessex Basin. *This volume*.

TALMA, A. S. & NETTERBERG, F. 1983. Stable isotope abundances in calcretes. *In*: WILSON, R. C. L. (ed.). *Residual Deposits*, pp. 221–33. Special Publication of the Geological Society of London, 11. Blackwell Scientific Publications, Oxford.

TAYLOR, S. R. 1983. A stable isotope study of the Mercia Mudstones (Keuper Marl) and associated sulphate horizons in the English Midlands. *Sedimentology*, **30**, 11–31.

WALTON, N. R. G. 1982. A detailed hydrogeochemical study of groundwaters in the southwest England Triassic Sandstone aquifer. *Report Series*, 81/5. Institute of Geological Sciences, London.

WIGLEY, T. M. L., PLUMMER, L. N. & PEARSON, F. J. 1978. Mass transfer and carbon isotope evolution in natural water systems. *Geochimica et Cosmochimica Acta*, **42**, 1117–40.

A. H. BATH & A. E. MILODOWSKI, Fluid Processes Reserach Group, British Geological Survey, Keyworth, Nottingham NG12 5GG, UK.

B. SPIRO, Mineral Science and Isotope Geology Research Group, British Geological Survey, 64 Gray's Inn Road, London WC1X 8NG, UK.

Appendix 1

Descriptive notes on carbonate cements in samples of Triassic sandstones from the Wessex Basin (estimates of carbonate contents by cathodoluminescence microscopy)

No.	Sample	Notes (including National Grid Reference)
Western outcrop		
1, 2	HW03	Holywell Quarry, Milverton (ST 1269 2703); Pebble Beds conglomerate with limestone pebbles, zoned early calcite cements
3	MV01	Milverton (ST 1160 2465); Pebble Beds; calcite on fracture surface
4	MV01	Milverton; Pebble Beds; *c.* 35% calcite, virtually all early calcite with later calcite filling rims
5	WS02	Wolston Quarry (ST 0945 4015); Pebble Beds; pure vein calcite with two zones: earlier is brecciated followed by later luminescent calcite
6	BL01	Bishops Lydyard (ST 1699 3021); Otter Sandstone; *c.* 30% calcite, some pellets; early:later calcite ≈50:50
7	LB01	Ladram Bay (ST 0975 8520); Otter Sandstone; highly calcareous concretion nodule, major part is luminescent calcite
8	LB02	Ladram Bay; Otter Sandstone; *c.* 25% calcite, virtually all non-luminescent calcite with occasional rims of brighter calcite

9	LB03	Ladram Bay; Otter Sandstone; calcrete sandstone, *c.* 20–25% calcite mostly later later : early $\approx$ 80 : 20
10	LBV01	Langford Budville (ST 1235 2218); Otter Sandstone; mostly calcrete and few pellets, rims of later calcite; early : later $\approx$ 80 : 20
Winterborne Kingston drillcore (*SY 8470 9796*)		
11	WK01	2316.3 m. Total carbonate *c.* 10%. Carbonate pellets with minor later carbonate; mostly replacive and pore-filling dolomite; <5% anhydrite cement
12	WK06	2321.2 m *c.* 50–60% carbonate; early detrital calcite cores (50%) surrounded by later calcite (*c.* 30%) and replacive dolomite (*c.* 20%); <5% anhydrite cement
13	WK09	2324.7 m. *c.* 10% total carbonate; mainly dolomite with minor later calcite; dol : calc. $\approx$ 95 : 5; <5% anhydrite
14	WK11	2328.1 m. *c.* 20% carb; micrite pellets with early calcite cement rims and abundant later calcite partly replaced by dolomite; pellets : late-calcite : dolomite $\approx$ 40 : 40 : 20; 5–10% anhydrite
15	WK18	2428.2 m. *c.* 60% carbonate, all dolomite replacing clasts and cement (50 : 50)
16	WK19	2428.7 m *c.* 50% carbonate, all dolomite replacing clasts and cement
17	WK22	2434.4 m. Layered calcareous mudstone with total replacement by dolomite (70–80% of total)
Marchwood drillcore (*SU 3991 1118*)		
(E samples from Upper Unit—feldspathic; MW samples from Lower Unit—lithic)		
18	E59535	1684.9 m. *c.* 25% luminescenet later calcite, very small amount of non-luminescent earlier calcite
19, 20	E59536	1684.4 m. 25–30% calcite, virtually all earlier calcite with some pellet calcite (80 : 20)
21	E59538	1690.5 m $\leq$10% of calcite, mostly earlier calcite
22	MW4	1696.3 m. 25–30% calcite; multiple zones earlier calcite and later calcite (20 : 80)
23	MW9	1699.7 m. *c.* 30% calcite; dominant earlier and later calcites (80 : 20); fine-grained
24	MW19	1707.3 m. *c.* 10% calcite, all later calcite; fairly coarse-grained
25	MW22	1709.6 m. *c.* 30% calcite, mainly earlier plus later calcites (80 : 20)
26	MW23	1714.2 m. *c.* 5% later calcite
27	MW24	1715.2 m. Abundant carbonate-cemented pellets, mainly earlier calcite cement
Western Esplanade drillcore (*SU 4415 1120*)		
28	E59553	1735.2 m. Highly calcareous (*c.* 25% total calcite); mainly later calcite cement with pellets and earlier cement (70 : 15 : 15)
29	E59557	1740.4 m. Carbonate clasts with stellate earlier non-luminescent calcite and later calcite cement (50 : 50)
30	E59559	1740.9 m. Cornstone pellets with stellate early and later intergranular calcite, patchy; 25 : 35 : 40; *c.* 10% total carbonate
31	E59562	1752 m. *c.* 15% total calcite; mainly later calcite cement with earlier calcite and pellets (80 : 20)
32	E59564	1754.1 m. *c.* 25% total calcite, well-cemented, pellets abundant and often stellate; later : early calcite and pellets $\approx$ 50 : 50
33	E59565	1757.9 m. *c.* 25–30% total calcite, mostly earlier cement; early : later calcite $\approx$ 70 : 30
Wytch Farm X14 drillcore (*SY 9804 8526*)		
34	E59540	1598.4 m. Mostly anhydrite cemented (*c.* 10%) with sparse (*c.* 5%) calcite cement; mostly later with early calcite and some pellets (80 : 20)
35	E59542	1601.3 m. *c.* 10–15% calcite and 10–15% anhydrite; mostly calcite pellets with some later and earlier calcite; (70 : 25 : 5)
36	E59544	1611.4 m. 5–10% calcite and <5% anhydrite cement; later calcite on clast cores (50 : 50)
37	E59546	1631.2 m. *c.* 10% total calcite and 5–10% anhydrite; clasts with clast calcite cores and later calcite rims (60 : 40); very minor earlier calcite rims
38	E59551	1649.8 m. *c.* 10% total calcite; carbonate-cemented pellets, probably later calcite; no detectable anhydrite

Appendix 2

Descriptive notes on carbonate cements in samples of Permo–Triassic sandstones from the East Yorkshire–Lincolnshire Basin (estimates of carbonate contents by cathodoluminescence microscopy)

No.	Sample	Notes (including National Grid Reference)
Cleethorpes drillcore (TA 30237 07090); Sherwood Sandstone		
1	CL(SH)1A	1111.9 m. 5–10% dolomite replacing ferruginous clasts and rare replacive late calcite; patchy minor anhydrite
2	CL(SH)1B	1111.9 m. Nodule of late calcite-cemented sandstone from CL(SH)1A
3	CL(SH)6	1115.5 m. 5–10% replacive dolomite; patchy later calcite cement, 2% and anhydrite
4	CL(SH)7	1116.3 m. Silty sandstone with minor anhydrite/gypsum; 25–30% replacive and pore-filling dolomite
5	CL(SH)13	1208.5 m. 10–15% replacive dolomite
6	CL(SH)20	1317.0 m. 10% replacive dolomite
Cleethorpes drillcore (TA 30237 07090); Basal Permian Sands		
7	CL(BP)3	1869.7 m. Mostly early non-ferroan dolomite (10%) overgrown by minor later ankerite
8	CL(BP)4	1870.7 m. Abundant (20–40%) replacive and pore-filling dolomite cements with ankeritic overgrowths
9	CL(BP)9	1884.0 m. Ferruginous siltstone with interstitial fine-grained dolomite, *c.* 10%.
10	CL(BP)11	1890.7 m. Similar to CL(BP)4; 10–15% vuggy replacive calcite
Wistow drillcore (SE 57433 35619); Sherwood Sandstone		
11	WSS-2	39.7 m. *c.* 5% replacive dolomite; rarer later poikilotopic calcite (2–5%)
12	WSS-3	49.1 m. 10% replacive dolomite
13	WSS-7	75.8 m. 10% dolomite; patchy later calcite (*c.* 1%) often enclosing or replacing dolomite
14	WSS-9	83.8 m. 10–15% dolomite replacing clasts; localized later poikilotopic calcite
15	WSS-14	106.6 m. Ferruginous siltstone with 10% interstitial dolomite rhombs
Outcropping Permo–Triassic sandstones		
16	789-1V	SK 606 898. Coarse siltstone, 10–15% interstitial dolomite and trace late calcite
17	178-1HX1	SK 7063 9972. *c.* 8% replacive dolomite, patchy late calcite *c.* 3%
18	1207-1V	Se 6451 2432. *c.* 10% dolomite replacing detrital clasts
19	526-H121	SE 4583 5587. *c.* 5% dolomite and sporadic nodules of late calcite

Six million year diagenetic history, North Coles Levee, San Joaquin Basin, California

J. R. Boles

SUMMARY: North Coles Levee is an uppermost Miocene turbidite sandstone reservoir in the San Joaquin Basin. The sands have undergone progressive burial and a complex history of siderite, dolomite and calcite cementation. Precipitation of calcite cement bands occurred between 40 and 80°C with the last cementation event corresponding to about 2.5 Ma BP. Between this time and the present, extensive dissolution of plagioclase and calcite resulted from influx of acid during kerogen maturation. Emplacement of 500 million barrels of oil closely followed or was contemporaneous with these events. Very late stage compaction effects including albitization of plagioclase and dolomite crystallization in crushed biotite have resulted from cement removal and/or fluid pressure drops during release of the gas cap.

This paper is a brief synthesis of the diagenetic history of a late Tertiary oil field in California. Results and discussions presented here are given in greater detail in earlier papers (Boles & Johnson 1983, Boles 1984 and Ramseyer & Boles 1986) and papers in preparation (Boles & Ramseyer, in press). A significant aspect of the history is the relatively small time interval (<2.5 Ma) which encompasses dissolution of framework grains and cements as well as hydrocarbon emplacement. These effects are consistent with the idea that porosity enhancement in the reservoir results from chemical evolution of the porewaters during or slightly prior to hydrocarbon emplacement (Surdam *et al.* 1984). The short time interval for these events is surprising considering the amount of mass transfer involved.

Geological setting

The San Joaquin Basin contains more than 7 km of largely upper Miocene and younger shale and sandstone (Fig. 1). The North Coles Levee oil field is located in the central basin area where the section has subsided continuously since Upper Miocene time. In this area the depositional sequence changes from deep marine facies in the Upper Miocene to shallow marine and non-marine facies in the Plio-Pleistocene (Fig. 2). The basin has proven recoverable hydrocarbon reserves in excess of 7.8 billion barrels of oil and 11.2 trillion cubic feet of gas in more than 40 producing fields (Callaway 1971). The young age of the basin, relatively simple structural history, and abundant subsurface data allows considerable control to be placed on timing and physical conditions of diagenesis.

North Coles Levee

The North Coles Levee (NCL) reservoir is a series of deep marine turbidite sandstones (Webb 1981). Foraminifera in surrounding mudstones indicate the sandstones were deposited at bathyal depths during the latest Mohonian stage (Lagoe, in press). Thus the diagenetic history described in this paper occurred within the past 7 Ma.

The main producing interval, between 2.6 and 2.9 km, is a series of amalgamated medium to coarse sandstones containing an estimated 500 million barrels of oil in place at the time of discovery. Average reservoir porosity is about 15% based on core plug measurements. Average permeability in core plugs is about 10 md with values as high as 150 md in some zones. The reservoir sands have quartz:feldspar ratios of 1:1 and plagioclase:K-feldspar ratios >1:1 (Ramseyer & Boles 1986). Sand isopach maps and facies analyses indicate they were derived from the Jurassic-age Sierran granodiorite batholith, 60 km to the NE (Webb 1981). Plagioclase/K-feldspar ratios in the sands are similar to Sierran granodiorites although the sands are enriched in quartz relative to Sierran sources, presumably due to weathering and transport processes.

The reservoir at 2.6 km burial depth originally produced fluids at 285 bars pressure at 105°C. Pressures today are maintained at higher levels due to water injection in the central part of the field (Fig. 3). Original formation water at the time of field discovery was NaCl rich with total dissolved solids somewhat less than in sea water (Table 1). There is no evidence of meteoric water influx into this part of the section and the lower salinities of these pore waters relative to sea water is attributed in part to clay dehydration

From: MARSHALL, J. D. (ed.), 1987, *Diagenesis of Sedimentary Sequences*, Geological Society Special Publication No. 36, pp. 191–200.

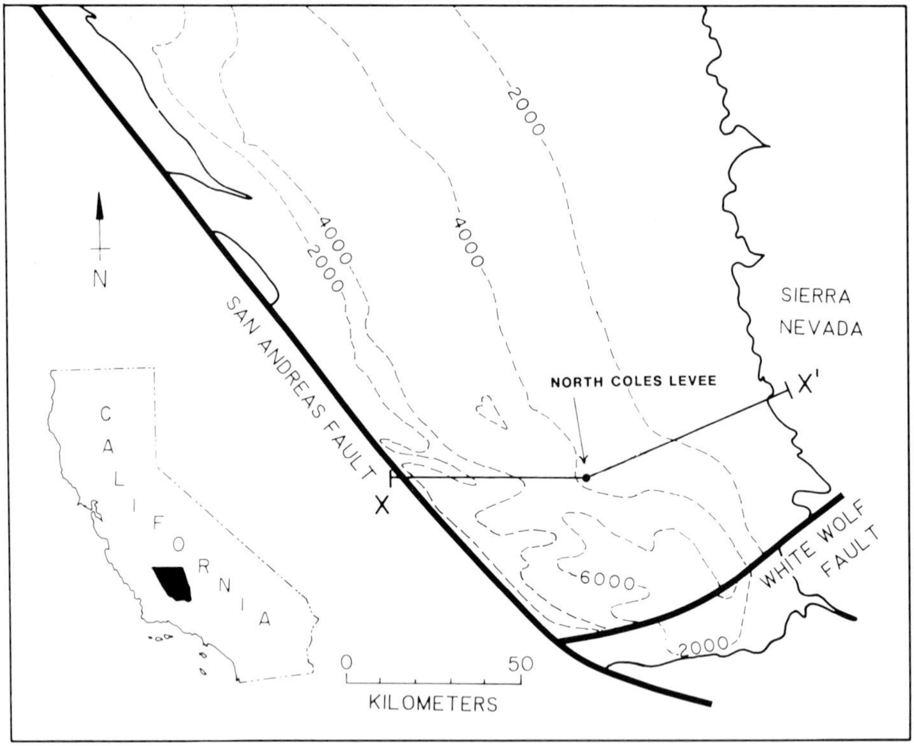

FIG. 1. Location of San Joaquin Basin and North Coles Levee oil field. Dashed lines show Tertiary sedimentary rock thickness (metres) from Callaway (1971). Cross-section X–X′ shown in Fig. 2.

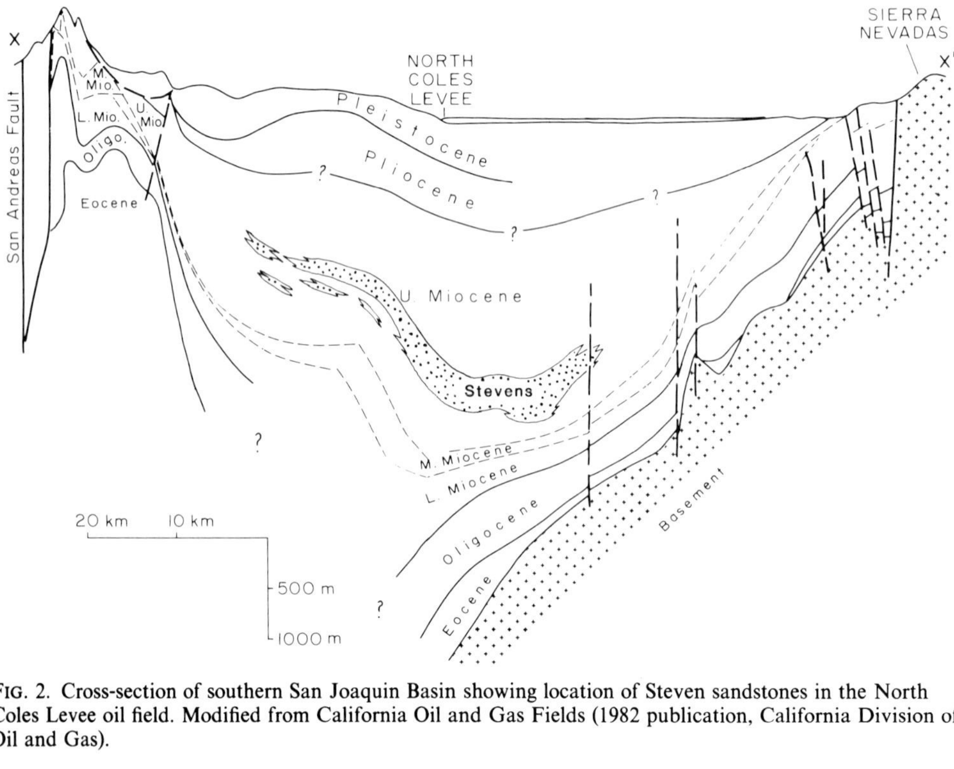

FIG. 2. Cross-section of southern San Joaquin Basin showing location of Steven sandstones in the North Coles Levee oil field. Modified from California Oil and Gas Fields (1982 publication, California Division of Oil and Gas).

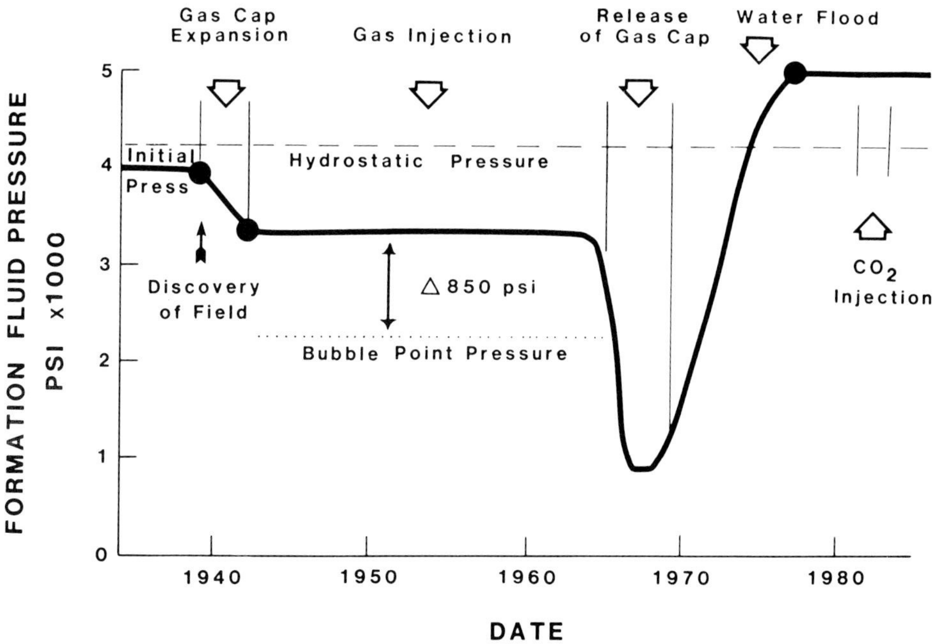

FIG. 3. Fluid pressure history of Stevens sandstone (2591 m) at North Coles Levee. Data from Arco, Bakersfield.

reactions (Ramseyer & Boles 1986). An unusual aspect of the water is that alkalinity is dominated by organic acids rather than carbonic acid species. The organic acids are believed to form complexes with metals (e.g. Al, Si) and hence provide a mechanism for moving low solubility components (e.g. Surdam *et al.* 1984, Crossey *et al.* 1986).

TABLE 1. *Comparison of seawater composition with evolved porewater at North Coles Levee (mg l^{-1}). Seawater composition from Drever (1982). Porewater HCO_3, acetate and propionate from R. C. Surdam Laboratory at University of Wyoming. Remaining analysis by ARCO Research Laboratory*

	Sea water	Well NCL 14-31 at 2641 m
Na^+	10 760	9300
K^+	399	200
Ca^{++}	411	790
Mg^{++}	1290	21
B^{+3}	4.5	120
Si^{+4}	0.5–10	44
Fe	<1	1.9
Mn	<1	0.7
Sr	8	49
Cl^-	19 350	14 000
NO_3^{--}	n.d.	120
SO_4^{--}	2710	35
HCO_3^-	142	96
Acetate	n.d.	3730
Propionate	n.d.	896
pH	8.2–8.4	6.94
TDS	35 000	29 403*

* Other elements analysed but not reported here.

Diagenetic history

The following section briefly describes the diagenetic phases in Stevens sandstones in their order of occurrence. These phases have formed

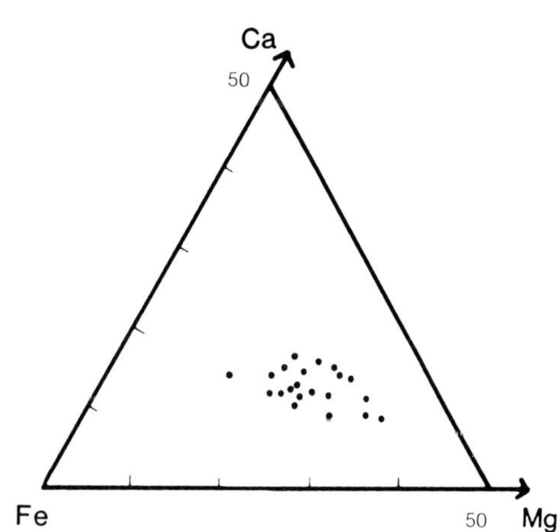

FIG. 4. Siderite compositions (mole %) from North Coles Levee. Data from microprobe analyses of five samples in well no. NCL 488-29 between 2727 and 2756 m depth.

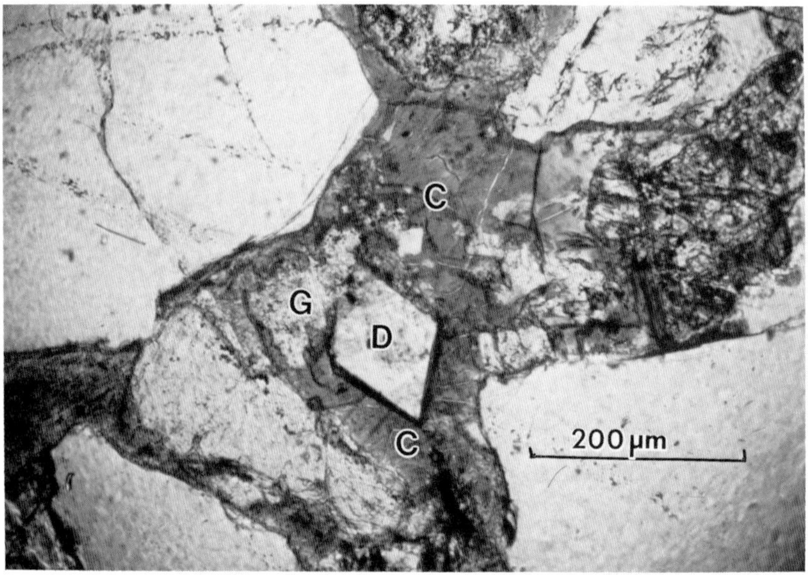

FIG. 5. Dolomite rhomb (D) attached to detrital grain (G) and surrounded by stained calcite (C). Well no. NCL 488-29, 2727 m.

during a single burial cycle at conditions ranging from those at the late Miocene marine sediment-water interface to present burial depths.

Siderite is one of the earliest diagenetic phases occurring as 10 μm yellowish crystals attached to detrital grains and as scattered crystals in pore spaces and frequently enclosed by later carbonate cements. Microprobe analyses indicate appreciable Mg substitution in the siderites (Fig. 4), similar to that observed in marine siderites from other areas (Matsumoto & Iijima 1981).

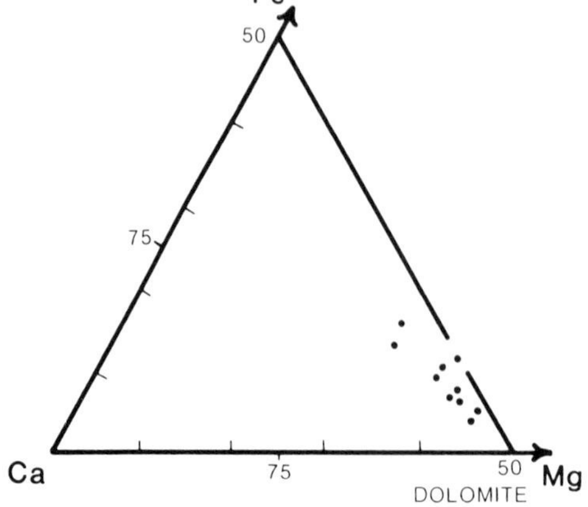

FIG. 6. Early dolomite compositions (mole %) from North Coles Levee cores. Data from microprobe analyses of 10 samples in well no. NCL 488-29 between 2716 and 2768 m depth.

Dolomite rhombs attached to detrital grains appear to post-date the siderite and predate most calcite cement (Fig. 5). Dolomite is usually a minor cement in the reservoir but several intervals (less than 1 m thick) contain 15–30 volume %. The dolomite has from 5 to 15 mole % Fe substitution and like many Tertiary dolomites is Ca rich with Ca : Fe + Mg > 1 : 1 (Fig. 6).

Calcite is the major pore-filling cement in the reservoir and is concentrated into thin zones (0.2–1 m thick) with up to 30% calcite. Where occurring with dolomite, calcite surrounds the euhedral dolomite crystals implying calcite formed after dolomite (Fig. 5). The calcite has Fe : Mg ratios of about 1 : 1 and from 1 to 10 mole % Fe + Mn + Mg substitution for Ca (Fig. 7). Detailed trace element and isotopic analyses of these and other carbonates is presented in a forthcoming paper where it is shown that the high Fe and Mg content of calcites from NCL is distinctive relative to calcite compositions from other depositional settings in the San Joaquin Basin (Boles & Ramseyer, in press).

Temperatures for NCL calcite precipitation are calculated from their oxygen isotopic composition and estimated values of pore fluid oxygen isotopes (see Boles & Ramseyer, in press). The $\delta^{18}O$ values for 18 calcites range from +21.2 to +25.9 (mean = +23.3). Pore water oxygen values were estimated with the assumption that $\delta^{18}O$ varied linearly between an initial value of 0 at the marine sediment–water interface (4–10°C) to the present-day value of +2.8 at 2600 m burial

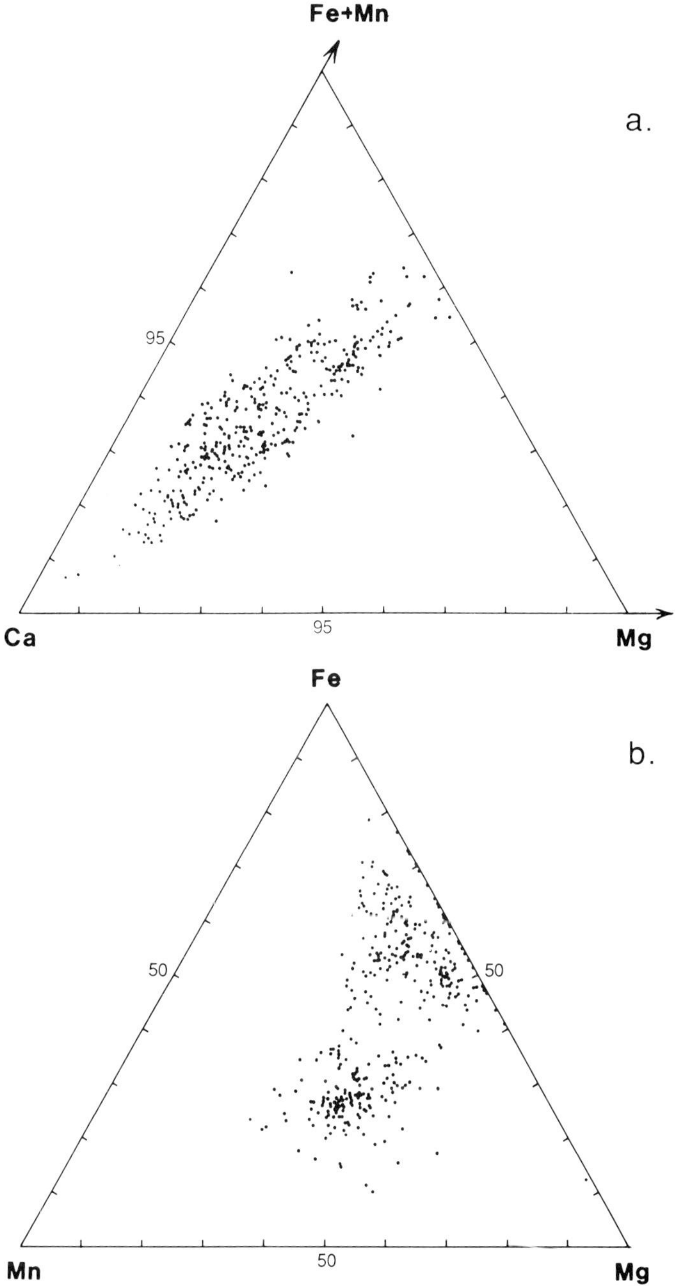

FIG. 7. Calcite compositions (mole %) from North Coles Levee cores. Data from microprobe analyses of 20 samples in wells no. NCL 88-29, 487-29, and 488-29 between 2069 and 2768 m depth.

depth (105°C). Results indicate that calcite precipitation occurred between about 40 and 80°C. Carbon isotopes for these calcites are generally light (average $\delta^{13}C_{PDB} = -4.2$) indicating precipitation in the zone of catagenesis in accordance with the calculated temperature range (e.g. Curtis *et al.* 1986). Fluid inclusions are generally very small in these calcites but a sample from 2727 m (present temperature 107°C) has inclusions with fluid-gas homogenization temperatures of 57–62°C and an isotopic precipitation temperature of 70–79°C. This agreement is reasonable considering the isotope temperature estimate is an average for the calcite in the rock and the samples contain a complex cementation history judging from compositional variation from closely spaced microprobe analyses.

Illite/smectite (I/S) mixed-layer clays are abundant in some sandstone intervals and in surrounding shales. Much of this clay was probably a smectite-rich detrital component of the turbidity currents. The I/S clays generally show expandibilities from about 30 to 50% smectite layers with up to 80% expandibility occurring in some samples (Ramseyer & Boles 1986). Expandibilities tend to be higher (up to 10% smectite layers) in calcite-cemented sandstones than uncemented samples, reflecting slowing of the smectite to illite reaction at low permeability conditions (see fig. 5 of Ramseyer & Boles 1986).

Porosity enhancement occurred after carbonate cementation. Calcic plagioclase and calcite were the important phases to be dissolved, whereas albite, K-feldspar and dolomite were unaffected. Plagioclase porosity forms up to one-third of the total porosity in the rock and has developed after carbonate cementation (Boles 1984). Plagioclase in calcite cemented rocks is not replaced by calcite nor does it have dissolution porosity, whereas plagioclase in adjacent uncemented sandstones exhibits extensive porosity (Fig. 8). Calcite cements adjoining pore space have ragged surfaces indicating dissolution has occurred but it is difficult to estimate the volume of calcite cement removed. It is probable that the more porous horizons were never completely cemented with calcite, and intergranular porosity has been enhanced there by dissolution of some calcite.

The dissolution of plagioclase and calcite appear to be synchronous events. Authigenic kaolinite is found in sandstones containing porous feldspars and is believed to conserve Al from plagioclase dissolution on the scale of a thin section (Boles 1984). The kaolinite is absent in tightly cemented sandstones but occurs in porous samples as interstitial crystal clusters. Calcite cement in contact with these crystals has ragged grain boundaries suggesting the kaolinite now occupies space once cemented with carbonate (Fig. 9). It is difficult to imagine a long residence time of aluminum in solution from a dissolving plagioclase, hence the kaolinite is believed to form contemporaneously with plagioclase dissolution. This implies calcite dissolution has also occurred contemporaneously or prior to these events. Acids derived from the maturation of organic matter have been suggested as the cause of porosity enhancement in sandstones (e.g. Curtis 1978, Surdam *et al.* 1984, Crossey *et al.* 1986). It is interesting to note that the dissolution event at NCL occurred between 80 and 105°C and this temperature range corresponds closely with a marked increase in organic acids in porewaters of the San Joaquin Basin (Carothers

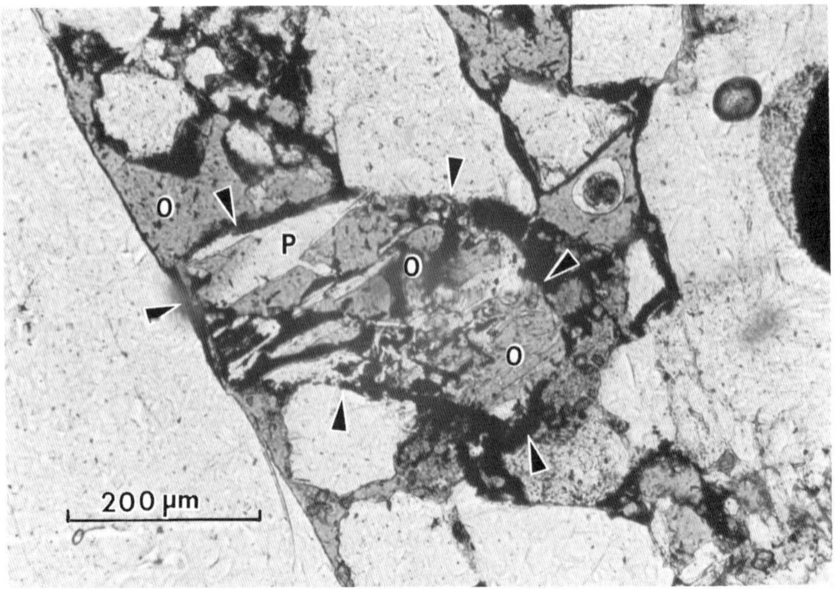

FIG. 8. Plagioclase remnant (P) with opaque residual hyrocarbon. Arrows denote original grain boundary. Well no. NCL 488-29, 2727 m. Plane light. Note extensive porosity 'O'.

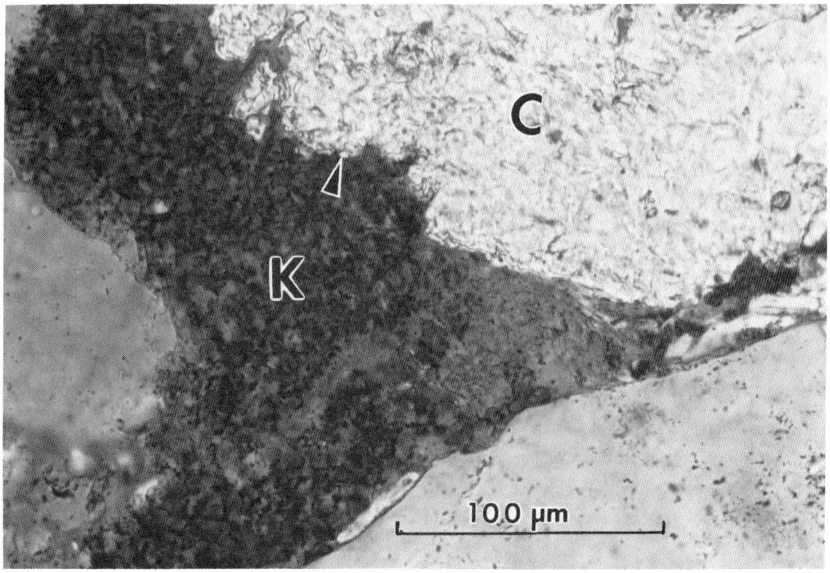

FIG. 9. Kaolinite (K) with calcite cement (C). Note ragged texture on calcite cement (arrow) suggesting kaolinite has filled pore space after some calcite dissolution. Well no. NCL 488-29, 2727 m, partially crossed nichols. Photograph taken in porous sandstone within 2 mm of cemented-uncemented interface.

& Kharaka 1978). The overlap of the peak in organic acid content of these porewaters with the apparent timing of aluminosilicate and carbonate dissolution at NCL shows the probable influence of kerogen maturation on porosity enhancement.

Authigenic pyrite, although not volumetrically important, is seen in fractures which cut across detrital grains and kaolinite. A possible source for the Fe may be calcite dissolution (see Fig. 7) although this source would also release Mg, the sink for which has yet to be identified. The

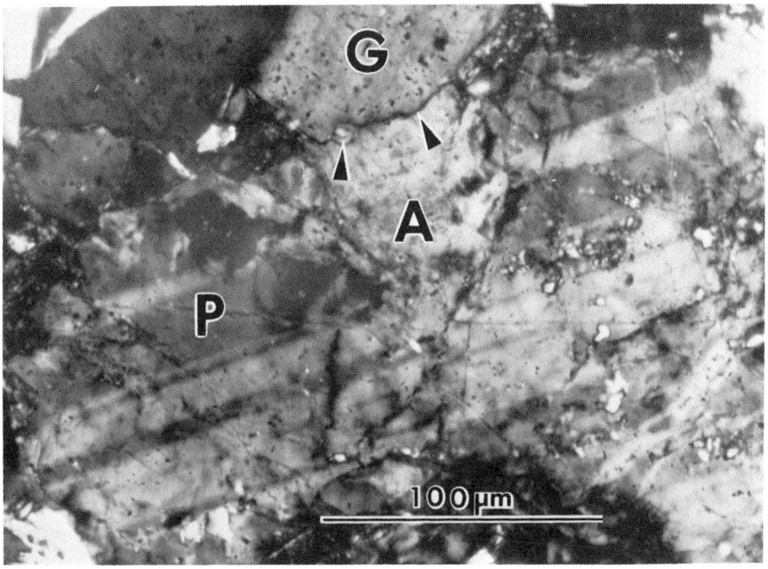

FIG. 10. Stress-induced albitization. Fresh plagioclase (P) is albitized (A) in halo about stressed contact (arrow) with resistant grain (G). Bedding approximately parallel to feldspar twinning. Well no. NCL 487-29, 2752.5 m, crossed nicols.

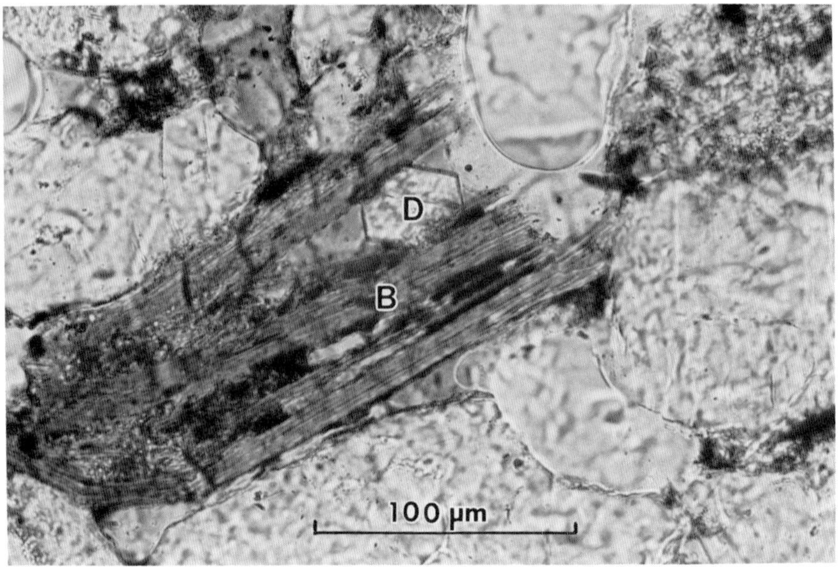

FIG. 11. Euhedral very late stage dolomite rhomb (D) in crushed biotite (B). Well no. NCL 488-29, 2738.8 m.

sulphur for this late-stage pyrite is probably sourced from hydrocarbons and associated products of kerogen maturation.

Late-stage compaction is responsible for very late diagenetic changes in these sandstones including albitization of plagioclase along stressed grain contacts and growth of dolomite rhombs along cleavage flakes of crushed biotites (Boles & Johnson 1983, Boles 1984). These effects are restricted to porous sandstones that have undergone compaction. The 'stress-induced' albitization occurs as an albitization halo in plagioclase grains in contact with other resistant detrital grains (Fig. 10). These contacts are usually concavo-convex and the resistant grain is commonly quartz. Presumably high silica activities at the stressed interface promote the albitization reaction.

The dolomite-biotite occurrence is explained by development of a high pH halo adjacent to biotite cleavage surfaces as a result of H^+ attraction to the mica surface (Boles & Johnson 1983). This carbonate association with the crushed biotite is remarkably selective as in a number of samples it is the only carbonate occurrence in the rock (Fig. 11). The dolomite has a $\delta^{18}O$ SMOW of +19.9. Assuming that the dolomite formed at the present reservoir temperature of 107°C the calculated pore fluid composition from which it precipitated is between +2 and −0.4 depending on the assumed dolomite-water fractionation curve (Katz & Matthews 1977 and Fritz & Smith 1970). The oxygen isotopic composition in the present formation water where these carbonates occur is not known precisely because the water in the central reservoir is an injected mixture of original formation water ($\delta^{18}O = +2.8$) and a small volume of water from shallow wells ($\delta^{18}O$ about −8). Nevertheless the similarity between these values and those predicted from the isotopic calculations demonstrate the recent origin of these dolomites.

Timing of diagenesis

A burial history curve for the Stevens sandstone at North Coles Levee shows the very recent timing of the above described diagenetic events (Fig. 12). The major carbonate cementation (calcite) occurred between 40 and 80°C which corresponds to 4.6 to 2.5 Ma BP. Thus the dissolution event, kaolinite crystallization, hydrocarbon maturation, and late stress effects are compressed into a time interval of less than 2.5 Ma. The hydrocarbon volume alone involves an estimated 500 million barrels emplaced into a reservoir with an average effective permeability of 1–10 md. The key point here is that the carbonate cements in these sandstones are relatively late events. Thus it follows that subsequent diagenetic events are compressed into a rather narrow time window.

The stress-induced diagenetic processes are clearly post-carbonate cementation and result from reservoir compaction. Removal of the calcite cement during the dissolution phase may have caused the collapse of the reservoir at

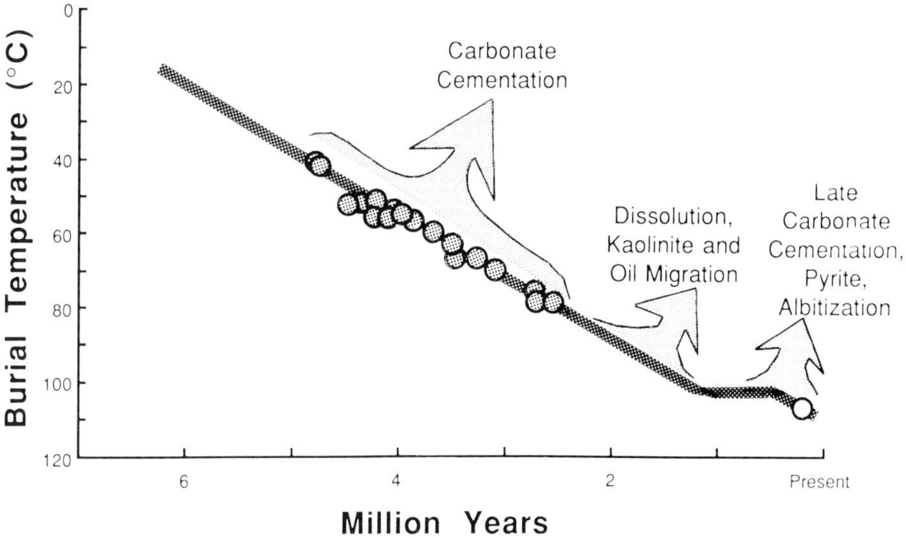

FIG. 12. Burial history (temperature) of Stevens sandstone at North Coles Levee in relation to diagenetic events. Burial history based on stratigraphy shown in Fig. 2 and present geothermal gradient. Carbonate cementation temperatures (circles) based on isotopic data and assumptions on isotopic fluid composition (see text). Exact sequence of events from 80 to 105°C is uncertain.

hydrostatic fluid pressure conditions. An alternative explanation is that these stress effects are even more recent, possibly human-induced. A fluid pressure history curve for North Coles Levee reveals a very low fluid pressure condition (1000 psi) recently existed (about 1965) for a short time period when the gas cap was released from the reservoir (Fig. 3). Most of the cores used in this study were cut after 1965 and these show the most pronounced compaction effects. Hence the possibility exists that the observed stress-related processes of dolomite precipitation and albitization occurred within the last 20 years!

ACKNOWLEDGMENTS: This paper is taken from a large study encompassing a number of oil fields in the San Joaquin Basin. The North Coles Levee study could not have been done without the complete cooperation of ARCO Oil and Gas Co., in particular ARCO geologist Bill Bazeley and engineers Art Tinnemeyer, Don Puckett, Ken Kuch and Tony Marino. Funding for this project includes ARCO Geoscience Group and NSF Grant EAR77-13705 (Boles). The assistance of UCSB personnel in preparing this manuscript including Dave Pierce (photographic expertise), Karen McCormack (computer support), Dianne Griffin (drafting), and Richard Coffman (proof-reading) is gratefully acknowledged.

References

BOLES, J. R. 1984. Secondary porosity reactions in the Stevens sandstone, San Joaquin Valley, California. *In*: MCDONALD, D. A. & SURDAM, R. C. (eds). *Clastic Diagenesis*, pp. 217–24. American Association of Petroleum Geologists Memoir, 37.

—— & JOHNSON, K. S. 1983. Influence of mica surfaces on pore-water pH. *Chemical Geology*, 43, 303–17.

—— & RAMSEYER, K. In press. Diagenetic carbonate in a Miocene sandstone reservoir, San Joaquin Basin, California. *American Association of Petroleum Geologists Bulletin*.

CALLAWAY, D. C. 1971. Petroleum potential of San Joaquin Basin, California. *In*: CRAM, I. M. (ed.). *Future Petroleum Provinces of the United States—their Geology and Potential*, pp. 239–53. American Association of Petroleum Geologists Memoir, 15.

CROSSEY, L. J., SURDAM, R. C., & LAHANN, R. 1986. Application of organic/inorganic diagenesis to porosity prediction. *In*: GAUTIER, D. L. (ed.). *Roles of Organic Matter in Sediment Diagenesis*, pp. 147–55. Society of Economic Paleontologists and Mineralogists Special Publication, 38.

CURTIS, C. D. 1978. Possible links between sandstone diagenesis and depth related geochemical reactions occurring in enclosing mudstones. *Journal of the Geological Society of London*, **135**, 107–17.

——, COLEMAN, M. L. & LOVE, L. G. 1986. Pore water evolution during sediment burial from isotopic and mineral chemistry of calcite, dolomite and siderite concretions. *Geochimica et Cosmochimica Acta*, **50**, 2321-34.

DREVER, J. I. 1982. *The Geochemistry of Natural Waters*. Prentice-Hall, Englewood Cliffs, NJ.

FRITZ, P. & SMITH, D. C. W. 1970. The isotopic composition of secondary dolomite. *Geochimica et Cosmochimica Acta*, **14**, 1875–7.

KATZ, A. & MATTHEWS, A. 1977. The dolomitization of $CaCO_3$: an experimental study at 252–295°C. *Geochimica et Cosmochimica Acta*, **41**, 297–308.

LAGOE, M. B. In press. The record of sea level and climatic fluctuations in an active margin basin: The Stevens sandstone, Coles Levee area, California. *Palaios*.

MATSUMOTO, R. & IIJIMA, A. 1981. Origin and diagenetic evolution of Ca-Mg-Fe carbonates in some coalfields of Japan. *Sedimentology*, **28**, 239–59.

RAMSEYER, K. & BOLES, J. R. 1986. Mixed-layer smectite/illite minerals in Tertiary sandstones and shales, San Joaquin basin, California. *Clays and Clay Minerals*, **34**, 115–24.

SURDAM, R. C., BOESE, S. W. & CROSSEY, L. J. 1984. The chemistry of secondary porosity. *In*: MC-DONALD, D. A. & SURDAM, R. C. (eds). *Clastic Diagenesis*, pp. 127–75. American Association of Petroleum Geologists Memoir, **37**.

WEBB, G. W. 1981. Stevens and earlier Miocene turbidite sandstones, southern San Joaquin Valley, California. *American Association of Petroleum Geologists Bulletin*, **65**, 438–65.

J. R. BOLES, Department of Geological Sciences, University of California, Santa Barbara, CA 93106, USA.

Trace-element source and mobility during limestone burial diagenesis—an example from the Middle Jurassic of eastern England

D. Emery

SUMMARY: Two phases of burial calcite from the Lincolnshire Limestone (Bajocian) of central Lincolnshire have been analysed for their trace element (Fe, Mn, Mg, Sr) and strontium isotopic compositions.

The earliest, non-ferroan, brightly luminescent phase has the following mean trace element concentrations: Fe— 919 ppm; Mn— 116 ppm: Mg— 1778 ppm; Sr— 327 ppm. The later, ferroan, dully luminescent phase has the following mean compositions: Fe— 3580 ppm; Mn— 225 ppm; Mg— 1400 ppm; Sr— 318 ppm. The strontium isotopic composition of the two phases is indistinguishable, with a mean $^{87}Sr/^{86}Sr$ ratio of 0.70820. This value is considerably more radiogenic than host Bajocian marine carbonate ($^{87}Sr/^{86}Sr = 0.70725$).

The Fe and Mn of the burial calcites were sourced by iron and manganese oxyhydroxides associated with shales bounding (and internal to) the Lincolnshire Limestone. The relatively shallow maximum burial depths (550 m) and low temperatures (maximum of 42°C) attained by the formation in the area of study preclude major clay or sandstone diagenetic transformations as potential sources for Fe and Mn. These trace elements moved into the Limestone in solution via compaction driven cross-formational flow from bounding shale aquicludes.

The Sr and Mg trace element components of the burial cements were derived from: (i) remobilized Bajocian marine carbonate, and (ii) remobilized Carboniferous marine carbonate. These sources satisfy the requirements of a linear Sr versus Mg relationship observed for the burial cements, that is, a constant Sr/Mg ratio with time, but highly variable absolute Sr and Mg contents. These sources also satisfy the near constant $^{87}Sr/^{86}Sr$ composition of the burial calcites.

The burial hydrology required by the trace element budget above is a convective flow system with cross-formational elements as normal bleeder faults connecting the Carboniferous Limestone and the Lincolnshire Limestone. Additional compaction-driven flow occurred from bounding shale formations (chiefly the underlying Grantham Formation and to a lesser extent the Upper Estuarine 'Series') to the Lincolnshire Limestone. Up-dip flow within the fully confined Limestone provides an enformational limb to the convective system.

During burial of sedimentary sequences, reactions can occur such that large quantities of cations (and anions) become redistributed in new and modified phases throughout the sequence. With regard to limestone burial diagenesis, the predominant newly formed phase is sparry calcite cement often with subordinate dolomite. In many carbonate sequences it is apparent that a large proportion of this burial calcite can be derived from the host depositional (usually marine) carbonate, by pressure dissolution or by leaching. It is not possible in many cases, however, to account for *all* the calcite cement by remobilization of pre-existing material. Furthermore, a source external to the depositional carbonate is necessitated for (i) trace elements such as iron and manganese shown by many authors to be present in burial calcite phases (Oldershaw & Scoffin 1967, Scholle & Halley 1985); (ii) strontium, as the strontium isotopic composition of burial cement phases described by Stueber *et al.* (1984), Faure & Szabo (1986) and Emery *et al.* (1987) is considerably more radiogenic than that of the host marine carbonate.

Many potential sources for these cations exist within bounding or intraformational clastic or clay-rich sequences; the relationship between ferroan calcite and ferroan dolomite burial cements and shale sequences bounding the carbonate of interest has been noted by many authors (Oldershaw & Scoffin 1967, Scholle & Halley 1985, Schofield & Adams 1986). Other studies (notably Stueber *et al.* 1984) have invoked the breakdown of rubidium-containing clastic phases (K feldspar and mica) to explain the higher $^{87}Sr/^{86}Sr$ ratios of the burial calcites relative to their host carbonate.

Whilst the circumstantial evidence for extra-formational derivation of diagenetic components

From: MARSHALL, J. D. (ed.), 1987, *Diagenesis of Sedimentary Sequences*, Geological Society Special Publication No. 36, pp. 201–217.

is difficult to dispute, it is clear that a more specific relationship between cation source, transport pathways and sink needs to be established. To this end, the burial history of the sediments of interest and its bounding formations needs to be well-constrained and thereby its thermal evolution and the attendant importance (or otherwise) of clastic and clay mineral diagenesis (e.g. Curtis 1983, Burley, Kantorowicz & Waugh 1985).

This type of source-specific chemical modelling can be applied to the Lincolnshire Limestone, a Middle Jurassic carbonate which, in the study area, underwent shallow burial to 550 m. This limestone displays a distinctive suite of burial carbonate phases, from earliest non-ferroan calcite to ferroan dolomite and latest ferroan calcite. The calcite phases, on which the most data are available, will be discussed relative to a burial budget of the following elements: Fe, Mn, Mg, Sr. It is important to emphasize that this is not a study on the relationship between porewater chemistry and precipitated phase based on distribution coefficients. Such work is only of practical value in Recent or Pleistocene carbonates, where the aquifer system and its attendant flow properties can be evaluated in detail, and where the dissolving and precipitating phases within the aquifer can be unequivocally identified. For details of such work the reader is referred to Veizer (1983) for an excellent review.

Geological setting

The Lincolnshire Limestone (Bajocian) is the major carbonate formation in the Middle Jurassic of eastern England, stretching from North Humberside, through its thickest development of 40 m at Boothby Pagnell in Lincolnshire to a feather-edge, probably determined by a combination of original thinning and pre-Bathonian

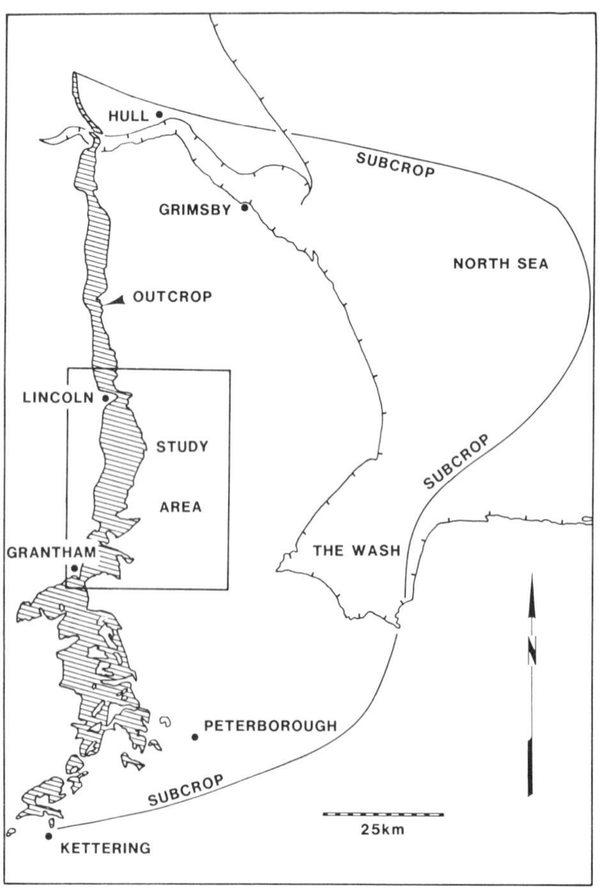

FIG. 1. Outcrop (cross-hatched) and subcrop of the Lincolnshire Limestone in England.

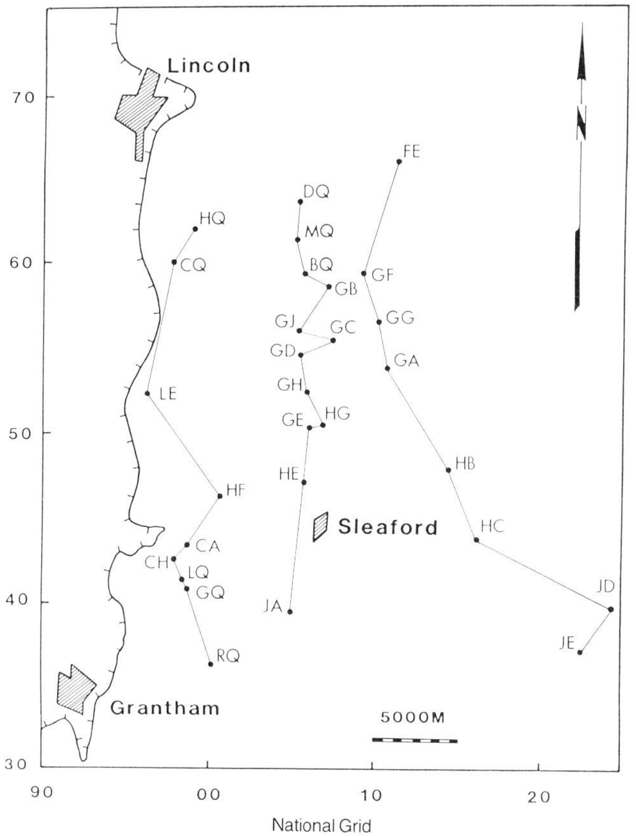

FIG. 2. Distribution of core and quarry material used in the present study. Hatched line indicates western limit of outcrop. HQ, Harmston Quarry; CQ, Coleby Quarry; LE, Leadenham Quarry; HF, North Rauceby Core; CA, Castle Quarry, Ancaster; CH, Copper Hill Quarry, Ancaster; LQ, Lyte's Quarry; GQ, Gregory's Quarry; RQ, Ropsley Quarry; DQ, Dunston Quarry; MQ, Metheringham Quarry; BQ, Blankney Quarry; GB, Kirby Green Core; GJ, Ashby Core; GC, Digby Core; GD, Bloxholm Core; GH, Cottage Farm Core; HG, Pear Tree Farm Core; HE, Leasingham Core; JA, Aunsby Core; FE, Dunston Fen Core; GF, Timberland Core; GG, Walcot Core; GA, Dorrington Core; HB, Asgarby Core; HC, Great Hale Core; JD, Bicker Core; JE, Donington Core.

erosion, in mid-Northamptonshire (Fig. 1). The study area for the present work and the distribution of the core and outcrop material is illustrated in Fig. 2. The Lincolnshire Limestone shows a very complex internal stratigraphy and a wide range of carbonate lithologies (Ashton 1980, Emery 1986). These can be interpreted as representing the landward migration of an offshore barrier-bar complex (now represented by the upper Lincolnshire Limestone) across a protected lagoon and back-barrier (lower and middle Lincolnshire Limestone: Ashton 1977 and in Tucker 1985). This in turn transgresses across non-marine coastal marsh deposits of the underlying Grantham Formation (Fig. 3).

After the deposition of the Limestone there was a regional regression and the Formation was tilted and eroded. Deposition resumed during the Bathonian, initially with the non-marine deposits of the Upper Estuarine 'Series', but later with renewed marine transgression. The unconformity between the Lincolnshire Limestone and Upper Estuarine 'Series' is represented by six ammonite Zones.

The structure of the Lincolnshire Limestone is simple, dipping gently to the east at an average of about 1° beneath later Jurassic strata, themselves overlain by Pleistocene and Holocene Fen deposits. It is an important aquifer in an area of low rainfall and intensive agriculture. Its hydrogeology has been studied by Downing *et al.* (1977) and Edmunds & Walton (1983) and cores taken for this purpose have been sampled in the present study, as well as available outcrops.

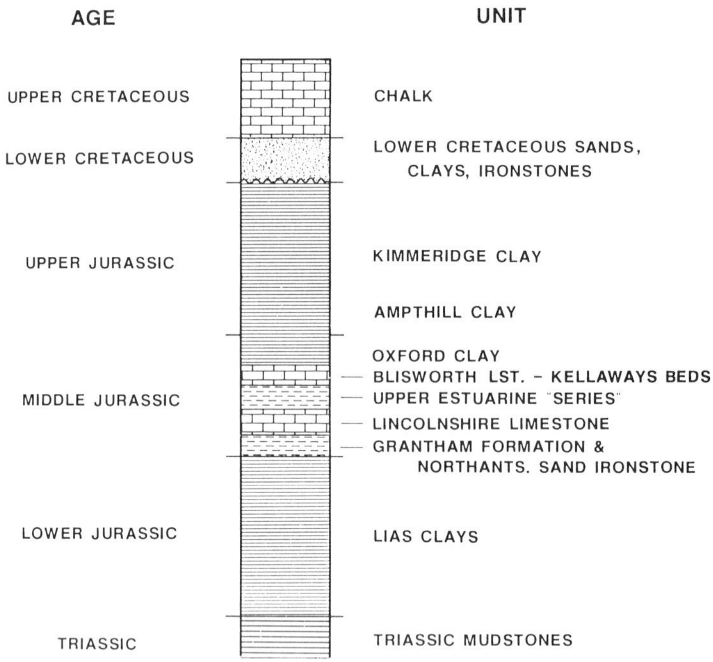

FIG. 3. Simplified stratigraphic column for the Mesozoic of the East Midlands.

Pre-burial diagenesis of the Lincolnshire Limestone

Diagenesis of the Lincolnshire Limestone began in the marine environment, with abundant microboring and precipitation of a variety of cements including major hardground cements (Marshall & Ashton 1980, Emery 1986). Pyrite was the major non-carbonate precipitate within the marine realm. Localized syndepositional subaerial exposure of the Formation gave rise to major aragonite dissolution and precipitation of a distinctive phase of meteoric cements (Emery 1986).

Post-depositional diagenesis commenced during folding and exposure of the Limestone in the Bajocian, with the establishment of a largely unconfined meteoric aquifer system with attendant aragonite dissolution and precipitation of a Formation-wide correlatable cement sequence, consisting of euhedral, distinctively zoned (under cathodoluminescence) low magnesium calcite as echinoderm overgrowths, in biomouldic porosity and rarely as intergranular fringe cements (Emery 1986). Thereafter the Limestone became entirely confined beneath the Upper Estuarine 'Series' and moved into the realm of burial diagenesis as defined by Scholle & Halley (1985), i.e. below the influence of shallow groundwater circulation and above the zone of low-grade metamorphism.

The burial history of the Lincolnshire Limestone

The Lincolnshire Limestone became buried underneath further Middle Jurassic shales and minor limestones (Fig. 3), the majority of which were deposited in the marine environment. During the Upper Jurassic, major thicknesses of the Oxford, Ampthill and Kimmeridge Clays accumulated, with particularly rapid deposition of the Kimmeridge Clay (Fig. 4). Minor uplift and erosion of the Kimmeridge Clay blanket probably occurred in pre-Cretaceous times, but the precise amount of emergence is difficult to evaluate (Fig. 4). Slow burial of the sequence under Lower Cretaceous sands, shales and ironstones persisted until deposition of the chalk, when rapid, differential burial occurred (Fig. 4). With the termination of Chalk deposition, the sequence underwent uplift related to end-Cretaceous inversion tectonism (Glennie & Boegner 1981). The Lincolnshire Limestone was ultimately re-exposed during the Miocene (Peach 1984).

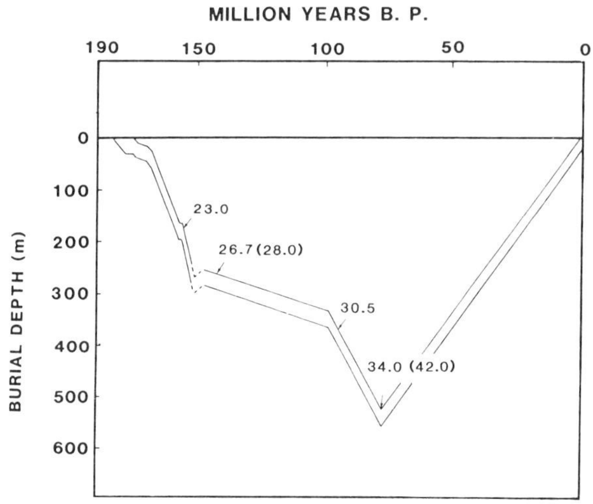

FIG. 4. Burial curve for the Lincolnshire Limestone within the study area. Broken lines represent uncertain emergence. Temperatures were calculated from equation 2 of Andrews-Speed *et al.* (1984). Temperatures in parentheses were calculated from oxygen isotopic composition of the burial calcites (Emery 1986). All temperatures are in °C.

The burial curve of Fig. 4 depicts temperatures attained by the Limestone in the study area during burial, calculated according to equation 2 in Andrew-Speed, Oxburgh & Cooper (1984). These temperatures represent likely maxima for the Limestone in the study area *due to heat conduction alone* through the sedimentary pile overlying the Limestone. The two temperatures in parentheses are those attained by the Limestone assuming heat transfer by fluid convection through the Limestone during burial. These latter temperature determinations are based on $\delta^{18}O$ compositions of the burial cement phases; the rationale and calculations behind these estimates are detailed in Emery (1986).

In summary, the maximum depth of burial of the Lincolnshire Limestone in the study area did not exceed 550 m, the maximum temperature attained by the Limestone was probably 42°C, and it is clear that the Formation was confined between major clay-rich sequences during burial. The most rapid burial of the formation occurred (i) during deposition of the Kimmeridge Clay and (ii) during deposition of the Chalk.

Burial cements of the Lincolnshire Limestone

The cements of the Lincolnshire Limestone precipitated in the burial regime consist of (i) non-ferroan calcite, (ii) ferroan dolomite and (iii) ferroan calcite.

Non-ferroan calcite

This phase forms up to 5% of the total rock by volume. It displays an irregular distribution, occurring as isolated poikilotopic crystals, impersistent lenses at least of the dimensions of individual thin-sections and contiguous bands. The crystals form both drusy and poikilotopic fills to intra- and intergranular porosity.

This phase stains a characteristic pink with mixed potassium ferricyanide and Alizarin Red-S stain (Dickson 1966), indicative of a low iron content (Dickson 1966), although a mauve tinge may be present in some examples. Under luminescence non-ferroan calcite is generally brightly luminescent with complex intracrystalline zonation in rare examples.

This calcite clearly post-dates major aragonite dissolution and compaction of allochems and earlier cement generations.

Ferroan dolomite

Dolomite is present both as cement and replacive phase and forms a minor component (approximately 2%) of the total rock by volume in the study area. Its origin intervening major phases of calcite precipitation is problematic and beyond the scope of the present paper.

Ferroan calcite

Ferroan calcite is the most volumetrically significant diagenetic precipitate in the Limestone, forming up to 25% of the whole rock by volume. It forms poikilotopic and drusy fills to intra- and intergranular porosity and is of ubiquitous distribution.

This phase stains a characteristic blue (Dickson 1966), and may display weakly ferroan or non-ferroan complex intracrystalline zonation. It is dully luminescent, with patchy intracrystalline non-luminescent zones; however, no concentric zonation, suggestive of a change in fluid chemistry with time, is observed.

The ferroan calcite precipitation post-dates further compaction of the Limestone, as evidenced by fracture of the non-ferroan calcite phase. The absolute timing of cement precipitation is considered to have been during deposition of the Upper Chalk (Emery 1986). This assertion is based chiefly on stable isotopic evidence and burial history reconstructions. The important feature of this cement is the relatively short timespan of cement precipitation and the absence of concentric intracrystalline chemical and isotopic zonation.

The non-ferroan calcite is also considered, albeit more speculatively, to have been precipitated over a relatively short timespan, and the above point also applies to this phase.

Analytical methods and results

(i) Inductively-coupled argon plasma spectrometry (ICP). Samples of calcite were extracted using a stout needle from thin-sections and slabs. 0.1 g of sample were dissolved in 10 ml of quartz-distilled 1.75 M HCl. The samples were analysed on a Phillips 50 MHz ICP. Results obtained (see Table 1 for mean and range) are expressed with a +/− 3% confidence. Details of this technique are available in Thompson & Walsh (1983).

(ii) Strontium isotopic compositions were determined by P. C. Smalley at the University of Oslo. Full details of this technique are available in Emery (1986) and Emery *et al.* (1987). Analysed $^{87}Sr/^{86}Sr$ ratios were adjusted to a value for the NBS 987 standard of 0.71014 to allow direct comparison with the data of Burke *et al.* (1982) which, while employing this very low normalization value for the NBS 987 standard, remains the most comprehensive data set for the Jurassic.

The results from ICP analysis are summarized in Table 1. Full, tabulated results of the Sr isotopic data are given in Emery (1986) and Emery *et al.* (1987).

TABLE 1. *All values are in parts per million. NFC—non-ferroan calcite; FC—ferroan calcite*

	Fe	Mn	Mg	Sr
Mean concentrations				
NFC	920	120	1780	330
FC	3580	230	1400	320
Range				
NFC	140–1680	80–230	600–2770	90–840
FC	2030–4830	160–390	540–2890	80–650

Discussion

Iron concentrations

The mean values and ranges for iron concentrations in the burial cement phases are given in Table 1.

The presence of iron within calcite cements depends on at least the following independent factors; the Eh and pH of the environment of precipitation and the supply of iron. An important result in *all* the Lincolnshire Limestone burial cements is the invariable presence of iron. If this iron is substituting for Ca^{2+} in the calcite lattice, the environment of precipitation of all the burial calcites must have been reducing. A control sample of (Recent?) speleothem calcite from the Lincolnshire Limestone was analysed with the burial calcites and, as anticipated, contained no iron, at least justifying contamination-free sampling. The variation of iron content in the burial cements is thus ascribed to a variation in the iron supply to the growing crystal faces, not simply the total iron supply. This qualification is necessary because the total iron supply to the porewaters may have been available to other minerals besides calcite. Co-precipitation of pyrite in a reducing environment will selectively remove Fe^{2+} from the porewaters, with consequent precipitation of an iron-free or iron-poor calcite (Curtis 1977).

Non-ferroan calcite

There is no textural evidence for co-precipitation of pyrite with the non-ferroan calcite phase, but such evidence may be difficult to ascertain, particularly in the absence of intergrowth fabrics. Indirect evaluations of the importance of pyrite precipitation at this stage can be considered through the sulphate budget of the aquifer and the carbon isotopic composition of the non-ferroan cements:

During the precipitation of the non-ferroan calcite, the Lincolnshire Limestone is shown to have been a confined aquifer, probably buried beneath Upper Jurassic Clays (Fig. 4). Under such circumstances it would be difficult to replenish the supply of sulphate to the porewaters. Any marine-derived sulphate would probably have been long exhausted by bacterial reduction and subsequent pyrite formation on shallow burial within the marine realm (Berner 1980). Supply of sulphate from meteoric waters is probably untenable; the presence of sulphate in the present Lincolnshire Limestone porewaters (Edmunds & Walton 1983) is due to a combination of pyrite oxidation at or near recharge and agrichemical poisoning of the aquifer. Neither of these situations would have pertained to the Lincolnshire Limestone aquifer during precipitation of the non-ferroan calcite.

Consideration of the mean and range $\delta^{13}C$ of the non-ferroan calcite demonstrates that it differs little from that of the marine carbonate (Fig. 5). There appears to have been no input of ^{13}C depleted organogenic carbonate to the calcite, as would be expected during bacterial sulphate reduction (Berner 1980, Irwin *et al.* 1977).

It is thus concluded that sulphate was absent from the diagenetic system at this time, and pyrite co-precipitation with the non-ferroan calcite did not occur. The absence of major quantities of iron at this time is further interpreted as a function of low overall iron supply to the porewaters.

Ferroan calcite

The iron content of both burial cement phases must be sourced from outside the carbonate-dominated portion of the Limestone, or indeed from outside the Limestone itself. External sources for iron in ferroan calcites have been suggested by many authors, notably Oldershaw & Scoffin (1967). These workers demonstrated that ferroan calcites precipitated in limestone sequences adjacent to shale bands displayed a decrease in Fe content away from the shales. My own general observations of British Jurassic carbonate sequences (Blisworth Limestone, Corallian) show a predominance of pore filling ferroan cements, presumably related to juxtaposed argillaceous sequences. By contrast, Moore & Druckman (1981) described non-ferroan burial cements from a clay-poor carbonate sequence. Bounding the Lincolnshire Limestone are two clay-rich units, the Grantham Formation and the Upper Estuarine Series (Fig. 3), and there are considerable thicknesses of clay within the Limestone itself (e.g. the Kirton Shale Member; Ashton 1980). There appears to be no difficulty regarding clay-rich sources for the iron of the burial cements. XRD analysis of clay-rich sequences within the Limestone (no analyses of the clastic formations outside the Limestone have been made) show that its clay mineralogy consists chiefly of kaolinite, but there is certainly sufficient smectite, illite and biotite present, all of which will contain structurally bound iron. Furthermore, smectite and illite have high cation exchange capacities resulting from charge imbalances in the crystal structure or at its surface (Carroll 1959, Rosler & Lange 1972). Dissolution or recrystallization of these phases could potentially yield iron to the porewaters of the Lincolnshire Limestone aquifer as Fe^{2+} under reducing conditions. However, conditions favourable for clastic diagenesis as above must also have pertained during burial of the Limestone. Such diagenetic events would not be anticipated

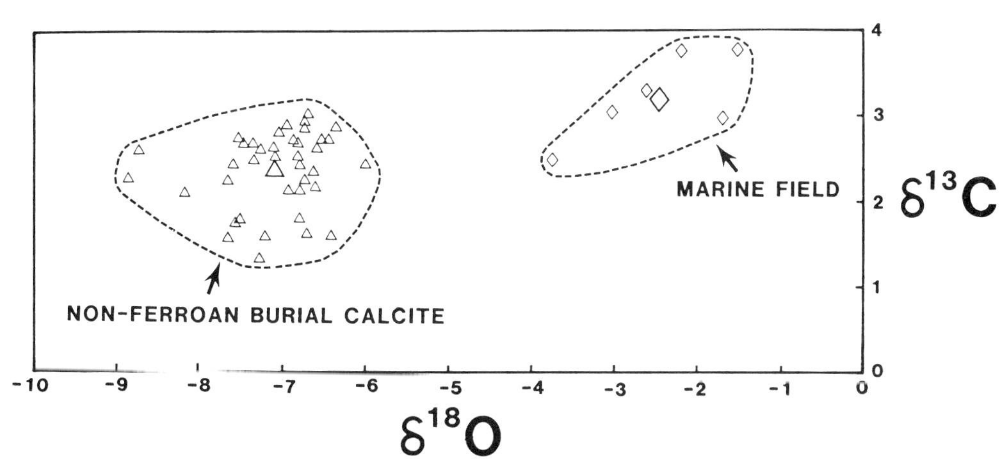

FIG. 5. The stable isotoic composition of the non-ferroan burial calcites compared to a marine field (defined by diagenetically unaltered bivalves) from the Lincolnshire Limestone. Note the significant depletion in ^{18}O of the burial calcites compared to the marine field, but the negligible depletion of ^{13}C. Data from Emery (1986).

at the relatively shallow burial depths and low temperatures attained by the Limestone during burial, according to the scheme of Curtis (1983) and Burley *et al.* (1985). It is concluded that clastic and clay mineral diagenetic events were of limited importance in cation supply to the burial porewaters of the Lincolnshire Limestone.

Another potential source, also associated with clay minerals, may be as iron oxide films on the surface of the clays, or as poorly ordered, amorphous iron oxy-hydroxides. These sources, suggested as possible sources for iron in sedimentary ironstones, are particularly attractive as they may also provide a complementary source of manganese as oxide films, also present on clay surfaces, as well as reactive, amorphous manganese oxy-hydroxides (Curtis & Coleman 1986). Furthermore, reducing conditions will liberate the iron and permit its transport as Fe^{2+}; no clastic or clay mineral diagenetic events are required. It is thus suggested that the iron and manganese of the burial cements was supplied from oxide films on the surface of clay minerals and/or from poorly ordered amorphous oxy-hydroxides.

Manganese concentrations

The mean value and ranges for manganese concentrations in the Lincolnshire Limestone burial cements are detailed in Table 1, and display relatively low concentrations compared to abundant published data (e.g. Fairchild 1983, Ten Have & Heijnen 1985). Incorporation of manganese into the calcite structure necessitates its reduction to the divalent state, Mn^{2+}. This again demands reducing conditions, but not to the same low Eh as required for iron incorporation into calcite (Oglesby 1976, Froelich *et al.* 1979). However, the invariable presence of reducing conditions as demonstrated by the iron evidence will have maintained manganese in the divalent state throughout the precipitation of the burial calcites, and since sulphate reduction is assumed to have been minimal, Mn in the porewaters will have been incorporated into the calcites.

Implications for the luminescence of burial cements

Figure 6 depicts a cross-plot for Fe versus Mn concentrations in the late spars. Open triangles are samples of non-ferroan calcite which exhibit bright luminescence. Not all the samples of non-ferroan cements were analysed for luminescence characteristics. No ferroan calcites (solid triangles) displayed bright luminescence. A simple correlation between low iron content, higher Mn/Fe ratio and bright luminescence exists. Similar correlations have been described within individual studies by a number of authors (see Fairchild 1983), but an aggregation of all results reveals a random scatter (Dickson, personal communication, 1986), albeit with brightly luminescent cements displaying higher Mn/Fe ratios. Such data scatter could be a function of variable instrumentation, or subjective attribution of the terms bright and dull. Despite this unavoidable subjectivity, it is widely accepted that high Mn/Fe ratios are prerequisite for bright luminescence of carbonates, although Ten Have & Heijnen (1985) have suggested that luminescence intensity is controlled by the Mn concentration, and not by the Mn/Fe ratio. Both these possibilities are borne out by the data on Lincolnshire Limestone burial cements.

Paired behaviour of iron and manganese

The Fe versus Mn cross-plot of Fig. 6 provides valuable information besides its implications for calcite luminescence. A broadly linear correlation between Fe and Mn of the non-ferroan and ferroan burial cements is exhibited. Trends for both the non-ferroan and ferroan calcites are parallel and the regression line drawn through the points intercepts the Mn concentration axis before the Fe concentration axis. The simplest interpretation of this result is that Mn^{2+} will be incorporated into the calcite lattice before Fe^{2+}, as Mn is more readily reduced from its tri- or tetravalent state than iron is reduced from its trivalent state (Curtis & Coleman 1986). This assertion predicts that any iron-free calcite precipitated will have manganese concentrations of ~ 70 ppm, although the presence of iron in all burial phases analysed implicates conditions of sufficiently low Eh during burial such that ferrous iron was variably mobilized.

The broadly linear trend for Fe versus Mn of Fig. 6 demonstrates that for each additional portion of calcite precipitated the Mn/Fe ratio therein was constant. This further implies that the Mn/Fe ratio of the porewaters was constant with time, as the concentration of a cation in calcite is a linear function of its concentration in the precipitating fluid assuming equilibrium partitioning (McIntire 1963).

The constancy of the Mn/Fe ratio of the porewaters with time was maintained, however, not by an unchanged absolute supply of Fe and Mn, but by an increased rate of supply of both cations which was balanced to maintain the Mn/Fe ratio, according to the simple linear relationship. There are two possible explanations for this observed relationship: (a) The Fe and

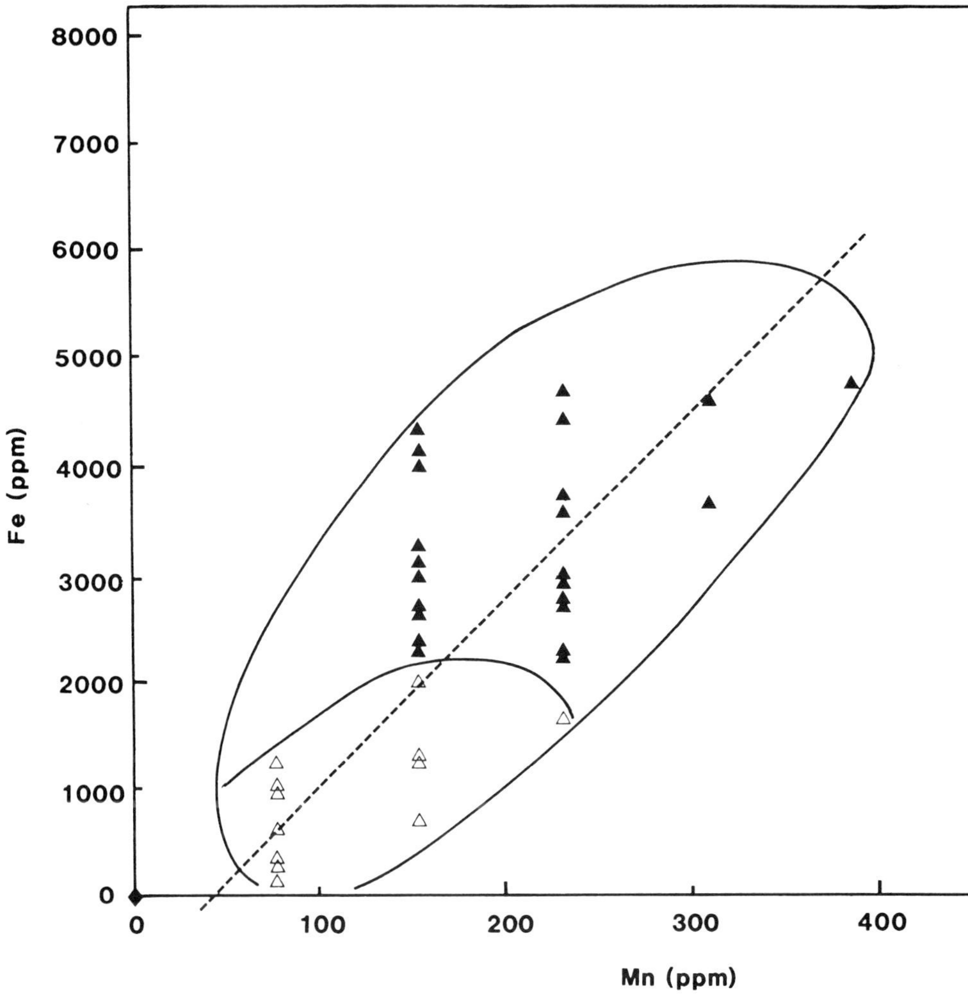

FIG. 6. Cross-plot of Fe versus Mn concentrations in the burial calcites. Open triangles represent non-ferroan burial calcites, solid triangles represent ferroan burial calcites.

Mn of oxide films and/or amorphous oxyhydroxides, the proposed source for these cations, were increasingly mobilized on progressive burial. (b) The mobilization rate of Fe and Mn was constant, but porewaters to which these cations were being contributed (i.e. argillaceous units within and/or outside the formation) formed an increasing proportion of the total Limestone porewaters through increasing cross-formational flow from the shale bands to the Limestone as a function of progressive burial.

Both possibilities relate to this simple cation budget route:

$$\text{Mn/Fe oxides, hydroxides}_{(\text{Fe, Mn})} \longrightarrow \text{Porewaters}_{(\text{Fe, Mn})} \longrightarrow \text{Calcite}_{(\text{Fe, Mn})}$$

Distinction between the two possibilities is difficult; however, progressive burial of the sequence is demonstrated both by petrography and burial curve reconstruction (Fig. 4) and, in the absence of contradictory data, option (b) is preferred.

Magnesium concentrations

The mean value and ranges for magnesium concentrations in the burial cements are shown in Table 1. Magnesium will be incorporated into the calcite lattice independently of redox potential of the environment of precipitation, as it exists naturally as a divalent ion of an appropriate ionic radius. It is an important component of marine calcite precipitates where it may substitute up to 18 mol% $MgCO_3$ in the lattice as high magnesium calcite. High magnesium calcite is generally considered to contain from 4–18 mol%

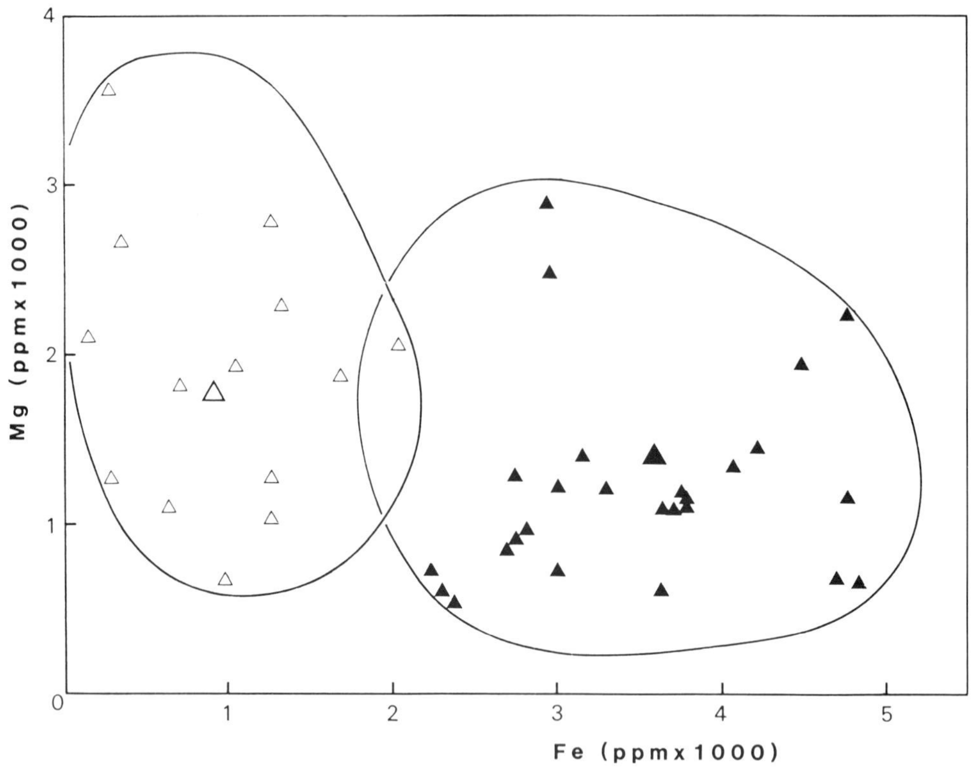

FIG. 7. Cross-plot of Mg versus Fe for the burial calcites. Large triangles represent mean concentrations, key as for Fig. 6.

$MgCO_3$ (Chave 1954). Magnesium may also substitute into the aragonite lattice. Since the Lincolnshire Limestone is composed dominantly of marine carbonate, a ready internal source of Mg was available for incorporation into the burial cements on remobilization, an important distinction from iron and manganese where an external source of cations was necessary. This is not to imply that high magnesium calcite persisted into the burial regime; earlier diagenetic events, particularly meteorically influenced diagenesis, transformed much (and possibly all) of the original marine high magnesium calcite into low magnesium calcite. However earlier cements and marine components still contain an appreciable component of magnesium, up to 7000 ppm in some echinoderm fragments analysed by energy-dispersive electron microprobe (Emery 1986).

The magnesium content of the burial cements is variable and is best illustrated by a magnesium versus iron cross-plot (Fig. 7) which shows no distinctive correlation between the two cations. However, for the relatively limited range in Fe of the non-ferroan calcites, compared to the ferroan calcites, the range in Mg is proportionally greater. The mean Mg concentration of the non-ferroan calcites is also slightly greater than that for the ferroan calcites (Table 1).

Consideration of the magnesium concentrations in isolation, however, does not permit a satisfactory evaluation of their source. Circumstantial evidence suggests that Mg could be at least partially derived from the host marine carbonate; this assertion is best verified by comparison of the Mg concentrations with Sr.

Strontium concentrations

Strontium is potentially of great utility in the chemical modelling of carbonate diagenesis (Veizer 1983), and has been used with success particularly in modelling diagenesis in the meteoric realm (Gavish & Friedman 1969, Kinsman 1969). Like magnesium, it exists as a divalent ion, and is incorporated into the calcite lattice independent of redox potential of the environment of precipitation. A further important property is its abundance in marine carbonate phases, particularly aragonite. A profusion of data exist on Sr concentrations in recent marine carbonates, notably after Kinsman (1969) and

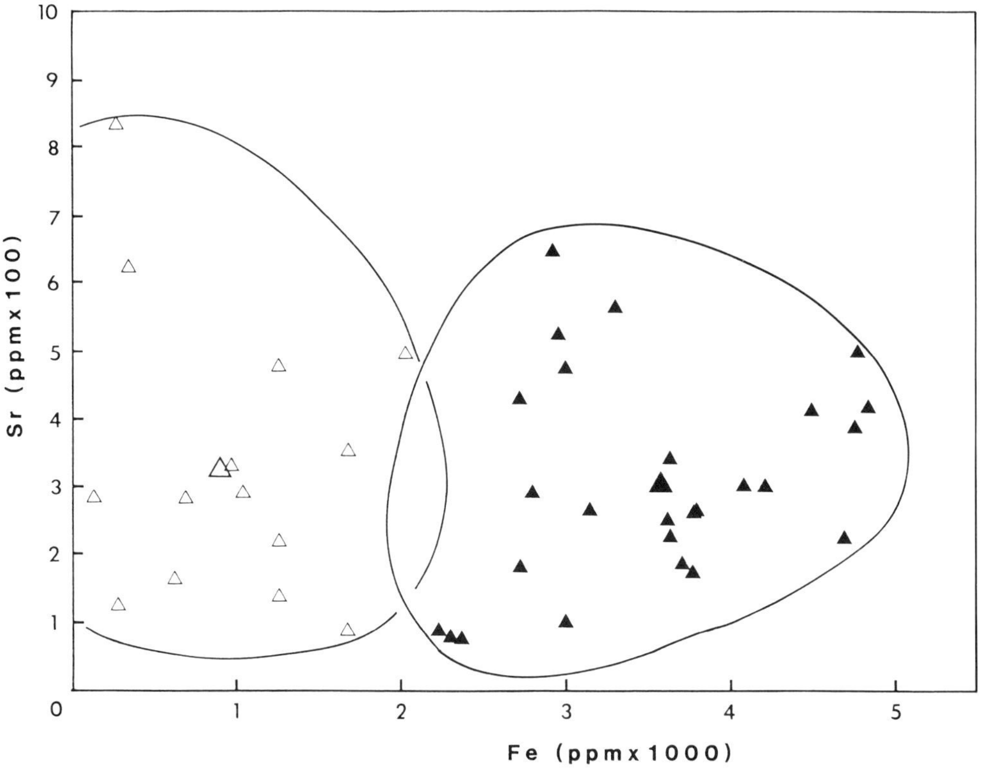

FIG. 8. Cross-plot of Sr versus Fe for the burial calcites. Large triangles represent mean concentrations, key as for Fig. 6.

reviewed by Veizer (1983) and it is clear that despite massive Sr loss from depositional aragonite, largely by dissolution in the meteoric realm, Sr would still be of sufficient concentration in marine carbonates and earlier cement phases to provide a major potential source of strontium for the burial cements. Figure 8 depicts an Sr versus Fe cross-plot for the burial calcites. From the form of this plot, it is clear that there is no distinctive correlation between Sr and Fe. Table 1 further emphasizes the similarity of the mean and range of Sr concentrations for the non-ferroan and ferroan calcites. The form of Fig. 8 is, however, very similar to that of Fig. 7. A direct plot of Sr versus Mg (Fig. 9) clearly demonstrates the origin of this similarity, in that Sr and Mg describe a positive linear correlation.

The regression line through the points of Fig. 9 has the simple formula Sr(concentration) = 0.21 Mg(concentration). This relationship has several significant implications: (a) The intercept of the regression line is at the origin. This would be expected for two elements whose incorporation into calcite is independent of redox potential. (b) The linear relationship between the two elements implies a constant Sr/Mg ratio in the burial porewaters with time, and that Sr and Mg were probably sourced by the same type of material. By contrast with the Fe versus Mn cross-plot, there is no absolute increase in the Mg and Sr contents with time (see also Figs 7 and 8). (c) Any material(s) ultimately sourcing the Sr and Mg of the burial calcites must have exhibited an invariant Sr/Mg with time, although the absolute concentrations of Sr and Mg may have shown considerable, unsystematic variation with time.

The above constraints require a Sr and Mg source for the burial calcites of constant Sr/Mg ratio, but one which would allow the absolute concentrations of both these cations to vary with time. The original marine carbonate of the Lincolnshire Limestone apparently satisfies these constraints; furthermore, it has clearly been removed by pressure-dissolution and normal dissolution processes during burial diagenesis.

If Bajocian marine carbonate is indeed the source of calcite for the burial cements, then the $^{87}Sr/^{86}Sr$ of the burial cements ought to be identical to that of original, diagenetically unaltered, Bajocian marine carbonate, i.e. their

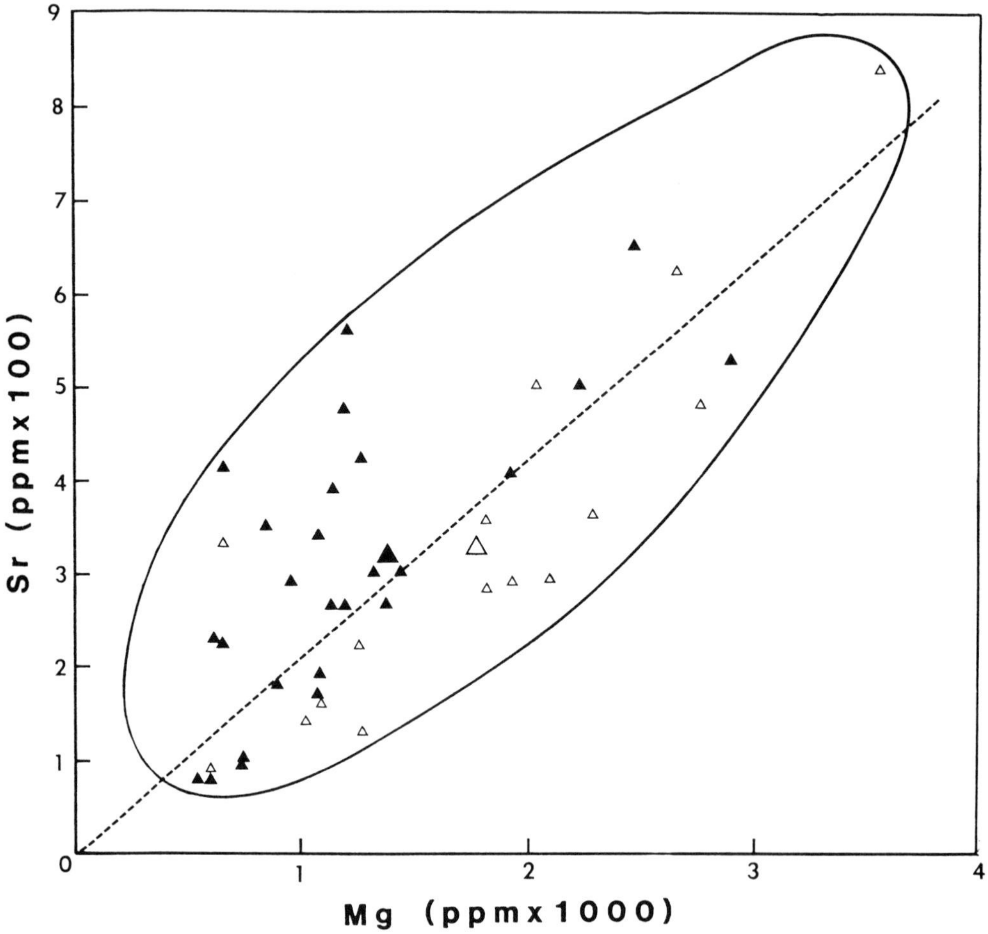

FIG. 9. Cross-plot of Sr versus Mg for the burial calcites.

mutual $^{87}Sr/^{86}Sr$ ratio should be approximately 0.70725 (Burke *et al.* 1982).

Strontium isotopes

The details of the Sr isotopic composition of the Lincolnshire Limestone are presented in Emery (1986) and Emery *et al.* (1987). For the purposes of the present argument, Fig. 10 clearly demonstrates that there is a considerable separation between the Sr isotopic composition of Bajocian marine material, as represented by brachiopods from the Lincolnshire Limestone (mean $^{87}Sr/^{86}Sr = 0.70725$), and the two phases of burial calcites (mean $^{87}Sr/^{86}Sr = 0.70820$). Furthermore, the strontium isotopic composition of the two phases of burial calcite is almost indistinguishable.

The model involving remobilization of Bajocian marine carbonate alone is clearly inappropriate; a source of additional radiogenic Sr is required in the first instance. However, this additional radiogenic Sr must still satisfy the criteria required by the Sr and Mg concentrations of the burial cements, in that its Sr/Mg ratio must be constant and identical to that of the remobilized Bajocian marine constituent. Furthermore, the absolute Sr and Mg concentrations of the calcite must be allowed to vary considerably, yet the near-constant $^{87}Sr/^{86}Sr$ ratio of the cements must remain unaffected.

These constraints probably rule out clay or clastic minerals as a source of Sr and Mg for the burial cements. Analysis of the Sr isotopic composition of clay mineral phases in the Lincolnshire Limestone (Smalley *et al.* 1985), and consideration of the Sr isotopic composition of associated K feldspars (Emery 1986) shows that these potential contributors are too radiogenic, and that the Sr/Mg ratios of these phases is

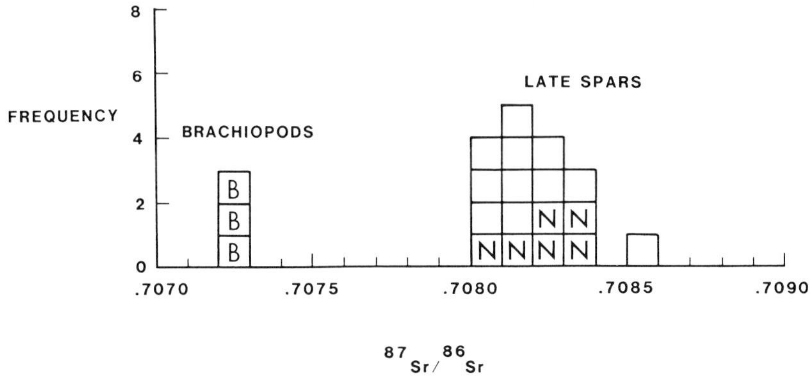

FIG. 10. The Sr isotopic composition of Lincolnshire Limestone brachiopods and burial cements.

probably inappropriate. In corroboration, it has already been stated that major clay and clastic mineral diagenesis is probably inappropriate at the burial depths and thermal regime experienced by the Lincolnshire Limestone in the study area.

However, in the more deeply buried area to the east, the Limestone may have attained temperatures as high as 70°C, with similarly elevated temperatures affecting the bounding shale formations. Under such circumstances, clay mineral transformation and clastic mineral breakdown reactions may have been expected to contribute ions to the burial porewaters of the Limestone. However, the present data on all elemental concentrations in the burial cements, particularly the Sr-isotopic information, does not support this. At least the following possibilities exist to account for this: (i) Clay mineral transformation and clastic mineral breakdown reactions in bounding the intraformational shales were unimportant at these elevated temperatures. (ii) Mineral sinks other than burial calcites are preferable to the cations of interest at these temperatures. (iii) Cations generated from the shales at these temperatures enter the Limestone porewaters but are circulated out of the Limestone without entering the study area. In the absence of data on burial cement phases further down-dip to the east it is impossible to constrain these possibilities.

A possible alternative source for Sr and Mg may be another marine carbonate within the stratigraphic column of the East Midlands. Such

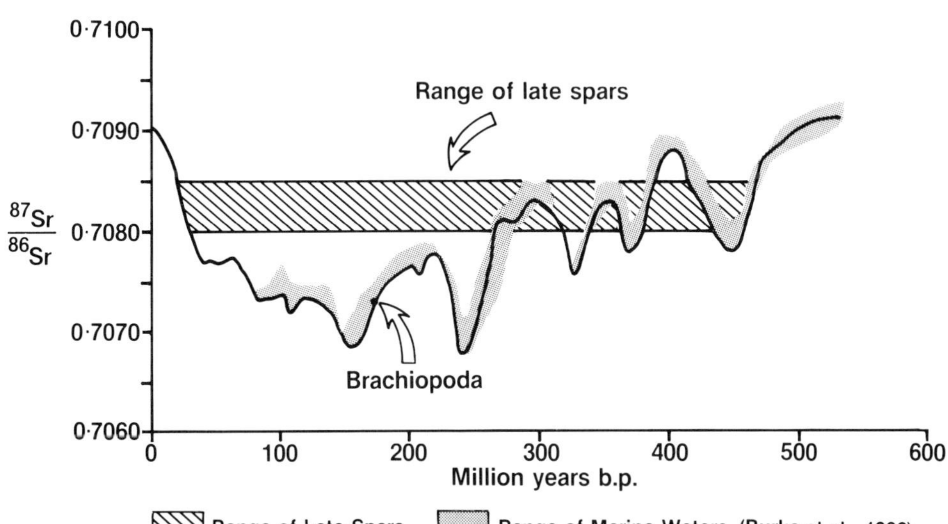

FIG. 11. The variation of $^{87}Sr/^{86}Sr$ in Phanerozoic marine waters (after Burke *et al.* 1982). Brachiopod and late spar results from this study are illustrated.

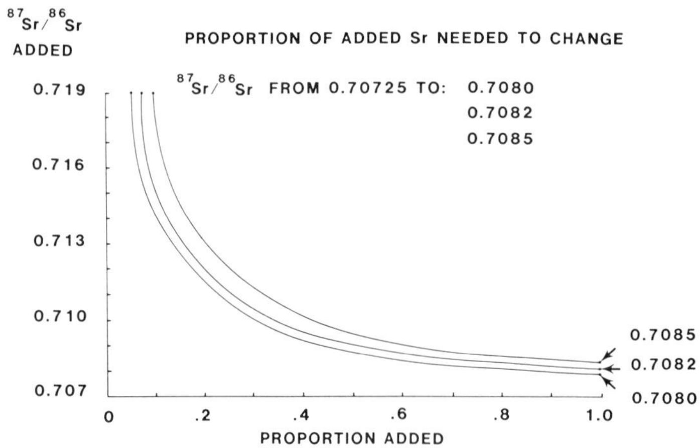

FIG. 12. Proportion of additional radiogenic Sr required to change the $^{87}Sr/^{86}Sr$ from 0.70725 (marine components) to burial calcite $^{87}Sr/^{86}Sr$ compositions.

material would satisfy the requirement of an Sr/Mg ratio which is near-identical to Bajocian marine carbonate. Figure 11 illustrates the evolution of the $^{87}Sr/^{86}Sr$ of marine waters with Phanerozoic time (Burke *et al.* 1982). The range in the strontium isotopic composition of the burial calcites has been added to this curve. It is evident from this construction that Carboniferous marine carbonate and other Lower Palaeozoic marine carbonates may have an $^{87}Sr/^{86}Sr$ ratio which is sufficiently elevated to account for the Sr isotopic composition of the calcite.

A further constraint which must be satisfied is one which allows the absolute concentrations of Sr and Mg in the burial calcites to show considerable variation whilst maintaining a broadly constant Sr isotopic composition. This can also be satisfied by remobilizing Palaeozoic marine carbonate, in that a large variation in the contributions of Bajocian and Palaeozoic carbonate will still maintain the required Sr isotopic composition of the burial calcites (Fig. 12: see also Emery 1986, Emery *et al.* 1987).

Remobilization of older carbonate may also account for a component of the calcium carbonate required by the burial cements. This mass-balance problem (Bathurst 1975) is a recurrent one in carbonate diagenesis; remobilized carbonate from elsewhere in the stratigraphic column on the scale envisaged above may be one solution to the difficulty.

In summary, it therefore appears possible to account for all the criteria pertinent to the Mg and Sr chemistry, and the Sr isotopic composition of the Lincolnshire Limestone burial cements by invoking contributions of these trace elements from host Bajocian marine carbonate and Palaeozoic marine carbonate. The oft-quoted problem of carbonate supply to buried Limestones may also be overcome by these sources.

A scheme for trace-element mobility during Lincolnshire Limestone burial diagenesis

Figure 13 depicts the essential elements of the model. Outstanding problems remain with regard to the identity of the Palaeozoic carbonate contributing Sr and Mg, and its essential hydraulic continuity with the buried Lincolnshire Limestone. The most obvious and proximal source of Palaeozoic carbonate in the East Midlands stratigraphic column is Carboniferous (Dinantian) marine carbonate, which satisfies all the requisite chemical and isotopic criteria. Seismic sections across the East Midlands Basin into the southern North Sea demonstrate the presence of normal faults cutting both Carboniferous Limestone and Lincolnshire Limestone. On burial, these faults would probably have acted as major fluid conduits, as fluid expelled from the sedimentary sequence attempted to bleed upwards along zones of high permeability. A tentative suggestion is that fluid migration up faults comprises a cross-formational limb of a convective system. Once in the permeable Lincolnshire Limestone, the fluid would have migrated up-dip, hence an enformational limb to the system which can also be demonstrated by the oxygen isotopic composition of the burial cements (Emery 1986). Briefly, as Fig. 4 demonstrates, the temperatures recorded by the precipitated ferroan calcite phase during maximum

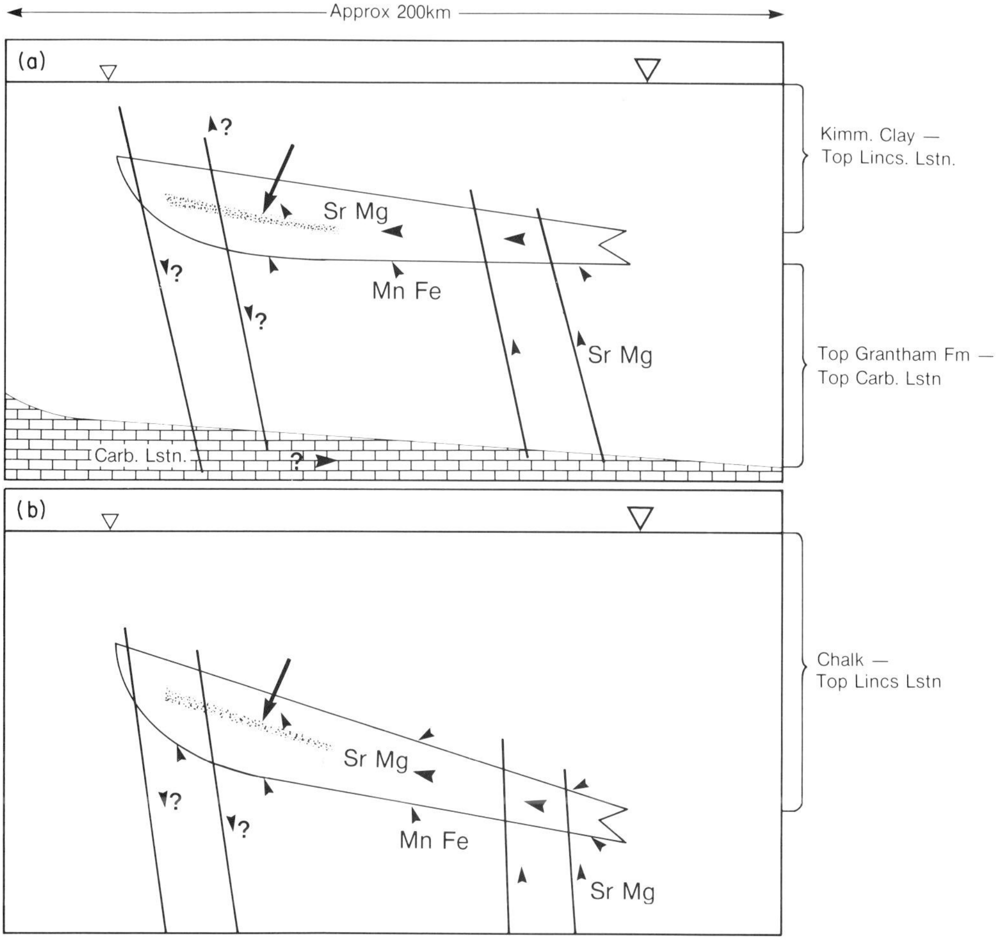

FIG. 13. A model for trace element source and mobility during precipitation of Lincolnshire Limestone burial cements. Small to large triangles represent increasing burial. Arrowheads represent increasing contributions of trace elements. Light shading illustrates shale interval (Kirton Shale Member) within the Lincolnshire Limestone. (a) Cation source for non-ferroan cements. (b) Cation sources for ferroan cements.

burial are considerably elevated over those due to that conduction through the overlying sedimentary pile alone. The most effective way of enhancing precipitation temperatures above the ambient conductive temperature is by up-dip fluid flow through the Limestone; this is further confirmed by the progressive down-dip ^{18}O depletion of both burial calcite phases. Whether an enformational limb was present within the Carboniferous Limestone during Lincolnshire Limestone burial calcitization is pure speculation.

Another component of cross-formational flow incorporated into the model is from bounding (and intraformational) shales, chiefly the underlying Grantham Formation and to a lesser extent the Upper Estuarine 'Series'. This flow component would account for transfer of Fe and Mn from the shales into the Lincolnshire Limestone aquifer during burial diagenesis.

Conclusions

(i) Two phases of burial calcite are present within the Lincolnshire Limestone, a non-ferroan early phase and a later ferroan phase.

(ii) The Fe and Mn trace element component of the burial cements is considered to be derived from Fe and Mn oxy-hydroxides associated with bounding and intraformational shales. Burial depths and temperatures associated with Lincolnshire Limestone are too low to account for derivation of these cations from clastic or clay mineral diagenesis.

(iii) Redox conditions during burial are believed to have been consistently reducing, permitting the mobilization of Fe and Mn as divalent cations and their subsequent incorporation into the burial calcites.

(iv) Increasing Fe and Mn contents of the later ferroan calcite phase was the result of increasing cross-formational flow from the bounding and intraformational shales on increasing burial.

(v) Sulphate was probably largely absent from the burial diagenetic regime.

(vi) Mg and Sr concentrations in both the burial calcites display a positive linear relationship. No absolute increase in the concentration of either cation is observed in the burial calcites with time.

(vii) The $^{87}Sr/^{86}Sr$ ratio of both burial calcites is indistinguishable (=0.70820). This isotopic composition is considerably more radiogenic than host Bajocian marine carbonate.

(viii) The Sr and Mg of the burial calcites is believed to have been derived from (a) remobilized Bajocian marine carbonate and (b) remobilized Carboniferous marine carbonate which mixed in variable proportions. These carbonate sources may be sufficient to provide the necessary calcite of the burial cements, thereby overcoming the recurrent problem of carbonate source during burial diagenesis.

(ix) A speculative fluid circulation system implied by the cation source evidence requires fluid movement from the Carboniferous Limestone up bleeder faults (the cross-formational limb of a convective system) into the Lincolnshire Limestone (the enformational limb of the system). A further cross-formational compactional flow component is provided by fluid flow from shales bounding (and internal to) the Limestone.

ACKNOWLEDGMENTS: The author is grateful to the following institutions for their time and use of equipment: The Department of Geology, King's College, London; The Institute for Energy Technology, Kjeller, Norway; The National Laboratory for Mass-Spectrometry, University of Oslo, Norway.

The author would also like to thank the following individuals for technical assistance and useful discussion: Dr J. A. D. Dickson, Dr P. C. Smalley, Dr N. Walsh, Cathy Lewin, Carol Birney, Alison Searl and Julian ('Sid') Fowles.

I am grateful for the support of a NERC studentship held at the University of Cambridge, and wish to thank Texaco Inc., Houston, for support given to the Cambridge carbonate group.

References

ANDREW-SPEED, C. P., OXBURGH, E. R. & COOPER, B. A. 1984. Temperature and depth dependent heat flow in the western North Sea. *Bulletin of the American Association of Petroleum Geologists*, **68**, 1764–81.

ASHTON, M. 1977. The stratigraphy and carbonate environments of the Lincolnshire Limestone (Bajocian) in Lincolnshire and parts of Leicestershire. *Unpublished PhD thesis*, University of Hull.

—— 1980. The stratigraphy of the Lincolnshire Limestone Formation (Bajocian) in Lincolnshire and Rutland (Leicestershire). *Proceedings of the Geologists' Association*, **91**, 203–23.

BATHURST, R. G. C. 1975. *Carbonate Sediments and their Diagenesis*. Elsevier, Amsterdam.

BERNER, R. A. 1980. *Early Diagenesis, a Theoretical Approach*. Princeton University Press.

BURKE, W. H., DENISON, R. E., HETHERINGTON, E. A., KOEPNICK, R. B., NELSON, H. F. & OTTO, J. B. 1982. Variation of seawater $^{87}Sr/^{86}Sr$ throughout Phanerozoic time. *Geology*, **10**, 516–19.

BURLEY, S. D., KANTOROWICZ, J. D. & WAUGH, B. 1985. Clastic diagenesis. *In*: BRENCHLEY, P. J. & WILLIAMS, B. P. J. (eds). *Sedimentology: Recent Developments and Applied Aspects*, pp. 189–226. Geological Society of London Special Publication, 18. Blackwell Scientific Publications, Oxford.

CARROLL, D. 1959. Ion exchange in clays and other minerals. *Bulletin of the American Association of Petroleum Geologists*, **70**, 754–79.

CHAVE, K. E. 1954. Aspects of the biochemistry of magnesium. *Journal of Geology*, **62**, 266–83.

CURTIS, C. D. 1977. Sedimentary geochemistry: environments and processes dominated by the involvement of an aqueous phase. *Philosophical Transactions of the Royal Society of London*, **286A**, 353–72.

—— 1983. Geochemical studies on the development and destruction of secondary porosity. *In*: BROOKS, J. (eds). *Petroleum Geochemistry and Exploration of Europe*, pp. 113–25. Geological Society of London Special Publication, 12. Blackwell Scientific Publications, Oxford.

—— & COLEMAN, M. 1986. Controls on the precipitation of early diagenetic calcite, dolomite and siderite concretions in complex depositional sequences. *In*: GAUTIER, D. L. (ed.). *Roles of Organic Matter in Sediment Diagenesis*, pp. 23–34. Society of Economic Paleontologists and Mineralogists Special Publication, 38.

DICKSON, J. A. D. 1966. Carbonate identification and genesis as revealed by staining. *Journal of Sedimentary Petrology*, **36**, 491–505.

DOWNING, R. A., SMITH, D. B., PEARSON, F. J. MONKHOUSE, R. A. & OTLET, R. L. 1977. The age of groundwater in the Lincolnshire Limestone, England, and its relevance to the flow mechanism. *Journal of Hydrology*, **33**, 201–16.

EDMUNDS, W. M. & WALTON, N. R. G. 1983. The Lincolnshire Limestone—hydrochemical evolution over a ten-year period. *Journal of Hydrology*, **61**, 201–11.

EMERY, D. 1986. The diagenesis of the Lincolnshire Limestone (Bajocian) in Lincolnshire. *Unpublished PhD thesis*. University of Cambridge.

——, DICKSON, J. A. D. & SMALLEY, P. C. 1987. The strontium isotopic composition and origin of burial cements in the Lincolnshire Limestone (Bajocian) of central Lincolnshire, England. *Sedimentology*, **34**, 795–806.

FAIRCHILD, I. J. 1983. Chemical controls of cathodoluminescence of natural dolomites and calcites: New data and review. *Sedimentology*, **30**, 579–83.

FAURE, G. & SZABO, Z. 1986. Isotopic studies of mineral cements in North American Sandstones: the Beria Sandstone of Ohio. *In*: RODRIGUEZ CLEMENTE, R. & FENOLL HACH-ALI, P. (eds). *Abstracts, Geochemistry of the Earth Surface and Processes of Mineral Formation*. Granada.

FROELICH, P. N., KLINKHAMMER, G. P., BENDER, M. L., LUEDTKE, N. A., HEATH, G. R., CULLEN, D., DAUPHIN, P., HAMMOND, D., HARTMAN, B. & MAYNARD, V. 1979. Early oxidation of organic matter in pelagic sediments of the eastern Equatorial Atlantic: suboxic diagenesis. *Geochimica et Cosmochimica Acta*, **43**, 1075–90.

GAVISH, E. & FRIEDMAN, G. M. 1969. Progressive diagenesis in Quaternary to late Tertiary carbonate sediments: sequence and time scale. *Journal of Sedimentary Petrology*, **39**, 1–32.

GLENNIE, K. W. & BOEGNER, P. L. E. 1981. Sole Pit inversion tectonics. *In*: ILLING, L. V. & HOBSON, G. D. (eds). *Petroleum Geology of the Continental Shelf of North-West Europe*, pp. 110–20. Heyden, London.

IRWIN, H., CURTIS, C. D. & COLEMAN, M. 1977. Isotopic evidence for the source of diagenetic carbonate during burial of organic rich sediments. *Nature*, **269**, 209–13.

KINSMAN, D. J. J 1969. Interpretation of Sr^{2+} concentrations in carbonate minerals and rocks. *Journal of Sedimentary Petrology*, **39**, 486–508.

MARSHALL, J. D. & ASHTON, M. 1980. Isotopic and trace-element evidence for submarine lithification of hardgrounds in the Jurassic of eastern England. *Sedimentology*, **27**, 271–89.

MCINTIRE, W. L. 1963. Trace element partition coefficients—a review of theory and application to geology. *Geochimica et Cosmochimica Acta*, **27**, 1209–64.

MOORE, C. H. & DRUCKMAN, Y. 1981. Burial diagenesis and porosity evolution, Upper Jurassic Smackover, Arkansas and Louisiana. *Bulletin of the American Association of Petroleum Geologists*, **65**, 597–628.

OGLESBY, T. W. 1976. A model for the distribution of manganese, iron and magnesium in authigenic calcite and dolomite cements in the Upper Smackover Formation in East Mississippi. *Unpublished MSc thesis*. University of Missouri.

OLDERSHAW, A. E. & SCOFFIN, T. P. 1967. The source of ferroan and non-ferroan calcite cements in the Halkin and Wenlock Limestones. *Journal of Geology*, **5**, 309–20.

PEACH, D. 1984. Some aspects of the hydrology of the Lincolnshire Limestone. *Unpublished PhD thesis*. University of Birmingham.

ROSLER, H. J. & LANGE, H. 1972. *Geochemical Tables*. Elsevier, Amsterdam.

SCHOFIELD, K. & ADAMS, A. E. 1986. Burial dolomitization of the Woo Dale Limestones Formation (Lower Carboniferous) Derbyshire, England. *Sedimentology*, **33**, 207–19.

SCHOLLE, P. A. & HALLEY, R. B. 1985. Burial diagenesis—out of sight, out of mind! *In*: SCHNEIDERMANN, N. & HARRIS, P. M. (eds). *Carbonate Cements*, pp. 309–34. Society of Economic Paleontologists and Mineralogists Special Publication, 36.

SMALLEY, P. C., RAHEIM, A., DICKSON, J. A. D. & EMERY, D. 1985. A strontium isotopic study of the Jurassic Lincolnshire Limestone Formation, England. *American Association of Petroleum Geologists. Research Conference Abstracts*. New Orleans, 1986.

STUEBER, A. M., PUSHKAR, P. & HETHERINGTON, E. A. 1984. A strontium isotopic study of Smackover brines and associated solids, Southern Arkansas. *Geochimica et Cosmochimica Acta*, **48**, 1637–49.

TEN HAVE, T. & HEIJNEN, W. 1985. Cathodoluminescence activation and zonation in carbonate rocks: an experimental approach. *Geologie en Mijnbouw*, **64**, 297–310.

THOMPSON, A. & WALSH, N. 1983. *A Handbook of Inductively Coupled Plasma Spectrometry*. Blackie, London.

TUCKER, M. J. 1985. Shallow-marine carbonate facies. *In*: BRENCHLEY, P. J. & WILLIAMS, B. P. J. (eds). *Sedimentology: Recent Developments and Applied Aspects*, pp. 147–69. Geological Society of London Special Publication, 18. Blackwell Scientific Publications, Oxford.

VEIZER, J. 1983. Chemical diagenesis of carbonates: theory and application of trace-element technique. *In*: *Stable Isotopes in Sedimentary Geology*. Society of Economic Paleontologists and Mineralogists. Short Course, 10.

D. EMERY, Department of Earth Sciences, University of Cambridge, Downing Street, Cambridge CB2 3EQ, UK. Present address: Sedimentology Branch, BP Petroleum Development, Britannic House, London EC2Y 9DU, UK.

Wilcox sandstone diagenesis, Texas Gulf Coast: a regional isotopic comparison with the Frio Formation

Lynton S. Land & R. Stephen Fisher

SUMMARY: The burial diagenesis of Eocene Wilcox sandstones differs significantly from the burial diagenesis of other Gulf Coast terrigenous clastic formations. Differences between the onshore Wilcox and Oligocene–Miocene Frio Formations are larger than regional variations within either formation. Younger, offshore units have undergone less diagenetic alteration than either of the older, onshore units, whereas Mesozoic sandstones are generally more extensively altered.

Cementation by quartz and calcite was the first diagenetic modification of volumetric significance to affect either the Wilcox or Frio Formations. The average $\delta^{18}O$ of Wilcox quartz cement is approximately +25‰ SMOW, in contrast to +31‰ for the Frio. In both formations, quartz cement is more abundant in the overpressured zone. Calcite has an average $\delta^{18}O$ of −10.8‰ PDB in the Wilcox Formation (compared to −7.2‰ for the Frio). $^{87}Sr/^{86}Sr$ values suggest that calcite was derived from coeval nannofossils, carbonate rock fragments of Mesozoic age, or by local input of fluids from underlying Mesozoic carbonates. Most calcite cement (and therefore quartz which pre-dates calcite cement) was apparently emplaced prior to extensive alteration of detrital silicates which would have released strontium with a high $^{87}Sr/^{86}Sr$ ratio.

The most common carbonate cement in the Wilcox Formation is ankerite which replaces calcite with increasing depth. The $\delta^{13}C$ of ankerite is essentially identical to that of calcite at the same depth, and is also similar to Frio and younger carbonate cements. In contrast, $\delta^{18}O$ values of Wilcox ankerite (average −9.8‰) indicate emplacement at higher temperature or from more depleted water than was true of Wilcox calcite. $^{87}Sr/^{86}Sr$ values for Wilcox ankerite are very radiogenic (>0.7100), indicating that ankerite emplacement occurred during or after active silicate diagenesis. Both ankerite and dolomite are very uncommon phases in Frio and younger sandstones despite massive smectite stabilization, suggesting that the conversion of smectite to illite was not the source of iron and magnesium for the late cements.

The volume of secondary porosity is similar in both Wilcox and Frio sandstones. Albitization and K-feldspar removal from both formations are essentially complete below 3000 m and essentially no unaltered detrital feldspar occurs below that depth.

Differences between the Gulf Coast formations are attributed to different geothermal gradients, differences in the basinal sediments over which the units prograded, and to the changing nature of connate fluids in the units themselves. Diagenesis of the Wilcox Formation, like other Gulf Coast terrigenous wedges, is understandable only in terms of interaction with underlying units during the large-scale evolution, in both time and in space, of the Gulf Coast diagenetic system.

The Eocene Wilcox and Oligocene–Miocene Frio Formations constitute the two largest clastic units beneath the on-shore Texas Gulf coastal plain (Fig. 1). The extensive collection of sandstones amassed by the Burea of Economic Geology, University of Texas at Austin, as part of the geothermal energy programme (Loucks *et al.* 1984) formed the basis for diagenetic studies of the Frio Formation (Land & Milliken 1981, Milliken *et al.* 1981, Land 1984). This same collection formed the basis for this study, supplemented by detailed study of the central part of the Texas Gulf Coast (Fisher 1982, Fisher & Land 1986) and previously published data (Boles 1978, Boles & Franks 1979, Boles 1981). In this paper we present the results of isotopic analysis of Wilcox cements and compare cementation conditions in the Wilcox and Frio Formations.

The emphasis of this paper, as in the regional study of the Frio Formation, is on the volumetrically important diagenetic reactions: quartz cementation, carbonate cementation, feldspar stabilization, and secondary porosity development. Clay cements other than kaolinite are of lesser volumetric significance in most Wilcox sandstones, and clays could not be isolated from the epoxy-impregnated samples available to this study.

In addition, we attempt to explain the observed regional differences between the diagenesis of Wilcox and younger formations. We briefly

From: MARSHALL, J. D. (ed.), 1987, *Diagenesis of Sedimentary Sequences*, Geological Society Special Publication No. 36, pp. 219–235.

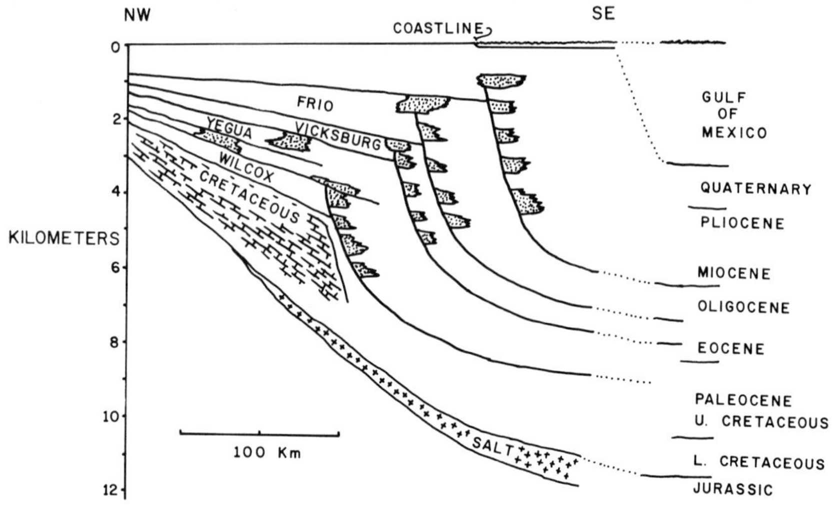

FIG. 1. Very schematic cross-section through the central part of the Texas Gulf Coast. We have modified the cross-section constructed by Bebout *et al.* (1982) by extrapolating the Gulf basinal sediments documented by Winker & Buffler (in press) beneath the continental Tertiary section. The (locally drastic) effects of salt and growth-fault tectonics have been ignored. This cross-section schematizes the relationship between thick sands (stippled) and growth faults, and between the offlapping Wilcox and Frio Formations and the thick underlying basinal Mesozoic marls, especially when the 23 × vertical exaggeration is taken into account.

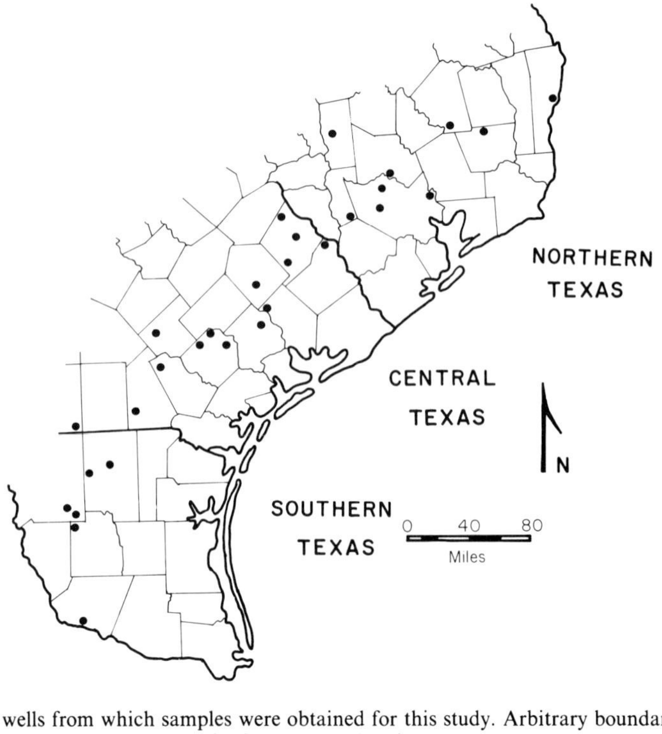

FIG. 2. Location of wells from which samples were obtained for this study. Arbitrary boundaries between 'northern', 'central' and 'southern' Coastal Gulf Texas are also shown.

compare the Wilcox Formation to underlying and overlying units to relate differences to the overall diagenetic evolution of the Gulf Coast clastic wedge.

Methods

Thin section stubs of 75 samples from 29 wells (Fig. 2) formed the sample base. Strontium for isotopic analysis was isolated from carbonate cements by reacting washed, powdered rock samples in 0.2 M HCl for less than five minutes, to minimize leaching of strontium from the silicate phases, followed by conventional cation exchange chromatography. Strontium isotope ratios were normalized to 0.7080 for the Eimer and Amend standard. Analytical precision is better than 0.0002. Otherwise, the analytical techniques used were identical to those outlined by Land (1984). The same cautions which were advanced in that study with regard to the possible biases introduced by the use of this sample suite apply to this study as well. Wells are from areas of active hydrocarbon exploration and areas of geothermal prospects, and may not be representative of Wilcox sands from other areas. Sampling of the available core may also have been biased toward the more altered (cemented or porous) rocks. Samples shallower than about 1500 m are generally unavailable from either formation.

Diagenetic reactions

Diagenetic reactions are discussed in the temporal order in which they occurred in the sandstones as determined by petrographic relationships. Quartz, calcite and kaolinite precipitation preceded the development of secondary porosity, some of which was subsequently occluded by ankerite and another generation of kaolinite.

Quartz cementation

The isotopic composition of Wilcox quartz overgrowth cements was determined by bulk analysis of the sand-sized quartz fraction of sandstones having varying amounts of quartz cement (Fig. 3). Data are presented in Table 1 (sandstones from which both quartz and feldspars were analysed isotopically), Table 2 (sandstones analysed solely for quartz), and in Fisher & Land (1986).

Wilcox sand containing no quartz overgrowths, or sand from which quartz overgrowths were removed with HF (Qz sand — cmt in Table 1) ranges from +10.2 to +16‰ in $\delta^{18}O$ (Fig. 3). This range is larger than is characteristic of younger units (10.6 to 14‰), but the average value of 13‰ for detrital Wilcox quartz is similar for all Tertiary units we have analysed (Gold 1984, Land 1984, Milliken 1985).

If Wilcox quartz sand is assumed to have a $\delta^{18}O$ of +13‰, then statistical analysis of all

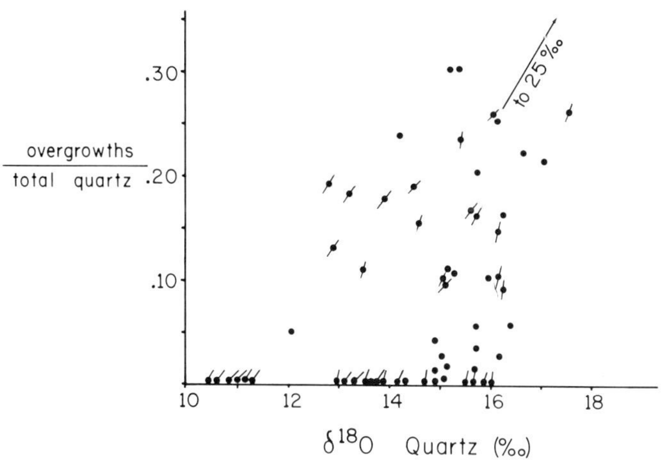

FIG. 3. The fraction of quartz present as quartz overgrowths, determined from point-counts, plotted versus the $\delta^{18}O$ of sand-sized quartz. Points along the bottom of the diagram include both sands having no overgrowths, and values obtained by analyzing the residue after 'stripping' quartz sands of their overgrowths with HF. Short line segments extrapolate between quartz sand including overgrowths (Qz sand), and quartz sand minus cement (Qz sand — cmt) from the same samples to values for pure overgrowths. The extrapolation from average sand having a $\delta^{18}O$ of 13‰ to pure overgrowths (25‰ + 4‰) includes data from this study (Tables 1 and 2), and from Fisher & Land (1986).

TABLE 1. *Summary of quartz and feldspar isotopic analyses*

Depth (m)	$\delta^{18}O$ total qz	$\delta^{18}O$ qz + feldspar sand	$\delta^{18}O$ qz sand	$\delta^{18}O$ qz sand —cmt	Per cent qz cmt	$\delta^{18}O$ feldspar*	Per cent feldspar	Per cent ab	$^{87}Sr/^{86}Sr$ feldspar
2357	15.2	15.7	16.0		10.2	11.8	7.1	67	0.7227
3008	15.1	17.1					15.9	81	0.7142
3031	15.7	16.8	16.2	15.7	10.4	21.2	11.9	90	
3194	16.0	17.6	17.5	15.9	26.3	18.0	19.0	99	0.7127
3363	14.7	15.1	15.1	13.3	9.6	15.1	26.1	97	
3445	14.7	16.1	15.8	13.8	16.3	19.4	8.3	97	
3656	14.6	15.5	15.0	14.3	10.2	16.8	28.0	98	
3976	15.0	15.6	15.1		11.1	22.4	6.8	98	
4006	14.7	15.9	16.2	16.0	9.0	14.1	14.5	98	
4112	14.7	16.1	15.4	14.7	23.8	19.1	18.8	98	0.7146
4297	14.6	16.7	16.2	15.6	14.9	19.7	14.2	99	0.7166
4344	13.3	15.1	14.6	13.5	15.7	17.1	20.2	98	0.7163

* Calculated.

$\delta^{18}O$ values are in ‰ relative to SMOW. Total quartz was determined after fusion of the whole rock in $NaHSO_4$ and removal of feldspars in H_2SiF_6 (Syers *et al.* 1968). Fine-grained quartz from rock fragments is thus included. In order to exclude such fine-grained quartz, the >62 μm (sand-sized) fraction of an aliquot of the rock was isolated by sieving following $NaHSO_4$ digestion, analysed for $\delta^{18}O$ before removal of feldspar (qz + feldspar sand), and after removal of feldspars (qz sand). The $\delta^{18}O$ feldspar was calculated from these two values and the % feldspar obtained by electron microprobe analysis of fused beads of the qz + feldspar sand fraction. Quartz overgrowths were then stripped from the qz sand fraction in HF, to yield samples of quartz sand presumably representative of the sand before cementation (qz sand — cmt). From these values, and the % qz cmt determined by point counts, the $\delta^{18}O$ of the overgrowths was calculated by material balance.

available data permits extrapolation to an average value for Wilcox quartz overgrowths of +25‰ (±4‰, SMOW). An average of +24‰ (±5‰) is obtained if only those values (Table 1; Fisher & Land 1986) obtained by stripping individual samples of their quartz overgrowths (see Milliken *et al.* 1981, for the procedure) are averaged.

TABLE 2. *Summary of isotopic analyses of quartz*

Depth (m)	$\delta^{18}O$ total quartz	$\delta^{18}O$ quartz sand	% quartz cement
1591	12.7	13.6	0
1619	11.1	(12.0)	4.0
1819	14.1	15.1	0.5
2552	14.3	15.0	1.3
2572	13.8	14.7	0
2603	14.1	14.9	4.1
2656	14.7	15.1	2.9
2718	13.2	14.1	0
2750		13.5	11.1
2796	15.5	16.3	3.6
2914	15.4	(16.3)	16.0
2943	14.7	15.3	10.9
2950	14.7	15.6	5.2
2975	14.7	(15.6)	20.0
2998	14.2	15.6	1.8
3035	13.8	14.7	2.0
3045	14.7	15.6	5.0
3270	15.5	16.4	5.0
3350		16.2	24.0
3421	15.7	(16.6)	22.0
3429		14.9	2.2
3591		17.0	21.5
3658	14.1	(15.0)	30.0
4608	14.3	(15.2)	30.0

Notation as in Table 1. Values for quartz sand in parentheses were calculated from the average difference (0.92‰) between total quartz and quartz sand for samples ($n=25$) from which both analyses were available.

Interpretation of quartz cementation in both the Frio and Wilcox Formations is complicated by the fact that the cements are zoned. Alternating zones of subtle orange luminescence with euhedral outlines are barely visible through the light microscope when excited by 15 kV electrons on the electron microprobe, and document a complex history of pore-filling. The cause of the luminescent zoning and whether or not it is accompanied by isotopic zoning, is unknown. Some euhedral overgrowths terminate in pore space and therefore quartz overgrowth cementation could have taken place over an extended period of time, and minor amounts could even be taking place from contemporaneous fluids.

Despite more variability in data from the Wilcox Formation as compared to the Frio Formation, it is clear that Wilcox quartz overgrowths are significantly depleted in ^{18}O relative to the value of +31‰ (Table 3) obtained from Frio sandstones (compare Fig. 3 with fig. 3 in Land 1984). Therefore Wilcox quartz overgrowths must have formed at a somewhat different temperature and/or from water of somewhat different $\delta^{18}O$ than was observed in the Frio Formation.

The average amount of quartz cement in Wilcox sandstones (4.6%) is approximately twice that found in average Frio sandstones (2.6%). The average percentage of quartz cement increases with depth (Fig. 4), as is also true of the

TABLE 3. *Comparison of Wilcox and Frio diagenetic phases*

	Wilcox	Frio
Quartz cement		
Volume % of rock	4.6 ± 2.5 (462)	2.6 ± 1.4 (645)
$\delta^{18}O$ SMOW	25 ± 4‰ (60)	31 ± 1.5‰(27)
Carbonate cement		
Dominant mineralogy	Ankerite	Calcite
% calcite of total carbonate	13	88
Other minerals	Calcite	Dolomite
Volume % of rock	3.3 ± 1.2 (462)	5.3 ± 2.9 (445)
$\delta^{18}O$ calcite PDB	−10.8 ± 2.1‰ (21)	−7.2 ± 1.1‰ (68)
$\delta^{18}O$ ankerite PDB	−9.8 ± 1.1‰ (70)	—
$\delta^{13}C$ carbonate PDB	−4.6 ± 2.0‰ (91)	−4.0 ± 2.6‰ (68)
Secondary porosity		
Volume % of rock	10	12
Primary feldspar		
% total feldspar of rock	23	27
% K-feldspar of total feldspar	40	30
%An in unaltered plagioclase	5	20
Diagenetic feldspar		
$\delta^{18}O$ albite SMOW	17 ± 3.0‰ (11)	17‰

± values are given as one standard deviation, with the number of samples in parenthesis. Data from the Wilcox Formation illustrated in Figs 3 to 8, and presented in Tables 1 and 2. Data from the Frio Formation from Milliken *et al.* (1981) and Land (1984).

Frio Formation. In the case of the Frio Formation, an increase in the degree of quartz cementation corresponds approximately with the top of the present-day overpressured zone (fig. 5 in Land 1984). In the Wilcox Formation, the relationship between the degree of quartz cementation and overpressure is not as sharp, but the gross relationship still exists. As in the

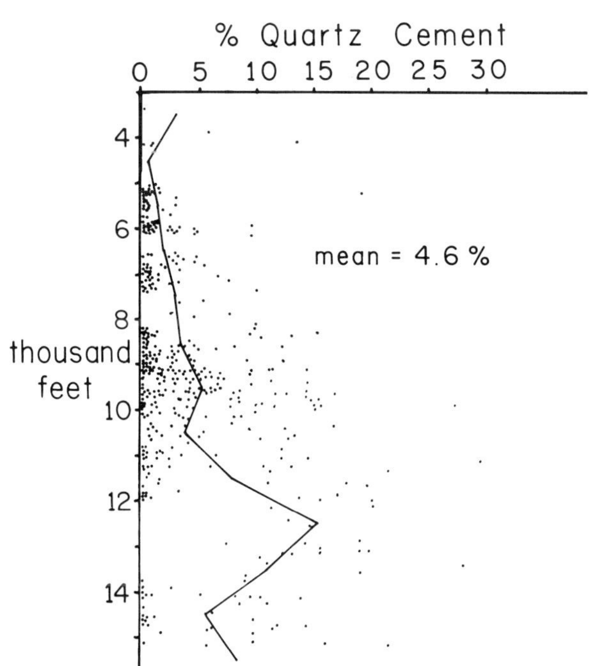

FIG. 4. Percentage quartz cement versus depth. Quartz cementation, as in the Frio Formation, is more common below modern geopressure which ranges between about 2400 m (8000 feet) and 3900 m (13 000) feet) in the study area. The line is a moving average of the raw data of Loucks *et al.* (1984).

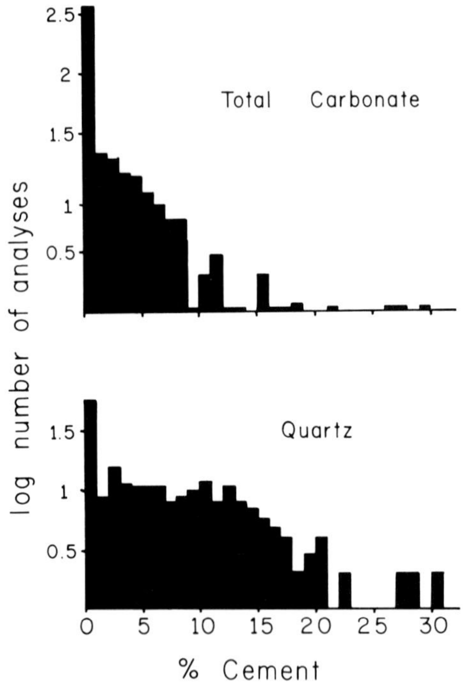

FIG. 5. Log of the number of point-counts versus the percentage total carbonate (top) and quartz (bottom) cement in Wilcox sandstones. Unlike the Frio Formation, quartz cementation is more uniform in Wilcox sandstones. Like the Frio, however, most sands, especially the matrix-rich ones, are uncemented, and very few sands are pervasively cemented. Raw data from Loucks *et al.* (1984).

Frio Formation, most sands, especially matrix-rich ones, contain little quartz cement, and very few sands in either Formation are extensively quartz-cemented (Fig. 5). The distribution of quartz cement in Wilcox sands is somewhat more uniform than in Frio sands (compare Fig. 5 with fig. 2 in Land 1984). More uniformly cemented rocks are produced as cementation proceeds, probably because as cement begins to occlude porosity and reduce permeability, fluid flow is directed into more permeable (less cemented and matrix-rich) sands. More intense early cementation by a more dynamic flow system, or a longer duration of cementation, may account for the more uniform and more intense quartz cementation in the Wilcox as compared to the Frio Formation.

Calcite cementation

Calcite is the dominant carbonate cement in shallow Wilcox sandstones, whereas ankerite is more characteristic of deeper sandstones. The presence of ankerite as shallow as 1800 m and calcite as deep as 3300 m document a more gradational boundary between the two types of cement in this regional study than was observed at 2700 m by Boles (1978) in a three county area in South Texas. We agree with Boles that ankerite characteristically replaces calcite in the deeper sands. We do not agree that a 'reaction front' occurs regionally over a narrow depth interval which corresponds to the transformation of smectite to illite. Nor do we agree on the source of components for the ankerite cement.

Electron microprobe analyses of Wilcox calcite cements indicate that they are relatively pure, containing more than 98% $CaCO_3$, similar to Frio calcite cements. Variations in composition are small, and do not correlate with depth. As in the case of calcite cements in the Frio Formation, luminescence is uniform and zoning is rarely observed. No systematic variation in the degree of carbonate cementation as a function of depth is apparent (Fig. 6). In this respect as well, Frio and Wilcox sandstones are quite similar, but the average amount of carbonate cement in Wilcox sandstones (3.3%) is slightly less than the amount found in average Frio sandstones (5.3% — Table 3).

Oxygen isotopic values for Wilcox calcite average −10.8‰ (Fig. 7), compared with an average of −7.2‰ observed for the Frio Formation (Table 3). As in the case of the Frio Formation (fig. 10 in Land 1984), no significant relation between depth and $\delta^{18}O$ is observed on a regional basis.

Carbon isotopic values for Wilcox calcite and ankerite cements are indistinguishable from each other (Fig. 8), and are similar to those found in the Frio Formation (Table 3). Values between −2 and −6‰ are typical of deeper sandstones, whereas shallower sandstones range between about −3‰ and −14‰. The reason for this distribution of values in both formations is not entirely clear, and several alternative scenarios can be constructed (Lundegard & Land 1986). If the source of $CaCO_3$ for sandstone cements is carbonate nannofossils in contemporeous shales, or $CaCO_3$ from older formations as suggested by the strontium isotopic compositions, then calcite cements should have a $\delta^{13}C$ near 0‰. Because the cements are depleted in ^{13}C with respect to inorganic carbon, between 10 and 50% organic carbon having a $\delta^{13}C$ of −26‰ must have been introduced into the diagenetic system in order to yield the measured $\delta^{13}C$ values. The exact organic reaction responsible cannot be identified with assurance, however. Lundegard *et al.* (1984) showed that decarboxylation reactions could not provide sufficient CO_2, or provide CO_2 of the appropriate isotopic composition in the correct

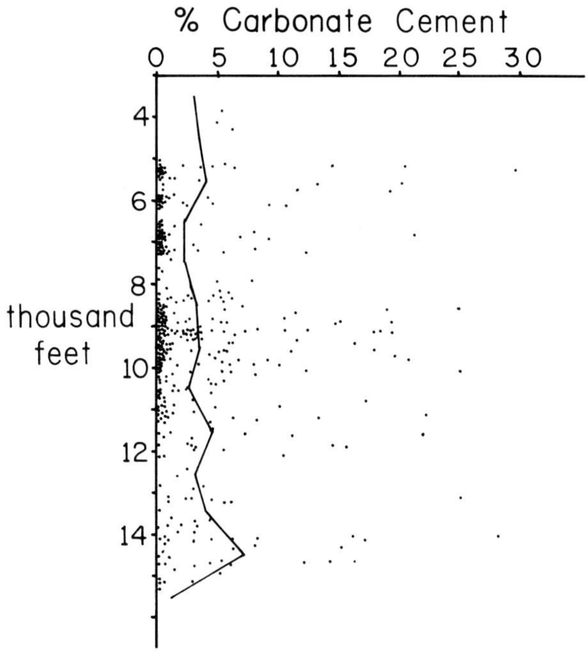

FIG. 6. Per cent carbonate cement (calcite + ankerite) versus depth. Similar to the Frio Formation, no systematic relationship with depth is observed. The line is a moving average based on the raw data of Loucks *et al.* (1984).

proportions. Hydrous pyrolysis in underlying formations might supply the carbon found in carbonate cements (−6‰) as the result of the release of ^{13}C-depleted methane (−46‰) from kerogen (−26‰). Simultaneous oxidation of kerogen and reduction of ferric iron and/or sulphate might also have released ^{13}C-depleted H_2CO_3 into the evolving pore water. Decomposition of dissolved acetate (Carothers & Kharaka 1978) is another possible source for ^{13}C-depleted carbon in the shallower samples.

Strontium isotopic analyses of Wilcox calcite cements range from 0.7070 to 0.7091 compared to values for Eocene sea water between 0.7076 and 0.7078 (Burke *et al.* 1982). The measured values strongly suggest that most of the $CaCO_3$ has been derived from coeval nannofossils, from Mesozoic carbonate rock fragments (rare in these rocks), or from fluids equilibrated with underlying Mesozoic strata (Fig. 9). We cannot account for $^{87}Sr/^{86}Sr$ ratios less than those characteristic of Eocene sea water (about 0.7077 — Burke *et al.* 1982) unless material has been derived from older marine carbonate or evaporite formations. Young volcanic debris which might contribute strontium with low $^{87}Sr/^{86}Sr$ ratios is rare in Wilcox sandstones. Based on currently available data, the reservoir of strontium from silicates having an $^{87}Sr/^{86}Sr$ ratio lower than that of Eocene sea water does not appear to be sufficiently large to account for the measured values. Most analyses of bulk feldspar or clay from the Gulf Coast yield $^{87}Sr/^{86}Sr$ ratios greater than about 0.71 (Perry & Turekian 1974, Morton 1983, Table 1 and work in progress).

Some calcite samples in the Wilcox and Frio Formations are slightly more radiogenic (enriched in ^{87}Sr) than Eocene sea water. Therefore, some calcite emplacement could have taken place from solutions in which silicate reactions, such as the transformation of smectite to illite, or feldspar dissolution or replacement, had begun. The generally non-radiogenic nature of Wilcox (and Frio) calcite indicates that either extensive silicate diagenesis had not yet taken place at the time calcite was emplaced, or that pore water strontium was dominated by strontium derived from the dissolution of contemporaneous nannofossils, Mesozoic rock fragments, or formation waters from underlying units (Morton & Land 1987).

Feldspar diagenesis

The conversion of all detrital feldspar to albite, or its dissolution to form secondary porosity, is grossly similar in both Wilcox (Fig. 10) and Frio sandstones (fig. 11 in Land 1984). The zone in

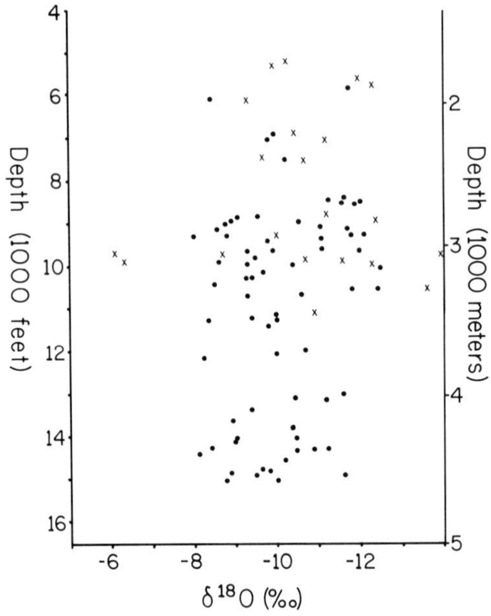

FIG. 7. $\delta^{18}O$ of Wilcox calcite (crosses) and ankerite (dots) cements versus depth. Calcite cements are more variable in their isotopic composition (standard deviation is 2.1‰) than ankerite (standard deviation is 1.1‰). Average Wilcox calcite (−10.8‰) is depleted relative to Frio calcite cement (−7.2‰, Table 3). Wilcox ankerite (−9.8‰) must have formed under conditions significantly different from Wilcox calcite, assuming a 3.8‰ fractionation between the CO_2 extracted from calcite and the CO_2 extracted from a co-genetic ankerite. Data include those of Boles (1978) and Fisher (1982). The scatter is not significantly reduced if present-day temperature, rather than depth, is plotted.

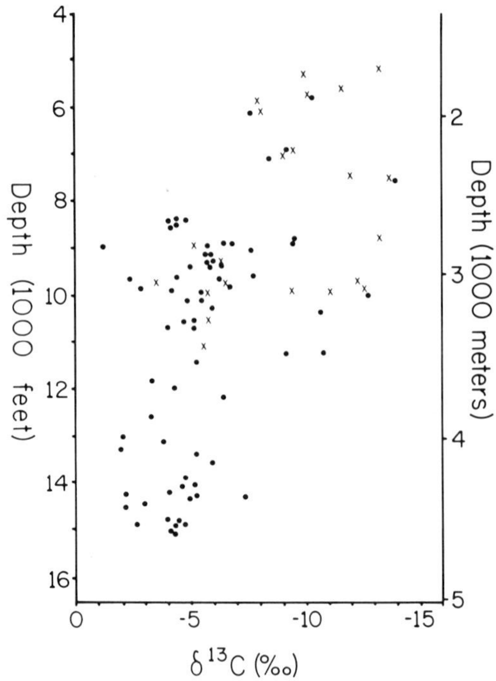

FIG. 8. $\delta^{13}C$ of Wilcox calcite (crosses) and ankerite (dots) cements versus depth. As in the case of the Frio Formation, a minimum value of about −2‰ is observed, and shallower cements tend to have a wider range in isotopic composition and to be more depleted in ^{13}C.

which K-feldspar is dissolved and plagioclase albitized extends between about 2500 and 3500 m, approximately 500 m shallower than was observed in the Frio Formation. The amount of primary detrital feldspar in the two formations is similar (Loucks *et al.* 1984, their figs 14 and 16), the only major difference being that Wilcox plagioclase was initially much more albitic, having an average composition of An_5 as compared to An_{20} in the Frio. This difference probably reflects not only somewhat different source rocks for the two formations, but the long time interval available for depositional recycling of Wilcox detritus on the craton prior to the Laramide orogeny. Table 1 presents data from which $\delta^{18}O$ values for authigenic albite have been calculated [$\delta^{18}O$ feldspar = ((100 * $\delta^{18}O$ qz + feldspar sand]-((100-%feldspar) * $\delta^{18}O$ qz sand))/%feldspar]. Recognizing the relatively large errors involved in these sorts of extrapolations, a value of +17‰ (±3 ‰) will be assumed. Wilcox and Frio authigenic albite (Milliken *et al.* 1981) is apparently similar in isotopic composition, suggesting that both formations may have approached similar conditions of temperature and $\delta^{18}O$ water at the time of albitization.

Ankerite

The oxygen isotopic composition of Wilcox ankerite cement (−9.8 ± 1‰, Table 3) is less variable than is true of calcite (−10.2 ± 2‰). The $\delta^{18}O$ of ankerite cement, as in the case of calcite cement in both the Frio and Wilcox Formations, displays no systematic variation with depth (Fig. 7). Our observations confirm the ionic compositional trend in ankerite, but not the decrease in $\delta^{18}O$ with increasing depth observed locally by Boles (1978, his fig. 8). In contrast to calcite cements in both the Frio and Wilcox Formations, significant variations in ankerite ionic composition do occur with increasing depth. Based on statistical analysis of 32 samples, two trends have been found.

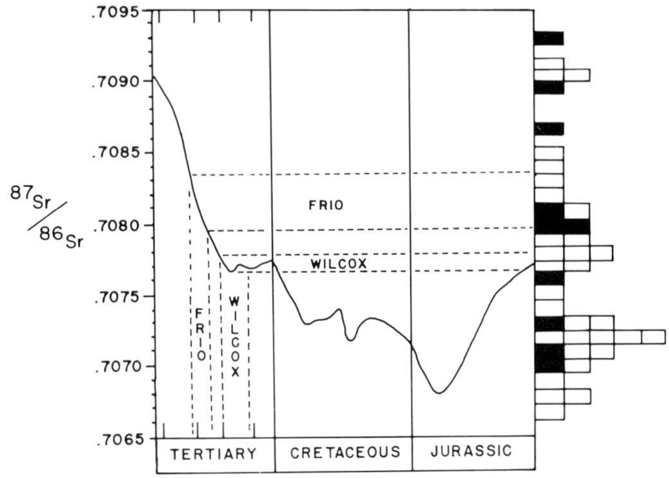

FIG. 9. $^{87}Sr/^{86}Sr$ of Wilcox (black bars) and Frio (open bars) calcite cements superimposed on the post-Triassic variation of $^{87}Sr/^{86}Sr$ in sea water presented by Burke *et al.* (1982). We interpret these data to indicate that some of the cations for cementation of both formations were derived from Mesozoic marine carbonate strata, either by remobilization of rock fragments in the sandstones, or as the result of fluids which transported components into the Tertiary strata. Calcite (and quartz) emplacement seems to have taken place prior to extensive silicate (feldspar and clay) stabilization reactions which would have released radiogenic strontium (^{87}Sr).

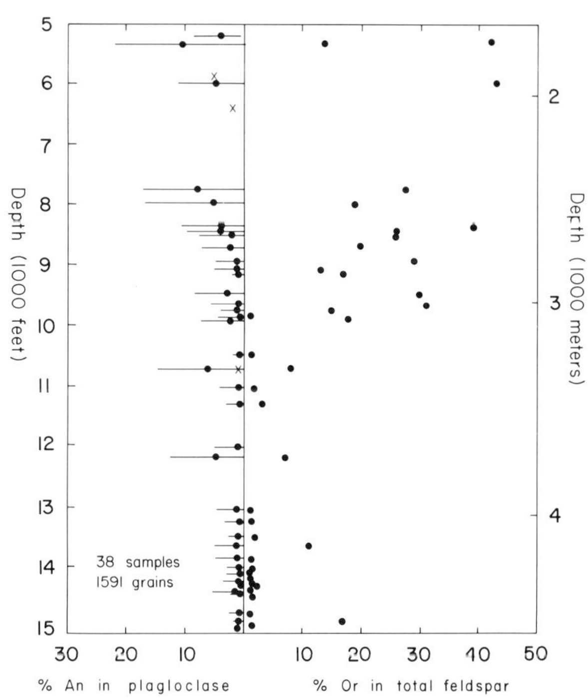

FIG. 10. Composition of Wilcox feldspars as a function of depth. Lines through plagioclase compositions are $\pm$ 1 standard deviation in An content based on about 80 analyses per thin section. All sands below about 3000 m (10 000 feet) are at least partially albitized and have suffered extensive K-feldspar dissolution. Most sands above about 2600 m (8500 feet), in contrast, are essentially unaffected by feldspar stabilization processes. Crosses from Boles (1981).

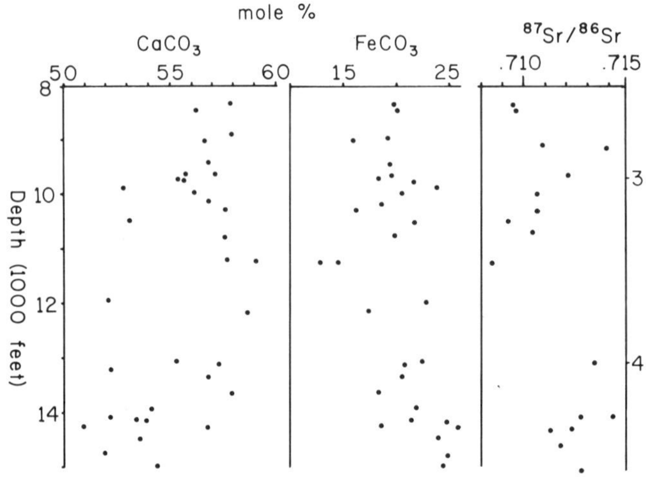

FIG. 11. Stoichiometry and $^{87}Sr/^{86}Sr$ of Wilcox ankerite as a function of depth. Less stoichiometric and less radiogenic phases are more typical of shallower samples, although considerable scatter exists.

First, deeper ankerite tends to be more stoichiometric ($Ca_{1\cdot0}Fe_{0\cdot52}Mg_{0\cdot48}(CO_3)_2$) and more enriched in ^{87}Sr, whereas shallower ankerite tends to be less stoichiometric ($Ca_{1\cdot20}Fe_{0\cdot24}Mg_{0\cdot56}(CO_3)_2$) and less enriched in ^{87}Sr (Fig. 11). Boles (1978) attributed the variations in ankerite composition to precipitation at different temperatures; less stoichiometric phases were hypothesized to be precipitated at lower temperatures, as is well documented for dolomite. If this explanation applies on a regional basis, then the lack of correlation between ankerite composition and $\delta^{18}O$ requires that the large temperature differences between 2500 and 4500 m (where temperatures differ by approximately 70° C today) must have been just offset by large variations in $\delta^{18}O$ water (see Fig. 12) to account for ankerite of similar isotopic composition over a large depth range. Alternatively, vertical changes in pore-fluid chemistry may have been involved. The less radiogenic nature of shallower ankerite might be explained by progressive dilution of radiogenic strontium by less radiogenic strontium (from the calcite which was being replaced) as rising fluids supplied the iron and magnesium from underlying units. Evolution of the pore fluids toward more calcium-rich, less radiogenic, ^{18}O-depleted (because of precipitation of ^{18}O-enriched mineral phases) compositions along vertical flow paths could also account for the compositional variations observed in Wilcox ankerite.

The source of radiogenic strontium incorporated in the ankerite is undoubtedly from the reaction of detrital K-bearing silicate minerals, but the relative importance of feldspar versus clay mineral stabilization reactions has not yet been determined. Almost all ankerite cement (Fig. 11) is less radiogenic than available analyses of either diagenetic illite or detrital feldspars (Perry & Turekian 1974, Morton 1983, Table 1 and work in progress). Therefore strontium isotopic compositions may reflect a mixture of the limited amount of strontium available from the silicate phases relative to the abundant strontium available from dissolution or recrystallization of marine carbonates or calcite cement.

Synthesis

Figure 12 shows the locus of temperature and $\delta^{18}O$ water values calculated for average Wilcox quartz and calcite, and for average ankerite and albite. As was the case in the regional study of the Frio Formation (Land 1984), the use of epoxy-impregnated thin section stubs precluded any attempt to obtain isotopic values for authigenic kaolinite (but see Fisher & Land 1986) or chlorite.

Quartz and calcite were the first volumetrically important cements to form, and based on petrographic relations, quartz commonly precedes calcite (Fig. 13). The isotopic data on both quartz and calcite suggest precipitation under similar conditions of temperature and $\delta^{18}O$ water, in agreement with the paragenetic relations seen in thin section. The ^{18}O-depleted nature of both Wilcox quartz and calcite cements

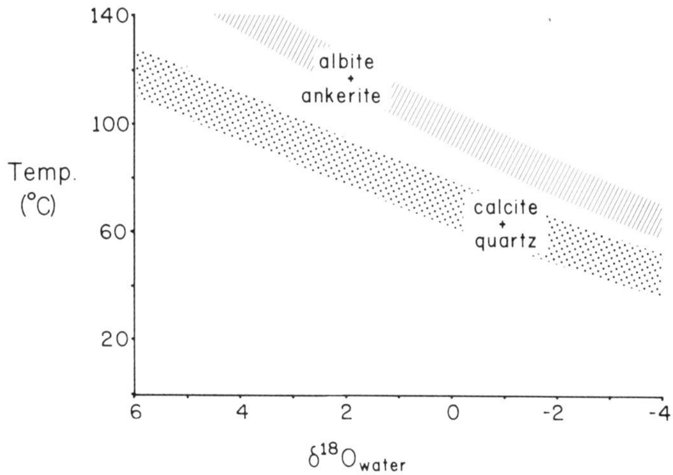

FIG. 12. Locus of temperature and $\delta^{18}O$ water values for average Wilcox quartz and calcite, and ankerite and albite. Because of the range of about $\pm 4‰$ in values for quartz, more variable conditions of emplacement are indicated for some of the quartz cement, but the average isotopic chemistry of quartz and calcite cement is in accord with the petrographic relationships (Fig. 13). The shallowest depth of occurrence of calcite and quartz today is similar, at approximately 85° C, and although ankerite occurs at shallower depths, it becomes dominant at depths corresponding to a temperature of about 120° C. The preferred conditions of emplacement are indicated by the positions of the mineral-pair labels.

relative to Frio cements (Table 3) must be due either to emplacement of Wilcox cements at higher temperature or to water depleted in ^{18}O (or both). Application of the quartz-calcite geothermometer to the averages listed in Table 3 predicts co-precipitation of Wilcox cements at a temperature of 68° C from water having a $\delta^{18}O$ of $-1‰$.

The geothermal gradient in the Wilcox Formation today, approximately 43° C km^{-1} in southern Gulf Texas (Bodner 1985) and approximately 33° C km^{-1} in central and northern Gulf Texas, is considerably higher than in the Frio Formation (approximately 25° C km^{-1}. These gradients have probably prevailed throughout most of the history of both Formations. The effects of the thermal event which accompanied rifting of the Gulf in Jurassic time would presumably have dissipated during the more than 10^8 years that elapsed prior to the onset of Wilcox deposition. A Tertiary geothermal gradient in the Wilcox Formation higher than that observed at present is currently unsupported.

If quartz and calcite cements precipitated from sea water buffered by clay mineral reactions in shales, then the $\delta^{18}O$ of the water must have been within about $2‰$ of $+5‰$. This conclusion is substantiated both by direct analysis of diagenetic illite and present-day formation water from the Frio Formation (Fig. 14), and by quantitative modelling (compare Fig. 14 with fig. 9 in Suchecki & Land 1983). Because only eight water samples from geopressured wells in the Wilcox Formation have been obtained (Fig. 14), we are forced to argue by analogy with the Frio Formation. If analogy with the Frio Formation is justified, and water having a $\delta^{18}O$ of approximately $+5‰$ was involved, then quartz and calcite in Wilcox sandstones were both emplaced at temperatures 20 – 30° C hotter than was true in the Frio Formation (compare Fig. 12 with fig. 4 in Land 1984). Higher thermal gradients in Wilcox sands could also account for the shallower emplacement of quartz cement in the Wilcox Formation, and in later albitization at shallower depths.

If the reaction of smectite to illite controlled the $\delta^{18}O$ of the porewater, then the strontium isotopic composition of the porewater would have been affected by that reaction as well. But calcite cements are considerably less radiogenic than the detrital clays. Therefore, clay transformations either postdated calcite (and quartz) cementation, or the large amount of strontium (thousands of ppm) released from contemporaneous nannofossils, and provided by water from underlying strata, overwhelmed the small amount (hundreds of ppm) of radiogenic strontium involved in the reaction of smectite to illite and the stabilization of detrital feldspars.

The alternative to emplacement of Wilcox quartz and calcite cements at higher temperatures

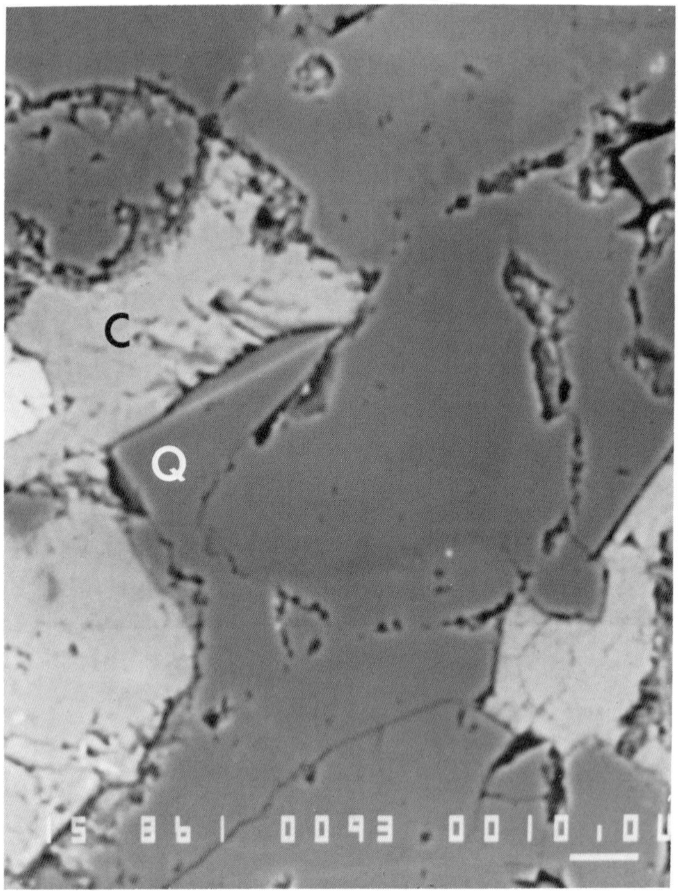

FIG. 13. Back-scattered electron image of calcite cement (C) which post-dates quartz overgrowth cement (Q). The $^{87}Sr/^{86}Sr$ of the calcite cement is 0.7082, suggesting that it was derived primarily from dissolution of coeval nannofossils. The fact that non-radiogenic calcite post-dates quartz in this sample (from 2251 m, Karnes County) also suggests that extensive silicate reactions, which should involve radiogenic strontium having an $^{87}Sr/^{86}Sr$ ratio no lower than 0.711, were not the source of SiO_2 for quartz cementation. Scale bar is 10 μm.

than was characteristic of the Frio Formation, is that the water which emplaced Wilcox quartz and calcite was several per mil lighter than the water which emplaced Frio cements. Meteoric water is unlikely to have gravitationally recharged Wilcox deltaic sediments because it is doubtful that sufficient head could have been generated in the low relief Wilcox deltaic plain. Deep, continuous recharge is inconsistent with the almost certain early establishment of geopressure in the thick, marine, pro-delta muds (Sharp & Domenico 1976). The initial burial of connate meteoric water is difficult to reconcile with the deposition of these dominantly marine sediments, but the involvement of large volumes of hydrothermally circulated meteoric water in quartz cementation of Mesozoic sands (Dutton 1986, McBride *et al.* in press) suggests that meteoric water may have replaced connate sea water prior to the development of over-pressure, if the hydrothermal systems which dominated the early diagenesis of Mesozoic sands continued into the Eocene.

Alternatively, water near SMOW in composition may have been derived by compaction of thick carbonate-rich Mesozoic strata over which the Wilcox Formation prograded (Fig. 1). Connate porewater in underlying strata would need to have been compacted away before extensive high-temperature calcite-water reactions took place, however, because such reations enrich the water in ^{18}O even more than the $+5‰$ observed

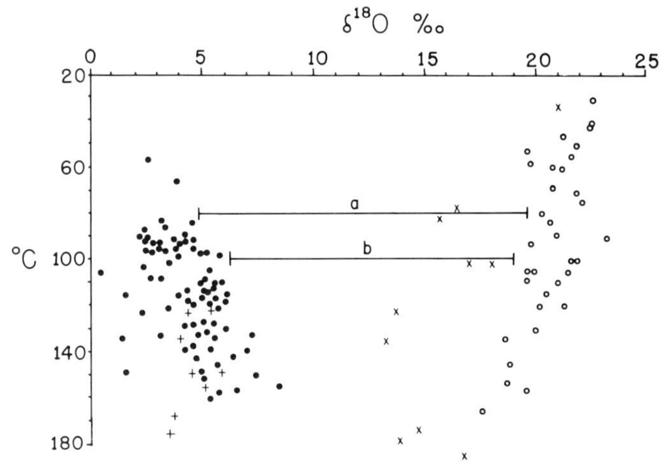

FIG. 14. $\delta^{18}O$ of modern Wilcox formation water below hard geopressure (crosses), Frio Formation water from all pressure regimes (dots), and the <1 μm fraction (essentially pure smectite-illite) of Frio shales (circles) and Wilcox shales (crosses). Data from Kharaka *et al.* (1977), Yeh & Savin (1977), Morton (1983), Jackson (in preparation) and Morton & Land (1987). Horizontal lines a and b indicate the equilibrium fractionation between illite-smectite and water at 80° and 100° C (14.8 and 12.7‰, respectively) using the equation of Eslinger & Savin (1973). At all depths, the most depleted Frio clays are nearly in equilibrium with the most enriched waters observed today at the same depth, even above present-day geopressure (approximately line b in many areas). Wilcox diagenetic illite, in contrast, is depleted in ^{18}O approximately 4‰ relative to Frio diagenetic illite, as is also observed for calcite and quartz cements in the respective sandstones (Table 3).

in terrigenous units (Land & Prezbindowski 1981).

Jackson (in preparation) has found Wilcox diagenetic illite to have a $\delta^{18}O$ of approximately +14‰ compared to values no lighter than +18‰ for the Frio Formation (Fig. 14). Illite values of +14‰ are difficult to explain unless a source of depleted water is invoked. The data on quartz and calcite cements in the sandstones, and diagenetic illite in the shales, therefore currently support Wilcox quartz and calcite cements being emplaced by water having a $\delta^{18}O$ near SMOW in composition. In contrast, Frio cements were apparently emplaced by smectite-buffered sea water having a $\delta^{18}O$ near +5‰.

The mechanism for cement emplacement was probably silica and CO_2 loss from ascending (cooling and devolatilizing) pore fluids. The volumes of cements involved could not have been transported on a 'once through' basis by the limited volume of water available from compaction of underlying strata. The volumes of quartz and calcite cement involved in Wilcox burial diagenesis are similar to those involved in Frio burial diagenesis, and the calculations presented by Land (1984) will not be repeated here. If those assumptions and calculations are correct, then at least one litre of water was required to emplace the volume of cement observed in each cubic centimetre of average Wilcox sandstone. Free convection appears to be the only viable mechanism available to transport such large volumes of water. Free convection in the rapidly deposited, undercompacted, sands and muds, at temperatures near 80° C, with water having $\delta^{18}O$ near 0‰ can account for the isotopic chemistry of Wilcox quartz and calcite cements. Free convection appears to be feasible in thick sandy sections pressurized to near lithostatic pressures (Blanchard & Sharp 1985), and in more shale-rich strata if extensive microfracturing occurs. Some components for both Frio and Wilcox calcite cements seem to have been derived from underlying Mesozoic marine carbonates.

The relative isotopic and minor element homogeneity, and lack of distinct trends with depth in both Wilcox and Frio calcite cements also requires an explanation. Cements having relatively uniform oxygen isotopic and trace element chemistry might be emplaced near the top of a convecting geopressured system, at relatively constant temperature (corresponding to the temperature at which geopressure is established) and $\delta^{18}O$ water (corresponding to a mix of rock-buffered connate water, rock-buffered mineral dehydration water, and water expelled from underlying formations). In each unit, the geopressured zone would undoubtedly have swept up through the thick clastic wedges as burial took place with time (Sharp & Domenico 1976). Cements could have been deposited near the top of the cell, where maximum temperature gradients could induce quartz precipitation and maximum pressure gradients could induce calcite precipitation (Wood & Hewett 1984). The cements would have been buried more deeply within the geopressured cell as sedimentation continued. Because they remained essentially in ionic (but not isotopic) equilibrium with the pore fluids, they did not react until either acids were introduced into the system so that wholesale dissolution occurred, or until the ionic composition of the pore fluids changed drastically so that replacement took place. In the case of the Wilcox Formation, if a more stable detrital clay suite required higher temperature to initiate reaction (Bruce 1984), and/or if water isotopically depleted in oxygen was characteristic of the connate porefluids, then the more depleted cements in the Wilcox Formation relative to cements in younger formations can be explained. More Wilcox cements are now found above the top of the present overpressured zone than is true of the Frio Formation, possibly because the pressure is beginning to dissipate in the older unit.

The major mineralogical (as opposed to isotopic) difference between Wilcox and younger sandstones is the late emplacement of ankerite in the Wilcox Formation. The reason for this difference may be variations in the history of the underlying strata, not in the history of co-genetically deposited basinal sediments. The Wilcox Formation is the only Tertiary unit in the Texas Gulf Coast in which ankerite is a major carbonate cement. Ankerite is extremely rare in post-Eocene sandstones relative to the amount of calcite present (Gold 1984, Land 1984, Milliken 1985, Land *et al.* 1987) despite the immense mass of smectite which has been converted to illite in the younger units.

Therefore it is questionable whether the reaction of smectite to illite provided iron and magnesium for ankeritization (Boles & Franks 1979). It is possible, of course, that Wilcox smectite was different from smectite deposited in the younger formations. The more volcanic-rich, westerly source for major south Texas Frio deposition contrasts with the more northerly cratonic source for the Wilcox (Winker 1982, Loucks *et al.* 1984). Chlorite and pyrite occur as late diagenetic phases in both formations, but their importance as sinks for iron and magnesium have not yet been quantitatively assessed in either. The reaction of a single phase containing radiogenic strontium (smectite to illite) could not possibly account for both an early generation of non-radiogenic cements (calcite, and by inference, quartz) and a late generation of radiogenic, ^{18}O-depleted ankerite as postulated by Boles & Franks (1979, their fig. 9).

It is tempting to invoke fluids derived from underlying Mesozoic strata to account for the replacement of calcite by ankerite in the immediately overlying Wilcox Formation. In some ways the late burial diagenesis of the underlying Jurassic and Cretaceous platform carbonates (Moore & Druckman 1981, Prezbindowski 1985, Woronick & Land 1985) and the diagenesis of Mesozoic sandstones (McBride 1981, Suchecki 1983, Dahl 1984, Dutton 1986, McBride *et al.* in press) more closely resembles some aspects of Wilcox sandstone diagenesis than does the diagenesis of younger Gulf Coast terrigenous units. All the Mesozoic units contain somewhat radiogenic late ankerite or ferroan dolomite cements and replacement phases which are essentially absent in post-Eocene formations, and the source for which is not clear. Weaver & Beck (1971) hypothesized the transport of iron, magnesium (and potassium) from underlying units into the Tertiary section. Fluids derived from the Mesozoic carbonates (which have $^{87}Sr/^{86}Sr$ ratios similiar to late phases in the Mesozoic strata — Stueber *et al.* 1984, Woronick & Land 1985) might also have supplemented hydrocarbons generated *in situ* from Wilcox shales (Jenden & Kaplan 1984).

Conclusion

Wilcox and Frio sandstones differ significantly in their diagenetic history. Although systematic diagenetic variation can be documented within each formation from north to south along the Texas Gulf Coast (e.g. Loucks *et al.* 1984), greater differences exist between the two Formations than exist within either. In general, younger Gulf Coast units are less diagenetically altered than either the Wilcox or Frio Formations (Gold 1984, Milliken 1985). In contrast, older clastic units, although not as volumetrically extensive or as widely distributed, are more pervasively altered and are altered at shallower depths. A growing number of isotopic analyses (Suchecki 1983, Franks & Forester 1984, Dutton 1986, McBride *et al.* in press, work in progress) suggest that the cements in older units are commonly even more depleted in ^{18}O than are the cements present in Wilcox sands. Apparently, both increased temperature and deeper penetration of meteoric water characterized the diagenesis of Mesozoic sandstones early in the diagenetic history of the Gulf sediment wedge.

Currently available data all suggest that the Gulf has evolved toward less extensive alteration of younger sands, alteration of younger sands at progressively greater burial depths, and toward generally lower temperature diagenesis with time (Land *et al.* 1987). The changing abundance and chemical nature of the authigenic phases, and the potential for large cross-formational fluxes of material (as suggested by the ^{87}Sr data and the necessity for vertical transport of liquid hydrocarbons into immature strata, for example), demand that diagenetic studies of Gulf Coast units, and sedimentary basins in general, be placed in a regional context. The burial diagenesis of any particular unit can only be understood as part of a much larger scale (in both time and space) diagenetic system. Considerably more data on sandstone, carbonate, evaporite and shale diagenesis will be needed before the diagenetic evolution of the Gulf Coast syncline can be understood to the point that it can be used for predictive purposes, as a model for other basins, or applied to uplifted and exhumed sedimentary wedges.

ACKNOWLEDGMENTS: We wish to thank Bob Loucks and Bob Morton of the Bureau of Economic Geology for amassing the samples on which this study was based and permitting us access to the point-count data and the thin section stubs. We have relied almost exclusively on the point-count data obtained by Loucks and his colleagues for Figs 4, 5 and 6. Numerous people have contributed to this study with their critical comments and encouragement. We especially want to thank the members of FOG (Friends of the Gulf), a seminar group of the last few years. Paul Blanchard, Bill Galloway, Paul Gold, Tim Jackson, Paul Lundegard, Wendy Macpherson, Earle McBride, Kitty Milliken and Jack Sharp were all active in directing our thinking and/or improving our prose. Steve Franks provided an especially constructive review, as did several other reviewers, including Fred Longstaffe, Sam Savin, Jim Boles, and Steve Crowley.

Laboratory support for this work was provided by the National Science Foundation, EAR-7824081 and EAR-8121009. Larry Mack and Rosemary Capo did most of the strontium isotopic analyses, and continuing strontium isotopic analyses of Gulf Coast rocks and waters constitutes Larry's PhD dissertation. Additional support was provided by the Mobil's Dallas Research Laboratory, Texaco USA, and the Geology Foundation of the University of Texas at Austin.

References

BEBOUT, D. G., WEISE, B. R., GREGORY, A. R. & EDWARDS, M. B. 1982. Wilcox sandstone reservoirs in the deep subsurface along the Texas Gulf Coast. *Bureau of Economic Geology, University of Texas, Austin, Report of Investigations,* **117**.

BLANCHARD, P. E. & SHARP, J. M. JR 1985. Possible free convection in thick Gulf Coast sandstone sequences. *Southwest Section American Association of Petroleum Geologists Transactions,* **11**, 6–12.

BODNER, D. 1985. Heat variations caused by groundwater flow in growth faults of the South Texas Gulf Coast basin. *MA thesis,* University of Texas at Austin.

BOLES, J. R. 1978. Active ankerite cementation in the subsurface Eocene of southwest Texas. *Contributions to Mineralogy and Petrology,* **68**, 13–22.

—— 1981. Active albitization of plagioclase, Gulf Coast Tertiary. *American Journal of Science,* **282**, 165–80.

—— & FRANKS, S. G. 1979. Clay diagenesis in Wilcox sandstones of southwest Texas: implications of smectite diagenesis on sandstone cementation. *Journal of Sedimentary Petrology,* **49**, 55–70.

BRUCE, C. H. 1984. Smectite dehydration — its relation to structural development and hydrocarbon accumulation in the Northern Gulf of Mexico basin. *American Association of Petroleum Geologists Bulletin,* **68**, 673–83.

BURKE, W. H., DENISON, R. E., HETHERINGTON, E. A., KOEPNICK, R. B., NELSON, H. F. & OTTO, J. B. 1982. Variation of seawater $^{87}Sr/^{86}Sr$ throughout Phanerozoic time. *Geology,* **10**, 516–9.

CAROTHERS, W. W. & KHARAKA, Y. K. 1978. Aliphatic acid anions in oil-field waters and their implications for the origin of natural gas. *American Association of Petroleum Geologists Bulletin,* **62**, 2441–53.

DAHL, W. M. 1984. Progessive burial diagenesis in lower Tuscaloosa sandstones, Louisiana and Mississippi (Abstract). *Clay Minerals Society Annual Meetings,* p. 42. Louisiana State University Department of Geology, Baton Rouge, Louisiana.

DUTTON, S. P. 1986. Diagenesis and burial history of the Lower Cretaceous Travis Peak Formation, East Texas. *PhD Dissertation,* University of Texas at Austin.

ESLINGER, E. & SAVIN, S. M. 1973. Mineralogy and oxygen isotope geochemistry of the hydrothermally altered rocks of the Ohaki–Broadlands New Zealand geothermal area. *American Journal of Science,* **273**, 240–67.

FISHER, R. S. 1982. Diagenetic history of Eocene Wilcox sandstones and associated formation waters, South-Central Texas. *PhD Dissertation,* University of Texas at Austin.

—— & LAND, L. S. 1986. Diagenetic history of Eocene Wilcox sandstones, South-Central Texas. *Geochimica et Cosmochimica Acta,* **50,** 551–62.

FRANKS, S. G. & FORESTER, R. W. 1984. Relationships among secondary porosity, pore-fluid chemistry and carbon dioxide, Texas Gulf Coast. *In*: MCDONALD, D. A. & SURDAM, R. C. (eds). *Clastic Diagenesis,* pp. 63–80. American Association of Petroleum Geologists Memoir, 37.

GOLD, P. B. 1984. Diagenesis of middle and upper Miocene sandstones, Louisiana Gulf Coast. *MA thesis*, University of Texas at Austin.

JACKSON, T. J. in preparation. Diagenesis of (Eocene) Wilcox sandstones and shales, Texas Gulf Coast. *PhD Dissertation*, University of Texas at Austin.

JENDEN, P. D. & KAPLIN, I. R. 1984. Maturation of organic matter in Paleocene-Eocene Wilcox group, South Texas: Relationship to clay diagenesis and sandstone cementation (Abstract). *American Association of Petroleum Geologists Bulletin*, **68**, 492.

KHARAKA, Y. K., CALLENDER, E. & CAROTHERS, W. W. 1977. Geochemistry of geopressured waters from the Texas Gulf Coast. In: *Proceedings of the Third Geopressured-Geothermal Energy Conference*, pp. 121–65. University of Southwest Louisiana, Lafayette, Louisiana.

LAND, L. S. 1984. Diagenesis of Frio sandstones, Texas Gulf Coast: a regional isotopic study. *In*: MCDONALD, D. A. & SURDAM, R. C. (eds). *Clastic Diagenesis*, pp. 37–62. American Association of Petroleum Geologists Memoir, 37.

—— & MILLIKEN, K. L. 1981. Feldspar diagenesis in the Frio formation, Brazoria County, Texas Gulf Coast. *Geology*, **9**, 314–8.

——, —— & MCBRIDE, E. F. 1987. Diagenetic evolution of Cenozoic sandstones, Gulf of Mexico sedimentary basin. *Sedimentary Geology*, **50**, 195–225.

—— & PREZBINDOWSKI, D. R. 1981. The origin and evolution of saline formation water, Lower Cretaceous carbonates, South-Central Texas, U.S.A. *Journal of Hydrology*, **54**, 51–74.

LOUCKS, R. G., DODGE, M. M. & GALLOWAY, W. E. 1984. Regional controls on diagensis and reservoir quality in Lower Tertiary sandstones along the Texas Gulf Coast. *In*: MCDONALD, D. A. & SURDAM, R. C. (eds). *Clastic Diagenesis*, pp. 15–46. American Association of Petroleum Geologists Memoir, 37.

LUNDEGARD, P. D. & LAND, L. S. 1986. Carbon dioxide and organic acids: Their origin and role in diagenesis, Texas Gulf Coast Tertiary. *In*: GAUTIER, D. L. (ed.). *The Roles of Organic Matter in Diagenesis*, pp. 129–46. Society of Economic Palentologists and Mineralogists Special Publication, 38.

——, —— & GALLOWAY, W. E. 1984. The problem of secondary porosity: Frio formation (Oligocene), Texas Gulf Coast. *Geology*, **12**, 399–402.

MCBRIDE, E. F. 1981. Diagenetic history of Norphlet Formation (Upper Jurassic), Rankin County, Mississippi. *Transactions of the Gulf Coast Association of Geological Societies*, **31**, 347–51.

——, LAND, L. S. & MACK, L. E. In press. Diagenesis of feldspathic eolian and fluvial sandstones, Norphlet Formation (Upper Jurassic), Rankin County, Mississippi, and Mobile County, Alabama. *American Association of Petroleum Geologists Bulletin*.

MILLIKEN, K. L. 1985. Petrology and burial diagenesis of Plio–Pleistocene sediments, northern Gulf of Mexico. *PhD Dissertation*. University of Texas at Austin.

——, LAND, L. S. & LOUCKS, R. G. 1981. History of burial diagenesis determined from isotopic geochemistry, Frio formation, Brazoria County, Texas. *American Association of Petroleum Geologists Bulletin*, **65**, 1397–413.

MOORE, C. H. & DRUCKMAN, Y. 1981. Burial diagenesis and porosity evolution, Upper Jurassic Smackover, Arkansas and Louisiana. *American Association of Petroleum Geologists Bulletin*, **65**, 597–628.

MORTON, J. P. 1983. Rb-Sr dating of clay diagenesis. *PhD Dissertation*. University of Texas at Austin.

MORTON, R. A. & LAND. L. S. 1987. Regional variations in formation water chemistry, Frio Formation (Oligocene), Texas Gulf Coast. *American Association of Petroleum Geologists Bulletin*, **71**, 191–206.

PERRY, E. A. JR & TUREKIAN, K. K. 1974. The effects of diagenesis on the redistribution of strontium isotopes in shales. *Geochimica et Cosmochimica Acta*, **38**, 929–35.

PREZBINDOWSKI, D. R. 1985. Burial cementation, is it important? A case study, Stuart City Trend, South Central Texas. *In*: SCHNEIDERMANN, N. & HARRIS, P. M. (eds). *Carbonate Cements*, pp. 241–64. Society of Economic Paleontologists and Mineralogists Special Publication, 36.

SHARP, J. M. JR & DOMENICO, P. A. 1976. Energy transport in thick sequences of compacting sediments. *Geological Society of America Bulletin*, **87**, 390–400.

STUEBER, A. M., PUSHKAR, P. & HETHERINGTON, E. A. 1984. A strontium isotopic study of Smackover brines and associated solids, southern Arkansas. *Geochimica et Cosmochimica Acta*, **48**, 1637–49.

SUCHECKI, R. K. 1983. Isotopic evidence for large-scale interaction between formation waters and clastic rocks (Abstract). Geological Society of America Annual Meeting, Indianapolis, Indiana, p. 701.

—— & LAND, L. S. 1983. Isotopic geochemistry of burial-metomorphosed volcanogenic sediments, Great Valley sequence, northern California. *Geochimica et Cosmochimica Acta*, **47**, 1487–99

SYERS, J. K., CHAPMAN, S. L., JACKSON, M. L., REX R. W. & CLAYTON, R. N. 1968. Quartz isolation from rocks, sediments and soils for determination of oxygen composition. *Geochimica et Cosmochimica Acta*, **32**, 1022–5.

WEAVER, C. E. & BECK, K. C. 1971. Clay water diagenesis during burial: How mud becomes gneiss. *Geological Society of America Special Paper*, 134.

WINKER, C. D. 1982. Cenozoic shelf margins, Northwestern Gulf of Mexico. *Transactions of the Gulf Coast Association of Geological Societies*, **32**, 427–48.

—— & BUFFLER, R. T. in press. North-south cross-section, Western Gulf of Mexico. *In*: *Gulf of Mexico Ocean Margin Drilling Program, Regional Atlas Series*, No. 6. Marine Science International, Woods Hole, Massachusetts.

WOOD, J. R. & HEWETT, T. A. 1984. Reservoir diagenesis and convective fluid flow. *In*: MC-

DONALD, D. A. & SURDAM, R. C. (eds). *Clastic Diagenesis*, pp. 99–110. American Association of Petroleum Geologists Memoir, 37.

WORONICK, R. E. & LAND, L. S. 1985. Late burial diagenesis, Lower Cretaceous Pearsall and Lower Glen Rose formations, South Texas. *In*: SCHNEIDERMANN, N. & HARRIS, P. M. (eds). *Carbonate Cements*, pp. 265–75. Society of Economic Paleontologists and Mineralogists Special Publication, 36.

YEH, H. W. & SAVIN S. M. 1977. Mechanism of burial metamorphism of argillaceous sediments: 3. O-isotopic evidence. *Geological Society of America Bulletin*, **88**, 1321–30.

LYNTON S. LAND. Department of Geological Sciences, University of Texas at Austin, Austin, Texas 78713, USA.

R. STEPHEN FISHER. Bureau of Economic Geology, University of Texas at Austin, Austin, Texas 78713, USA.

Regional dolomitization of subtidal shelf carbonates: Burlington and Keokuk Formations (Mississippian), Iowa and Illinois

David C. Harris & William J. Meyers

SUMMARY: Cathodoluminescent petrography of crinoidal limestones and dolomites from the Mississippian (Osagean) Burlington and Keokuk Formations in Iowa and Illinois has revealed a complex diagenetic history of calcite cementation, dolomitization, chertification and compaction. Dolomite occurs abundantly in subtidal, open-marine facies throughout the study area. Three luminescently and chemically distinct generations of dolomite can be recognized regionally. Dolomite I, the oldest generation, is luminescent, thinly zoned, and occurs mainly as a replacement of lime mud. Dolomite II has dull red unzoned luminescence, and occurs mainly as a replacement of dolomite I rhombs. Dolomite III is non-luminescent, and occurs as a syntaxial cement on, and replacement of, older dolomite I and II rhombs.

Petrography of these dolomite generations, integrating calcite cement stratigraphy, chertification and compaction histories has established the diagenetic sequence. Dolomites I and II pre-date all calcite cements, most chert, intergranular compaction and stylolites. Dolomite III precipitation occurred within the calcite cement sequence, after all chert, and after at least some stylolitization. The stratigraphic limit of these dolomites to rocks older than the St Louis Limestone (Meramecian) suggests that dolomitization took place before or during a regional mid-Meramecian subaerial unconformity.

A single dolomitization model cannot reasonably explain all three generations of dolomite in the Burlington and Keokuk limestones. Petrographic and geochemical characteristics coupled with timing constraints suggest that dolomite I formed in a sea water–fresh water mixing zone associated with a meteoric groundwater system established beneath the pre-St Louis unconformity. Dolomite II and III may have formed from externally sourced warm brines that replaced precursor dolomite at shallow burial depths. These models therefore suggest that the required Mg for dolomite I was derived mainly from sea water, whereas that for dolomites II and III was derived mainly from precursor Burlington—Keokuk dolomites through replacement or pressure solution.

A fundamental problem in carbonate diagenesis has been the origin of widespread dolomites, particularly those in subtidal, open marine depositional facies. In recent years research on dolomites has become increasingly more sophisticated, using a wide range of geochemical and crystallographic approaches (e.g. Zenger *et al.* 1980, Reeder 1981, Saller 1984, Banner *et al.* in press). Essential to using these approaches is the establishment of a detailed petrographic and stratigraphic framework. We report herein one such regional petrographic study of dolomites in the Osagean Burlington and Keokuk Formations in southeastern Iowa and adjacent Illinois, a study which is part of a broader project on the geochemistry and petrology of regional dolomites in these strata (Harris 1982, Smith 1984, Cander 1985, Kaufman 1985, Prosky & Meyers 1985, Banner 1986, Daniels 1986, Banner *et al.* in press).

The main approach in this petrographic study has been the use of cathodoluminescence, which has allowed us to: (1) identify three major, regionally extensive stages of dolomite precipitation, (2) establish the relative ages of the three dolomite types, and (3) to establish the timing of the dolomites relative to other diagenetic events, which in turn constrains their absolute ages. The study is based on 175 thin sections chosen from about 385 samples collected from 25 measured sections throughout the outcrop belt of southeastern Iowa and adjacent Illinois (Fig. 1). These samples were spaced at about 0.3–1 m intervals throughout the preserved thickness of the Burlington and Keokuk Formations, which averaged about 20 m (range = 8–50 m).

Stratigraphy and depositional facies

In southeastern Iowa, the main units investigated were the Osagean Burlington and Keokuk Formations (Fig. 1). Additionally, the underlying Kinderhookian and overlying Meramecian units were sampled in order to put constraints on timing of calcite cements and dolomites.

Burlington and Keokuk Formations

The Burlington and Keokuk Formations were part of a broad shallow shelf that flanked the south side of the Transcontinental Arch during

From: MARSHALL, J. D. (ed.), 1987, *Diagenesis of Sedimentary Sequences*, Geological Society Special Publication No. 36, pp. 237–258.

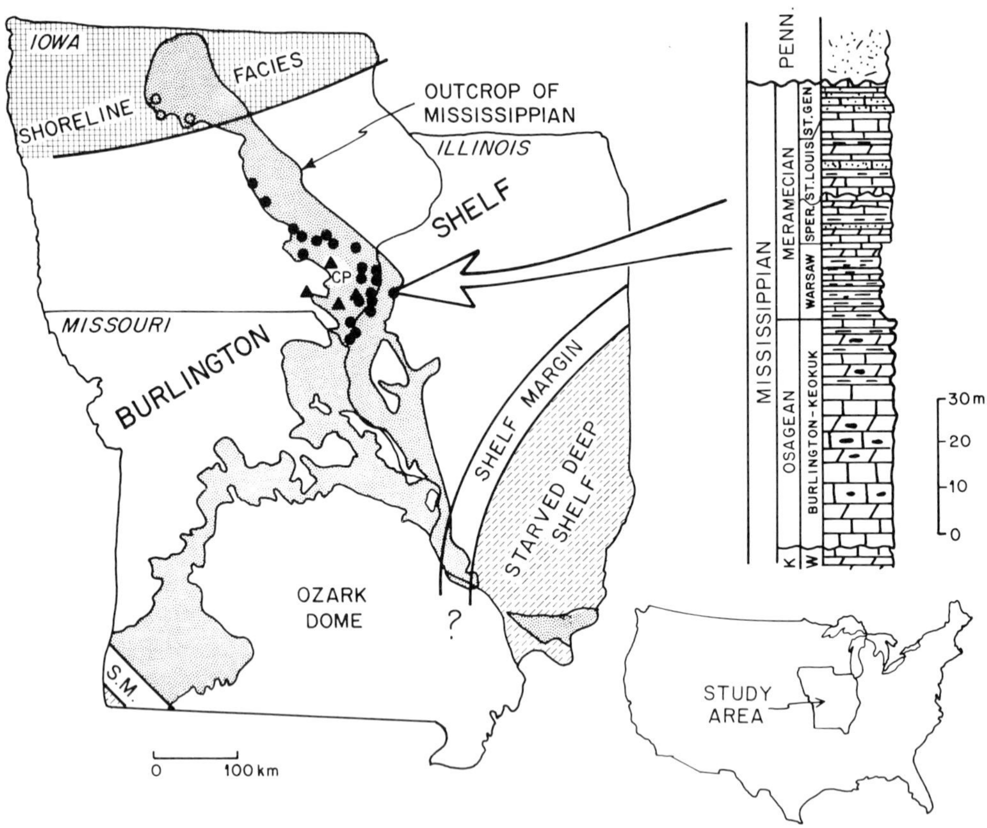

FIG. 1. Palaeogeographic map of the Burlington Shelf, and stratigraphic section of Osagean and Meramecian rocks in Iowa and western Illinois. The outcrop belt of Mississippian carbonates is indicated by the shaded pattern. Black circles and triangles indicate locations of measured sections in the Burlington and Keokuk Formations and in Meramecian strata, respectively, used in this study. A possible Burlington–Keokuk equivalent shoreline facies, studied by Sixt (1983), occurs as indicated to the NW of the study area.

Osagean time. These shelf carbonates extend westward into Kansas, and thin toward the SE in Illinois (Fig. 1), and toward the south in Arkansas, Kansas, and Oklahoma to a deeper starved shelf facies comprising a condensed sequence of phosphatic and cherty carbonates and shales (Harris & Parker 1964, Lane 1978, Lane & DeKeyser 1980).

The Burlington and Keokuk Formations in the study area range up to about 50 m thick and comprise skeletal limestones and dolostones. The Keokuk differs from the Burlington mainly by containing more shaly carbonates, but comprises essentially the same depositional facies. These carbonates are dominated by crinoids, fenestrate bryozoans and brachiopods, and vary in depositional textures from grainstones to mudstones. Locally they contain glauconite pellets and fish teeth horizons. Chert nodules and lenses are common throughout the Burlington and Keokuk, especially in mud-supported rocks. Bedding typically varies from about 30 to 80 cm thick and is defined by stylolites, shaly partings, and contacts between contrasting depositional textures.

The Burlington and Keokuk therefore comprise strictly subtidal shelf, open-marine facies, lacking ooids, stromatolites, fenestral carbonates and other indicators of sea-level/peritidal facies. The nearest possible peritidal facies are the Lower Humboldt beds of the Gilmore City Formation in north-central Iowa recently interpreted to be shoreline equivalents of the Burlington (Sixt 1983) (Fig. 1). Relevant to the diagenetic history of the Burlington and Keokuk, there is no substantive physical evidence for intraformational subaerial disconformities, nor for subaerial exposure at the Burlington–Keokuk contact. We also see no evidence for submarine hardgrounds within grain or mud-supported rocks.

Younger Mississippian formations

Meramecian rocks in the study area are up to about 45 m thick, and consist in ascending order of the Warsaw Shale, Spergen Formation, St Louis Formation and the St Genevieve Limestone (Fig. 1). These in turn are overlain unconformably by Pennsylvanian (Desmoinesian) strata which cut down through progressively older units towards the east.

The Warsaw conformably overlies the Keokuk and consists of fossiliferous calcareous shales and shaley dolomites. The Spergen Formation conformably overlies the Warsaw and consists of limestones and dolomites that are locally shaly, sandy and oolitic. The St Louis Formation consists of fine-grained, generally unfossiliferous lime mudstones containing collapse(?) breccias and shale beds. The breccias are interpreted as evaporite solution-collapse breccias based on the presence of evaporites in the subsurface in Illinois (Carlson 1979).

The contact between the Spergen and St Louis has been interpreted as an unconformity based on the apparent overlap of St Louis rocks on to Kinderhookian strata in north-central Iowa, and the absence of the Warsaw and Spergen in this area (Carlson 1979). The recent work of Sixt (1983) implies that this angular discordance is not real, because the rocks previously identified as St Louis in north-central Iowa are probably Burlington equivalents. In spite of this there is a regional unconformity at the base of the St Louis (Lane, personal communication) in southeastern Iowa, which may represent the Meramecian eustatic sea-level drop shown on published global sea-level curves (Vail *et al.* 1977).

In summary, the Spergen and St Louis Formations represent shallowing, and restriction of the former open-circulation deeper shelf of the Burlington–Keokuk sediments. Relevant to the diagenetic history there were at least two unconformities that could have resulted in freshwater diagenesis of the Burlington–Keokuk rocks: the pre-St Louis and the pre-Pennsylvanian unconformities.

Dolomite petrography

Burlington–Keokuk dolomites consist mainly of euhedral to subhedral rhombs that average 70–100 μm in size (range = 30–200 μm) (Fig. 2a, b). The rhombs have unit extinction and are usually inclusion-poor, or have inclusion-rich cores and clear rims. These inclusions comprise opaque phases, $CaCO_3$, and fluid inclusions (Smith 1984).

The dolomite selectively replaced lime mud and bryozoans, and only rarely replaced crinoids and brachiopods. Dolomite pervasively replaced lime mud in mud-supported textures. In cement-free skeletal packstones, dolomite occurs as a 'matrix' between grains (Fig. 2c). In cement-bearing packstones, dolomite occurs as geopetal fabrics on tops of grains and at the bottoms of axial canals of crinoids (Fig. 2d). This geopetal distribution reflects the original distribution of lime mud. In some cement-bearing grain rocks, dolomite occurs as clumps of rhombs near the centres of inter-crinoid interstices, and as isolated single rhombs or clusters of a few rhombs encased in calcite cements. In the first case, the clumps of rhombs near the centres of intergranular interstices seemingly float in calcite cement (Fig. 3a), a distribution that could be interpreted as late, post-cement precipitation of dolomite. However, based on dolomite-cement timing discussed below, these clumps of dolomite are interpreted as replacements of micritic grains, possibly peloids or bryozoans. In the second case, rhombs encased in calcite cement occur in contact with fossil grains in both geopetal and non-geopetal distributions, and as rhombs completely encased in calcite cement (Fig. 3b). These rhombs may be dolomite cements anchored on grains (or other rhombs) in a third dimension, or may be replacements of former microcrystalline calcite grains or mud.

Porosity in the Burlington–Keokuk carbonates is restricted to dolostones and to clumps of dolomite within the grain-supported textures, mainly as intercrystalline porosity (Figs 2a, b, 4c, 5c, 8b), as fossil moulds, and more rarely as intra-rhomb pores.

Cathodoluminescence petrography

Cathodoluminescence petrography shows that there are three major regionally extensive generations of dolomite—dolomite I, II and III—each having a distinct cathodoluminescent signature and distinct geochemistry. All three occur in mud- and grain-supported textures.

Dolomite I is the single most abundant dolomite type and is characterized by finely zoned orange luminescence (Fig. 4a, b). The thin concentric zoning is due to variations in intensity and colour of luminescence from yellowish-orange to brownish-orange. Although the fine zoning in dolomite I occurs throughout the study area, the details of the zonal stratigraphy vary between measured sections. Dolomite I is Ca-rich (54.5–56.5 mole% $CaCO_3$), contains 800–1500 ppm Mn, 400–8000 ppm Fe and has $^{87}Sr/^{86}Sr$ near Mississippian sea water (0.7076) (Prosky & Meyers 1985, Banner *et al.* in press). Stable isotopes of dolomite I are relatively heavy,

FIG. 2. Dolomite from the Burlington and Keokuk Formations. (a) Euhedral to subhedral, clear rhombs characteristic of Burlington dolomite. P = intercrystalline pores. Scale = 250 µm. (b) SEM micrograph of Burlington dolomite. Scale = 100 µm. (c) Dolomitized crinoidal packstone in which original lime mud matrix has been dolomitized (D), but crinoid grains (CR) are preserved as calcite. Scale = 500 µm. (d) Geopetal distribution of dolomite (G) in a crinoidal packstone reflecting original distribution of lime mud.

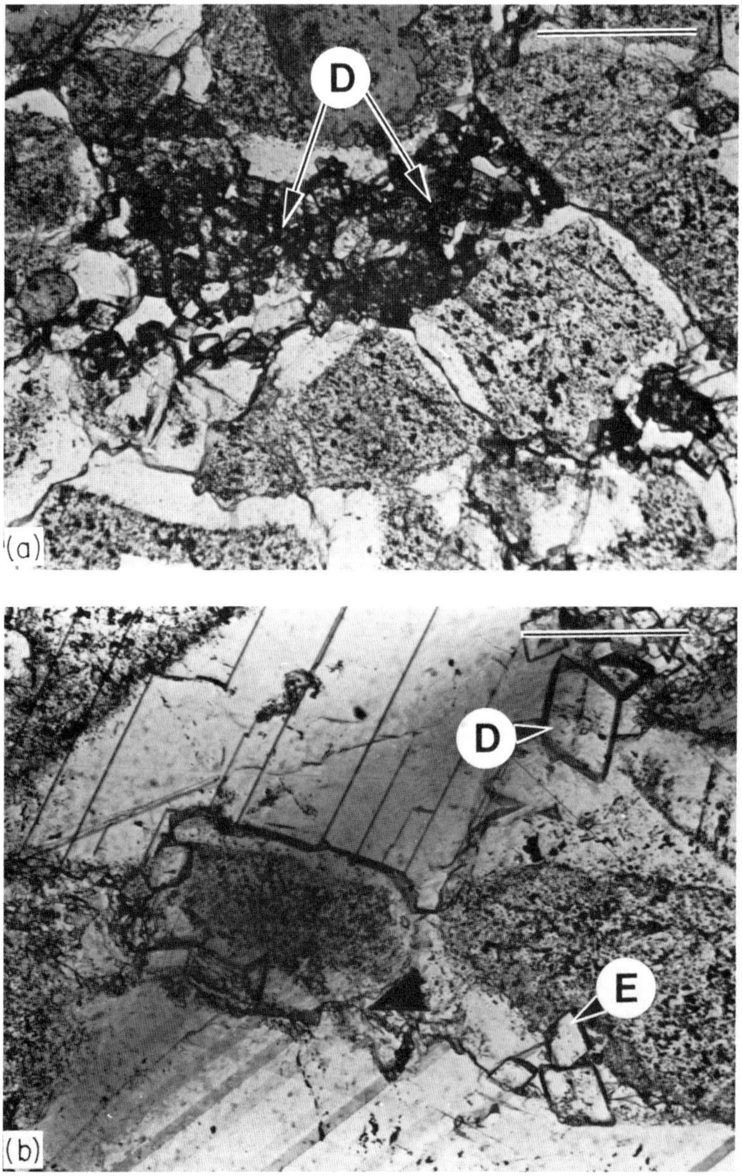

FIG. 3. Dolomite from the Burlington and Keokuk Formations. (a) Dolomite occurring as clumps of rhombs between grains (D) interpreted as a replacement of micritic skeletal or peloidal grains. Scale = 500 μm. (b) Dolomite rhombs (D) encased and isolated in calcite cements in crinoidal grainstone. Rhombs may be either cements, or replacements of small amounts of micritic calcite. E = euhedral rhomb, part of which has replaced crinoid. Scale = 250 μm.

with mean $\delta^{18}O$ of $-0.4‰$ PDB ($n = 33$, range $= -2.2$ to $+2.5‰$), and mean $\delta^{13}C$ of $+2.5‰$ PDB ($n = 33$, range $= -0.9$ to $+4.0‰$) (Banner 1986).

Dolomite II, the second most abundant generation, is characterized by unzoned red luminescence that varies from moderate red to dark red-brown (Fig. 4c, d). Dolomite II most commonly occurs as the interiors of rhombs that have dolomite I rims, in which case the contacts between the two are irregular and cut across zones within dolomite I (Fig. 4c). The geometries of the dolomite II cores are often irregular, and volumetrically range from a small percentage to

FIG. 4. Dolomite from the Burlington and Keokuk Formations under cathodoluminescence. (a) Zoned, luminescent dolomite I rhombs (I) surrounded by calcite cement. CR = crinoid. E = euhedral portions of rhomb that has replaced crinoid. Scale = 150 μm. (b) Dolostone composed of luminescently zoned dolomite I rhombs. Scale = 150 μm. (c) Unzoned dolomite II (II) occurring as replacements of cores of dolomite I rhombs. Zoned rims (I) are remnants of dolomite I. Note presence of some pure dolomite I rhombs. Scale = 150 μm. (d) Dolomite II rhomb (II) with dull red luminescence, and with brightly luminescent patches similar to dolomite I (I). Note deflection of calcite cement zones (numbered 2–4) by rhomb at arrows (X). III = thin dolomite III. Dolomite is encased in calcite cement 5 (5).

occupying nearly the entire rhomb (Fig. 4c). Dolomite II also occurs as rhombs without dolomite I rims, in which cases it commonly contains small brightly luminescing patches (Figs 4d, 5d), some of which are similar in luminescence colour to dolomite I. Geochemical characteristics of dolomite II differ markedly from those of dolomite I. Dolomite II is nearly stoichiometric (51–52 mole% $CaCO_3$), and contains 1000–2600 ppm Mn and 1600–43 000 ppm Fe, the darker luminescing varieties having the greater Fe and Mn contents (Prosky & Meyers 1985). Dolomite II contains radiogenic Sr ($^{87}Sr/^{86}Sr = \sim 0.7094$), mean $\delta^{18}O$ of $-3.8‰$ PDB ($n = 31$, range $= -0.2$ to $-6.6‰$ PDB), and mean $\delta^{13}C$ of $+2.7‰$ PDB ($n = 30$, range $= -1.0$ to $+4.1‰$ PDB) (Banner *et al.* in press, Banner 1986).

The irregular geometries of the dolomite I–dolomite II contacts could conceivably be explained by either dolomite I or dolomite II being the younger. Specifically, zoned dolomite I rims could be interpreted as syntaxial overgrowths on corroded dolomite II rhombs or, conversely, zoned dolomite I rims could be interpreted as the chemically more resistant relics after replacement of cores by dolomite II. We favour the second interpretation, that dolomite II replaced dolomite I, an interpretation first proposed by Prosky (in preparation) and Banner (1986), for the following reasons. First, some of the small brightly luminescing patches within dolomite II (Fig. 4d) are most likely relics of dolomite I. Secondly, dolomite II rhombs are about the same size as dolomite I rhombs, even where they occur in the same sample (Fig. 4c), and dolomite II dolostones have porosities comparable to those of dolomite I dolostones. The presence of rhombs composed totally of dolomite I in a dolomite II dolostone implies that dolomite I fluids moved through these rocks for the same length of time as through dolomite I dolostones. If, in the same sample, dolomite I rims were syntaxial overgrowths on corroded dolomite II cores, then these rims should be equal in thickness to about one-half of the width of the co-existing totally dolomite I rhombs. In such a case, the dolomite II plus dolomite I overgrowths should be significantly larger than the co-existing dolomite I rhombs, and the dolostones should have lower intercrystalline porosity than dolomite I dolostones. Therefore, the fact that dolostones dominated by dolomite II have about the same rhombs sizes and porosities as those dominated by dolomite I is best explained by dolomite II being a replacement. Third, dolomite I is more Ca-rich than dolomite II, and commonly the cores of dolomite I are more Ca-rich than their rims (Prosky, in preparation). Fourth, most intrarhomb pores (hollow cores and selectively dissolved zones) occur in dolomite I, and are rare in dolomite II. These features imply that dolomite I was less stable than dolomite II. Thus, there was a chemical 'motive' for the replacement of dolomite I by dolomite II, and for the common selective replacement of dolomite I cores.

The third volumetrically most important generation, dolomite III, is non-luminescent, is strongly ferroan, and is unzoned (Fig. 5a). Dolomite III is present on most dolomite I rhombs (Fig. 5a) to some degree, and occurs more rarely on dolomite II (Fig. 5d). Most commonly dolomite III comprises syntaxial rims on dolomites I or II, with the contact being either conformable (Fig. 5a) or irregular (Fig. 5a, d). The irregular contacts with dolomite I cut across the fine zoning in dolomite I, and commonly deeply embay the dolomite I rhombs (Fig. 5a). In some cases dolomite III occurs in the centres and as zones within dolomite I rhombs (Fig. 5b), in which cases their contacts are irregular and cut across the zones of dolomite I. In the latter case, both the inner and outer contacts of dolomite III zones with dolomite I host are unconformable (Fig. 5b). In some samples whole rhombs of dolomite III are present, some of which contain irregular remnants of zoned dolomite I (Fig. 5c). Geochemically, dolomite III is the most Fe- and Mn-rich, with about 52 800–89 400 ppm Fe and 2500–8200 ppm Mn, and contains about 53 mole% $CaCO_3$ (Prosky & Meyers 1985).

We interpret the above described irregular contacts as dolomite III having replaced dolomite I or II (Fig. 5a, d), and the conformable contacts as dolomite III having grown as syntaxial cements on dolomite I (Fig. 5a) or on dolomite II (Fig. 4d). The dolomite III rhombs containing relics of dolomite I and the associated dolomite III rhombs (Fig. 5c) are interpreted as dolomite III having completely replaced dolomite I rhombs.

In summary, the three major dolomite types represent progressive replacement; dolomite I was replaced by dolomite II, and was replaced and overgrown by dolomite III; dolomite II was replaced and overgrown by dolomite III (Fig. 6). The most extensively replaced dolomite, dolomite I, is the least stoichiometric of the three. Furthermore, the replacement was accomplished by progressively more iron-rich dolomites. In the above described cases where a younger dolomite generation has replaced an older dolomite, we see no definitive evidence for the older dolomites having experienced a major moulding stage. For example, we see no evidence of collapsed hollow dolomite I rhombs pre-dating replacement by either dolomite II or dolomite III.

FIG. 5. Dolomite from the Burlington and Keokuk Formations under cathodoluminescence. (a) Dolomite III (III) occurring as syntaxial rims and replacements of zoned dolomite I (I) rhombs. Dolomite rhombs are surrounded by calcite cement 5 (5). Scale = 150 μm. (b) Dolomite I (I) containing a zone selectively replaced by dolomite III (III). Note irregular inner and outer contacts of dolomite III with dolomite I. Scale = 150 μm. (c) Whole rhombs of dolomite III (III), some of which contain irregular relicts of dolomite I (I). 2, 3, 5 = calcite cement zones. P = pores. Scale = 150 μm. (d) Dolomite III (III) rims on dolomite II (II) rhombs in a skeletal mold cemented by

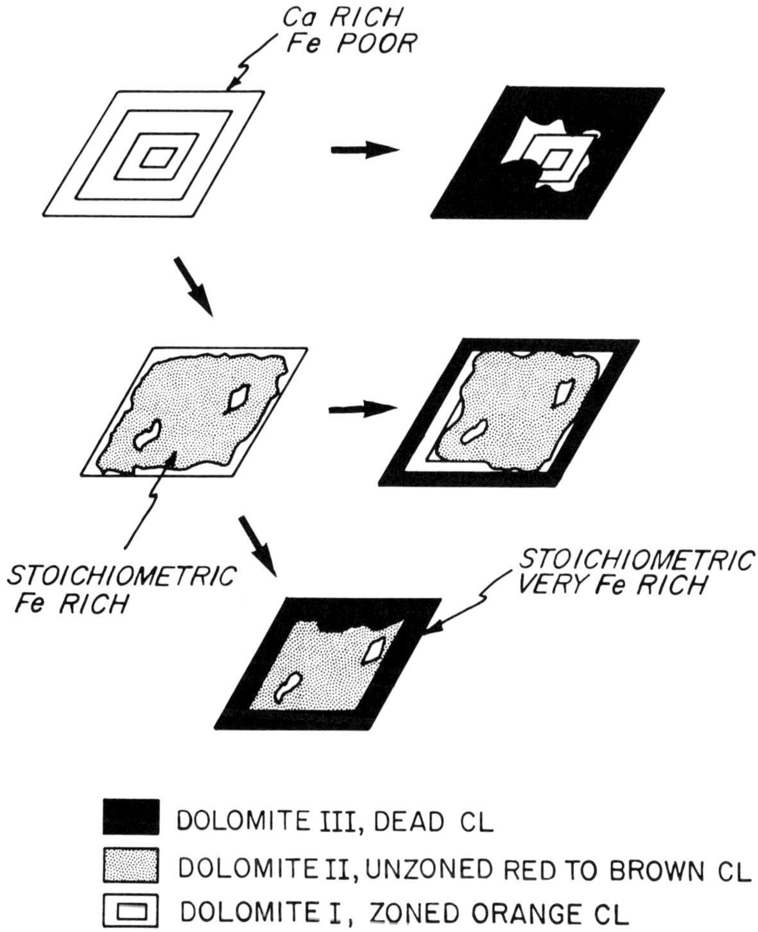

FIG. 6. Summary of Burlington–Keokuk dolomite petrography. Dolomites II and III represent progressive replacement and overgrowth of older dolomite I or II rhombs by younger, more stoichiometric and ferroan dolomite phases. Disconformities between one dolomite generation, and a younger generation may represent a discrete period of dolomite dissolution, or may have resulted directly from the replacement process.

Timing of dolomitization

Critical to constraining the possible diagenetic environments of dolomitization is determining the age of each dolomite generation relative to the diagenetic and burial history of the Burlington–Keokuk strata. This was accomplished through four approaches: (1) timing of dolomites relative to calcite cements, (2) stratigraphic distribution of dolomites, (3) timing of dolomites relative to compaction, and (4) timing of dolomites relative to chertification.

Dolomite–calcite cement relationships

Cathodoluminescence petrography has established a detailed and regionally extensive calcite cement stratigraphy in the skeletal packstones and grainstones of the Burlington–Keokuk rocks (Harris 1982, Harris & Meyers 1985, Cander 1985, Kaufman 1985, Daniels 1986, Kaufman *et al.* in press). This cement stratigraphy is essential to identifying the paragenetic sequence, including the timing of the dolomites. The cements are dominated by crinoid-syntaxial calcites that contain as many as six major zones (Fig. 7a, b, c, d). From oldest to youngest, these comprise: zone 1, non-ferroan, moderate to bright luminescence with many subzones; zone 2, non-ferroan, non-luminescent with luminescent hairline subzones; zone 3, non-ferroan, moderate luminescence; zone 4, non-ferroan, non-luminescent with luminescent hairline subzones; zone 5, ferroan, dull fairly uniform luminescence with no or few

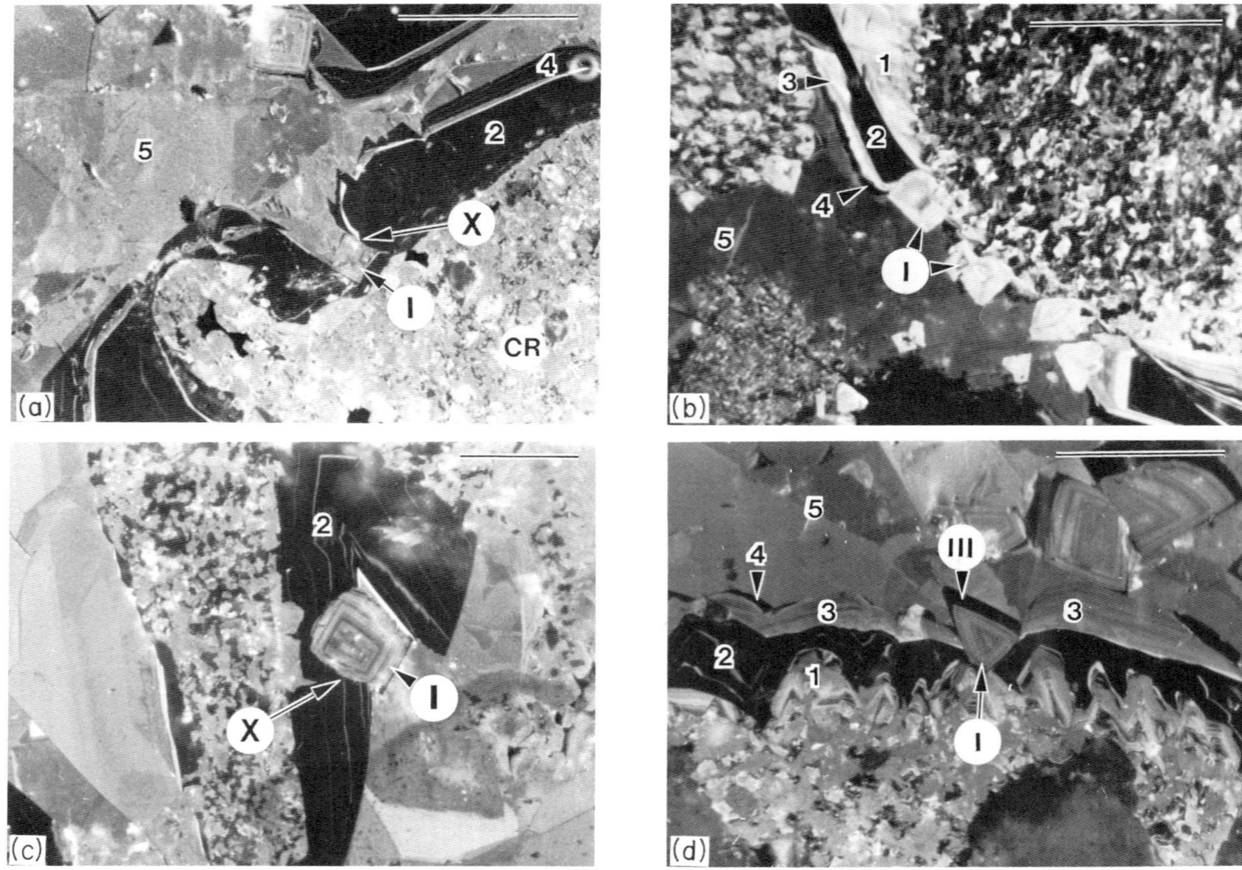

FIG. 7. Dolomite-calcite cement relationships under cathodoluminescence. (a) Calcite cement zones 2–5 occur as syntaxial cements on crinoid grain (CR). Cement zones 2, 3, and 4 are strongly deflected by (X) and pinch out adjacent to dolomite I rhomb (I). Calcite cement 5 overgrew the dolomite rhomb and re-entrant in the early cements. The interpretation is that dolomite I pre-dated calcite 2. Scale = 500 μm. (b) Dolomite I rhombs (I) occur in contact with crinoid grain, with calcite cements 1–4 pinching out adjacent to dolomite rhombs and absent on crinoid where rhombs occur. The interpretation is that dolomite I pre-dated calcite cement zone I. Scale = 150 μm. (c) Zone 2 calcite cement (2) surrounds part of a dolomite I rhomb (I) with irregular contact between the two (X). Note the sharp truncation of hairline zones in calcite cement zone 2 at the rhomb contact (X). Since calcite cements 1–4 are deflected and pinch out against dolomite I in many other cases, this relationship is interpreted as resulting from poikilitic growth of calcite zone 2 around the rhomb. Scale = 100 μm. (d) Dolomite III rims (III) on dolomite I rhombs (I) occur only on surfaces not in contact with calcite cement

broad subzones; zone 6, non-ferroan, moderate uniform luminescence distinctly brighter than zone 5.

These six zones have been interpreted as non-marine cements based on their petrography and geochemistry. They lack the characteristics of shallow warm-water marine cements, such as columnar or prismatic crystal morphologies, microdolomite inclusions, and post-cement marine internal sediment. Furthermore, they are low in Mg (mean Mg for zone 1 = 1600 ppm; zone 2 = 725 ppm; zone 3 = 635 ppm; zone 4 = 320 ppm; zone 5 = 890 ppm; zone 6 = 600 ppm; Grams, in preparation). Additionally, cement-rich rocks contain no intraclasts of grainstones even within beds that experienced slow deposition (glauconitic and fish bone horizons). Finally, the sequence of cathodoluminescent zones, probably reflecting fluctuating Eh during precipitation, is consistent over large areas and throughout relatively large stratigraphic intervals. These features are difficult to reconcile with a model of cementation by marine porewaters during progressive burial.

The six cement zones have been interpreted as being pre-St Louis Formation in age. This age determination is based on the presence of a cement stratigraphy within St Louis grain rocks that differs from that in the Burlington–Keokuk strata, and that is found syntaxially on zones 5 and 6 in a few places in Burlington–Keokuk rocks. Had the Burlington–Keokuk rocks been completely uncemented during cementation of the St Louis Formation, the St Louis cements should have been precipitated directly on Burlington–Keokuk crinoids.

Having established the calcite zonal stratigraphy, the relative age of each dolomite can be interpreted from the geometric relationships between cement zones and dolomite rhombs. For example, luminescent zones in calcite cements are commonly deflected or pinch out near a dolomite I rhomb (Fig. 7a) and, in many cases, where rhombs are in direct contact with crinoids they completely prevent the growth of cement (Figs 4d, 7b). In these cases rhombs were not covered with cement until the late zone 5 or 6 precipitated. These geometric relationships are interpreted to indicate that dolomite I deflected or blocked the growth of early cements (zones 1 to 4), and therefore pre-dated calcite cement zone 1. Early cements did not grow over the rhombs, presumably due to an inhibition of calcite nucleation on dolomite substrates. Another possible interpretation is that the dolomite I rhombs replaced grains (peloids, bryozoans) that deflected the cements, in which case dolomite I could be younger than cement zone 4. Arguing against this is the absence of undolomitized relics of the grains (Fig. 7b), and the generally excellent fit of the dolomite rhombs to the re-entrants in the early cements (Figs 4d, 7a, c).

In some instances, rhombs sharply truncate major cement zones and hairline zones (Fig. 7c), with no deflection or pinchout of zones adjacent to rhombs. These rhombs have the same luminescent signature as rhombs which block calcite growth in other areas within the same samples, and therefore the apparent truncation geometries are interpreted as dolomite pre-dating calcite cements. The apparent truncation of calcite cement zones by dolomite rhombs can be interpreted as dolomite either pre-dating, or post-dating and replacing cement. On the other hand, the deflection of cement zones by dolomite rhombs is readily explained by dolomite pre-dating cement, and is difficult to explain by dolomite post-dating cement. The irregular contact between dolomite I and non-luminescent cement in Fig. 7(c) is best interpreted as the cement peripherally replacing dolomite I. If dolomite I had replaced the non-luminescent cement, the dolomite rhomb would be likely to be euhedral where it embays the calcite crystal, as is the usual geometry where dolomite replaces single-crystal crinoids (Figs 3b, 4a).

The geometric relationships between dolomite II and the calcite cement zones are the same as those between dolomite I and the cements (Fig. 4d). However, because dolomite II is a replacement of dolomite I, we cannot definitively establish its age relative to the early cements (zones 1 to 4). Dolomite II commonly is encased in calcite cement zone 5 in grain rocks (Fig. 4d) and in dolostones, and was rarely replaced by calcite cement 5. These relationships date dolomite II as older than calcite cement 5. Although we favour dolomite II being older than cement zone 4, we cannot rule out dolomite II having replaced dolomite I after cement zone 4 and before cement zone 5.

Several lines of evidence indicate a post-zone 4, pre-zone 5 age for dolomite III. Dolomite I rhombs partially encased in early calcite cements have dolomite III rims developed *only* on surfaces not in contact with cement zones 1 to 4 (Fig. 7d). These dolomite III rims are surrounded by cement zones 5 or 6 (Fig. 7d). The dolomite III rims thus precipitated before zone 5 cements, otherwise the necessary porosity would not have been available. There is no evidence to suggest that the dolomite III rims are a mould-fill or a calcite cement replacement after zone 5 cementation. Dolomite III is often encased in calcite 5 where it replaced and overgrew dolomite I and II in grain rocks (Figs 4d, 5a, c) and in dolostones (Fig. 5d). Dolomite III also post-dates the

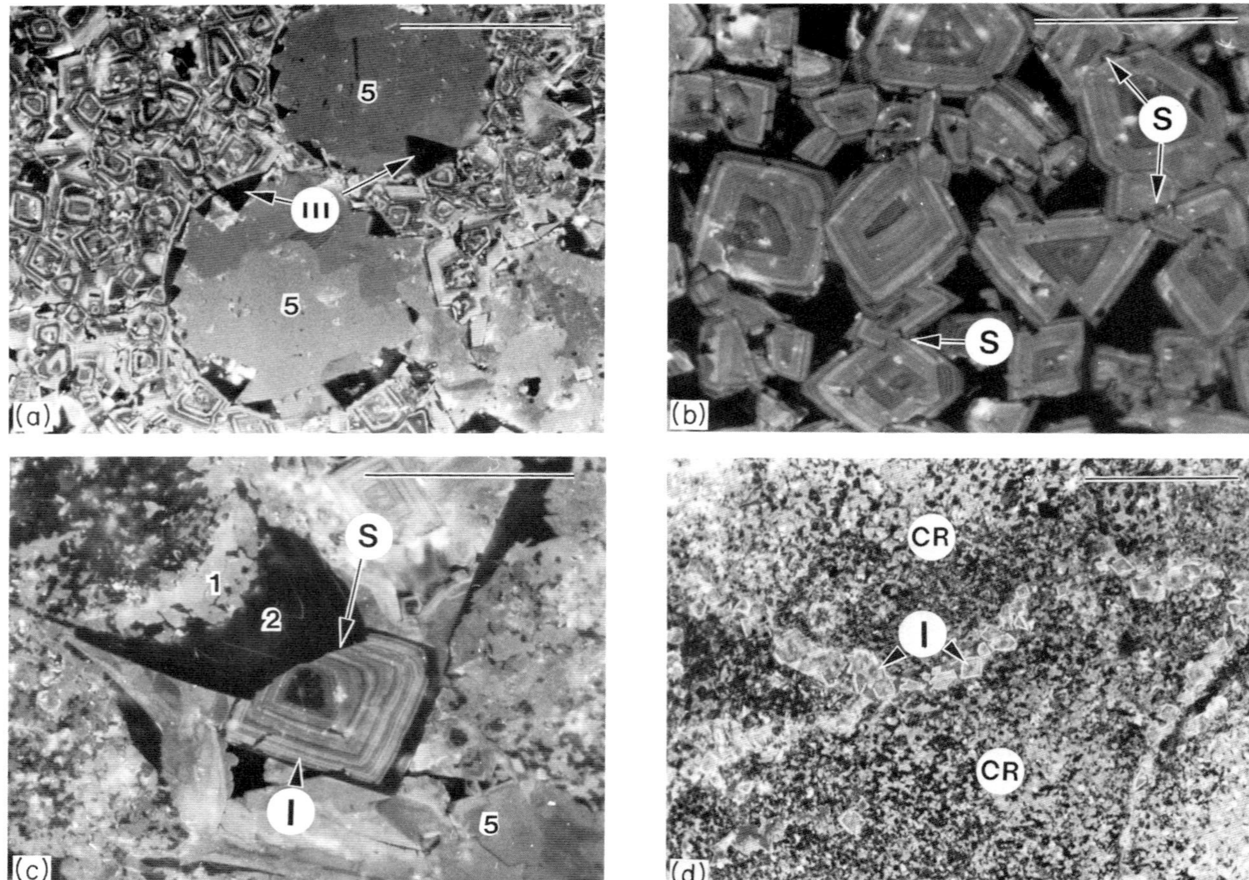

FIG. 8. Cathodoluminescence petrography of dolomites. (a) Dolomite III occurs as syntaxial rims (III) on dolomite I rhombs bordering skeletal molds. Dolomite III post-dates mould formation and pre-dates calcite cement 5 (5). Scale = 150 μm. (b) sutured contacts between dolomite I rhombs (S) resulting from pressure solution. Scale = 150 μm. (c) Irregular contact (S) between dolomite I rhomb (I) and calcite cement zone 2 (2). Calcite zone 2 embays rhomb. This is interpreted as pressure solution between rhomb and cement after zone 2 cement. Scale = 150 μm. (d) Dolomite I rhombs (I) occur along pressure solution contacts between crinoids (CR) in chemically compacted [illegible] having formed before compaction. Scale = 500 μm

formation of mouldic porosity in mud-supported dolomite textures, where it forms overgrowths on dolomite I rhombs which border crinoid(?) moulds (Fig. 8a). In these cases, dissolution of the fossil preceded the growth of dolomite III rims, and later zone 5 cements filled the mould.

Stratigraphic distribution of dolomite

Since all calcite cements in the Burlington–Keokuk are interpreted to have formed before mid-Meramecian St Louis time, the three dolomite generations are also interpreted to have formed before this time. Dolomite from the Spergen Formation at Coppock Quarry (CP in Fig. 1), which immediately underlies the pre-St Louis unconformity, contains rhombs with luminescence identical to dolomite II in the Burlington–Keokuk rocks. St Louis and St Genevieve lime mudstones above the unconformity in the same quarry are undolomitized. Their lack of dolomite when compared to totally dolomitized muds in the Burlington–Spergen sequence implies that dolomitization of the latter formations occurred before St Louis time.

Dolomite-compaction relationships

The above observations imply that dolomitization was one of the earliest major diagenetic events affecting Burlington–Keokuk rocks. If this is correct then the dolomites would be expected to pre-date some of the compaction processes. The relationship between mechanical compaction and dolomites is ambiguous, but commonly dolomite was involved in chemical compaction, as shown by sutured contacts between rhombs of dolomite I (Fig. 8b) and between rhombs of dolomite II. We have not observed sutured contacts between dolomite III rhombs. Pressure-solution contacts also occur between dolomite I rhombs and calcite cements (Fig. 8c). In the latter, the calcite cement has embayed the rhomb. The exact correspondence of zone 2 with the embayed part of the dolomite rhomb makes a growth relationship interpretation highly unlikely, and dates the pressure solution as post-zone 2.

Pressure solution also occurred between rhombs and crinoid grains. Compaction of dolomitized packstones has resulted in fitted, pressure-solved contacts between crinoids, with dolomite rhombs caught between grains (Fig. 8d). In these cases, dolomite rhombs embay the adjacent crinoids along pressure solution contacts, and involve both dolomite I (Fig. 8d) and dolomite II, but never dolomite III. Dolomitization of lime mud after chemical compaction of crinoids is ruled out since lime mud would have been removed by pressure solution between grains.

Stylolites always truncate dolomite I and II rhombs, but at least some are older than dolomite III. Evidence for this is the presence of dolomite III rims on truncated surfaces of dolomite I (or II) rhombs at stylolites (Fig. 9a).

The above features indicate that dolomite I and II rhombs have undergone inter-rhomb pressure solution, and were therefore formed early in the burial history of the Burlington–Keokuk strata. This is consistent with the interpretation that they were one of the earliest diagenetic events. Dolomite III post-dates intergranular pressure solution and at least some stylolitization.

Dolomite-chert timing

Timing of chertification has been interpreted from the petrographic relationships between cherts and calcite cements in grainstone and packstone beds. Most commonly, chertification followed incipient calcite cementation, as indicated by chalcedony coating and pseudomorphing dog-tooth calcite crystals on crinoids. These calcite cements are as old as zone 1; rarely, outside chert nodules but in close proximity to them, chalcedony coats cements as young as zone 4. Fractures that cut chert nodules are filled with cement zone 5. Rarely, chert nodules show no evidence of cementation before chertification. Therefore, chertification began before calcite cementation in a few cases, but in most nodules began after incipient cementation. Chertification occurred as late as zone 4 calcite cementation outside of nodules, but in no instance post-dates zone 5. Therefore, any dolomite that pre-dates chertification must pre-date zone 5.

With regard to dolomite–chert relationships, most information comes from mud-supported textures. Within chert nodules, a complete spectrum from undolomitized to totally dolomitized textures are seen frozen by microquartz replacement. Least common are nodules that contain no evidence of dolomite rhombs in microquartz. These samples are from the oldest part of the largest chert nodules and represent lime mud that was replaced by microquartz before any dolomitization. Most commonly chert contains scattered dolomite rhombs floating in microquartz. In most cases rhombs are partially preserved (Fig. 9b), but others are ghosts, having been nearly totally replaced by microquartz; the only dolomite remaining is minute dolomite inclusions (Fig. 9c) which show up under cathodoluminescence. In some dolostones micro-

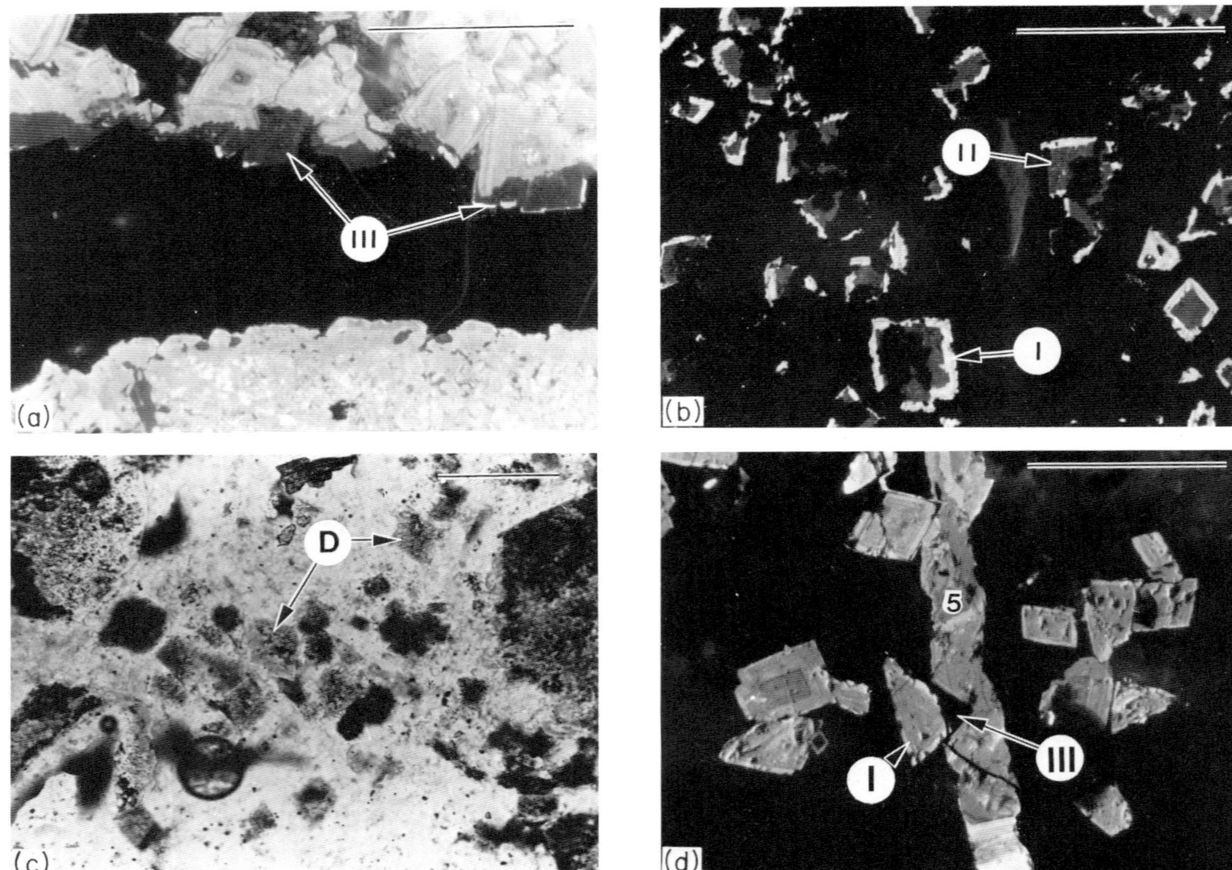

FIG. 9. Dolomite-compaction and dolomite-chert relationships under cathodoluminescence. (a) Dolomite I rhombs are truncated at stylolite (now slightly opened), and are overgrown by dolomite III (III). Dolomite III therefore post-dates the stylolite. Scale = 150 μm. (b) Rhombs composed of dolomite I rims (I) and dolomite II (II) cores are irregularly embayed by microquartz chert (non-luminescent). The interpretation is that dolomites I and II were partially replaced by and therefore pre-date the chert. Scale = 150 μm. (c) Relict inclusions of dolomite in microquartz chert define ghosts of former rhombs (D). Transmitted plane polarized light. Scale = 250 μm. (d) Fracture in chert nodule truncates dolomite I rhombs (I), which are syntaxially overgrown with dolomite III (III). Calcite cement zone 5 (5) fills fracture. Non-luminescent portions are microquartz chert. These relationships indicate that dolomite III post-dates chert and pre-dates

quartz chert filled inter-rhomb pores and replaced some rhombs, indicating that chertification followed complete dolomitization.

The fact that dolomite rhombs occur in cherts does not in itself prove that dolomitization preceded chertification, and could be interpreted conversely. The presence of rhombs pseudomorphically replaced by microquartz (Fig. 9c) does, however, prove that chertification was the later event. Under cathodoluminescence, rhombs in microquartz matrix are irregular, have corroded edges, and are deeply embayed, the microquartz-dolomite contacts cutting across zoning (Fig. 9b). These geometries are best explained as chert having partially replaced dolomite rhombs. Note that this chertification of rhombs involved both dolomite I and II, but never dolomite III.

Dolomite III occurs in cherts only in fractures, where it has syntaxially overgrown fracture-truncated dolomite I and II rhombs (Fig. 9d). These fractures are filled with zone 5 calcite cement, which encases the dolomite III syntaxial overgrowths (Fig. 9d). Dolomite III is not seen on rhombs encased in chert, and because it forms on rhombs truncated by chert fractures, it is interpreted as being a post-chertification event. This timing correlates well with dolomite III–calcite cement relationships since it has been shown that dolomite III post-dates zone 4 cement.

In summary, chert–dolomite relationships show that dolomite I and II formed before most chertification and therefore before zone 5 calcites, a timing relationship that agrees with that determined from the dolomite-cement zones relationships. This also implies that some chertification of lime muds began before dolomitization, and argues against dolomitization beginning on the seafloor.

Diagenetic sequence and summary of timing of dolomitization

The diagenetic sequence of the Burlington–Keokuk carbonates is summarized here in order to place dolomitization into the overall diagenetic history, and thus to constrain the possible environments and models for dolomitization. The diagenetic sequence is summarized in Figs 10, 11 and 12. Figure 10 shows the relative timing of each diagenetic event; the length of events and the time between events is arbitrary. Not all events can be precisely fitted into the sequence, and their placement is marked by question marks. Figures 11 and 12 are diagrams summarizing diagenetic sequences in grain-supported and mud-supported rocks respectively.

Based on the foregoing evidence, dolomitization was one of the earliest diagenetic events to affect the Burlington–Keokuk sediments, preceded only by some chertification. Most dolomitization (I, II) occurred before all of the major calcite cementation, and before most of the chertification. Its early age is consistent with dolomite being involved in inter-rhomb pressure solution. Based on our sampling of younger Mississippian strata, the stratigraphic distribution of dolomite and post-dolomite cements implies a pre-St Louis age for the dolomites. Other features that are consistent with this

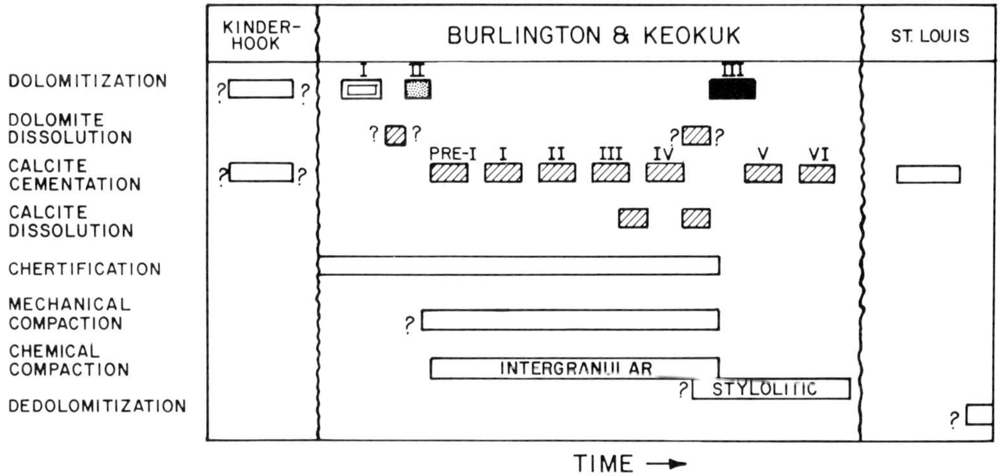

FIG. 10. Sequence of diagenetic events for Burlington and Keokuk rocks in the study area. The relative timing of each event is shown, and uncertain relationships are indicated by question marks. The length of events and the time between events are arbitrary.

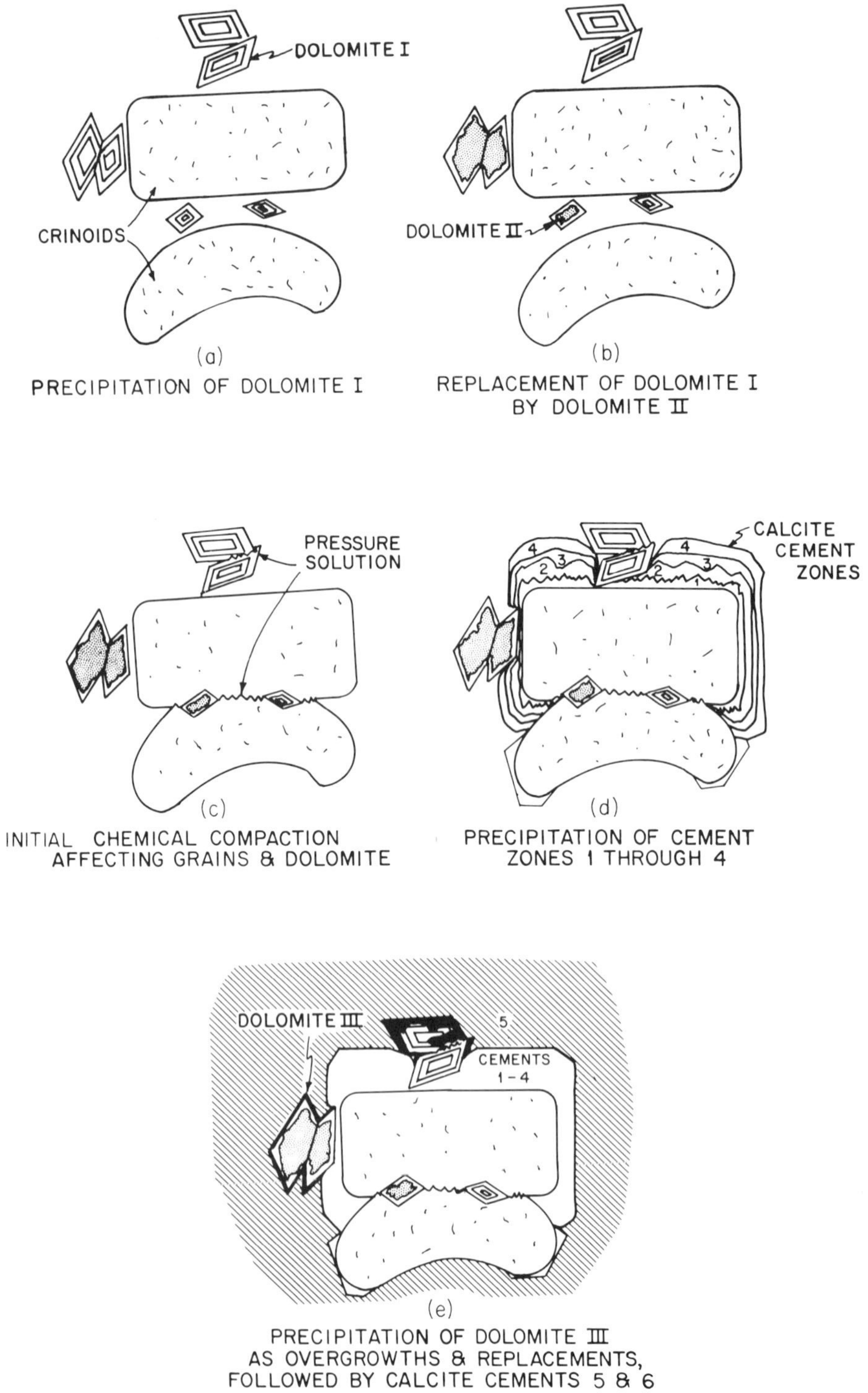

FIG. 11. Summary of diagenetic sequence in crinoidal grainstones. In (a) and (b), dolomite rhombs may either be cements, or may have replaced microcrystalline components such as small peloid or bryozoan grains, or in some cases small amounts of lime mud.

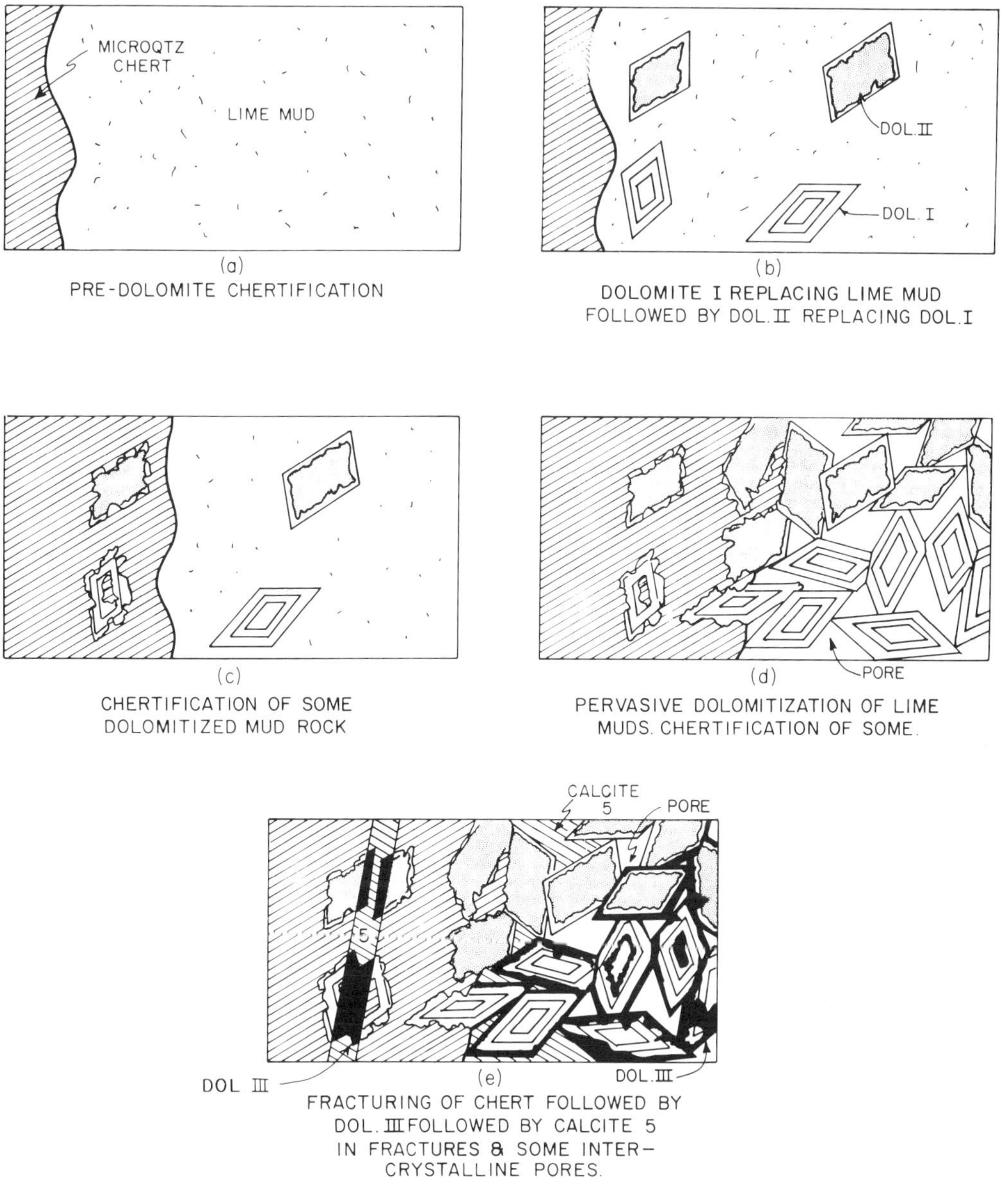

FIG. 12. Summary of diagenetic sequence in mud-supported rocks. In (d) and (e), undolomitized lime mud has been removed by dissolution during or after initially more pervasive dolomitization. It should be noted that some beds remained completely composed of dolomite I, while others were nearly entirely replaced by dolomite II before chertification.

Mississippian age for cherts and dolomites are the presence of chert and dolomite clasts in pre-Pennsylvanian palaeo-karst in the Burlington–Keokuk rocks of central Missouri (Daniels 1986). Petrography of calcite cements associated with these palaeokarst features has led Daniels (1986) and Kaufman *et al.* (in press) to interpret cement zones 1 to 4 as pre-karst and zones 5 and 6 as post-karst, which of course differs from the timing of zones 5 and 6 proposed here.

If our pre-St Louis age is correct it constrains burial depths to less than 100 m during dolomitization of the Burlington–Keokuk strata and to less than about 30 m during dolomitization of the Spergen sediments. If the dolomites were post-Spergen but pre-Pennsylvanian, it would imply burial depths less than about 500 m, based on the thickness of Meramecian and Chesterian strata from central western Illinois (Willman *et al.* 1975).

Models for dolomitization

Accepting the pre-St Louis age for the dolomitization, we rule out deep burial models involving elevated burial temperatures. Similarly, the absence of peritidal facies within the Burlington–Keokuk rules out synsedimentary tidalite models. Such models are particularly difficult to invoke considering that proposed Burlington-equivalent shoreline facies of the Lower Humboldt beds in north-central Iowa are largely non-dolomitic (Sixt 1983). We consider below four other models: meteoric-marine mixing zone waters, post-depositional hypersaline waters, marine or modified marine waters, and deep subsurface brines that migrated into shallow burial settings. For background on these and other dolomite models, the reader should see the reviews by Morrow (1982) and Machel & Mountjoy (1986).

Mixing-zone model

Several facts support a mixing-zone origin for Burlington–Keokuk dolomites. Most apparent is the dolomitization of subtidal, open-marine sediments which contain no supratidal, hypersaline indicators, and which is regional in extent. Also consistent with this model is the early timing of dolomitization. Dolomite probably formed after some burial as implied by some of the chert pre-dating dolomite I. This is based on analogy with deep sea cherts, which do not form on the seafloor, but begin forming only after some shallow burial (von Rad & Rosch 1974, Wise & Weaver 1974). On the other hand, dolomites (at least dolomite I) formed before calcite cementation. This in turn suggests that dolomitization occurred before establishment of a regional fresh groundwater system, since the calcite cements are interpreted as freshwater precipitates (Harris 1982). During the subaerial exposure that resulted in the calcite cements, a mixing-zone would have formed during flushing of the sediments by fresh water, and this brackish water would have swept through the sediments ahead of the freshwater lens. Thus, this mixed water would have been the first non-marine water to permeate the sediments. The fact that there was no calcite cementation during precipitation of dolomite I and II is consistent with this, as is the absence of dolomitized fossils (they are preserved as moulds). Significant freshwater systems could have been established with only modest relative sea-level drops. Water table elevations of only about 2.5 m could have generated a freshwater lens 100 m thick (based on the Ghyben–Herzberg relation), thick enough to extend stratigraphically through the Burlington. During late Mississippian, the proposed fresh waters conceivably could have had recharge areas in the La Salle Arch, or the Transcontinental Arch. Considering that shallow water facies are common in the Spergen and younger strata, there could also have been short-lived emergences during their deposition that have not been clearly recognized, and that could have resulted in flushing of the underlying formations with fresh water.

Although the mixing-zone model has many attractive aspects, it does not explain the three different generations of dolomite. It is not at all clear why the three chemically distinct dolomites I, II and III should all form in the same general diagenetic environment, nor why they should show a progressive increase in iron. The C, O and Sr isotopic compositions of dolomite I is consistent with a mixing-zone model, but the depleted O and radiogenic Sr isotopic compositions of dolomite II are difficult to reconcile with this model. It is possible that each episode of dolomitization represents a separate sea-level event, separated by periods of sedimentation. Thus, the younger the dolomite, the greater the burial depth during mixing zone dolomitization.

Another problem with invoking the mixing-zone hypothesis is the presence of two-phase fluid inclusions in the dolomites and calcite cements. These have been studied by Smith (1984) and Smith *et al.* (1984) who interpreted them as primary inclusions. Homogenization temperatures are in the range of 90–100°C, and final melting temperatures of frozen inclusions indicate bulk salinities in the range of 100 000–200 000 ppm. If these are truly pristine primary inclusions, they clearly rule out a simple seawater–freshwater-mixing-zone.

Hypersaline reflux model

Another possible model invokes the 'refluxing' of brines from the St Louis or other post-Keokuk Mississippian units. The St Louis Limestone can be eliminated as a likely source of brines, even though it contains solution collapse breccias which are interpreted as former evaporites. Since Burlington–Keokuk dolomite is restricted to pre-St Louis formations, and St Louis rocks are undolomitized in the study area, this model would imply that dolomitizing fluids were formed during or immediately after Warsaw or Spergen time. We have seen no evidence of evaporitic or restricted marine depositional conditions in the Warsaw. On the contrary, the Warsaw contains open marine fossils. We cannot rule out evaporitic facies having been present in the Spergen Formation. Although we have not seen evidence of evaporites in the Spergen, we have not studied the unit in detail, and it is possible that evaporites existed in upper Spergen rocks and were removed by pre-St Louis erosion. A check for 'occult' gypsum or anhydrite in Burlington and Keokuk dolomites, using the methods of Beales & Hardy (1980), found no evidence of gypsum or anhydrite inclusions. If indeed refluxing hypersaline brines were the main dolomitizing agent then we would expect to see some gypsum or anhydrite moulds or calcite pseudomorphs in the pervasively dolomitized mud rocks of the Burlington and Keokuk. An argument for a brine reflux model involves the interpretation of Chowns & Elkins (1974) of geodes from Warsaw and upper Keokuk strata as replacements of anhydrite nodules. The strongest evidence they present for this is their recognition of abundant length-slow chalcedony (a common replacement of evaporites; Folk & Pittman 1971), and relict anhydrite inclusions in quartz crystals in the geodes.

The hypersaline model is consistent with the early timing of the dolomites, and with the high salinity fluid inclusions. The luminescent zoning of dolomite I is probably compatible with precipitation from proposed refluxing and down-soaking brines which could have had low enough Eh's to incorporate Mn and Fe. As with the mixing-zone model, dolomites II and III are not readily explained by the reflux model. It is not clear what might have led to the multiple replacements. Furthermore, the high Fe contents of dolomite II and III argue against their precipitation in high sulphate, low Eh brines. Had this been the situation, most of the Fe probably would have been tied up in sulphides (Frank *et al.* 1982). Additionally, the wide range in O isotope values for dolomite I, some of them depleted in ^{18}O, are difficult to reconcile with hypersaline evaporitic waters. Similarly, the radiogenic Sr in dolomite II is difficult to reconcile with the reflux model, since the bulk of the dissolved Sr in a refluxing brine would likely be from Mississippian sea water.

Marine or modified marine waters

The early pre-cement and pre-compaction age for dolomites I and II, and the Osagean 'marine' Sr isotopes of dolomite I are consistent with dolomitization on or near the seafloor. On the other hand, the presence of some chertification of lime muds before dolomitization suggests that dolomitization took place under some shallow burial, assuming analogy with extant deep sea nodular cherts (e.g. von Rad & Rosch 1974). If this is true, then dolomite I could have formed under a few metres of sediment in sea water of normal salinity. Marine porewaters in lime muds tend to be reducing due to bacterial decay of organic matter, and likewise tend to involve bacterial reduction of sulphate with attendant production of Fe sulphide. Baker & Kastner (1981) have suggested that the removal of sulphate catalyses dolomite formation. This chemical setting would conceivably result in Fe-poor, Mn-bearing dolomites such as dolomite I. It would also involve bacterial decay of organic matter and therefore generate a wide range of $\delta^{13}C$ values, including some very light values (Irwin *et al.* 1977, Hudson 1977). We do not see light C isotopic signatures for either dolomite I or II. Furthermore, the wide range of O isotope values for dolomites I and II, many depleted in ^{18}O, cannot be explained by a seawater model at shallow burial, nor can the radiogenic Sr in dolomites II. Another problem with a simple marine model is the replacement of dolomite I by dolomites II and III. A marine model would imply rather uniform porewater chemistry, yet these dolomites probably precipitated from waters chemically different from one another.

A model invoking marine waters with slightly elevated salinities refluxing from younger formations, such as proposed by Simms (1984), would be consistent with the early timing of dolomites, but not with the ranges of O isotopes for dolomite I and for dolomite II, nor with the radiogenic Sr in dolomite II.

Basinal brine model

The recognition of fluid inclusions containing highly saline brine and having elevated homogenization temperatures raises the possibility that extraneous hot brines moved through the Burlington–Keokuk sediments and dolomitized them (Smith 1984). Possible sources for these are subsurface brines expelled from nearby basinal areas, such as the Illinois Basin, by compaction, gravity, or tectonic related processes (Cathels & Smith 1982, Leach *et al.* 1984, Bethke 1986). These brines were possibly similar to and genetically related to brines thought to have formed Mississippi Valley type Pb-Zn deposits (review by Sverjensky 1984). In support of this idea is the widespread occurrence of vugs in Burlington, Keokuk and Warsaw strata, many of which contain sphalerite, quartz, coarse calcite, pyrite and, more rarely, kaolinite and saddle dolomites, minerals that are common in Mississippi Valley-type ore deposits. In Iowa we do not know the precise age relationships between these vug minerals and the Burlington–Keokuk dolomites and calcite cements, however, work in Illinois and Missouri has shown these vug-filling phases to cut cements as young as calcite 5 (Cander 1985, Kaufman 1985, Kaufman *et al.* in press). This young age is consistent with the occurrence of sphalerites in lower Pennsylvanian strata in the Missouri–Iowa–Illinois region (Leach 1973) and with the late Pennsylvanian–early Permian age proposed for the Mississippi Valley-type ores (Wu & Beales 1981).

Considering the above data, this model would require an early phase of brine expulsion, before most of the mineralization, and would require that high temperature fluids moved over long distances at shallow burial depths (less than 100 m). The marine Sr isotope and relatively heavy O isotope signatures of dolomite I are not consistent with this model. On the other hand, the radiogenic Sr and ^{18}O-depleted isotope signatures of dolomite II are consistent with a brine model. Similarly, the increasing Fe contents of the younger dolomites (II and III) might be explained by pulses of progressively more Fe-rich brines. With regard to the fluid inclusions, two-phase fluid inclusions similar to those in the dolomites are found in all the calcite cement zones, and the two-phase fluid inclusions in dolomite I are similar to those in dolomite III (Smith 1984, Smith *et al.* 1984). Considering the differences in timing, cathodoluminescence and geochemistry between many of these phases, it raises the possibility that the fluids in the inclusions, and their homogenization temperatures are secondary. In summary, fluid inclusions provide evidence that high temperature saline brines moved through the Burlington–Keokuk sediments, but it is still an open question whether these are samples of the same fluids that formed the calcite cements and dolomites.

Conclusions

(1) The distribution of dolomite in the Burlington and Keokuk limestones is controlled primarily by distribution of lime mud-rich depositional textures. Lime mud is pervasively replaced throughout the study area.

(2) Three regionally extensive generations of dolomite can be recognized by their cathodoluminescent characteristics. Dolomite I, the oldest generation, is luminescent, usually thinly zoned, and was responsible for the replacement of most of the lime mud. Dolomite II has a dull red luminescence, is unzoned, and occurs mainly as a replacement of dolomite I rhombs. Dolomite III is non-luminescent, and occurs as a syntaxial cement on, and replacement of dolomites I and II.

(3) Dolomite petrography indicates that dolomite I and probably dolomite II pre-date all calcite cements. Luminescent zoning within calcite cement crystals is deflected and inhibited by the presence of these rhombs. Dolomite I and II pre-date most chert, intergranular chemical compaction, and stylolites. Dolomite III post-dates cement zone 4 and pre-dates zone 5. It formed after all chert, after the formation of moldic porosity in dolomite I, and after some stylolites. The stratigraphic restriction of dolomite to pre-St Louis Formation units suggests that dolomitization occurred before or during the pre-St Louis unconformity.

(4) A single dolomitization model cannot reasonably explain all three generations of Burlington–Keokuk Formation dolomites. Timing constraints coupled with petrographic and geochemical characteristics suggest that dolomite I formed in a sea water–freshwater mixing-zone associated with a meteoric groundwater system established beneath the pre-St Louis unconformity. Dolomite II and III, on the other hand, may have formed from externally sourced brines at moderate temperatures. These proposed brines became more Fe-rich with decreasing age, and the required Mg was derived mainly from the precursor dolomite that was replaced.

ACKNOWLEDGMENTS: This paper is derived from a MS thesis by the senior author while a graduate student at the State University of New York at Stony Brook. Financial support from Amoco Production Company, Research Center, Tulsa, for fieldwork and thin sections is gratefully acknowledged. We would like to thank Brian F. Glenister, R. C. Hager and the Iowa Geological Survey for their help during fieldwork. Raid Brothers Quarries (Burlington, Iowa) and Kaser Construction Company (Des Moines, Iowa) granted permission to work in their quarries. Special thanks go to A. C. Kendall and W. J. Meyers for their initial conception of the Burlington project, and their continued support and encouragement. We thank John Miller, Jim Marshall, and an anonymous reviewer for the thoughtful and helpful suggestions for improving the manuscript.

This paper does not necessarily represent the views of the Standard Oil Production Company.

References

BAKER, P. & KASTNER, M. 1981. Constraints on dolomite formation. *Science*, **213**, 214–6.

BANNER, J. L. 1986. Geochemical constraints on the origin of Burlington–Keokuk dolomites. *PhD Thesis*. State University of New York at Stony Brook.

——, HANSON, G. N. & MEYERS, W. J. in press. Determination of initial Sr isotope compositions of dolostones from the Burlington–Keokuk Fms. of Iowa, Illinois and Missouri: constraints from cathodoluminescence and glauconite paragenesis. *Journal of Sedimentary Petrology*.

BEALES, F. W. & HARDY, J. W. 1980. Criteria for the recognition of diverse dolomite types with an emphasis on studies of host rocks for MVT ore deposits. *In*: ZENGER, D. H., DUNHAM, J. B. & ETHINGTON, R. L. (eds). *Concepts and Models of Dolomitization*, pp. 197–213. Society of Economic Paleontologists and Mineralogists Special Publication, 28.

BETHKE, C. M. 1986. Hydrologic constraints on the genesis of the Upper Mississippi Valley mineral district from Illinois Basin brines. *Economic Geology*, **81**, 233–49.

CANDER, H. S. 1985. Diagenetic history of the Burlington–Keokuk Limestone, Illinois. *MS Thesis*. State University of New York at Stony Brook.

CARLSON, M. P. 1979. The Nebraska–Iowa region. *In*: CRAIG, L. C. & VARNES, K. L. (eds). *Paleotectonic Investigations of the Mississippian System in the United States,—Part I: Introduction and regional Analyses of the Mississippian System*, pp. 107–14. United States Geological Survey Professional Paper, 1010-F.

CATHLES, L. M. & SMITH, A. T. 1983. Thermal constraints on the formation of Mississippi Valley-type lead-zinc deposits and their implications for episodic basin dewatering and deposit genesis. *Economic Geology*, **78**, 983–1002.

CHOWNS, T. M. & ELKINS, J. E. 1974. The origin of quartz geodes and cauliflower cherts through the silicification of anhydrite nodules. *Journal of Sedimentary Petrology*, **44**, 885–903.

DANIELS, L. D. 1986. Diagenesis and paleokarst of the Burlington–Keokuk Formation (Mississippian), Missouri. *MS Thesis*, State University of New York at Stony Brook.

FOLK, R. L. & PITTMAN, J. S. 1971. Length-slow chalcedony: a new testament for vanished evaporites. *Journal of Sedimentary Petrology*, **41**, 1045–58.

FRANK, J. R., CARPENTER, A. B. & OGLESBY, T. W. 1982. Cathodoluminescence and composition of calcite cement in the Taum Sauk Limestone (Upper Cambrian), southeast Missouri. *Journal of Sedimentary Petrology*, **52**, 631–8.

GRAMS, J. in preparation. Trace element geochemistry of calcite cements in the Burlington–Keokuk Formations, Iowa and Illinois. *MS Thesis*. State University of New York at Stony Brook.

HARRIS, D. C. 1982. Carbonate cement stratigraphy and diagenesis of the Burlington Limestone (Miss.), S.E. Iowa, W. Illinois. *MS Thesis*, State University of New York at Stony Brook.

—— & MEYERS, W. J. 1985. Carbonate cement stratigraphy of Burlington Limestone (Osagean) of Iowa: evidence for Eh gradients in a regional Mississippian paleo-groundwater system. *Bulletin of the American Association of Petroleum Geologists*, **69**, 263 (Abstract).

HARRIS, S. E. & PARKER, M. C. 1964. Stratigraphy of the Osage Series in southeastern Iowa. *Iowa Geological Survey Report of Investigations I*.

HUDSON, J. 1977. Stable isotopes and limestone lithification. *Journal of the Geological Society of London*, **133**, 637–60.

IRWIN, H., CURTIS, C. & COLEMAN, M. 1977. Isotopic evidence for source of diagenetic carbonate formed during burial of organic rich sediments. *Nature*, **269**, 209–13.

KAUFMAN, J. 1985. Diagenesis of the Burlington–Keokuk Limestones (Mississippian), eastern Missouri. *MS Thesis*. State University of New York at Stony Brook.

——, CANDER, H. S., DANIELS, L. D. & MEYERS, W. J. in press. Calcite cement stratigraphy and cementation history of the Burlington–Keokuk Formation (Mississippian), Illinois and Missouri. *Journal of Sedimentary Petrology*.

LANE, H. R. 1978. The Burlington Shelf (Miss.) north-central United States. *Geologica et Palaeontologica*, **12**, 165–76.

—— & DE KEYSER, T. L. 1980. Paleogeography of the late early Miss. (Tournaisian 3) in the central and southwestern U.S. *In*: FOUCH, T. D. & MAGATHAN, E. R. (eds). *Paleozoic Paleogeography of West-Central United States, West-Central U.S. Paleogeography Symposium I*, pp. 149–62. Rocky Mountain Section of the Society of Economic Paleontologists and Mineralogists, Denver.

LEACH, D. L. 1973. A study of the barite-lead-zinc deposits of central Missouri and related mineral deposits in the Ozark region. *PhD Thesis*. University of Missouri, Columbia.

——, VIETS, J. G. & ROWAN, L. 1984. Appalachian–Ouachita orogeny and Mississippi Valley-type lead-zinc deposits. *Geological Society of America Program with Abstracts, 97th Annual Meeting*, p. 572 (Abstract).

MACHEL, H. G. & MOUNTJOY, E. W. 1986. Chemistry and environments of dolomitization—a reappraisal. *Earth-Science Reviews*, **23**, 175–222.

MORROW, D. W. 1982. Diagenesis II. Dolomite—part II: Dolomitization models and ancient dolostones. *Geoscience Canada*, **9**, 95–107.

PROSKY, J. L. in preparation. Trace element geochemistry of regional dolomites in the Burlington–Keokuk Formations (Mississippian) of Iowa and Illinois. *MS Thesis*. State University of New York at Stony Brook.

—— & MEYERS, W. J. 1985. Nonstoichiometry and trace element geochemistry of the Burlington–Keokuk dolomites. *Society of Economic Paleontologists and Mineralogists Annual Midyear Meeting Abstracts*, p. 73 (Abstract).

REEDER, R. J. 1981. Electron optical investigation of sedimentary dolomites. *Contributions to Mineralogy and Petrology*, **76**, 148–57.

SALLER, A. H. 1984. Petrologic and geochemical constraints on the origin of subsurface dolomite, Enewetak Atoll: an example of dolomitization by normal seawater. *Geology*, **12**, 217–20.

SIMMS, M. 1984. Dolomitization by groundwater–flow systems in carbonate platforms. *Transactions of the Gulf Coast Association of Geological Societies* **XXXIV**, 411–20.

SIXT, S. 1983. Depositional environments, diagenesis and stratigraphy of the Gilmore City Fm. (Miss.) near Humboldt, north central Iowa. *MS Thesis*. University of Iowa, Iowa City.

—— SMITH, F. O. 1984. A fluid inclusion study of the dolomite-calcite transitions in the Burlington and Keokuk Limestones (Mississippian), S.E. Iowa, W. Illinois. *MS Thesis*. State University of New York at Stony Brook.

——, REEDER, R. J. & MEYERS, W. J. 1984. Fluid inclusions in Burlington Limestone (Middle Mississippian)—evidence for multiple dewatering events from Illinois Basin. *Bulletin of the American Association of Petroleum Geologists*, **68**, 528–29 (Abstract).

SVERJENSKY, D. A. 1984. Oil field brines as ore-forming solutions. *Economic Geology*, **79**, 23–37.

VAIL, P. R., MITCHUM, R. M. & THOMPSON, S. 1977. Seismic stratigraphy and global changes of sea level. *In*: PAYTON, C. E. (ed.) *Seismic Stratigraphy—Applications to Hydrocarbon Exploration*, pp. 83–97. American Association of Petroleum Geologists Memoir, 26.

VON RAD, U. & ROSCH, H. 1974. Petrography and diagenesis of deep-sea chert from the central Atlantic. *In*: HSÜ, K. J. & JENKYNS, H. C. (eds). *Pelagic Sediments: on Land and Under the Sea*, pp. 273–99. Special Publication of the International Association of Sedimentologists, 1. Blackwell Scientific Publications, Oxford.

WILLMAN, H. B., ATHERTON, E., BUSCHBACH, T. C., COLLINSON, C., FRYE, J. C., HOPKINS, M. E., LINEBACK, J. A. & SIMON, J. A. 1975. *Handbook of Illinois Stratigraphy*. Illinois State Geological Survey Bulletin, 95.

WISE, S. & WEAVER, F. 1974. Chertification of oceanic sediments. *In*: HSÜ, K. & JENKYNS, H. (eds). *Pelagic Sediments: on Land and Under the Sea*, pp. 301–26. Special Publication of the International Association of Sedimentologists, 1. Blackwell Scientific Publications, Oxford.

WU, Y. & BEALES, F. W. 1981. A reconnaisance study by paleomagnetic methods of the age of mineralization along the Viburnam Trend, southeastern Missouri. *Economic Geology*, **76**, 1879–94.

ZENGER, D. H., DUNHAM, J. B. & ETHINGTON, R. L. (eds) 1980. *Concepts and Models of Dolomitization*. Society of Economic Paleontologists and Mineralogists Special Publication, 28.

DAVID C. HARRIS, Standard Oil Production Company, 5400 LBJ Freeway, Suite 1200, Dallas, TX 75240, USA.

WILLIAM J. MEYERS, Department of Earth and Space Sciences, State University of New York, Stony Brook, NY 11794, USA.

The diagenesis of the Great Estuarine Group, Middle Jurassic, Inner Hebrides, Scotland

J. D. Hudson & J. E. Andrews

SUMMARY: The Bathonian Great Estuarine Group consists of sandstones, silty shales and mainly shelly limestones, deposited in micro-tidal brackish lagoons in a warm, seasonal climate. The rocks show, in general, a lack of intense early diagenesis or deep-burial diagenesis, and are thus suitable for the study of shallow-burial diagenetic processes. Early diagenetic changes can be directly linked to depositional environment. They were only important on the lagoon margins, where cyanobacteria flourished in schizohaline 'algal marsh' settings and dolomite formed in response to evaporation of low-salinity lagoonal waters. During burial to a few hundred metres, and after initial compaction, cementation by ferroan calcite occurred. This was pervasive in limestones and formed large concretions in sandstones. Quartz and feldspar grains in sandstones were marginally corroded by the porewaters that precipitated calcite, and some feldspars had previously or concurrently suffered partial solution. Aragonitic mollusc shells were replaced by calcite in permeable rocks but remained unaltered in shales. Widespread volcanism in the Palaeocene buried the area beneath basaltic lavas and dykes were intruded. These events produced little metamorphic or burial-diagenetic change, except in close proximity to minor intrusive contacts or in the vicinity of plutons. The evolution of porewaters can be monitored in the carbon and oxygen isotopic composition of diagenetic calcites. After initial burial, they were dominantly of meteoric derivation. Vitrinite reflectance measurements confirm the mild thermal history of most of the sediments, and show the local effects of heating by igneous intrusions.

The Great Estuarine Group, a paralic sequence of sedimentary rocks, is an attractive subject for diagenetic studies because it is lithologically varied and fossiliferous, thus containing a wide variety of sedimentary particles of differing diagenetic potential. In most of its outcrop area, it has suffered only shallow burial, and some of the 'unstable' minerals have survived virtually unaltered (e.g. shell aragonite and smectitic clay). Elsewhere, the rocks have been subjected to higher degrees of diagenetic alteration and even, in some areas, thermal metamorphism. Because it was mainly deposited in waters of less than normal salinity, sub-littoral synsedimentary lithification was limited. It therefore provides a useful contrast to some other sequences where early marine cementation was pervasive. The name of the Group emphasizes that the deposition, as well as the diagenesis, of the sediments involved the interplay of at least two types of water, marine and meteoric. Although not all aspects of Great Estuarine Group diagenesis have been studied as thoroughly as we would like, we do now have enough diverse information to attempt a synthesis of the diagenesis of the Group as a whole, and to relate this to its burial history.

In this paper we consider the different stages in the evolution of these rocks: derivation, deposition, early diagenesis, burial diagenesis, metamorphism, uplift and erosion. Of these the middle two are the most important in the present context, but we want to stress the interdependence of all stages.

The Great Estuarine Group

The islands of the Inner Hebrides (Fig. 1) expose Mesozoic, mainly Jurassic, rocks that rest on Precambrian and Palaeozoic basement and are overlain by Tertiary volcanics: they are also cut by Tertiary intrusions. A similar succession is found in the submarine area of the Minch between the mainly Precambrian rocks of the Scottish mainland and the Outer Hebrides. Although direct evidence of palaeo-shorelines is lacking, we believe that the Inner Hebrides–Minch region approximates to a depositional Minch Basin, bordered on either side by swells that were, at least in part, fault-bounded (Hudson 1983). The basin is crossed by a major fault, the Camasunary Fault (Fig. 1), that itself had episodes of movement during the Mesozoic as well as subsequently, but was probably largely quiescent during the time that most concerns us. It divides the Minch Basin in the broad sense into the Sea of the Hebrides basin to the NW, and the Inner Hebrides basin to the SE separated by a palaeo-high in central Skye. The main Jurassic exposures follow the NNE–SSW trend of the basin and its margins and hence the

From: MARSHALL, J. D. (ed.), 1987, *Diagenesis of Sedimentary Sequences*, Geological Society Special Publication No. 36, pp. 259–276.

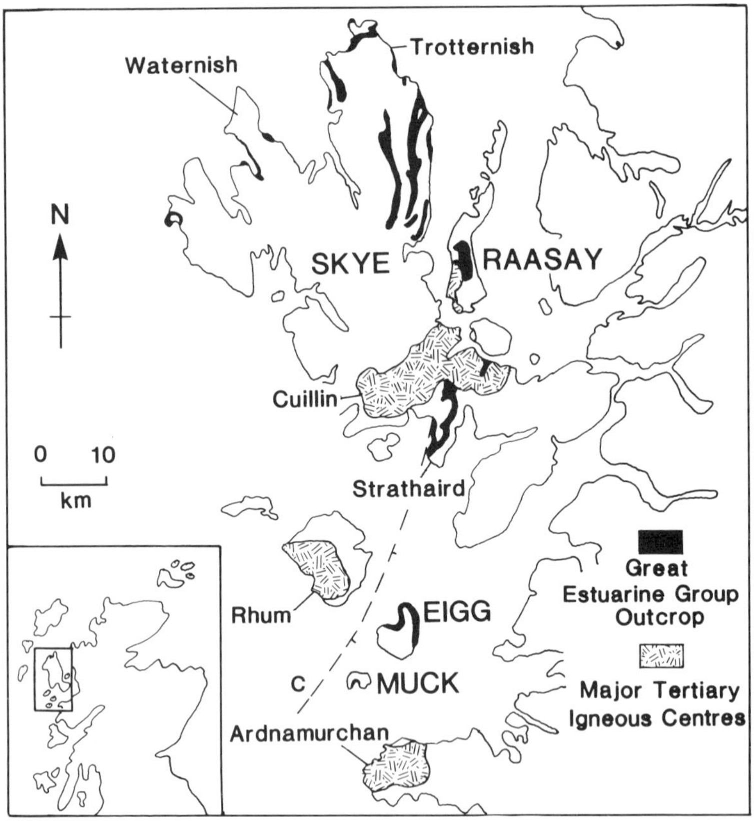

FIG. 1. Map showing outcrop of the Great Estuarine Group (black), and distribution of the major Tertiary igneous centres, northern Inner Hebrides. Fault line marked 'C' shows the position of the Camasunary fault.

depositional strike. There is limited evidence of westward thinning on Skye, and of southward thinning from a depocentre in North Skye.

The Great Estuarine Group (Fig. 2) is a paralic episode in an otherwise marine Jurassic succesion. During the accumulation of some 280 m of sediment (present, compacted thickness) the depositional surface was never more than a few metres from sea-level, yet despite the great variety of facies seen in vertical sequences, lateral variability is remarkably small (apart from deltaic sands) along the 90 km of outcrop from North Skye to Muck, and probably originally further. Because of the lack of good zone fossils the diachroneity of facies cannot be accurately assessed, but it was probably small (Hudson 1980).

The sediments

The sediments of the Great Estuarine Group are lithologically diverse but not well differentiated: most of the shales are more or less silty, many of the sands argillaceous and carbonaceous, the limestones sandy or marly. Nevertheless it is of interest to compare the approximate proportions of the major rock types. Based mainly on the logs presented by Hudson & Harris (1979) and Harris & Hudson (1980), the proportions for the main outcrop (Trotternish, Strathaird, Eigg) are given in Table 1. The Group is sandier than the world average of sediment types (e.g. Garrels & Mackenzie 1971, Pettijohn, Potter & Siever 1972), but is probably a reasonably representative sample of epicontinental, near-source sedimentary rock.

Derivation

The composition of the coarser sand detritus was investigated by Hudson (1964) and Harris (1984). In the Sea of the Hebrides basin much of the sand is feldspathic (5–25% of total grains). Orthoclase and microcline are roughly equal in average abundance, plagioclase only minor. In the Inner Hebrides basin feldspar forms less than 5% of the coarse detritus. Heavy minerals are dominated by the stable species zircon, tourma-

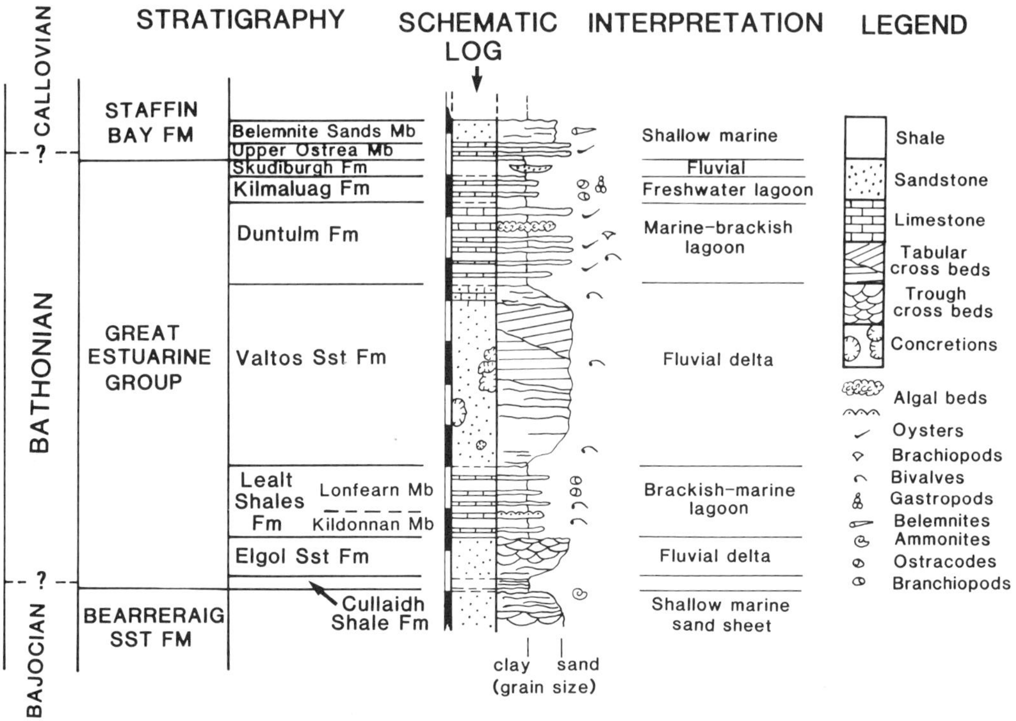

FIG. 2. The stratigraphy of the Great Estuarine Group (after Harris & Hudson 1980) including a schematic graphic log of the Group (Trotternish) and palaeoenvironmental interpretations. The black and white scale bar divisions represent ~25 m.

TABLE 1. *Proportions of different lithologies in the Great Estuarine Group*

	Total thickness (m)	Shale/silt %	Sandstone %	Limestone %
Trotternish	264	43	42	15
Strathaird	137	68	18	14
Eigg	129	50	39	11
Weighted mean (Composite thickness)	(530)	51	35	14

The weighted mean has been computed by adding total thicknesses, thus emphasizing the thicker Trotternish succession.

line and rutile, but with substantial proportions of metamorphic minerals. Garnet is most abundant in the north and staurolite in the south, with a few samples rich in epidote from NW Skye. These data indicate a dominant source from the Scottish mainland. A minor Lewisian source contributed in the NW, probably from an Outer Isles landmass (Hudson 1964). The rôle of intermediate sources between a metamorphic ultimate source and the Jurassic basin (Old Red Sandstone, Permo–Triassic?) remains unclear. Studies on Sm-Nd model ages of detritus in the Kilmaluag Formation of the Isle of Muck (Inner Hebrides basin) are consistent with a largely Dalradian source (Andrews *et al.* in press) while studies on Sr in the dolostones favour an important ancient limestone component in the hinterland. Studies of clay mineralogy have revealed an abundance of illite-smectite mixed layer clays (Andrews 1987), probably signifying a volcanic source. This is not seen in the sandstone-dominated formations. Most probably the source area was, from time to time, mantled by fine-grained volcanic ash, which was transported into the basin along with the fine-grained weathering products of the bedrock.

The sourcelands were thus within a few tens of kilometres of the basin of deposition. They were, for long periods, deeply weathered, with a soil mantle containing a volcanic-derived component, and well vegetated. At these times they delivered fine-grained sediment to the basin. At other times they were sharply uplifted, delivering coarse, unweathered sand with terrestrial plant debris as a minor but important part of the load, via short, rapidly flowing rivers and their deltas (Harris 1984).

Deposition

The detrital sediment, and factors such as depth of water, occasional exposure, conditions of turbulence, turbidity and rate of sedimentation were not very different between the Great Estuarine Group and many shallow-water marine formations. Yet the rocks are different, and this is because of the predominantly brackish-water deposition of the Group.

(1) The brackish waters favoured a narrow range of taxa of shelled invertebrates, mainly molluscs, but high productivity meant that these few flourished exceedingly (Hudson 1963, 1980). Some of these shells occur abundantly in shales and in rock-forming abundance in limestones, in almost monotypic shell beds. Apart from their obvious quantitative importance, they give excellent opportunities for comparative diagenetic studies, both between types of shell and between shales and limestones. Oysters (*Praeexogyra hebridica*) in the Duntulm Formation were primary low-Mg calcite. *Praemytilus strathairdensis* in the Kildonan Member, Lealt Formation, was composed of nacreous aragonite. *Neomiodon* spp., were composed of crossed-lamellar aragonite. High biological productivity is also revealed by the abundance of phosphatic fish and reptile remains and by an organic carbon content of 1–2% in many of the shales.

(2) The well-preserved mollusc shells, together with other evidence, allow some estimates to be made of the water types present during deposition, particularly their isotopic composition and temperatures (Tan & Hudson 1974, Hudson 1980). The results are summarized in Table 2. An important distinction between 'open' and 'closed' coastal lagoons (Hudson 1980) is worth re-stressing. In 'open' lagoons, which dominated the lower formations of the Group, free connection was at all times maintained between a freshwater supply and the open sea. Hence salinity varied between 0 and 35‰, and the chemistry of all but the most dilute waters was controlled by the marine component. In 'closed' lagoons the chief supply was fresh water, and its concentration by evaporation produced waters chemically and isotopically different from sea water, with different faunas and different diagenetic sequences. This became particularly important in the Kilmaluag Formation (Andrews 1985).

Early diagenesis

The conditions of temperature, salinity etc., that we have listed (Table 2) were fundamental in controlling those diagenetic reactions that took place on the lagoon floor or immediately beneath it, where pore waters were closely related to depositional waters and re-working resulted in the same particle repeatedly experiencing 'depositional' and 'diagenetic' effects. This stage corresponds to early, penecontemporaneous, synsedimentary, or eo-diagenesis as used by various authors. In the present case it is useful to distinguish between the lagoon floor, on which the sediments were under almost perpetual, though shallow, cover of water from lagoon margins where exposure was important.

As in most sedimentary sequences, the coarse-grained clastic sediments show the least effect of early diagenesis. Quartz (and even feldspar) have low diagenetic potential for alteration by low temperature dilute waters and relatively mobile, organic-poor porewaters did not precipitate authigenic minerals. For converse reasons the shales show more obvious diagenetic effects, and in the Great Estuarine Group they are linked with the limestones which commonly occur as thin beds within them. Nevertheless, on the lagoon floor early diagenetic effects were less obvious than in many contemporaneous marine sediments. In particular, cementation was inhibited: little early carbonate cement formed and no hardgrounds are found, nor was there much obvious dissolution of aragonite or calcite. On the lagoon margin, however, or where run-off

TABLE 2. *Setting of Great Estuarine Group deposition*

Palaeolatitude	35°–40°N
Temperature	15–25°C
Water depth	0–10 m (?); micro-tidal
Salinity 'open' lagoons	0 (fresh water) to 35‰ (sea water)
'closed' lagoons	0 (fresh water to evaporated low SO_4 waters
$\delta^{18}O$ of water (‰, SMOW)	−6 (fresh water) to −1 (sea water) (increased by evaporation)
$\delta^{13}C$ of ppd carbonate (‰, PDB)	−3 (fresh water) to +3 (sea water)

Compiled from papers referred to in the text, particularly Tan & Hudson (1974) and Hudson (1980).

and desiccation affected closed lagoons, more obvious changes took place: calcite and dolomite could form.

The lagoon floor: clays and silts

The main clay mineral assemblages were determined by derivation rather than diagenetic changes (Andrews 1987), but iron-rich minerals are strongly linked to depositional environment. This is shown by the colour of the rocks. Typically, marine-brackish mudrocks are pale to dark grey, due to pyrite formed by sulphate reduction in the porewaters, and to organic matter. Freshwater mudrocks are generally pale green to dark greenish or brownish-grey. Near absence of pyrite presumably allows the greenish tints to be revealed. Finally, and related to the filling of the lagoon, the silts of the Skudiburgh Formation are typical red-brown alluvial 'redbeds', with grey-green reduced mottlings (Andrews 1985).

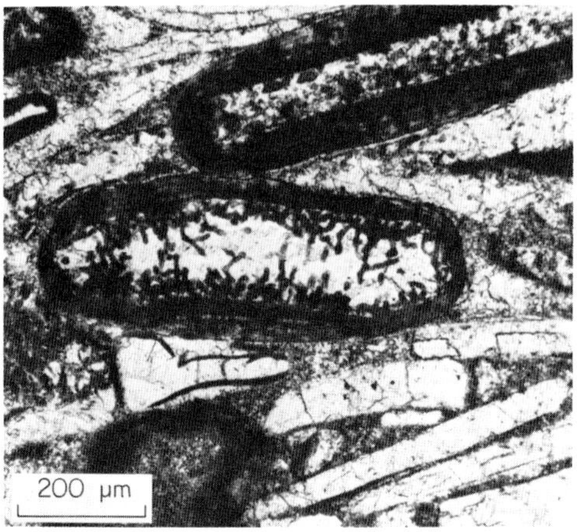

FIG. 3. Photomicrograph of a limestone from the Lonfearn Member, Lealt Formation, Trotternish, Skye. Molluscan shell fragments are variably coated with micrite envelopes or oolitic sheaths. That in the centre shows 'algal' borings; the one beneath it a fractured micrite envelope.

Lagoon floor carbonates

Most of the limestones are thin (<0.5 m) laterally extensive (tens of metres minimum, sometimes several kilometres) molluscan shell beds, interbedded with shales or silts. All transitions from shell layers to shell limestones can be found and the faunas of the limestones are the same as those of the shales that enclose them. This argues against long-distance transport. The shell beds to not show the characteristics of storm deposits (Aigner 1985): no erosional bases, no internal grading. They are probably the result of repeated winnowing events and illustrate the interplay of depositional and diagenetic processes. The *Neomiodon* shell beds of the Lealt Formation, and the finer-grained facies of the Valtos Formation, can be divided into different types in general similar to those from the Purbeck Formation of southern England described by El-Shahat & West (1983). Some of the major types are:

(1) Shales or silts with scattered shells, articulated but not in life position.
(2) Shales with shell layers, mostly whole but disarticulated valves, convex-up.
(3) Thin (few centimetres) shelly limestones containing both articulated shells and single valves, as well as shell fragments.
(4) Similar shelly limestones with a preponderance of broken and rounded shell fragments, and generally finer-grained.
(5) Limestones with rounded shell fragments and abundant coated grains (ooliths) (Fig. 3). These are confined to the Lonfearn Member, Lealt Formation.

The shells in the limestones are frequently intensely 'bored', probably by cyanobacteria. The borings are filled with either fine-grained calcite or pyrite. Shell beds of types 3, 4 and 5 often contain a mixture of marginally bored and pyritized shells with almost pristine ones, thus showing how winnowing mixed shells which had a long residence time on the lagoon floor, or even within the sulphate reduction zone of the sediments, with newly dead ones. Many of the bored fragments also show well-developed micrite envelopes or coatings, beneath which the deeper borings extend for a few tens of microns (Fig. 3). Oolitically coated grains in the Lonfearn Member, Lealt Shales, retain radial fabrics. These ooids are now composed of ferroan dolomite, and their high $\delta^{18}O$ values were interpreted by Tan & Hudson (1971) as showing that the dolomite is early diagenetic. By analogy with modern radial ooids from a broadly analogous depositional setting, the Laguna Madre, Texas (Land *et al.* 1979), it is likely that the Lonfearn Member ooids were Mg calcite originally; however, we note that radial fabric is not a single criterion with which to distinguish former carbonate mineralogy (Richter 1983). Dolomitization presumably took place beneath the sulphate reduction zone, allowing Fe^{2+} to become incorporated in the dolomite.

Coated grains also sparsely occur in the Duntulm Formation, Isle of Muck, where ?*Cor-*

bula shells display radial oolitic coatings of non-ferroan calcite in what appears to have been a storm winnowed horizon. $MgCO_3$ is *c.* 0.70 mol% enriched in the ooid cortex relative to the burial diagenetic ferroan spar which probably means that the ooids were originally composed of a Mg calcite.

Carbonate concretions

Septarian calcite concretions up to about 1 m across and 20 cm thick occur at several horizons, most notably near the middle of the Lealt Formation in the upper Kildonan and lower Lonfearn Members. The beds concerned are shale–limestone alternations with the concretions occurring in shales, but commonly close to limestone beds. The shales have faunas varying from bed to bed from fresh water to marine-brackish. The $\delta^{13}C$ values (−6‰) of the concretions are low but not extreme, and can be explained as derived from mixtures between sulphate reduction bicarbonate and bicarbonate derived from the dissolution of fossils incorporated in the shales. Their $\delta^{18}O$ values (−2‰) confirm their shallow-burial origin (Tan & Hudson 1974). Calcite which filled the septarian cracks was introduced later, but the timing of the opening of the cracks is unknown (cf. Astin 1986). More rarely, sideritic concretions a few centimetres long occur in the freshwater facies of the Kilmaluag Formation.

At this diagenetic stage the rocks, apart from concretions, were not lithified, nor had aragonite been transformed to calcite. The limestones were still porous and the muds little compacted. Pores were filled with water closely related to the water of deposition, with loss of much sulphate and some carbonate.

The open lagoon margin: algal marsh deposits, Duntulm Formation

If the lagoon floor was diagenetically quiet, the lagoon margin, whose deposits are only sparsely represented in the record, was very active. This is especially well shown in the 'algal beds' of the Duntulm Formation (Hudson 1970, Andrews 1986). The setting was that of a supra-littoral fringe to a micro-tidal marine-brackish lagoon: the fact that the open lagoon was not far away is shown by marine fossils washed on to the supra-littoral deposits. The setting was classically schizohaline. Most growth of the cyanobacterium (blue-green alga) *Cayeuxia* probably took place during periods of abundant fresh water flushing from rainwater and run-off, and during this time its sheath was calcified. Periodic inundation of the marsh by sea water from the adjacent lagoon, followed by desiccation, resulted in the precipitation of some gypsum and brecciation of the calcified algal 'heads' forming an 'algal gravel'. Intense bacterial diagenesis in organic-rich carbonate sediments produced a spectrum of diagenetic textures (cement fringes, spherulites, microspars) quite unlike anything seen in the subaqueous deposits of the lagoon floor.

The closed lagoon margin: dolostones of the Kilmaluag Formation

In the Great Estuarine Group depositional dolomite is virtually confined to the low-salinity, closed lagoon facies of the Kilmaluag Formation; a fact in good general accord with the sulphate-inhibition theory of dolomite formation (Baker & Kastner 1981). Within the Kilmaluag Formation, dolomite occurrences correlate strongly with evidence of evaporation and desiccation. Evidence from carbon and oxygen isotopes agrees with an origin from evaporated meteoric waters, with dolomite forming via dolomitization of a pre-existing calcite (probably Mg calcite) mud during seasonal episodes of evaporation (Andrews *et al.* in press). Once again, the variable conditions of the lagoon margin promoted more active diagenetic change than the constant conditions of the lagoon floor. Once formed, the aphanitic Kilmaluag dolomite, enclosed within fine-grained rocks, escaped substantial diagenetic change during further burial.

The final stage of lagoon closure: alluvial calcretes of the Skudiburgh Formation

Within the mottled clays of the Skudiburgh Formation white-grey nodular concretions, typically 2–8 cm in diameter occur. These are usually cracked in 'septarian style' and the ground mass of non-ferroan calcite microspar contains spherulitic structures of probable bacterial origin. These concretions have been interpreted as calcrete nodules (Andrews 1985). Their stable isotope geochemistry, $\delta^{13}C$ values *c.* −10‰, $\delta^{18}O$ values −3 to −4‰, suggest formation from meteoric derived soil-water which may have undergone some evaporation.

Burial diagenesis

The deposition and early diagenesis of the Great Estuarine Group certainly took place during Bathonian times on or just below the lagoon floor, so the time and place of the processes are well defined. With burial the sediments entered a more obscure phase in which the relative, and still more the absolute, timing and depth of burial at which different events happened are hard to discern. As most diagenetic changes are fluid-rock interactions, the nature of the porewaters is also critical. We can look for clues, mostly to relative order of events, in the sediments themselves, and to some extent reconstruct the burial history from independent evidence.

As any sediments are buried, certain changes take place which distinguish burial from near-surface diagenesis. Light is excluded, temperature stabilizes and then gradually increases, biological activity apart from bacteria ceases and compaction starts. Adjacent beds that may have experienced very different conditions at and immediately after formation are now subject to similar processes. Slow processes such as compaction or convection-driven water flow have time to take effect. It does not follow, however, that burial diagenesis is a one-way sequence of changes. Uplift can reverse the temperature trend, and new sources of water can traverse the sediment due, for instance, to hydrologic flow down aquifers in response to uplift of nearby land, or to changes in sea-level.

Because burial diagenesis is the hardest of the stages to define and understand, it is convenient to approach it by excluding other effects. As well as working forwards from deposition and early diagenesis, we can work backwards from the present. The effects of modern weathering on our rocks are rather easy to recognize and hence to avoid during sampling. The Tertiary igneous activity that affected our region may have had subtle effects related to heat flow and depth of burial—we discuss this below—but the influence of contact metamorphism near minor igneous intrusions is easy to recognize and avoid. The regional effects of the Skye plutonic centre and its hydrothermal aureole are also clearly distinct from burial diagenesis and well recognized from independent evidence (Taylor & Forester 1971). This process of elimination leaves us with the results of 'normal' burial diagenesis, which we believe were mostly completed during the period of initial burial of the sediment pile. In support of this conclusion, blocks of lithified Jurassic sediments, principally limestones, occur in Tertiary volcanic explosion breccias in North Skye, Strathaird and Muck. They do not differ in field appearance from equivalent lithologies in nearby *in situ* outcrops.

Burial history

Figure 4 shows a simplified version of the burial history of the Great Estuarine Group based mainly on North Skye. Although variations occur across the basin, it will apply in general terms throughout.

Deposition of the Group was terminated by the transgression of the Staffin Bay Formation in the latest Bathonian (Bradshaw & Fenton 1982). Slow but relatively steady subsidence ensued during the deposition of the Staffin Shale Formation through the Callovian, Oxfordian and Lower Kimmeridgian (Sykes 1975), to a total observed thickness of 120 m. We do not know how much sediment was deposited above this, and before an important phase of pre-upper Cretaceous erosion (Hudson 1983), but a maximum of 500 m seems generous. The Upper Cretaceous marine strata are thin (few metres) and discontinuous, and there is no evidence of former great thicknesses in the region. There was a further brief period of erosion before the onset of volcanism in the Palaeocene. The earliest volcanics are palagonite tuffs, but most of the lavas are continental, with frequent evidence of weathering between lava flows. The preserved thickness of lavas on Skye is about 600 m (Emeleus 1983); a former maximum thickness of 900 m (G. P. C. Walker, personal communication) is used on Fig. 4. Since high lava fields may have been built up, depth below land surface rather than sea-level is shown. Most of this activity was probably concentrated in a short period of time from about 60 to 58 Ma; the lavas on Eigg and Muck may be older, about 63.5 Ma (Dagley & Mussett 1986), and the Sgurr of Eigg pitchstone is the youngest well-dated igneous rock in the area at 52 ± 1 Ma, early Eocene (Dickin & Jones 1983, Dagley & Mussett 1986). This later activity was probably not widespread in its effects whereas the earlier volcanic episode could have involved a regional increase in heatflow (Emeleus 1983, p. 389).

The uplift and erosional history of the area after the igneous activity and prior to the Pleistocene glaciations is poorly known. Watson's (1985) review suggests that the Palaeocene was a time of general uplift in Scotland, and that substantial deformation of the lavas took place shortly afterwards. Oligocene non-marine sediments rest on lavas in the Canna basin. Most probably the present land outcrop of the lavas, and the associated plutonic centres, were uplifted during the same episode of deformation, that is prior to the late Oligocene.

Compaction

The finer-grained rocks have suffered substantial compaction and some of the shales of the Lealt Formation are well-laminated 'paper shales' containing crushed bivalve shells. More silty

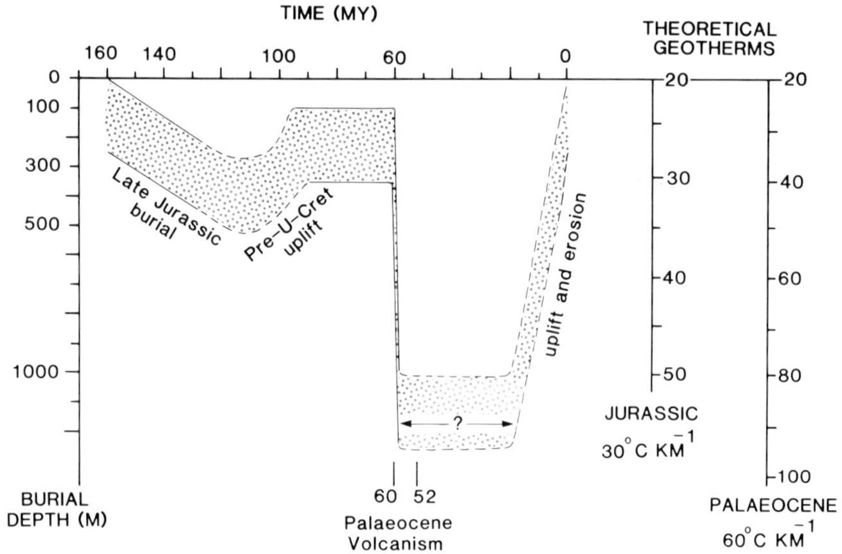

FIG. 4. Postulated burial history for the Great Estuarine Group, based on north Skye. For discussion, see text. Two geotherms are shown. The 'Jurassic' geotherm is a conventional 30°C km^{-1} assumption and probably applies except during Palaeocene volcanism. At that time the geothermal gradient was higher, and an arbitrary doubling to 60°C km^{-1} is shown.

and calcareous shales, on the other hand, show much less compaction, and all fine-grained rocks are soft, porous and friable when not close to igneous contacts. Meade (1966) and Rieke & Chilingarian (1974) (e.g. fig. 17, p. 42) show that a large range of compaction values are to be found in argillaceous rocks at depths of burial of a few 100 m, depending on grain size, composition and other properties of the shales concerned. Qualitatively, the rocks are similar to those of the Jurassic of the East Midlands of England, which have never been buried to a depth of more than about 500 m and are in a stable region not affected by volcanic activity. Limestones show some evidence of pressure solution, with microstylolites sometimes developed between adjacent allochems. More extensive stylolite development characterizes some of the 'algal beds' of the Duntulm Formation, where dissolution of *Cayeuxia* nodules has taken place adjacent to clay seams. Compaction phenomena in sandstones are discussed below.

Concretionary cementation of sandstones

By far the most striking diagenetic phenomenon in these rocks is the occurrence of very large (up to more than 1 m radius) ellipsoidal calcareous concretions (doggers, nodules) in the Valtos Sandstone Formation. These are well seen at the type locality in Trotternish but even better on Eigg.

Several observations bear on their interpretation.

(1) Sandstones less than 1 m thick are nearly always fully cemented whereas thicker ones are concretionary.

(2) In the upward-coarsening deltaic sand units that typify the Valtos Sandstone Formation, the lower parts of the sand bodies are flat-bedded, and contain carbonaceous and argillaceous partings. These are partially cemented by concretions bounded by bedding planes, though rounded in plan view. The classic ellipsoids occur in the upper parts of sand bodies where permeability is more homogeneous (Fig. 5a).

(3) The uppermost part of a sand body is often a coarse sand or granule conglomerate about 1 m thick, which is usually also rich in *Neomiodon* shell debris. This is typically fully cemented, and half-ovoid concretionary masses appear as though 'suspended' below it.

(4) There may be a more general correlation between the occurrence of shells in the sandstones and adjacent shales and the occurrence of concretions. Some of the more densely concretionary sands are also shelly, whereas the coarsest and thickest sand on Eigg (Division C, Hudson & Harris 1979) is only sparsely concretionary, and the Elgol Sandstone Formation, which is unfos-

FIG. 5. (a) Calcite cemented concretions in the Valtos Formation, Valtos cliff section, Trotternish, Skye. The concretions are about 1.5 m in diameter. Note contrast between 'flat' lower concretion and rounded upper one. (b) Compaction of bedding in uncemented sandstone around a concretion. Valtos. Hammer is 38 cm long.

siliferous and enclosed in sparsely fossiliferous shales above and below, is scarcely cemented by calcite at all.

(5) In some units, notably the upper part of the Valtos Sandstone Formation in Trotternish, concretions are very rich in whole or broken *Neomiodon* shells, whereas the uncemented sand outside the concretions is unfossiliferous, showing no trace of the shells that it must formerly have contained.

(6) Whereas sedimentary lamination and cross-bedding look similar at first, whether within concretionary masses or outside them, in some instances at least bedding is deflected around concretions. About 15–20% compaction has been observed in the few examples measured (Fig. 5b).

(7) In thin sections from within the concretions one commonly observes pressure-solution of shells against quartz and feldspar grains, and sometimes shells have been reduced to mere films (Fig. 6).

(8) In the field, on bright days such as are, surprisingly to some, common in the Hebrides, the calcite cements glisten with lustre-mottling, revealing crystal sizes up to centimetres across. This is confirmed in thin section, where poikilotopic calcite crystals may enclose hundreds of sand grains and several *Neomiodon* shells. Frequently, more than one shell forms part of a single large poikilotopic crystal continuous from cement to neopmorphic replacement.

(9) The calcite of the cements is always ferroan.

(10) The isotopic composition of the calcite is generally within the same range as that of limestones within the Great Estuarine Group (Tan & Hudson 1974).

Taken together, these observations clearly establish the burial diagenetic origin of the concretions. The pressure dissolution of shells must clearly have happened before the concretions formed (observation 7). Compaction continued after concretion growth also, at least around some of the concretions (observation 6). This phase of compaction was accompanied or preceded by dissolution of shells outside the concretions (observation 5). Petrographic observations (8) suggest that at the time of concretion formation *Neomiodon* shells were composed of aragonite, as they still are in adjacent shales. This would explain the scarcity of nucleation points for calcite (hence the large crystals) and the enclosing of several then-aragonitic shells in one poikilotopic calcite crystal. The chemistry of the concretions also agrees with a burial diagenetic origin, and contrasts with that of early-diagenetic septarian concretions of the shales in the Great Estaurine Group (Tan & Hudson 1974),

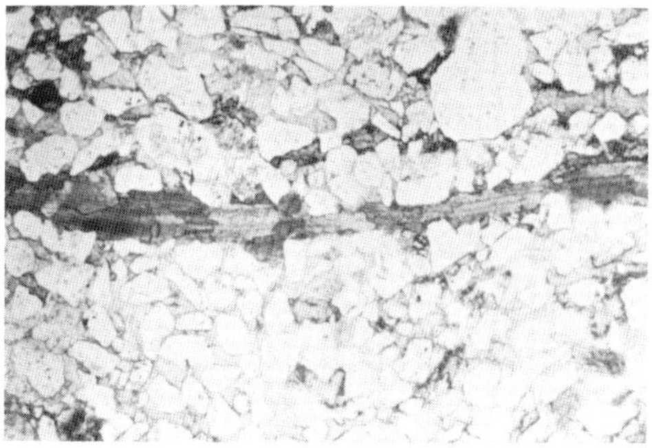

FIG. 6. Photomicrograph of a calcite-cemented sandstone from a concretion (cf. Fig. 3). A *Neomiodon* shell, about 0.4 mm thick, shows pressure solution against quartz grains; this must have happened before the sandstone was cemented. The replacement calcite in the shell is slightly pseudo-pleochroic. Valtos Formation. Plane-polarized light.

and with that of carbonates clearly related to Tertiary igneous activity (see below).

The formation of these concretions after initial burial but before deepest burial allows some further observations and speculation concerning their growth in relation to other events. Berner (1980, p. 124) showed that sands will pass through a concretionary stage on their way to becoming fully cemented, rather than being cemented via a 'cementation front': essentially because fluids can continuously flow past a growing concretion. A concretionary sand thus results when the supply of $CaCO_3$ runs out before cementation is complete: the more concretionary sand bodies in the Great Estuarine Group, such as Division E, Valtos Sandstone Formations, Eigg, are about 20% cemented, as recent work by M. Wilkinson (University of Leicester) shows. These considerations say nothing about the size of concretions (many small or few large ?): in these sandstones, as in many others, large concretions are the rule. Berner (1968) gives formulae for the rate of concretion growth, and using his recommended values for the various parameters involved (some of which, such as degree of $CaCO_3$ supersaturation, are poorly known), the time for formation of a spherical concretion of radius 1 m can be calculated. A pure $CaCO_3$ concretion growing by diffusion only in stationary porewater would take 4.55 Ma to form. If growth in ground-water flowing at 3 m yr^{-1} (Bathurst 1975) is considered the time is reduced to 1.44 Ma. These figures are only approximate, but they do indicate that these concretions could have taken in the order of 1 Ma to form. This is a small part of the time available, if that is reckoned to be the period of late Jurassic and early Cretaceous initial burial, say 40 Ma.

A very different geometry of cementation occurs alongside Tertiary basaltic dykes that cut the Valtos Formation sandstones on Eigg. The dykes are typically about 1 m wide, and themselves have now largely weathered away. Their baked margins, about 0.5 m wide, stand proud, and are fully cemented at the actual contact (Fig. 7). Away from this, cementation is concretionary, with numerous small, more or less coalescent concretions typically about 2 cm across, and gradually dying away into the surrounding sandstone. Evidently the short-lived intense heating associated with dyke intrusion induced rapid precipitation (or re-distribution) of calcite, and precipitation was in the form of many small concretions. Slow, steady precipitation during burial diagenesis produced, by contrast, the fewer large concretions described above. The isotopic composition of the tiny concretions is indicative of precipitation from heated waters (see below).

Mineral stability

Thin sections of the large concretions show marginal corrosion of quartz and feldspar grains against calcite cement, and cracks within grains are calcite-filled and possibly enlarged. Apart from this corrosion, many K-feldspar grains appear unaltered in thin section. However, recent SEM observations show that decomposed feldspar grains also occur, and many of these are plagioclase (M. Wilkinson, personal communication).

Certain heavy minerals show differential preservation in cemented versus uncemented sands: kyanite, staurolite, sphene, epidote (Hudson 1964, Harris 1984). The data may be summarized to give a stability ranking (Table 3). Comparison with Morton's (1984) data clearly shows that loss of these minerals is related to burial diagenesis rather than weathering. Because of the differential preservation in concre-

FIG. 7. (a) Cemented margins of a weathered basaltic dyke in Valtos Formation sandstone. Isle of Eigg. Dyke is about 1 m wide. (b) Detail of a dyke margin, showing small concretions in sandstone. Same locality. Hammer head 15 cm.

TABLE 3. *Stability ranking of heavy minerals under different conditions of alteration*

	Morton (1984)		This work
	'Deep burial' pH > 7, saline	'Weathering' pH < 7, fresh	Great Estuarine Group
Unstable:	Olivine, pyroxene	Olivine, pyroxene	—
	Andalusite, sillimanite	Amphibole	—
	Amphibole	Sphene	Amphibole
	Epidote	Apatite	Epidote
	Sphene	Epidote, garnet	Sphene
	Kyanite	Chloritoid, spinel	Kyanite
	Staurolite	Staurolite	Staurolite
	Garnet	Kyanite	Garnet
	{Apatite, Chloritoid, Spinel}	{Andalusite, Sillimanite, Tourmaline}	—
Stable:	Rutile, tourmaline, zircon	Rutile, zircon	{Rutile, Tourmaline, Zircon}

tions, most of this diagenesis must have taken place since concretion growth. The near-absence of the highly unstable minerals that must have been available at source, notably amphiboles, is best explained by loss during source weathering, transport, or pre-concretion diagenesis (Hudson 1964).

Cementation and neomorphism in shells and limestones

The commonest type of limestone in the Great Estuarine Group is a molluscan biosparite or biosparrudite, but finer-grained limestones also occur, and mollusc shells occur in sandstones and shales as well as in limestones. We summarize here some of the variations related to lithology and original shell mineralogy.

Shell preservation

(1) Shells enclosed in shales still generally retain their original mineralogy and fabric: foliate calcite in the case of *Praeexogyra hebridica*, nacreous high Sr (4800 ppm) aragonite in *Praemytilus strathairdensis*, crossed-lamellar aragonite, lower in Sr (3700 ppm) in *Neomiodon* spp, to name the three commonest bivalves involved.

(2) In thin, argillaceous 'earthy' limestones even *Neomiodon* may retain chalky aragonite (cf. El-Shahat & West 1983), *Praemytilus* and *Praeexogyra* retain mineralogy and fabric.

(3) In all purer limestones *Neomiodon* is replaced by sparry calcite, but the calcite may retain aragonite and organic inclusions which render it pseudopleochroic (Sandberg & Hudson 1983).

(4) *Praemytilus* commonly retains its aragonitic mineralogy and nacreous shell micro-structure even in fully-cemented sparites, although these are only a few centimetres thick and enclosed in shales. Evidently nacreous microstructure was more resistant to replacement than crossed-lamellar, possibly because of its higher organic content (Hudson 1967) and layered nature.

(5) *Praeexogyra* shells, being composed of low-Mg calcite, are very resistant to diagenetic alteration. Their foliate microstructure is always well-preserved, even when the shells have been compacted and cemented into a 'micro-breccia'.

Spar cements and shell neomorphism

As recorded above, shelly limestones generally lack early cements. The predominant cement is a post-compactional ferroan spar, resulting in fabrics identical to those of the 'compacted biosparrudites' of El Shahat & West (1983, fig. 6). Fabrics comparable to their 'early-lithified biosparrudites' (El Shahat & West 1983, fig. 7) are rare. The ferroan spar of the cement is essentially identical to that which replaces aragonitic molluscs, except that Sr is enhanced in the replaced shells. Single crystals commonly transect shell-margins and post-date fracturing of the shells (Sandberg & Hudson 1983). The isotopic and elemental compositions of the spar are those typical of burial cements (Table 4; Andrews 1986, Sandberg & Hudson 1983).

TABLE 4. *Chemical characteristics of burial spar cements*

Elemental composition (ppm)	
Mg	<2500
Sr	200–600
Mn	*c.* 1000
Fe	6000–13 000
Isotopic composition, ‰ PDB	
$\delta^{13}C$	0 to +2*
$\delta^{13}O$	−10 to −6

* More negative $\delta^{13}C$ values (to −4.6) occur in algal limestones in which early diagenetic calcites are also more negative (Andrews 1986). Data from Andrews (1986), Sandberg & Hudson (1983) and Tan & Hudson (1974).

Microspars

Most of the finer-grained limestones are microsparites rather than micrites. They have been most thoroughly studied from algal limestones of the Duntulm Formation (Andrews 1986). Microspar is typically composed of 4–30 μm, inclusion-rich calcite in equant to loaf-shaped crystals. Crystals are often separated by clay seams and patches (Andrews 1985, fig. 13). Molluscan shell debris, peloids and microfossils appear to 'float' in the microspar. It is undoubtedly of neomorphic origin (Bathurst 1975, pp. 484–6).

Both ferroan and non-ferroan microspars occur. The non-ferroan variety is particularly common in argillaceous limestones and probably formed in early diagenesis, perhaps within the sulphate reduction zone where Fe was incorporated into pyrite. It is often transitional to ferroan microspar which presumably formed under deeper-burial conditions where iron was available for incorporation in the carbonate. In its turn this microspar can pass into pseudo-spar, true void-filling spar, or into fibrous displacive calcite.

The $\delta^{18}O$ values of microspars (−5 to −9‰) overlap with those of true spars but are generally somewhat heavier. $\delta^{13}C$ values are lighter (−0.2 to −6.7‰) reflecting the incorporation of some organic-derived carbon (Andrews 1986) (cf. Figs 9a and 10).

Fibrous calcite veins

Many of the shales include beding-parallel layers of fibrous calcite ('beef'), a few millimetres to a few centimetres thick. The thicker examples may show cone-in-cone structure. Commonly, these layers occur above or below thin shelly limestones or calcareous sandstones, or they may occur within limestones along shale partings. In such cases transitions can be observed from normal microspar, to fibrous-and-displacive microspar, to 'beef' veins. This suggests that microspar neomorphism and 'beef' veining were contemporaneous, and the ferroan nature and carbon and oxygen isotopic composition of the calcite involved shows that both were burial diagenetic. Marshall (1982) and Stoneley (1983) discuss the possibility that such veins were formed during episodes of over-pressuring of the shales, to help account for their displacive growth.

Marshall's example from the Valtos Formation of Eigg showed a sharp change in chemistry and isotopic composition at a petrographically recognizable boundary, as well as gradual changes along the lengths of the fibres. These results are best explained by the introduction of new sources of porewaters, as well as gradual porewater evolution (Marshall 1982).

Most of the fibrous calcite veins appear to belong to 'normal' burial diagenesis. However it is noticeable that in Strathaird there are many such veins, some not parallel to bedding. These may relate to the Tertiary hydrothermal system of that region. The example from Strathaird studied by Marshall (1982), however, is a burial-diagenetic one. Its isotopic composition suggests it pre-dated the Tertiary heating and survived it without isotopic change.

The effects of Tertiary volcanism and burial

More than 1 m or so from minor intrusions such as basaltic dykes, and more than a few kilometres from the major plutonic centre of Skye, metamorphic effects are far from obvious, despite the fact that the whole area was buried beneath hundreds of metres of basaltic lavas, and that these contain zeolite minerals deposited by circulating hot waters (Walker 1970).

We have investigated some of the problems involved by studying rocks from outcrops categorized as follows:

(1) Outcrops with few or no minor intrusions, not close to major sills, and lacking signs of baking, e.g. Duntulm and Staffin Bay, Skye; Kildonan, Eigg.
(2) Outcrops in areas with many minor intrusions, including thick (10 m+) sills as well as dykes, in which there is pervasive baking of the country rocks, e.g. Lealt area, Skye.
(3) Strathaird, which is within 5 km of the margin of the Skye plutonic complex, and within its hydrothermal aureole (Taylor & Forester 1971).
(4) Samples from within about 0.5 m of an igneous contact.

In category 1, shales are soft and crumbly, smectitic clay minerals are preserved, and shale colours are either pale grey, greenish, or reddish-brown, depending on depositional facies. Aragonitic fossils are preserved in shales, and retain depositional $\delta^{18}O$ values.

In category 2, shales are harder and more brittle, but fissile, and are darker coloured. Former aragonitic fossils are normally either represented by moulds only, or replaced by calcite.

In Strathaird, category 3, all shales are dark grey or black, and have lost fissility. Aragonitic fossils are never preserved. The average $\delta^{18}O$ of limestones, especially fine-grained ones, is lowered by about 5‰ compared to their facies equivalents elsewhere (Tan & Hudson 1971, 1974). Smectitic clays are lacking (Andrews 1987).

In category 4, all these conditions are accentuated, and even calcitic *Praeexogyra* shells may be recrystallized, but these effects die away rapidly outwards from the intrusion in areas otherwise belonging to category 1 (Tan & Hudson 1971, p. 763, fig. 2). Smectitic clays are lacking, the crystallinity of illite is increased, and some chlorite is formed.

Vitrinite reflectivity

We have limited data on vitrinite reflectivity on samples from throughout the outcrop (measured by R. Lee and B. S. Cooper, University of Newcastle upon Tyne, in 1975). There is no obvious correlation with position in the basin, stratigraphical position, or inferred distance beneath the basalts. The results (Fig. 8) can most usefully be considered in the light of our alteration categories, and of the burial history (Fig. 4).

The R_0 values show most samples from category 1 below 0.5, with seven of them below 0.3. Some higher values, up to 0.8, are from outcrops also yielding values <0.3. With one exception, all category 4 samples have $R_0 > 2$, and those from Strathaird >1.5. Category 2 samples are, as expected, transitional, ranging from 0.5 to 1.3.

There is an interesting exception to these simple findings. The Duntulm Formation outcrop in Lon Ostatoin, Trotternish, has crumbly green shales with preserved aragonitic fossils (Tan & Hudson 1974, p. 120). Yet R_0 values range from 0.7 to 1.3 (Fig. 7), like those from category 2 shales. The section at Lon Ostatoin contains two sandstones cemented with calcite of $\delta^{18}O$ -15 and -19‰ (Tan & Hudson 1974), and is situated 30 m below a picrite-dolerite sill about 100 m thick.

There is much difference of opinion on the interpretation of R_0 values in terms of burial history and temperature. Many workers, e.g. Waples (1980) believe that R_0 is determined by a sediment's total history, with an approximate doubling of the rate of maturation for every 10°C rise in temperature (Lopatin theory). Price (1983), however, holds that maximum temperature reached effectively defines R_0 in sedimentary basins. A similar conclusion as regards water-dominated hydrothermal systems was reached by Barker (1983).

Lopatin calculations (Waples 1980) and comparison with data from the continuously subsiding basin of the North Sea (Lerche *et al.* 1984)

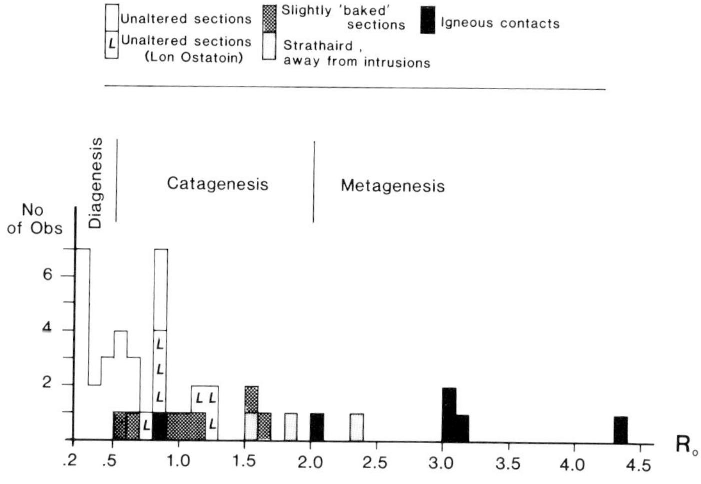

FIG. 8. Histogram of vitrinite reflectivity values (R_0, %) classified according to apparent degree of thermal alteration of shales. For discussion see text.

confirm the mild thermal history of our category 1 samples.

Barker's (1983) graph should be applicable to the hydrothermal aureole of the Skye plutonic centre (Taylor & Forester 1971). The R_0 values from Strathaird suggest temperatures of approximately 200–240°C. This agrees with an estimate of at least 200°C from spore colouration (Riding 1984).

Tertiary calcite cements

Although most cements in the limestones and sandstones are burial diagenetic products, there are several examples which we relate to Tertiary heating either by spatial association, like the small concretions along dykes mentioned above, or because they have unusually low $\delta^{18}O$ values. The lightest apparently 'normal-diagenetic' calcites are some calcite concretions at −13.7‰, but the overwhelming majority of diagenetic calcites are not lighter than −10‰. Values less than −15‰ are therefore regarded as probably related to Tertiary hot water. They are geographically widespread, from North Trotternish to Muck. They range from fossil fills (like those recorded from the underlying Bearreraig sandstone by Marshall 1981) to apparently normal cements of thin sandstones (like those in Lon Ostatoin associated with anomalous R_0 values, see above), and from dyke-margin concretions (above) to late cavity-filling dolomites and calcites (Andrews 1986). However none of these cements are volumetrically important, and several are merely the partial fillings of voids left over from earlier cementation. No doubt Tertiary hot waters traversed other porous rocks too, but without precipitating any carbonate.

The only region in which pre-existing calcites (but not dolomites) appear to have re-equilibrated with heated waters is Strathaird (Tan & Hudson 1971, 1974). This is consistent with the >200°C temperatures probably reached there.

Evolution of water chemistry

Since the formation and diagenesis of the reactive minerals represents a series of water-mineral interactions, with or without the intervention of organisms, it is useful to try to trace the evolution of the waters concerned. Most of our information concerns the carbonates. We can only work back from the carbonate to the water by making assumptions. In many cases we can only do it at all because some metastable carbonates have persisted through later events, with their compositions largely unchanged. The rocks in general are not, and never have been equilibrium assemblages.

The carbon isotopic composition of a carbonate (Fig. 9) should be a direct reflection of the

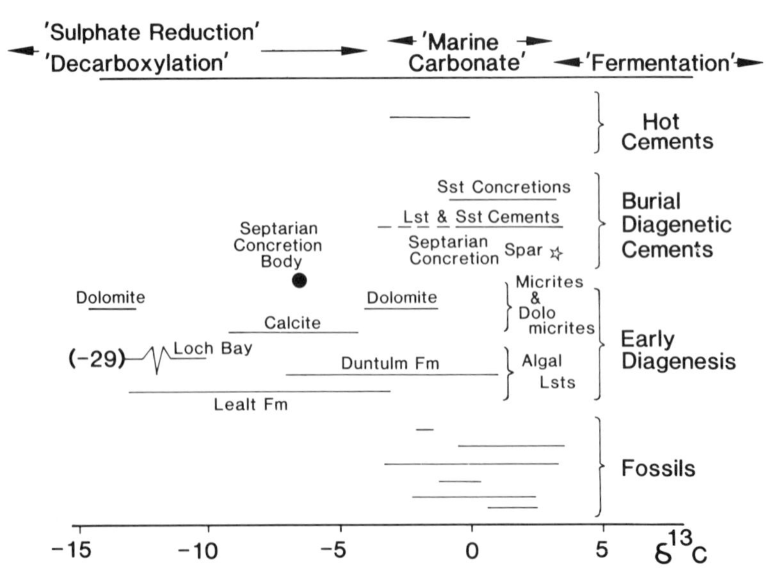

FIG. 9. Summary of carbon isotopic compositions of carbonate fossils and cements ($\delta^{13}C_{PDB}$). Characteristic $\delta^{13}C$ ranges for different types of carbonate (Curtis 1977) are shown. Most fossils have $\delta^{13}C$ values near the marine range. Early diagenetic calcites are lighter, reflecting 'soil gas' and 'sulphate-reduction' carbon sources. Burial diagenetic cements have $\delta^{13}C$ similar to fossils. Data from Tan & Hudson (1974) and Andrews (1986).

$\delta^{13}C$ of the dissolved carbonate, fractionations and temperature effects on them being small (Anderson & Arthur 1983). The oxygen isotopic composition depends on both $\delta^{18}O$ of water and temperature, so a unique solution is not normally possible, but as we discuss, interpretations can be constrained within reasonable limits.

During deposition, the main agents of carbonate formation within the lagoons were molluscs. Both aragonite and calcite were produced by different forms. $\delta^{13}C$ values are related to palaeoecology, being generally lighter in freshwater shells (Tan & Hudson 1974). $\delta^{18}O$ reflects salinity in 'open lagoonal' situations, both being determined by seawater–freshwater mixing, but the situation is again not simple (Tan & Hudson 1974). The molluscan faunas themselves are certainly a better guide to palaeosalinities than their isotopic compositions. However, molluscs from just above the Great Estuarine Group (Staffin Bay Formation and basal Staffin Shales) are fully marine and give apparently reliable palaeotemperatures using conventional assumptions (Tan *et al.* 1970). Assuming $\delta^{18}O$ of ocean water was $-1‰$ in the Jurassic (Shackleton & Kennett 1975), and allowing for possible seasonal growth (Hudson 1968), temperatures of shell deposition ranged between about 15 and 25°C. Applying these temperatures to other shell analyses allows the $\delta^{18}O$ of water in the lagoons to be calculated (Fig. 10). Early diagenetic calcites, probably precipitated during meteoric-water flushing on algal marshes of the lagoon margins (Andrews 1986) have $\delta^{18}O$ very similar to that of freshwater fossils. Both approaches suggest that the isotopically lightest fresh water in the lagoon had a $\delta^{18}O$ of about $-5‰$. The heaviest lagoonal waters were probably associated with evaporation, either on closed-lagoon margins where dolomites formed (Andrews *et al.*, in press) or possibly even in the lagoons themselves (some *Praeexogyra* analysed by Tan & Hudson 1974).

Early diagenesis just within the sediment was the time of maximum bacterial metabolism of organic matter, and this is reflected in the very wide range of $\delta^{13}C$ values in the precipitated carbonates. Algal limestones are consistently light in carbon (Tan & Hudson 1974, Andrews 1986). Septarian concretions also have light carbon and normal-depositional oxygen. Many of the early diagenetic calcites are relatively Mg-rich, suggesting that they were Mg calcites originally. The good preservation of these cal-

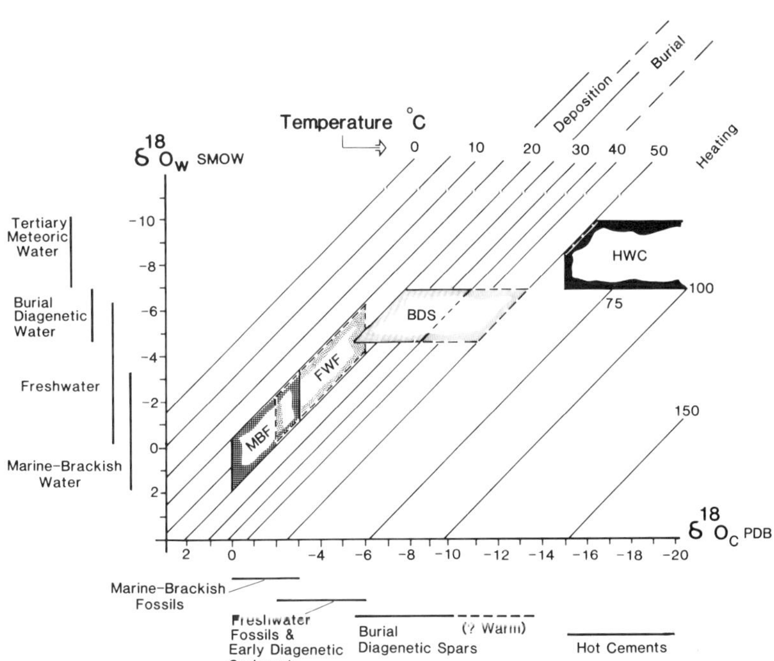

FIG. 10. Interpretation of oxygen isotopic composition of carbonates ($\delta^{18}O_c$, PDB) in terms of temperature and isotopic composition of water ($\delta^{18}O_w$, SMOW). Temperature lines correspond to equilibrium precipitates ($\delta^{18}O_c$) from water ($\delta^{18}O_w$) according to the standard palaeotemperature equation (Anderson & Arthur 1983). MBF = marine-brackish fossils, FWF = freshwater fossils, BDS = burial diagenesis spars, HWC = hot-water cements. Data from Tan & Hudson (1974) and Andrews (1986). Discussion in text.

cites, even when very fine-grained as in 'algal' tube calcifications, shows that they were not aragonite: indeed we have no evidence of inorganic aragonite precipitation. All the early calcites are non-ferroan, being formed either in oxic environments or in the sulphate reduction zone. Although these early diagenetic carbonates are very informative, they are volumetrically minor.

There ensued a period in which carbonates were evidently not precipitated, and this included initial compaction of the rocks, fracturing of shells, and pressure solution of aragonite against quartz in sandstones. Then, for unknown reasons, calcite precipitation started again, giving the main phase of burial cementation. This was pervasive throughout the Group affecting shales (except when very impermeable), limestones, and sandstones. The importance of cross-formational water flow is indicated by the ferroan nature of the calcites, because Fe (and Mn) cannot have been obtained from within the purer sandstones and limestones. The formidable nature of the cementation problem at this stage is indicated by Berner's (1980, p. 124) calculation that at least 300 000 pore volumes are needed. The $\delta^{13}C$ of the late spars is virtually identical to that of the commonest fossils, strongly suggesting that dissolution of fossil carbonates was the main source of carbon, rather than organic reactions (Fig. 9). The $\delta^{18}O$ values (Fig. 10), assuming a maximum temperature of formation of 35°C (500 m burial, 30°C km^{-1} geothermal gradient, 20°C surface temperature) imply $\delta^{18}O$ of porewaters of between -4.5 and $-7‰$ (Fig. 10). These overlap with meteoric water values calculated from freshwater fossils and early freshwater precipitates. The slightly lighter values are also reasonable because much of the meteoric water which entered the sediments during burial no doubt flowed down porous aquifers, having been precipitated on the hilly hinterland rather than in the basin itself. There are several potential aquifers within the Jurassic succession, both within the Great Estuarine Group itself and beneath it (Hudson 1983). The lightest supposed burial cements, from some of the sandstone concretions, have $\delta^{18}O$ values down to $-13.7‰$. These imply somewhat lighter, or warmer, formation waters. It should be emphasized that, although meteoric-derived, these burial diagenetic waters were probably by no means 'fresh'. They were certainly reducing, Fe-rich, and probably more or less saline. Polyphase growth of fibrous calcite veins shows that more than one groundwater type traversed these rocks.

Again we have an obscure interval before the heating associated with Palaeocene volcanism apparently triggered the precipitation of small quantities of calcite and dolomite cement. At this time the porewaters could have been isotopically lighter in oxygen, because Palaeocene rainwater probably had a $\delta^{18}O$ down to $-12‰$ (Taylor & Forester 1971). However by the time it got into the Jurassic rocks (and assuming it got there at all), it would probably already have exchanged with the heated rocks it traversed. Figure 10 assumes $\delta^{18}O$ between -7 and $-10‰$ for Palaeocene groundwater, and this implies tempertures around 100°C for the precipitation of the lightest calcites we observed. These calcites often only partially fill the pores in which they formed, and are quantitatively minor compared to normal burial cements.

After this, very little seems to have happened to the Great Estuarine Group rocks. They were uplifted, faulted, jointed and exposed to the modern erosional cycle. But the late Quaternary glaciations stripped off any weathering mantle that may once have enveloped them, and the rocks we now see are, apart from trivial recent weathering, probably much as they were in the Eocene. Indeed, as we have stressed how small the Tertiary igneous effects were, some of them are probably much as they were in the Cretaceous.

ACKNOWLEDGMENTS: We thank Jim Harris for use of information on heavy minerals from his thesis, and our several co-workers, particularly Francis Tan, Jim Marshall, Tony Fallick, Joe Hamilton, Rob Raiswell and Mark Wilkinson, for discussions in the field and laboratory. JEA thanks NERC for a studentship held at the University of Leicester and for the use of facilities at SURRC, East Kilbride and the University of Edinburgh. We thank Stuart Haszeldine and Gill Harwood for helpful reviews.

References

AIGNER, T. 1985. *Storm Depositional Systems*. Lecture notes in Earth Sciences, 3. Springer-Verlag, Berlin.

ANDERSON, T. F. & ARTHUR, M. A. 1983. Stable isotopes of oxygen and carbon and their application to sedimentologic problems. *In*: ARTHUR, M. A. *et al.* (eds). *Stable Isotopes in Sedimentary Geology*. Short Course Notes No. 10, Society of Economic Paleontologists and Mineralogists, Tulsa, Oklahoma.

ANDREWS, J. E. 1985. The sedimentary facies of a late Bathonian regressive episode: the Kilmaluag and Skudiburgh Formations of the Great Estuarine Group, Inner Hebrides, Scotland. *Journal of the Geological Society of London*, **142**, 1119–37.

—— 1986. Microfacies and geochemistry of Middle Jurassic algal limestones from Scotland. *Sedimentology*, **33**, 499–520.

—— 1987. Jurassic clay-mineral assemblages and their post depositional alteration: upper Great Estuarine Group, Scotland. *Geological Magazine*, **124**, 261–71.

——, HAMILTON, P. J. & FALLICK, A. E. In press. The geochemistry of early diagenetic dolostones from a low-salinity Jurassic lagoon. *Journal of the Geological Society of London*,

ASTIN, T. R. 1986. Septarian crack formation in carbonate concretions from shales and mudstones. *Clay Minerals*, **21**, 617–31.

BAKER, P. & KASTNER, M. 1981. Constraints on the formation of sedimentary dolomite. *Science*, **213**, 214–6.

BARKER, C. E. 1983. Influence of time on metamorphism of sedimentary organic matter in liquid-dominated geothermal systems, western North America. *Geology*, **11**, 384–88.

BATHURST, R. G. C. 1975. *Carbonate Sediments and their Diagenesis* (2nd edn). Elsevier, Amsterdam.

BERNER, R. A. 1968. Rate of concretion growth. *Geochimica et Cosmochimica Acta*, **32**, 477–83.

—— 1980. *Early Diagenesis: a Theoretical Approach*. Princeton University Press.

BRADSHAW, M. J. & FENTON, J. P. G. 1982. The Bajocian 'Cornbrash' of Raasay, Inner Hebrides: palynology, facies analysis and a revised geological map. *Scottish Journal of Geology*, **18**, 131–45.

CURTIS, C. D. 1977. Sedimentary geochemistry: environments and processes dominated by the involvement of an aqueous phase. *Philosophical Transactions of the Royal Society of London*, Series A, **286**, 353–72.

DAGLEY, P. & MUSSETT, A. E. 1986. Palaeomagnetism and radiometric dating of the British Tertiary Igneous Province: Muck and Eigg. *Geophysical Journal of the Royal Astronomical Society*, **85**, 221–42.

DICKIN, A. P. & JONES, N. W. 1983. Isotopic evidence for the age and origin of pitchstones and felsites, Isle of Eigg, NW Scotland. *Journal of the Geological Society of London*, **140**, 691–700.

EL-SHAHAT, A. & WEST, I. M. 1983. Early and late lithification of aragonitic bivalve beds in the Purbeck Formation (Upper Jurassic–Lower Cretaceous) of southern England. *Sedimentary Geology*, **35**, 15–41.

EMELEUS, C. H. 1983. Tertiary igneous activity. *In*: CRAIG, G. Y. (ed.). *Geology of Scotland*, pp. 357–97. Scottish Academic Press, Edinburgh.

GARRELS, R. M. & MACKENZIE, F. T. 1971. *Evolution of the Sedimentary Rocks*. Norton, New York.

HARRIS, J. P. 1984. Environments of deposition of Middle Jurassic sandstones in the Great Estuarine Group, N.W. Scotland. *Unpublished PhD Thesis*. University of Leicester.

—— & HUDSON, J. D. 1980. Lithostratigraphy of the Great Estuarine Group (Middle Jurassic), Inner Hebrides. *Scottish Journal of Geology*, **16**, 231–50.

HUDSON, J. D. 1963. The recognition of salinity-controlled mollusc assemblages in the Great Estuarine Series (Middle Jurassic) of the Inner Hebrides. *Palaeontology*, **6**, 318–26.

—— 1964. The petrology of the sandstones of the Great Estuarine Series, and the Jurassic palaeogeography of Scotland. *Proceedings of the Geological Association of London*, **75**, 499–528.

—— 1967. The elemental composition of the organic fraction, and the water content, of some recent and fossil mollusc shells. *Geochimica et Cosmochimica Acta*, **31**, 2361–78.

—— 1968. The microstructure and mineralogy of the shell of a Jurassic mytilid (Bivalvia). *Palaeontology*, **11**, 168–82.

—— 1970. Algal limestones with pseudomorphs after gypsum from the Middle Jurassic of Scotland. *Lethaia*, **3**, 11–40.

—— 1980. Aspects of brackish-water facies and faunas from the Jurassic of north-west Scotland. *Proceedings of the Geological Association of London*, **91**, 1 & 2, 99–105.

—— 1983. Mesozoic sedimentation and sedimentary rocks in the Inner Hebrides. *Proceedings of the Royal Society of Edinburgh*, **83b**, 47–63.

—— & HARRIS, J. P. 1979. Sedimentology of the Great Estuarine Group (Middle Jurassic) of North-West Scotland. *In*: *Sedimentation Jurassique W. Européen*, pp. 1–13. Association des Sedimentologistes Francais, Publ. Speciale no. 1.

LAND, L. S., BEHRENS, E. W. & FRISHMAN, S. A. 1979. The ooids of Baffin Bay, Texas. *Journal of Sedimentary Petrology*, **49**, 1269–78.

LERCHE, I., YARZAB, R. F. & KENDALL, C. G. STC. 1984. Determination of paleoheat flux from vitrinite reflectance data. *Bulletin of the American Association of Petroleum Geologists*, **68**, 1704–17.

MARSHALL, J. D. 1981. Zoned calcites in Jurassic ammonite chambers: trace elements, isotopes and neomorphic origin. *Sedimentology*, **28**, 867–87.

—— 1982. Isotopic composition of displacive fibrous calcite veins: reversals in pore-water composition trends during burial diagenesis. *Journal of Sedimentary Petrology*, **52**, 615–30.

MEADE, R. H. 1966. Factors influencing the early stages of the compaction of clays and sands—review. *Journal of Sedimentary Petrology*, **36**, 1085–101.

MORTON, A. C. 1984. Stability of detrital heavy minerals in Tertiary sandstones from the North Sea Basin. *Clay Minerals*, **19**, 287–308.

PETTIJOHN, F. J., POTTER, P. E. & SIEVER, R. 1972. *Sand and Sandstone*. Springer-Verlag, New York.

PRICE, L. C. 1983. Geologic time as a parameter in organic metamorphism, and vitrinite reflectance as an absolute palaeo-geothermometer. *Journal of Petroleum Geology*, **6**, 5–38.

RICHTER, D. K. 1983. Calcareous ooids: a synopsis. *In*: PERYT, T. M. (ed.). *Coated Grains*. Springer-Verlag, Berlin.

RIDING, J. B. 1984. The palynology of the Tobar Cearn Siltstone Member, Staffin Shale Formation (Jurassic, Callovian/Oxfordian), Strathaird, southern Skye. *Reports of the British Geological Survey*, **16** (10), 1–5.

RIEKE, H. H. & CHILINGARIAN, G. V. 1974. *Compaction of Argillaceous Sediments*. Developments in Sedimentology, 16. Elsevier, Amsterdam.

SANDBERG, P. A. & HUDSON, J. D. 1983. Aragonite relics in Jurassic calcite-replaced bivalves. *Sedimentology*, **30**, 879–92.

SHACKLETON, N. J. & KENNETT, J. P. 1975. Palaeotemperature history of the Cenozoic and the initiation of Antarctic glaciation: Oxygen and carbon isotope analyses in the DSDP sites 277, 279 and 281. *In*: *Initial Reports of the Deep Sea Drilling Project*, **29**, 743–55. US Government Printing Office, Washington.

STONELEY, R. 1983. Fibrous calcite veins, overpressures, and primary oil migration. *Bulletin of the American Association of Petroleum Geologists*, **67**, 1427–8.

SYKES, R. M. 1975. The stratigraphy of the Callovian–Oxfordian stages (Middle–Upper Jurassic) in northern Scotland. *Scottish Journal of Geology*, **11**, 51–78.

TAN, F. C. & HUDSON, J. D. 1971. Carbon and oxygen isotopic relationships of dolomites and co-existing calcites, Great Estuarine Series (Jurassic), Scotland. *Geochimica et Cosmochimica Acta*, **37**, 755–67.

—— & —— 1974. Isotopic studies of the palaeoecology and diagenesis of the Great Estuarine Series (Jurassic) of Scotland. *Scottish Journal of Geology*, **10**, 91–128.

——, —— & KEITH, M. L. 1970. Jurassic (Callovian) paleotemperatures from Scotland. *Earth and Planetary Science Letters*, **9**, 421–6.

TAYLOR, H. P. JR & FORESTER, R. W. 1971. Low-O^{18} igneous rocks from the intrusive complexes of Skye, Mull and Ardnamurchan, western Scotland. *Journal of Petrology*, **12**, 465–97.

WALKER, G. P. L. 1970. The distribution of amygdale minerals in Mull and Morvern (western Scotland). *In*: MURTY, T. V. V. E. R. K. & RAO S. S. (eds). *West Commemorative Volume*, pp. 181–94. University of Sagar, India.

WAPLES, D. 1980. Time and temperature in petroleum formation: application of Lopatin's method to petroleum exploration. *Bulletin of the American Association of Petroleum Geologists*, **64**, 916–26.

WATSON, J. V. 1985. Northern Scotland as an Atlantic–North Sea divide. *Journal of the Geological Society of London*, **142**, 221–43.

J. D. HUDSON, Department of Geology, University of Leicester, Leicester LE1 7RH, UK.

J. E. ANDREWS, School of Environmental Sciences, University of East Anglia, Norwich NR4 7TJ, UK.

Oxygen-isotope studies of clastic diagenesis in the Lower Cretaceous Viking Formation, Alberta: implications for the role of meteoric water

Fred J. Longstaffe & Avner Ayalon

SUMMARY: The oxygen-isotope compositions and paragenetic sequence of diagenetic minerals from the Lower Cretaceous Viking sandstone and conglomerate from south-central Alberta, Canada have been used to identify changes in porewater composition during diagenesis and to relate these changes to major geological events within the western Canada sedimentary basin. Large-scale influx of meteoric water has played an important role in diagenesis of this unit, especially following uplift of the basin in early Eocene time.

Diagenetic phases include early kaolinite (δ^{18}O SMOW = +24.9 to +28.2 ‰), siderite (δ^{18}O SMOW = +22.0 to +23.9 ‰; δ^{13}C PDB = −6.3 to −1.3 ‰), calcite (δ^{18}O SMOW = +26.0 ‰; δ^{13}C PDB = −0.4 ‰) and chlorite (δ^{18}O SMOW = +10.7 − +13.4 ‰), followed in order by dolomite (δ^{18}O SMOW = +19.3 − +22.1 ‰; δ^{13}C PDB = −5.9 to −2.6 ‰), calcite (δ^{18}O SMOW = +13.9 − +15.9 ‰; δ^{13}C PDB = −8.4 to −6.1 ‰), ankerite (δ^{18}O SMOW = +16.3 − +17.9 ‰; δ^{13}C PDB = −12.4 to −3.5 ‰), kaolin group minerals (δ^{18}O SMOW = +12.0 − +15.7 ‰), and illite (δ^{18}O SMOW = +14.2 − +15.7 ‰) and illite/smectite (δ^{18}O SMOW = +13.8 − +16.5 ‰). Quartz overgrowths (δ^{18}O SMOW = +16.2 − +27.4 ‰) began crystallizing relatively early during burial diagenesis and continued to form at least up to the onset of late diagenetic formation of clay minerals.

The interpretation of these results is that shallow diagenesis, early in the burial history (glauconite, pyrite, calcite, chlorite), occurred largely in the presence of seawater-derived fluids, although a freshwater influence is indicated where siderite and(or) early kaolinite cements are abundant. As compaction and burial diagenesis proceeded (diagenetic chlorite, quartz overgrowths, dolomite), the porewaters became enriched in ^{18}O due to water/rock interaction. Burial diagenesis was terminated in the early Eocene by uplift related to the major Laramide Orogeny. Recharge of the basin by low ^{18}O meteoric water occurred at this time. The meteoric water then became involved in the formation of diagenetic quartz, calcite and ankerite (and the dissolution and albitization of feldspar) at or near maximum burial temperatures, and in the crystallization at lower temperatures of kaolin group minerals, illite and illite/smectite as the post-Eocene erosion progressed.

In this paper we report the results of an oxygen-isotope study of diagenetic minerals from the Lower Cretaceous Viking Formation over an area of about 20 000 km^2 in southwestern Alberta, Canada (Fig. 1). Similar studies of diagenetic minerals from Upper Cretaceous sandstones in the western Canada sedimentary basin have shown that waters of meteoric origin have played an important role in their diagenesis (Longstaffe 1983, 1984, 1986, Ayalon & Longstaffe, in press). The purpose of this study is to determine whether meteoric water has been involved on a regional scale in the diagenesis of Lower Cretaceous sandstones from the Viking Formation, and to deduce how the oxygen-isotope composition of porewaters in these rocks varied in response to depositional environment, burial, and the subsequent uplift and erosion of the basin that resulted from the major Laramide Orogeny in early Eocene time.

The oxygen-isotope compositions of formation waters provide important information concerning their origin (e.g. sea water, meteoric water) and modification during diagenesis. Diagenetic minerals should obtain oxygen-isotope signatures characteristic of the porewater and the temperature at which the crystallization occurred. The δ^{18}O values of these minerals can then be used to reconstruct some aspects of porewater evolution throughout diagenesis, provided that the paragenetic sequence of diagenetic minerals can be determined, and that the oxygen-isotope compositions of the minerals have remained substantially unchanged since crystallization.

At temperatures typical of sedimentary environments, most minerals do not experience significant oxygen-isotope exchange with water. Such exchange occurs only during mineral dissolution and precipitation, the potential for which increases as temperature rises. Of the phases most commonly formed during diagenesis, quartz is the least affected by such re-equilibration (Yeh & Savin 1977). Exchange of oxygen

From: MARSHALL, J. D. (ed.), 1987, *Diagenesis of Sedimentary Sequences*, Geological Society Special Publication No. 36, pp. 277–296.

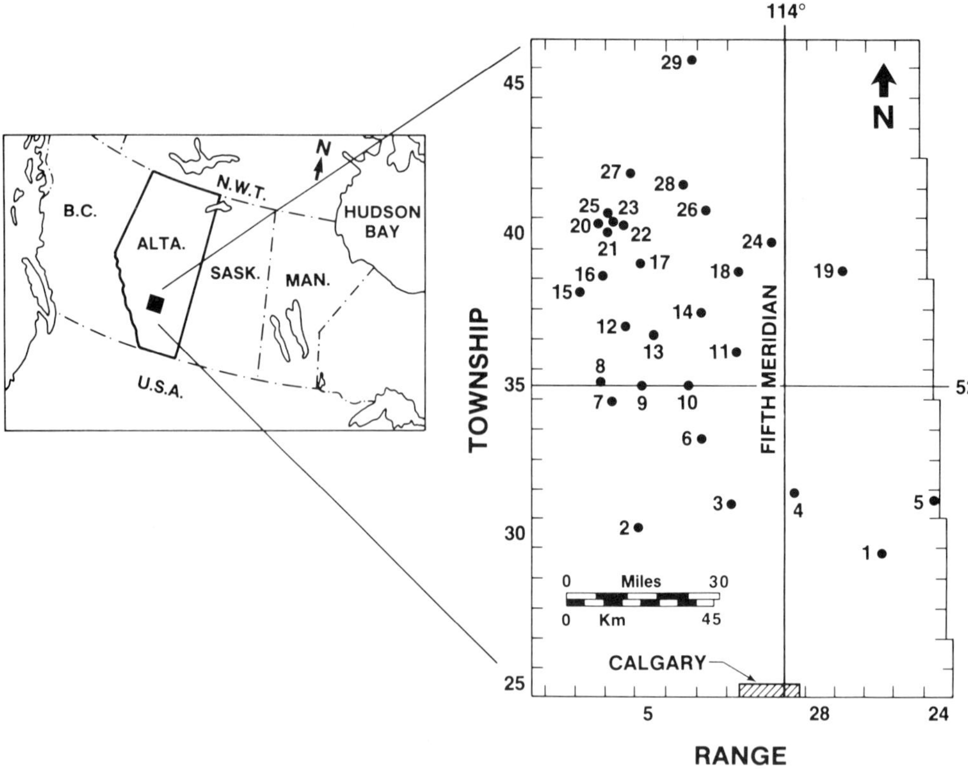

FIG. 1. Location of study area and sample locations. Numbers refer to samples listed in Tables 1 and 2.

between water and structural sites in most clay minerals also seems to be unimportant until temperatures well in excess of 100°C have been attained (Yeh & Savin 1976, 1977, Eslinger & Yeh 1981). Due to a greater ease of recrystallization, oxygen-isotope exchange can affect carbonate minerals (calcite more so than dolomite) at lower temperatures than silicate minerals (Clayton 1959); the possibility of isotopic exchange between calcite cement and porewater during subsequent stages of diagenesis should not be totally ignored. Within these constraints, however, the oxygen-isotope compositions of most diagenetic minerals should provide a valuable record of the temperatures and water $\delta^{18}O$ compositions prevailing during their formation.

Geological background

The Lower Cretaceous Viking Formation (Slipper 1918) has been extensively studied because of its importance as a hydrocarbon reservoir. Most recently, the sedimentology of the study area has been discussed by Grant (1985), Robb (1985), Dean (1986) and Hein *et al.* (1986); the reader is referred to these studies for a comprehensive listing of other important works concerning the Viking Formation, and for detailed discussions of the stratigraphy and sedimentology of the unit.

During the earlier part of late Albian time, the interior of western North America was inundated by transgressive seas from the north and, to a lesser extent, from the south (Williams & Stelck 1975, Weimer 1983), resulting in the deposition of marine shales (Joli Fou Formation). A regression towards the end of late Albian time marked the beginning of Viking deposition, the sediment being derived dominantly from the rising Cordillera to the west.

Hein *et al.* (1986) have characterized the Viking as a coarsening upward sequence of interbedded, fine-grained sandstone and shale, fine- or medium-grained sandstone, and chert-pebble conglomerate and pebbly sandstone. Within the study area, the Viking Formation can be up to 60 m thick. Three main depositional events have been proposed by Hein *et al.* (1986): (i) progradation of a shoreline-attached, clastic wedge during a regression, (ii) cut-and-fill of channels that have dissected the clastic wedge, and (iii) marine reworking of the sediment into sheet-like deposits of sandstones and conglomerates on the offshore shelf.

During the later portion of the early Turonian, a second major transgression occurred, ending Viking deposition; the overlying Lloydminster and Colorado shales were deposited at this time. The first phase of the Laramide Orogeny began in the late Cretaceous, causing thrusting and uplift along the eastern Cordillera (Taylor *et al.* 1964, Dickinson & Snyder 1978). To the east, continental deposition dominated through the late Cretaceous and into the early Tertiary as the sediment was shed from the Cordillera into the downwarped basin containing the Viking Formation (Beaumont 1984). Maximum burial of the Viking Formation probably occurred in the late Palaeocene or early Eocene (Taylor *et al.* 1964, Hitchon 1984). At this time the major pulse of the Laramide Orogeny (early Eocene) resulted in extensive deformation (overthrusting) in the eastern Cordillera and significant uplift of the sedimentary basin in Alberta; extensive erosion of the accumulated Tertiary and Upper Cretaceous rocks ensued (Taylor *et al.* 1964, Beaumont 1981).

Analytical procedures

Thirty-eight core samples of the Viking Formation have been examined (Fig. 1) using optical microscopy, scanning electron microscopy, energy dispersive spectrometry, X-ray diffraction, inductively coupled plasma spectrometry and oxygen- and carbon-isotope geochemistry. All of the analytical procedures, including the mineral separation techniques used to obtain subsamples for isotopic analyses of the diagenetic minerals (except quartz overgrowths), are discussed in Ayalon & Longstaffe (in press) and are not repeated here.

To separate the quartz overgrowths, samples were crushed using a mortar and pestle, and disaggregated in distilled water using an ultrasonic probe at low power. The samples were then dispersed in distilled water, and the <2 μm fraction removed using standard sedimentation techniques. The samples were dried and the 125–250 mesh size-fraction obtained by sieving. Quartz was extracted from this fraction following the method of Syers *et al.* (1968) and Jackson (1979). The quartz overgrowths were then isolated from the detrital quartz following the method of Lee & Savin (1985), and examined for authigenic morphology and purity using the scanning electron microscope.

The oxygen- and carbon-isotope data are presented in the usual δ-notation relative to Standard Mean Ocean Water (SMOW) for oxygen (Craig 1961) and the belemnite *Belemnitella americana* from the PeeDee Formation (PDB) for carbon (Craig 1957). To calculate the oxygen-isotope results for calcite and dolomite, values of 1.01025 and 1.01110, respectively, were employed for the total carbonate CO_2-phosphoric acid extracted CO_2 fractionation-factor (α) at 25°C (modified after Sharma & Clayton 1965). The calcite α was used in all calculations for siderite; using the α (25°C) recently reported by Rosenbaum & Sheppard (1986) for siderite would lower the values reported here by about 1.36‰. The dolomite α was used in all calculations for ankerite; using the new α (25°C) reported by Rosenbaum & Sheppard (1986) for ankerite would lower the values reported here by about 0.66 ‰. These differences are insufficient to affect the sense of the arguments and conclusions that follow. CO_2 was initially obtained from the siderite and ankerite by reaction at 25°C for 21 days; these experiments were repeated on fresh samples at 50°C for periods of 16, 24, 48 and 72 hr. While the CO_2 yields increased at the higher temperatures, no significant differences in the oxygen (temperature corrected) or carbon-isotope compositions of the carbonates were noted, an observation in agreement with Becker & Clayton (1972).

An oxygen-isotope CO_2-H_2O fractionation factor of 1.0412 at 25°C has been used throughout this study. During the course of the experiments a value for silica standard NBS-28 of $+9.62 \pm 0.12$‰ was obtained. Calcite standard NBS-20 gave an average $\delta^{18}O$ SMOW = $+26.50 \pm 0.02$‰ and $\delta^{13}C$ PDB = -1.17 ± 0.01‰. Except where noted otherwise, the analytical results reported are precise to better than ± 0.2 ‰. Higher uncertainties arise when the isotopic composition of a phase has been calculated from that of a binary mixture (e.g. chlorite and, in some samples, kaolinite; see Longstaffe 1986 and Ayalon & Longstaffe, in press).

Results

The petrology and diagenesis of the Viking Formation have been discussed by many workers, including Gammel (1955), Glaister (1958), Thomas & Oliver (1979), Amajor (1980), Foscolos *et al.* (1982), Dean *et al.* (1984), Grant (1985), Robb (1985), Dean *et al.* (1985) and Dean (1986). A detailed discussion of the diagensis of several oil fields within the Viking Formation is given in Dean *et al.* (in preparation). Preliminary stable isotope results have been reported by Longstaffe & Ayalon (1986).

Three main rock types have been studied: conglomerate, medium- and coarse-grained sandstone, and fine-grained sandstone; their framework-grain petrography is summarized here largely after Grant (1985), Robb (1985) and Dean (1986). The conglomerates can be classified as quartzarenites or sublitharenites (Folk 1968), and are comprised of framework grains (75–85%), cement (5–15%) and porosity (3–15%). Framework grains consist of chert fragments (45–80%), quartz (5–25%), rock fragments (mostly shale, trace–15%) and feldspar (mostly plagioclase, 1–12%). Ragged flakes of kaolinite (detrital?) are present in some samples. The dominant cements are quartz and clay minerals (kaolin group clay minerals > illitic clay minerals > smectite; Table 1). Minor siderite, dolomite and ankerite cements are also present.

The medium- and coarse-grained sandstones are quartz arenites or sublitharenites, and are composed of detrital grains (70–80%), cement (8–25%) and porosity (3–13%). Framework grains consist of quartz (5–50%), chert fragments (20–65%), rock fragments (mostly shale, 0–14%) and feldspar (mostly plagioclase, 2–30%). Detrital (?) kaolinite occurs in some samples. The main cements are quartz and clay minerals (kaolin group and illitic clays, minor smectite; Table 1).

The fine-grained sandstones are sublitharenites, and are composed of framework grains (55–80%), cements (10–50%) and porosity (trace–10%). The framework grains are quartz (35–60%), chert fragments (5–20%), rock fragments (mostly shale, 2–25%), and feldspar (mostly plagioclase, 1–6%). In some samples, dolomite grains are present. The cements are comprised of carbonate minerals (siderite, dolomite and ankerite; calcite is rare), quartz and clay minerals. Unlike the conglomerates and coarse-grained sandstones, diagenetic illitic clay minerals are more abundant than kaolin group minerals (Table 1).

Most of the detrital silica has $\delta^{18}O$ values between +14.0 and +16.9‰ (Table 2), characteristic of a sedimentary origin. Higher values (+17.6 to +20.0‰, Table 2) were obtained for a few samples of conglomerate and coarse-grained sandstone, probably reflecting the higher proportions of chert fragments versus quartz grains in these rocks.

The $\delta^{18}O$ values of the <2 μm clay fraction from two shale samples are +13.5 and +17.1‰ (Table 2); values of +17 to +20‰ have been reported for <2 μm clay fractions of Viking shales, mudstones and argillaceous sandstones located further to the east in Alberta (Longstaffe 1983). These values are similar to those obtained for other Cretaceous shales from the western Canada sedimentary basin (Longstaffe 1983, 1984, 1986, 1987), and are also typical of detrital clay mixtures from marine sediments (Savin & Epstein 1970a, b).

Of particular interest is that the <2 μm clay size-fractions from the Viking sandstones and

TABLE 1. *Relative percentage of clay minerals in the <2 μm size fraction*

Sample[1]	Depth (m)	Rock type	Kaolin group	Chlorite	Illite[2]	Smectite
2	2977.6	Shale	32	5	63	
4b	1870.6	Fsst[3]	13		87	
4c	1872.6	Fsst	5	10	85	
5a	1367.8	Congl[4]	47		18	35
5b	1373.4	Fsst	32		68	
7	2719.1	Fsst	18	6	76	
8a	2731.4	Csst[5]	60	4	35	1
8b	2734.6	Congl	63	3	23	11
10	2174.5	Shale	18	10	68	4
11	1903.5	Msst[6]	14	10	76	
12	2386.9	Msst	5	30	65	
13	2290.6	Fsst	21	10	69	
14b	1987.4	Fsst	15	10	75	
16b	2404.6	Fsst	5	5	90	
18	1904.5	Csst	60	5	23	12
19a	1544.6	Csst	75	5	20	
21	2257.8	Congl	91		9	
22	2229.7	Congl	40	15	45	
25	2294.3	Congl	31	5	60	4
26b	1923.3	Congl	47	8	30	15
28	1909.6	Fsst	30	8	62	

[1] See Fig. 1 for locations of samples. [2] Includes illite/smectite. [3] Fine-grained sandstone. [4] Conglomerate. [5] Coarse-grained sandstone. [6] Medium-grained sandstone.

conglomerates, which are comprised mostly of authigenic clay minerals and some silica, are also ^{18}O-rich (+15.1 to +20.8‰; Table 2). This result contrasts markedly with the much lower $\delta^{18}O$ values (+9 to +14‰) obtained for similar materials from Upper Cretaceous sandstones in Alberta; these low values have been interpreted to indicate the involvement of a sizeable fraction of low-^{18}O meteoric water in the crystallization of the diagenetic clays (Longstaffe 1983, 1984, 1986, 1987, Ayalon & Longstaffe, in press).

Diagenetic minerals

The Viking Formation has experienced a variety of diagenetic processes, including compaction, cementation, grain alteration and dissolution. Most pores have been at least partially filled by authigenic minerals. Diagenetic minerals that have been observed in this study include glauconite, iron oxide, siderite, pyrite, Fe-chlorite, quartz, calcite, albite, dolomite, ankerite, kaolin group minerals (kaolinite, dickite), illite, illite/smectite and smectite.

Iron oxide: Trace amounts of iron oxide occur as very thin coatings on some detrital grains. These 'dust rims' characteristically mark boundaries between detrital silica and authigenic quartz overgrowths and were probably inherited during sedimentation.

Glauconite: Glauconite is normally present in only minor quantities as rounded or oval-shaped grains (200–400 µm), and as matrix material compacted between rigid detrital grains. It is most likely of very early diagenetic origin. The glauconite occurs generally in offshore rather than nearshore portions of the study area.

Siderite: Siderite is most abundant in the finer-grained rocks, comprising up to 60% of some fine-grained sandstones; however it is patchy in distribution (Dean 1986). Where such relationships can be observed, most siderite appears to have formed rather early in diagenesis, prior to most other diagenetic cements (Fig. 2a). In some samples, siderite has filled all pore space, effectively preventing subsequent formation of other diagnetic phases. Framework grains in such samples are normally less compacted than in other areas (Dean 1986).

The $\delta^{18}O$ SMOW and $\delta^{13}C$ PDB values of the siderite range from +22.0 to +23.9‰, and −6.3 to −1.3‰, respectively (Table 2). Even lower $\delta^{18}O$ SMOW values (+18.1‰; $\delta^{13}C$ PDB = −4.6‰) have been reported by Dean (1986) for siderite from the southwesternmost portion of the study area, where a rooted sedimentary facies has been identified (Hein *et al.* 1986). Detailed stable isotope studies of the siderite are in progress (Connolly, personal communication, Staley, personal communication).

Pyrite: Pyrite is a common though volumetrically unimportant early diagenetic phase in some fine-grained sandstones. It occurs as framboid-like aggregates of small (0.5–1 µm) closely packed crystallites (Fig. 2b) on silica grains, or as individual, larger (4 µm) euhedra associated with the framboidal pyrite.

Chlorite: Authigenic, Fe-rich chlorite is volumetrically minor in the samples studied (Table 1). It occurs as small rosettes or fan-shaped clusters of pseudohexagonal crystals that have developed directly upon detrital grains, beginning early in the diagenetic history of the rock. Chlorite from <0.2 and (or) 0.2–0.5 µm mixtures with illite/smectite was calculated to have $\delta^{18}O$ values of +10.7 to +13.4‰ (Table 2) (for procedures, see Longstaffe 1986 and Ayalon & Longstaffe, in press); these values are precise only to ±1‰.

Calcite: Calcite is very uncommon in these rocks. It occurs both as a comparatively early cement between relatively uncompacted grains, coated by clay cements (Fig. 2c), and as a later phase, where it overlies earlier cements such as siderite and quartz, but is itself overlain by authigenic kaolin group minerals and illitic clay. The few samples of the later calcite have lower $\delta^{18}O$ SMOW (+13.9 to +15.9‰, Table 2) and $\delta^{13}C$ PDB (−8.4 to −6.1‰, Table 2) values than the one sample of early calcite ($\delta^{18}O$ SMOW = +26.0‰, $\delta^{13}C$ PDB = −0.4‰; Table 2, no. 13) that was analysed. Further isotopic studies of the calcite are in progress (Connolly, personal communication, Staley, personal communication).

Quartz: Silica is the most abundant authigenic cement in the Viking Formation; some pores have been completely filled by diagenetic quartz crystals, which can vary in length from 2 to 60 µm. Quartz overgrowth formation appears to have occurred throughout much of the diagenetic history of the rock. Some early kaolinite is engulfed by quartz overgrowths. An authigenic origin for much of this clay is suggested by its well crystallized morphology (Fig. 2d, e). In other samples, quartz overgrowths are post-dated by diagenetic kaolin group minerals (Fig. 2f) and illite or illite/smectite (Fig. 2g).

A wide variation in $\delta^{18}O$ was obtained for the diagenetic quartz (+16.2 to +27.4‰, Table 2). The fine- and medium-grained sandstones have $\delta^{18}O$ values in the lower end of this range (+16.2 to +20.8‰); the coarse-grained sandstones and

TABLE 2. *Stable isotope results*

Sample[1]	Depth (m)	Rock type	$\delta^{18}O$ SMOW ‰											$\delta^{13}C$ PDB ‰			
			<2 µm	AQ[6]	DQ	I/S	I	Chl	K	Calc	Dol	Ank	Sid	Calc	Dol	Ank	Sid
1a	1280.0	Fsst[2]									20.0				−5.0		
1b	1288.1	Fsst										17.9				−4.9	
2	2977.6	Shale	13.5														
3	2200.0	Csst[3]								15.9				−6.1			
										14.6				−7.2			
4a	1866.0	Fsst										17.3				−3.5	
4b	1870.6	Fsst		20.5	14.7	16.5	14.2				20.3				−2.6		
						15.4											
4c	1872.6	Fsst		17.1	16.0	14.7	15.3	10.7									
5a	1367.8	Congl[4]		25.6					15.7								
5b	1373.4	Fsst		19.5	15.0	14.1											
6	2200.0	Congl										16.3				−6.3	
7	2719.1	Fsst		16.9	15.3	14.2					19.3	17.0			−5.9	−7.7	
						14.3											
8a	2731.4	Csst	15.1	25.7	15.2	15.2											
8b	2734.6	Congl		22.6	15.8				13.0								
9	2468.0	Fsst									22.1				−2.8		
10	2174.5	Shale	17.1														
11	1903.5	Msst[5]	20.5		14.0								23.6				−2.8
12	2386.9	Msst		17.7	16.9	15.5		11.6									
						16.1		11.7									
13	2290.6	Fsst	15.7	16.2	14.3					26.0				−0.4			
14a	1981.0	Fsst											23.9				−1.3
14b	1987.4	Fsst		17.2	15.3	16.3											
15	2537.0	Fsst											22.0				−1.5

TABLE 2. *continued*

Sample[1]	Depth (m)	Rock type	δ¹⁸O SMOW ‰											δ¹³C PDB ‰			
			<2 µm	AQ[6]	DQ	I/S	I	Chl	K	Calc	Dol	Ank	Sid	Calc	Dol	Ank	Sid
16a	2393.0	Congl							12.4								
16b	2404.6	Fsst				14.4											
						13.8											
17	2201.0	Fsst											23.3				−6.3
18	1904.5	Csst		19.4			15.7	13.4	12.0								
									13.8								
19a	1544.6	Csst		26.5	16.7												
19b	1548.4	Csst										16.6				−12.4	
20	2289.2	Congl							12.6								
21	2257.8	Congl		18.1	19.2				24.9								
									25.4								
22	2229.7	Congl	16.3	21.1	16.3												
23	2276.0	Congl							28.2								
24	1684.6	Csst	17.0	27.4	20.0												
25	2294.3	Congl	20.8	17.7	14.8												
26a	1923.0	Congl							15.4								
26b	1923.3	Congl		26.9	17.6												
27	2122.0	Csst							28.1								
28	1909.5	Fsst		20.8	14.0				26.9								
29	1730.0	Congl								13.9				−8.4			

[1] See Fig. 1 for sample locations. [2] Fine-grained sandstone. [3] Coarse-grained sandstone. [4] Conglomerate. [5] Medium-grained sandstone. [6] AQ = authigenic quartz, DQ = detrital quartz, I/S = illite/smectite, I = illite, Chl = chlorite, K = kaolin group minerals, Calc = calcite, Dol = dolomite, Ank = ankerite, Sid = siderite; all phases are diagenetic, except where noted.

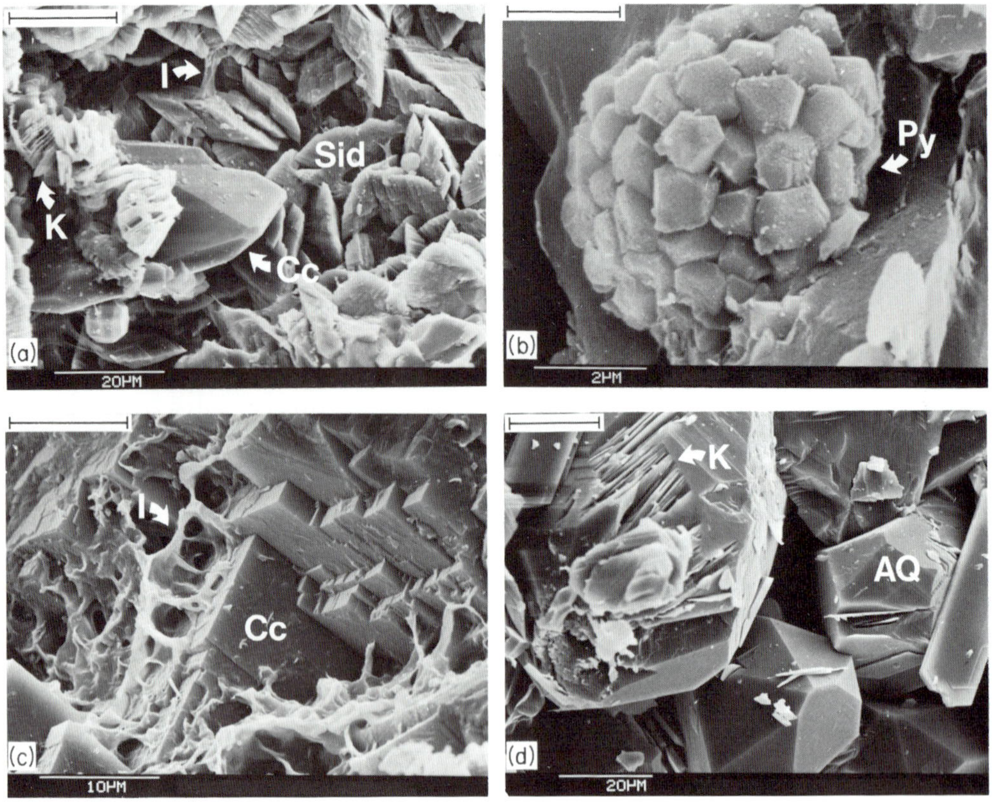

FIG. 2. (a) Diagenetic siderite (Sid), calcite (Cc), kaolinite (K) and illite/smectite (I). Scale = 20 µm. (b) Early diagenetic pyrite (Py) on detrital quartz. Scale = 2 µm. (c) Diagenetic calcite (Cc) filling pore and coated by illite and (or) illite/smectite (I). Scale = 10 µm. (d, e) Kaolin group minerals (K) engulfed by authigenic quartz (AQ). Scale = 20 µm. (f) Authigenic kaolin group minerals (K) filling pores; quartz overgrowths (AQ)

conglomerates tend to have higher $\delta^{18}O$ values (+17.7 to +27.4‰). This difference may indicate that some grain-size dependent condition affected silica diagenesis. That selective contamination by detrital quartz or chert fragments is wholly responsible for this shift in values is less likely, but cannot be ruled out entirely.

Feldspar: Diagenetic reactions involving feldspar (mostly plagioclase) include alteration, dissolution and the formation of albite overgrowths. Alteration includes the partial or complete replacement of feldspar by carbonate and (or) kaolinite; Dean (1986) has also noted albitization of plagioclase. Dissolution of plagioclase grains has also been observed, together with, in some cases, the appearance of authigenic albite overgrowths on the partially dissolved grain. The relative timing of plagioclase dissolution and authigenic albite crystallization is unclear; the two processes may be related (Boles 1982, Land 1984).

Dolomite/ankerite: Late diagenetic carbonate cement in the Viking Formation is comprised mostly of ankerite and (or) dolomite. Ankerite cement coats authigenic quartz, replaces detrital feldspar and dolomite, and fills pores. In turn, it is coated by authigenic clay minerals, typically illite or illite/smectite (Fig. 2h). In some samples, the Ca-rich nature of the ankerite suggests that it has directly replaced calcite (Dean 1986). In others, earlier calcite cement is replaced by diagenetic dolomite, which can be rimmed or partly replaced by ankerite. Further studies are in progress (Connolly, personal communication).

The ankerite has $\delta^{18}O$ SMOW and $\delta^{13}C$ PDB values of +16.3 to +17.9‰ and −12.4 to −3.5‰, respectively (Table 2). The diagenetic dolomite is slightly richer in both ^{18}O and ^{13}C ($\delta^{18}O$ SMOW = +19.3 to +22.1‰; $\delta^{13}C$ PDB = −5.9 to −2.6‰, Table 2). Unaltered detrital dolomite from the Viking Formation and overlying Upper Cretaceous units such as the Milk River sandstone has generally higher $\delta^{18}O$

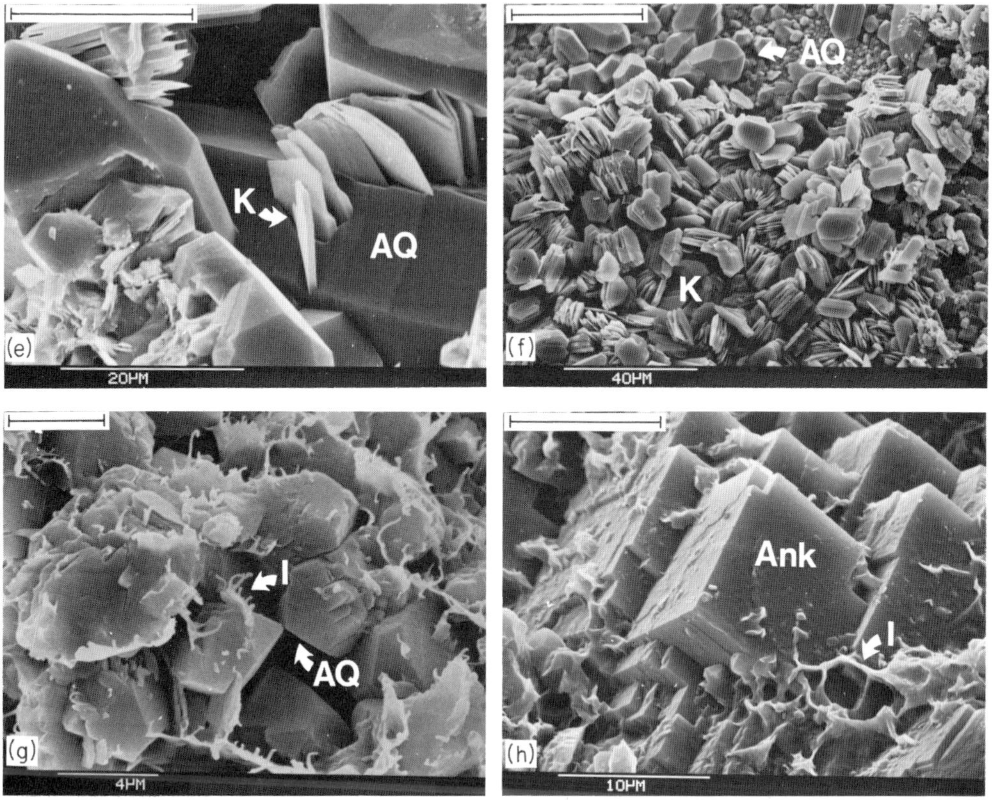

line the pore walls. Scale = 40 µm. (g) Authigenic illite/smectite and (or) illite (I) coating quartz overgrowths (AQ). Scale = 4 µm. (h) Authigenic ankerite (Ank) filling pore and coated by illite and (or) illite/smectite (I). Scale = 10 µm.

SMOW values (+26 to +28‰; Longstaffe 1983, 1984, unpublished data). Some contamination of the diagenetic dolomite samples by detrital dolomite may have occurred.

Kaolin group minerals: Kaolin group minerals (kaolinite and dickite) are particularly abundant as late diagenetic pore-fillers in the coarse-grained sandstones and conglomerates (Table 1), where they occur as loosely packed booklets of pseudohexagonal flakes (10–25 µm in diameter, Fig. 2f). In addition, as noted previously, some well crystallized, earlier formed kaolinite has been engulfed by quartz overgrowths (Fig. 2d, e). In other samples, more poorly crystallized (subhedral) flakes of kaolinite are associated with diagenetic quartz. These flakes may be detrital; alternatively, authigenic kaolinite can attain such a morphology due to dissolution during later stages of diagenesis (Hurst 1980).

Two populations of $\delta^{18}O$ values were obtained for the kaolinite group minerals (Table 2). The early kaolinite (mostly euhedral, but also some 'ragged' flakes) has $\delta^{18}O$ values of +24.9 to +28.2‰ (±0.5 where determined from quartz-kaolinite mixtures; Ayalon & Longstaffe, in press). The later diagenetic material has much lower values (+12.0 to 15.7‰; ±0.5). Some cross-contamination of samples between early and late kaolinite may have occurred.

Illite, illite/smectite and smectite: The 10 Å clay minerals are common diagenetic phases in the Viking Formation, especially in the fine- and medium-grained sandstones (Table 1). They occur as mixtures of illite (>90% non-expandable layers) and illite/smectite (65–80% non-expandable layers). Their most common modes of occurrence are as (i) webby to mat-like aggregations that coat grains and earlier cements (Fig. 2c), and (ii) filamentous material with curled projections that is attached to grains and earlier cements, and extends into pore space (Fig. 2g). The illitic clay minerals appear to be among the last major authigenic cements to form. They coat most previous cements, including some of the late diagenetic, pore-filling kaolin group minerals (see, e.g. Dean 1986). However, the presence of some earlier diagenetic illitic clays cannot be ruled out from petrographic data; such material can be very difficult to recognize, especially if it has become matted on grain surfaces. The illite/smectite and illite obtained from <0.2 or 0.2–0.5 μm size-fractions have $\delta^{18}O$ values of +13.8 to +16.5‰ and +14.2 to +15.7‰, respectively (Table 2).

Smectite is relatively uncommon, occurring mostly in the conglomerates or the coarse-grained sandstones as a late-stage pore-lining. It has the characteristic honeycomb-like mode of aggregation (Wilson & Pittman 1977).

Paragenetic sequence–summary: It is difficult to define the order of diagenetic mineral formation for an individual sample uniquely. However, the petrographic evidence from the core samples suggests that the paragenetic sequence proposed in Fig. 3 is generally applicable throughout the study area. Very early (synsedimentary) formation of glauconite was followed during shallow burial by the crystallization of pyrite and (or) siderite. Chlorite diagenesis also appears to have begun relatively early; its timing relative to siderite or pyrite crystallization is as yet unclear. Some calcite and kaolinite cement also formed early in diagenesis. Quartz crystallization occurred throughout much of diagenesis, followed by ankerite, authigenic kaolin group minerals and diagenetic illite, illite/smectite and (or) smectite. At least some diagenetic illitic clay postdates other authigenic clay minerals.

The timing of feldspar dissolution and the formation of albite overgrowths is difficult to bracket precisely from the petrographic record alone. However, most literature suggests that large-scale dissolution and albitization of feldspar begins at relatively high temperatures (*e.g.* 110–120°C, Boles 1982). The interval suggested in Fig. 3 is compatible with such observations (see below).

Discussion

The petrographic and oxygen-isotope results for the diagenetic minerals can be combined to deduce changes in porewater $\delta^{18}O$ during deposition, burial, uplift and erosion of the Viking sandstones and conglomerates. The relationship between the measured $\delta^{18}O$ of each diagenetic mineral and the possible combinations of water $\delta^{18}O$ and temperature are shown in Fig. 4(a, b). Our goal is to fit a porewater evolution curve to this grid in a manner that intersects the range of possible temperatures and water $\delta^{18}O$ values in an order consistent both with the observed paragenetic sequence (Fig. 3) and with geological events in the Viking Formation.

Some uncertainties in this approach should be acknowledged. The reliability of the limiting curves varies with our knowledge of the oxygen-isotope equilibrium fractionation factor (α) for mineral-water pairs at low temperatures. For minerals such as calcite it is well understood; much less is known for phases like Fe-chlorite. As discussed earlier, the potential for isotopic re-equilibration of phases like calcite (*i.e.* dissolution–reprecipitation), and the efficacy of mineral separation, could also affect the position of these curves.

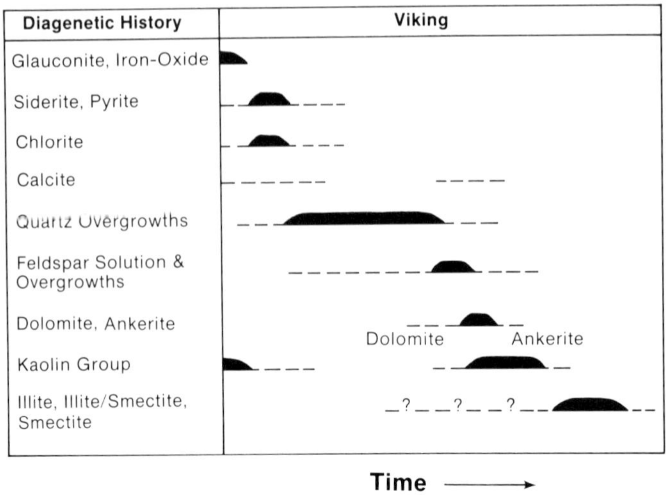

FIG. 3. General paragenesis of diagenetic minerals in the Viking Formation.

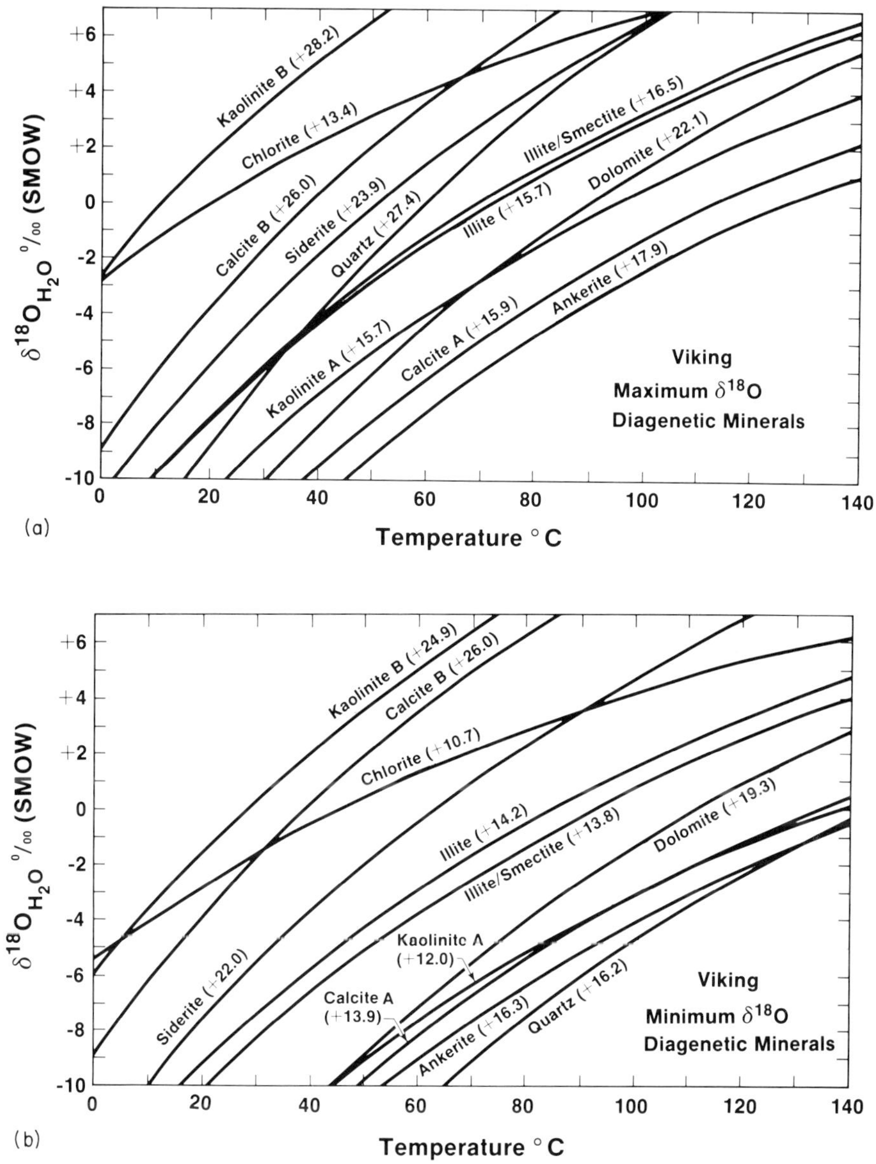

FIG. 4. $\delta^{18}O$ of porewater versus temperature for diagenetic minerals from the Viking Formation. The curves for the minerals are shown for the maximum and minimum $\delta^{18}O$ values of each phase using the following equations (T = degrees K): (1) $\Delta^{18}O$ chlorite-H_2O = 1.56 (10^6) T^{-2} − 4.7 (Wenner & Taylor 1971). (2) 10^3 ln α illite/smectite-H_2O = [(2.43 + 0.24E)](10^6) T^{-2} − 4.82 (Eslinger & Savin 1973, Yeh & Savin 1977; E = % expandable layers in illite/smectite). (3) 10^3 lnα calcite-H_2O = 2.78(10^6)T^{-2} − 2.89 (Friedman & O'Neil 1977 after O'Neil *et al.* 1969); also used for siderite; no correction for Fe-content was made. (4) 10^3 lnα kaolinite-H_2O = 2.5(10^6) T^{-2} − 2.87 (Eslinger 1971, Land & Dutton 1978). (5) 10^3 lnα quartz-H_2O = 3.38(10^6) T^{-2} − 2.90 (Friedman & O'Neil 1977, after Clayton *et al.* 1972). (6) 10^3 lnα dolomite-H_2O = 3.14(10^6) T^{-2} − 2.0 (Land 1983). (7) 10^3 lnα ankerite-H_2O = 2.78(10^6) T^{-2} + 0·32 (Dutton & Land 1985).

Present conditions in the Viking Formation provide an end-point for the porewater evolution curve. Available $\delta^{18}O$ values for formation waters range from −7 to −4 (Hitchon & Friedman 1969, Schwartz *et al.* 1981 and unpublished data). The present depth of the Viking sandstones and conglomerates analysed in this study ranges from about 1300 m in the east to 2700 m in the west. Assuming an average surface temperature of 5°C and an average geothermal gradient of about 24°C km^{-1} (Hitchon 1984), formation temperatures of 40–70°C can be calculated. These estimates are similar to corrected bottom-hole temperatures available for petroleum fields in the Viking Formation.

Estimation of the maximum temperatures reached during burial is more problematical. The palaeogeothermal gradient and (or) the thickness of sediment removed from the basin since maximal burial in the early Eocene has been examined by many workers, including Magara (1976), Hacquebard (1977), Beaumont (1981), Majorowicz & Jessop (1981), Hitchon (1984), Nurkowski (1984), Beaumont *et al.* (1985), Majorowicz *et al.* (1985) and Longstaffe (1986). According to Hitchon (1984), major changes in the geothermal gradient accompanied the second stage of the Laramide Orogeny; the geothermal gradient across Alberta ranged from ≈30°C km^{-1} in the west to ≈23°C km^{-1} in the east during the late Palaeocene. By the Eocene, the pattern was reversed, increasing from ≈21°C km^{-1} in the west to ≈27°C km^{-1} in the east, a subdued version of the present situation. In Hitchon's (1984) model, the major (i.e. second pulse) Laramide Orogeny produced a topographic high (maximum relief) in western Alberta.

The resulting high potentiometric surfaces led to regional recharge of cold meteoric water in the WSW, flowing to the ENE in the basin. This gravity-induced flow of cold meteoric water reversed the geothermal gradient pattern established by dominantly compactional flow prior to the early Eocene. Fluid flow, more so than the thickness (or rate of erosion) of the Phanerozoic section, became the dominant control upon the geothermal gradient and heat flow, at least for Cretaceous and Tertiary sandstones and conglomerates of the basin.

Hitchon's (1984) conclusions are based partly upon calculations of palaeogeothermal gradients made by Hacquebard (1977), using vitrinine reflectance (coal rank) and the equilibrium moisture contents of coals. More recently, Beaumont *et al.* (1985) studied organic reactions in shales through Alberta; they also concluded that since the early stages of the major Laramide Orogeny, the palaeogeothermal gradient increased systematically away from the Foothills, but noted that the gradient may have been more similar to that of the present day (27°C km^{-1} in the Foothills to 40°C km^{-1} in the Edmonton area). In general, however, no reversal in the heat-flow pattern has occurred since the early Eocene (Hitchon 1984, Majorowicz *et al.* 1985); the hydrodynamic regime in Alberta has remained fundamentally unchanged since that time.

From his data, Hacquebard (1977) calculated that the maximum sediment thickness eroded since the Tertiary ranged from 3000 m in WSW Alberta to 800 m in the ENE. Similar trends, though generally lower values (1900–900 m) were reported by Nurowski (1984) from his studies of the equilibrium moisture content of Upper Cretaceous coals. Using a crustal flexture model together with Hacquebard's (1977) data, Beaumont *et al.* (1985) suggested that between 1600 and 1900 m of overburden were removed from the study area.

From the available information on palaeogeothermal gradients, maximum burial depths, and estimated average early Eocene surface temperatures (+16°C; Piel 1971), a range of maximum burial temperatures between 100 and 150°C can be calculated for the study area. The optimum estimate lies in the range of 110°–130°C, with the temperature maximum decreasing to the ENE. This estimate is compatible with that indicated by the smectite to illite transformation in Viking shales from the study area (Dean 1986).

Early (shallow) diagenesis (stage I)

The porewater evolution curve proposed to explain the paragenetic sequence of diagenetic minerals, and their $\delta^{18}O$ values, is shown in Fig. 5, together with appropriate portions of the curves from Fig. 4. Given the shallow marine setting for most of the Viking sediments, the most probable starting composition for porewater during low-temperature formation (≈20°C) of early diagenetic phases (chlorite, calcite B; Fig. 5a) was sea water ($\delta^{18}O$ = 0‰). The water may have been slightly more ^{18}O-poor (e.g. $\delta^{18}O$ = −1‰), given possible early contributions by meteoric water (see below), potential variations in the primary $\delta^{18}O$ values of entrapped bottom waters (Savin & Yeh 1981) or, less likely, the effects of interaction and exchange with underlying material (Lawrence *et al.* 1975). The $\delta^{13}C$ value of the early calcite (−0.4‰) is typical of marine carbonates derived from CO_2 of inorganic origin (Land 1980).

The $\delta^{18}O$ values of the early kaolinite (kaolinite B, Fig. 5a) are compatible with crystallization from sea water ($\delta^{18}O$ = 0 ± 2‰ at 20°C). However, geochemically it is much more likely that the kaolinite formed in the presence of a significant fraction of fresh water (*e.g.* Bjørlykke *et al.* 1979, Hurst & Irwin 1982); waters with $\delta^{18}O$ values of −5 to −2‰ are indicated at 5°C (Fig. 4); by 15°C, water with $\delta^{18}O$ = −3 to 0‰ is required.

It is even more difficult to envisage the formation of the early diagenetic siderite from sea water alone (*i.e.* $\delta^{18}O$ = 0‰). First, crystallization temperatures in excess of 50°C would be required (still higher temperatures result, following Rosenbaum & Sheppard 1986). Second, formation of early diagenetic siderite is geochemically much simpler in brackish to freshwater environments (Berner 1971, Gautier 1982, Postma 1982).

The isotopic data support the involvement of meteoric water at low temperatures in the formation of the siderite. First, its relatively high $\delta^{13}C$ values (average $-3‰$; Table 2) suggest that isotopically very light CO_2 normally associated with thermal decarboxylation of organic matter (e.g. Irwin *et al.* 1977) was unimportant in siderite crystallization. Instead, the $\delta^{13}C$ values are typical of many early, shallow diagenetic siderites, and of CO_2 of dominantly inorganic origin. Some contribution by ^{13}C-rich CO_2 derived by bacterial fermentation at relatively low temperatures (<50–60°C) cannot be ruled out entirely (Dean 1986). Second, from the oxygen-isotope data, siderite crystallization at temperatures between 20 and 40°C would require water with $\delta^{18}O$ of -7 to $-2‰$ (Fig. 4). Even lower $\delta^{18}O$ values for the water would result following Rosenbaum & Sheppard (1986); only slightly higher values are anticipated from an experimentally determined equilibrium oxygen-isotope, siderite-water relationship. The even lower $\delta^{18}O$ for siderite reported by Dean (1986) from a rooted zone (i.e. barrier island, delta, shoreline) in the southwestern portion of the study area further indicates that fresh water has been important in siderite genesis.

A possible implication of these conclusions is that despite the shallow marine setting proposed by Hein *et al.* (1986) for the Viking sediments, fresh water was introduced into at least some sedimentary facies at certain localities, early in diagenesis. This influx may have been achieved through hydraulic connection to freshwater aquifers or by recharge of locally (and temporarily?) emergent zones; either process could help to account for the patchiness of siderite occurrence in the Viking sandstones.

Burial diagenesis (stage II)

With increasing depth of burial, the $\delta^{18}O$ of the dominantly marine porewater can be expected to increase due to dissolution of rock fragments, the possibility of isotopic exchange with phases like calcite and (or) contributions from the dewatering of clay minerals (Clayton *et al.* 1966, Hitchon & Friedman 1969, Suchecki & Land 1983). Following such a porewater evolution path with increasing temperature, the fields for early diagenetic calcite and chlorite are left and those for authigenic quartz and then dolomite are entered (stage II, Fig. 5b), in keeping with the paragenetic sequence (Fig. 3).

The extent to which the porewater became enriched in ^{18}O during compaction and burial is difficult to determine. Involvement of very ^{18}O-rich waters derived from Palaeozoic carbonates seems most unlikely; instead, the maximum $\delta^{18}O$ of the porewater probably did not exceed $\approx +6‰$, because of the buffering of the water composition by clay mineral reactions (*e.g.* Suchecki & Land 1983). Lower values, like those shown in Fig. 5(b) ($+2 \pm 1‰$) are even more likely. First, the field for authigenic dolomite cannot be intersected at maximum burial temperatures by water compositions much greater than $+3 \pm 1‰$; second, maximum $\delta^{18}O$ values for water of $\approx +3‰$ can be calculated from a combination of fluid inclusion temperatures and the oxygen-isotope compositions of associated diagenetic quartz from other Lower Cretaceous sandstones and conglomerates in Alberta which have reached much higher temperatures (≈190°C) during burial (Tilley & Longstaffe 1986).

Maximum burial and uplift (stage III)

While the oxygen-isotope requirements for dolomite can be satisfied at or near temperatures associated with maximum burial, the range in $\delta^{18}O$ of the diagenetic quartz and subsequent carbonate cements (calcite, ankerite) cannot be explained unless a significant lowering in porewater $\delta^{18}O$ occurred at, or shortly following, maximum burial of the unit (Fig. 5b). This change could only be accomplished by a pervasive influx of low-^{18}O meteoric water ($\delta^{18}O \approx -15‰$, see Taylor 1974). Such an occurrence is entirely consistent with the geological development of the basin; maximum burial coincided with maximum relief which, in turn, led to a major, regional influx of fresh water into the Cretaceous sandstones and conglomerates (Hitchon 1984). The exact slope of this portion of the porewater evolution curve (Fig. 5b, stage III) is idealized; a net cooling effect arising from uplift and the recharge of cold meteoric water can be expected.

The oxygen-isotope results for diagenetic quartz from the coarse-grained sandstones and conglomerates overlap much of this part of the porewater evolution curve, as well as the earlier stage involving compactional waters (stage II, Fig. 5b). In contrast, diagenetic quartz from the fine- and medium-grained sandstones has $\delta^{18}O$ values confined almost entirely to this low-^{18}O segment of the curve. The reason for this observation is not entirely clear; one explanation is that some of the finer grained rocks had better hydraulic connections to adjacent units (as

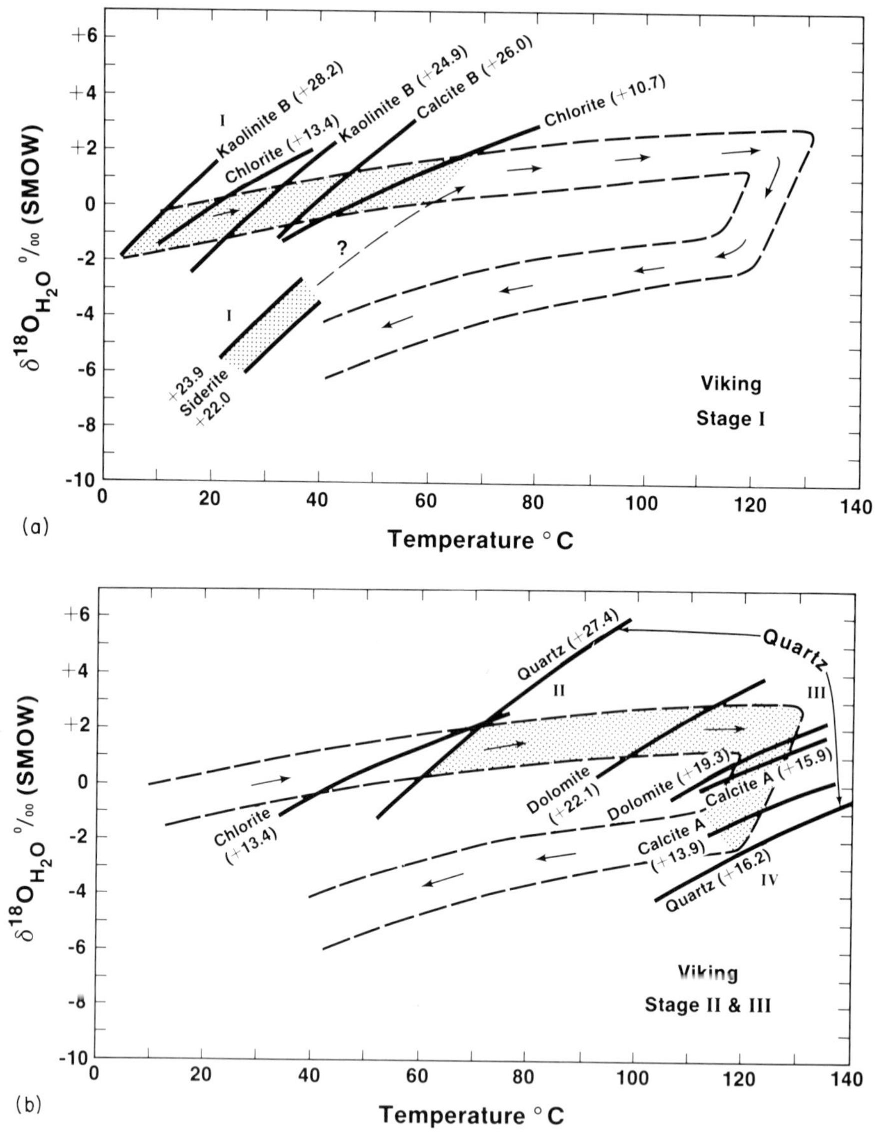

FIG. 5.

proposed earlier for siderite formation). Another explanation is that dissolution of chert fragments released abundant silica for diagenesis only in the coarser grained rocks. Quartz cementation in the finer grained sandstones may have required externally derived silica, carried into the unit by meteoric water. The Fe required for ankeritization of calcite could have been transported in the same manner. Still other scenarios are possible; no unequivocal explanation is available at present.

The carbon-isotope values of the carbonate cements show some interesting trends. Diagenetic dolomite, which is interpreted to have formed mostly from evolved, compaction-derived porewater at maximum burial, has $\delta^{13}C$ values (-5.9 to $-2.6‰$; average $= -4.1‰$; Table 2) very similar to the early diagenetic siderite. Such values are more weighted towards inorganic (*e.g.* pre-existing calcite) than organic carbon reservoirs. Low-^{13}C CO_2, such as can be derived by thermal decarboxylation of organic matter, appears to have been relatively unimportant. In contrast, the few samples of the later diagenetic calcite and ankerite, which apparently formed at high temperatures in the presence of a sizeable

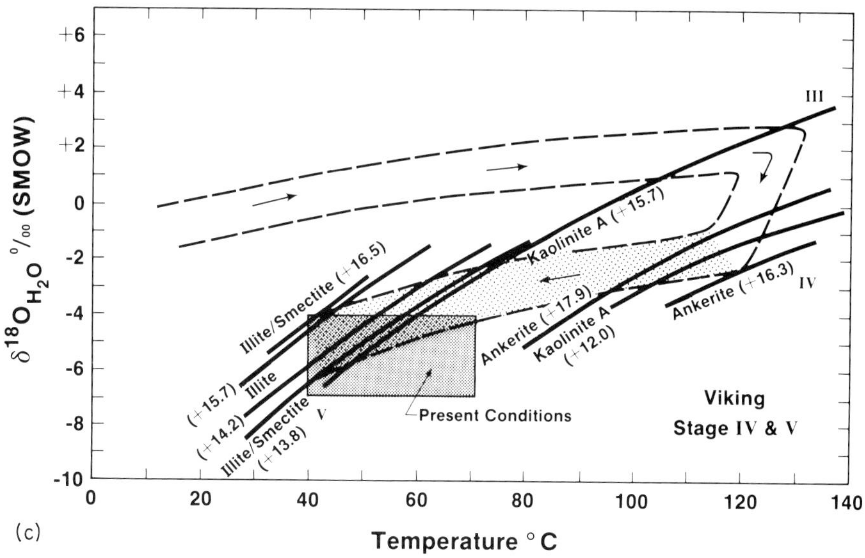

FIG. 5. Idealized porewater evolution path ($\delta^{18}O$ of water versus temperature) for the Viking sandstones and conglomerates. The curve has been deduced from the paragenetic sequence (Fig. 3) and the oxygen-isotope compositions (Fig. 4a,b) of the diagenetic minerals. (a) Stage 1, early (shallow) diagenesis; (b) stage II, burial diagenesis and stage III, maximum burial and uplift; (c) stage IV, erosion and stage V, present day.

fraction of meteoric water, have significantly lower $\delta^{13}C$ values (-12.4 to -3.5‰; average = -7.1‰) which demand involvement of organically derived CO_2. Meteoric water may have helped to deliver CO_2 (as well as Si, Fe?) which was released into shale porewaters by the maturation of organic matter, and expelled into the coarser grained units. Alternatively, the surface-derived meteoric water may have originally dissolved significant quantities of low-^{13}C soil CO_2.

Erosion (stage IV)

The crystallization of the late stage, authigenic kaolin group minerals (kaolinite A, Fig. 5C) and illitic minerals largely post-dates the formation of diagenetic quartz and associated carbonate cements (Fig. 3). To accommodate these observations and the $\delta^{18}O$ values of the authigenic clay minerals requires that the porewater substantially declines in temperature. The most logical trajectory is towards the present range of temperature and porewater $\delta^{18}O$ (stage V, Fig. 5c). Such cooling can be expected as a result of the substantial erosion that followed the early Eocene uplift of the Viking Formation.

The porewater evolution curve ends within the field known for present conditions in the study area; this range overlaps with much of the isotopic data for the authigenic illite and illite/smectite (stage V, Fig. 5c). The occurrence of illitic clays as the last major diagenetic phase (Fig. 3) makes it at least possible that their formation is continuing at present. Porewaters with suitable K^+/H^+, for example, could have evolved from the originally fresh water as water/rock interaction progressed (e.g. feldspar dissolution), and the ionic strength of the porewater increased. Compositional variations (e.g. % illite) in the illite/smectite are controlled by water chemistry rather than temperature in the Viking sandstones and conglomerates (Dean *et al.*, in preparation). However, the very fine-grained nature of the illitic clays, together with the present uncertainty over the exact mechanism by which the smectite to illite transformation occurs, leaves open the possibility that oxygen isotopic re-equilibration with present formation water has occurred. At least some of the diagenetic illitic clay may have formed earlier, at higher temperatures from more ^{18}O-rich porewater. Radiogenic isotope studies (Rb/Sr; K/Ar) to resolve the timing of illitic clay formation are in progress.

Concluding remarks

One objective of this paper is to demonstrate that the oxygen-isotope compositions of diagenetic minerals can provide a valuable record of the temperatures and porewater $\delta^{18}O$ compositions prevailing during their formation. Figure 6 summarizes the model deduced for diagenetic minerals from the Viking Formation.

We also wished to determine whether meteoric water had been involved on a regional scale in the diagenesis of the Viking sandstones and conglomerates and, more specifically, to show whether porewater compositions could be related directly to changing geological conditions during the evolution of the basin. Our conclusions are summarized in Fig. 7.

Shallow diagenesis, early in the burial history (stage 1; diagenetic glauconite, pyrite, calcite, chlorite) occurred mostly in the presence of seawater-derived pore fluids, although a fresh-water influence is indicated where siderite and early kaolinite cement are abundant. As compac-

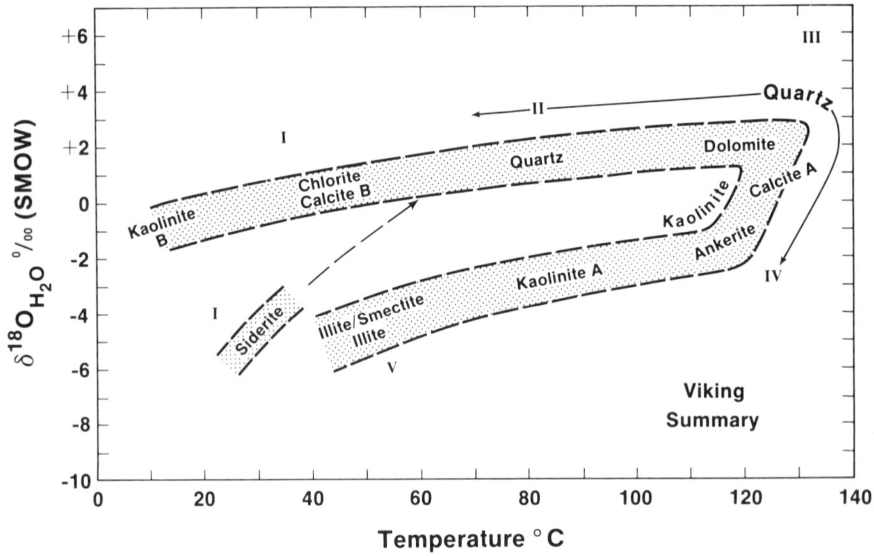

FIG. 6. Idealized porewater evolution path ($\delta^{18}O$ of water versus temperature) for the Viking sandstones and conglomerates: summary.

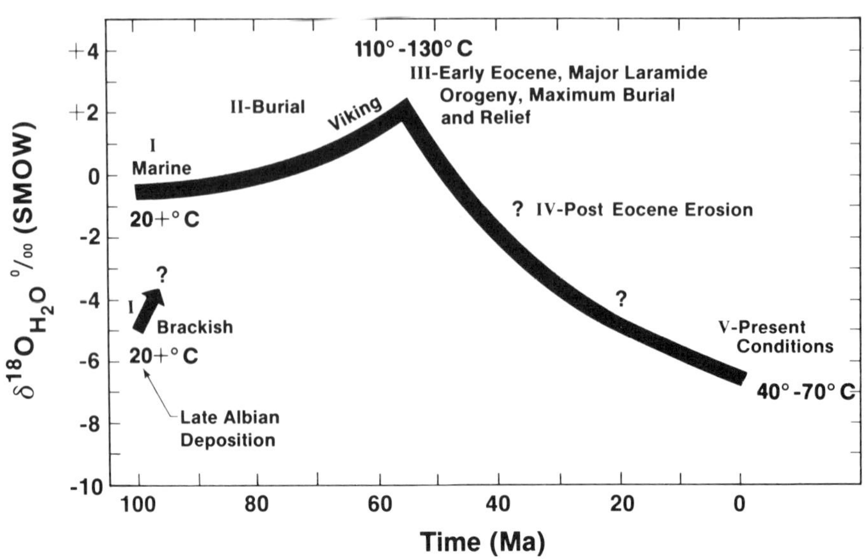

FIG. 7. Idealized porewater evolution path ($\delta^{18}O$ of water versus time) for the Viking sandstones and conglomerates. The exact shape of the curve from stage III to stage V is uncertain (see text).

tion and burial diagenesis proceeded (stage II, diagenetic chlorite, quartz overgrowths, dolomite), the porewaters became enriched in ^{18}O due to water/rock interaction. This diagenetic style was halted by the major Laramide Orogeny in the early Eocene. Maximum burial and relief (due to uplift) was accompanied by recharge of Tertiary and Cretaceous rocks in the basin by low-^{18}O meteoric water (stage III, feldspar dissolution, albitization, and diagenetic quartz, calcite and ankerite). The influx of surface-derived fresh water continued as the post-Eocene erosion progressed (stage IV, diagenetic kaolin group minerals, illite, illite/smectite) until the porewater attained its present temperature and composition (stage V, diagenetic illite, illite/smectite?).

The model proposed for the diagenesis of the Lower Cretaceous Viking sandstones and conglomerates implies large-scale movement of meteoric water, at least within the Cretaceous portion of the western Canada sedimentary basin. That meteoric water has been important on a basin-wide scale is further substantiated by the even lower $\delta^{18}O$ values of equivalent minerals from Upper Cretaceous sandstones within the same locality (Longstaffe 1983, 1986, Ayalon & Longstaffe 1985, in press). The more ^{18}O-rich nature (by about 3–5‰) of both early and late diagenetic minerals from the Viking Formation may have resulted from a combination of factors, including: (i) the starting porewater in the Viking sandstones and conglomerates and conglomerates was dominantly sea water, rather than brackish to fresh water (as was the case in some Upper Cretaceous units; Longstaffe 1983, 1986, Ayalon & Longstaffe 1985, in press); (ii) the fraction of low-^{18}O meteoric water mixed with pre-existing formation water was smaller in the Viking Formation than in the Upper Cretaceous units, and (or) (iii) meteoric water which penetrated the Viking Formation had greater opportunity for water/rock interaction and ^{18}O-enrichment (e.g. longer travel path, higher temperatures, more time) than did fresh water entering the stratigraphically higher units.

The relative importance of these possibilities and many others not considered here is not yet well understood. For example, the rate at which meteoric waters entered the subsurface following uplift of the basin and in turn became enriched in cations and ^{18}O due to water/rock interaction, is unknown. Hence, the exact shape of the $\delta^{18}O$ water versus time curve from stages III to V (Fig. 7) remains speculative. Radiogenic and deuterium-isotope studies of the illite and illite/smectite may provide greater insight into this portion of the porewater evolution curve in the western Canada sedimentary basin.

ACKNOWLEDGMENTS: The capable assistance of Ms D Caird in the laboratory is gratefully acknowledged as are numerous discussions with M. Dean. Grants in aid of this research were provided by the Natural Sciences and Engineering Research Council of Canada (NSERC A7387) to FJL. Helpful reviews of an earlier version of this paper were provided by K. Bjørlykke and P. J. Hamilton. Financial aid for travel to the Symposium from its organizers and industrial sponsors is acknowledged with great pleasure.

References

AMAJOR, L. C. 1980. Chronostratigraphy, depositional patterns and environmental analysis of subsurface Lower Cretaceous (Albian) Viking reservoir sandstones in central Alberta and part of southwestern Saskatchewan. *PhD thesis*, University of Alberta, Edmonton, Canada.

AYALON, A. & LONGSTAFFE, F. J. 1985. Diagenesis and pore-water evolution of the basal Belly River sandstone, Alberta: mineralogic, petrographic and stable isotope studies (abstract). *Geological Society of America, Program with Abstracts,* **17**, 516.

—— & —— in press. Oxygen-isotope studies of diagenesis and porewater evolution in the western Canada sedimentary basin: evidence from the Upper Cretaceous basal Belly River sandstone. *Journal of Sedimentary Petrology*.

BEAUMONT, C. 1981. Foreland basins. *Geophysical Journal of the Royal Astronomical Society,* **65**, 291–329.

BEAUMONT, E. A. 1984. Retrogradational shelf sedimentation: Lower Cretaceous Viking Formation, central Alberta. *In*: TILLMAN, R. W. & SIEMERS, C. T. (eds). *Siliciclastic Shelf Sediments,* pp. 163–77. Society of Economic Paleontologists and Mineralogists, Special Paper 34.

——, BOUTILIER, R., MACKENZIE, A. S. & RULLKOTTER, J. 1985. Isomerization and aromatization of hydrocarbons and the paleothermometry and burial history of Alberta foreland basin. *American Association of Petroleum Geologists Bulletin,* **69**, 546–66.

BECKER, R. H. & CLAYTON, R. N. 1972. Carbon isotopic evidence for the origin of a banded iron-formation in Western Australia. *Geochimica et Cosmichimica Acta,* **36**, 577–95.

BERNER, R. A. 1971. *Principles of Chemical Sedimentology*. Elsevier, New York.

BJØRLYKKE, K., ELVERHØI, A. & MALM, A. O. 1979. Diagenesis in Mesozoic sandstones from Spitsbergen and the North Sea—a comparison. *Geologisches Rundschau,* **68**, 1152–71.

BOLES, J. R. 1982. Active albitization of plagioclase, Gulf Coast Tertiary. *American Journal of Science,* **282**, 165–80.

CLAYTON, R. N. 1959. Oxygen isotope fractionation in the system calcium carbonate-water. *Journal of Chemical Physics*, **30**, 1246–50.

——, FRIEDMAN, I., GRAF, D. L., MAYEDA, T. K., MEENTS, W. F. & SHIMP, N. F. 1966. The origin of saline formation waters, 1. Isotopic composition. *Journal of Geophysical Research*, **71**, 3869–82.

——, O'NEIL, J. R. & MAYEDA, T. K. 1972. Oxygen isotope exchange between quartz and water. *Journal of Geophysical Research*, **77**, 3057–67.

CRAIG, H. 1957. Isotopic standards for carbon and oxygen and correction factors for mass-spectrometric analysis of carbon dioxide. *Geochimica et Cosmochimica Acta*, **12**, 133–49.

—— 1961. Standards for reporting concentrations of deuterium and oxygen-18 in natural waters. *Science*, **133**, 1833–4.

DEAN, M. E. 1986. Diagenesis of the Viking Formation, south-central Alberta. *MSc thesis*. University of Alberta, Edmonton, Canada.

——, LONGSTAFFE, F. J. & HEIN, F. J. 1984. Mineralogical controls on reservoir quality in the Harmattan Oil Field, Lower Cretaceous Viking Formation, south-central Alberta (abstract). *Geological Association of Canada, Program with Abstracts*, **9**, 57.

——, —— & —— 1985. Regional patterns of diagenesis in the Viking Formation, south-central Alberta (abstract). *Thirty-sixth annual technical meeting, Petroleum Society of CIM*, p. 50.

——, —— & —— In preparation. Diagenesis of the Viking Formation, south-central Alberta. *Bulletin of Canadian Petroleum Geology*.

DICKINSON, W. R. & SNYDER, W. S. 1978. Plate tectonics of the Laramide Orogeny. *In*: MATHEWS, W. (ed.). *Laramide Folding Associated with Basement Block Faulting in the Western United States*, pp. 355–66. Geological Society of America, Memoir, **151**.

DUTTON, S. P. & LAND, L. S. 1985. Meteoric burial diagenesis of Pennsylvanian arkosic sandstones, southwestern Anadarko Basin, Texas. *The American Association of Petroleum Geologists Bulletin*, **69**, 22–38.

ESLINGER, E. V. 1971. Mineralogy and oxygen isotope ratios of hydrothermal and low-grade metamorphic argillaceous rocks. *PhD dissertation*, Case Western Reserve University, Cleveland, United States.

—— & SAVIN, S. M. 1973. Mineralogy and oxygen isotope geochemistry of the hydrothermally altered rocks of the Ohaki-Broadlands, New Zealand geothermal area. *American Journal of Science*, **273**, 240–67.

—— & YEH, H. 1981. Mineralogy, O^{18}/O^{16} and D/H ratios of clay-rich sediments from Deep Sea Drilling Project site 180, Aleutian Trench. *Clays and Clay Minerals*, **29**, 309–15.

FOLK R. L. 1968. *Petrology of Sedimentary Rocks*. Published by Hemphill's Book Store, Austin, Texas.

FOSCOLOS, A. E., REINSON, G. E. & POWELL, T. G. 1982. Controls on clay-mineral authigenesis in the Viking sandstone, central Alberta, 1. shallow depths. *Canadian Mineralogist*, **20**, 141–50.

FRIEDMAN, I. & O'NEIL, J. R. 1977. Compilation of stable isotope fractionation factors of geochemical interest. *In*: FLEISCHER, M. (ed.). *Data of Geochemistry* (6th edn.). United States Geological Survey Professional Paper, **440-KK.**

GAMMEL, H. G. 1955. The Viking Member in central Alberta. *Bulletin of the Alberta Society of Petroleum Geologists*, **3**, 63-9.

GAUTIER, D. L. 1982. Siderite concretions: indicators of early diagenesis in the Gammon Shale (Cretaceous). *Journal of Sedimentary Petrology*, **52**, 859–71.

GLAISTER, R. P. 1958. Petrology of the Blairmore sandstone. *Bulletin of the Alberta Society of Petroleum Geologists*, **6**, 43–9.

GRANT, S. K. 1985. Geologic study of the Viking Formation, Harmattan East Field. *MSc thesis*. University of Alberta, Edmonton, Canada.

HACQUEBARD, P. A. 1977. Rank of coal as an index of organic metamorphism for oil and gas in Alberta. *In*: DEROO, G., POWELL, T. G., TISSOT, B. & MCCROSSAN, G. (eds). *The Origin and Migration of Petroleum in the Western Canadian Sedimentary Basin, Alberta*, pp. 11–22. Geological Survey of Canada Bulletin, **262**.

HEIN, F. J., DEAN, M. E., DEIURE, A. M., GRANT, S. K., ROBB, G. A. & LONGSTAFFE, F. J. 1986. The Viking Formation in the Caroline, Garrington and Harmattan East fields, western south-central Alberta: sedimentology and paleogeography. *Bulletin of Canadian Petroleum Geology*, **34**, 91–110.

HITCHON, B. 1984. Geothermal gradients, hydrodynamics, and hydrocarbon occurrences, Alberta, Canada. *The American Association of Petroleum. Geologists Bulletin*, **68**, 713–43.

—— & FRIEDMAN, I. 1969. Geochemistry and origin of formation waters in the western Canada sedimentary basin—1. Stable isotopes of hydrogen and oxygen. *Geochimica et Cosmochimica Acta*, **33**, 1321–49.

HURST, A. 1980. Occurrence of corroded authigenic kaolinite in a diagenetically modified sandstone. *Clays and Clay Minerals*, **28**, 393–6.

—— & IRWIN, H. 1982. Geologic modelling of clay diagenesis in sandstones. *Clay Minerals*, **17**, 5–22.

IRWIN, H., CURTIS, C. & COLEMAN, M. 1977. Isotopic evidence for the source of diagenetic carbonates formed during burial of organic-rich sediments. *Nature*, **269**, 209–13.

JACKSON, M. L. 1979. *Soil Chemical Analysis—Advanced Course* (2nd edn). Published by the author, Madison, Wisconsin.

LAND, L. S. 1980. The isotopic and trace element geochemistry of dolomite: the state of the art. *In*: ZENGER, D. H., DUNHAM, J. B. & ETHINGTON, R. A. (eds.). *Concepts and Models of Dolomitization*, pp. 87–110. Society of Economic Paleontologists and Mineralogists, Special Paper, **28**.

—— 1983. The application of stable isotopes to studies of the origin of dolomite and to problems of diagenesis of clastic sediments. *In*: ARTHUR, M. A. (organizer). *Stable Isotopes in Sedimentary Geology*, pp. 4.1–4.22. Society of Economic Paleontologists and Mineralogists, Short Course, **10**.

—— 1984. Frio sandstone diagenesis, Texas Gulf Coast: a regional isotopic study. *In*: SURDAM, R. &

MacDonald, D. (eds). *Clastic Diagenesis*, pp. 47–62. American Association of Petroleum Geologists, Memoir, **37**.

—— & Dutton, 1978. Cementation of a Pennsylvanian deltaic sandstone: isotopic data. *Journal of Sedimentary Petrology*, **48**, 1167–76.

Lawrence, J. R., Gieskes, J. M. & Broecker, W. S. 1975. Oxygen isotope and cation composition of DSDP pore waters and the alteration of layer II basalts. *Earth and Planetary Science Letters*, **27**, 1–10.

Lee, M. & Savin, S. M. 1985. Isolation of diagenetic overgrowths on quartz and grains for oxygen isotopic analysis. *Geochimica et Cosmochimica Acta*, **49**, 497–501.

Longstaffe, F. J. 1983. Diagenesis, IV. Stable isotope studies of diagenesis in clastic rocks. *Geoscience Canada*, **10**, 44–58.

—— 1984. The role of meteoric water in diagenesis of shallow sandstones: stable isotope studies of the Milk River aquifer and gas pool. *In*: Surdam, R. & MacDonald, D. (eds). *Clastic Diagenesis*, pp. 81–98. American Association of Petroleum Geologists, Memoir, **37**.

—— 1986. Oxygen isotope studies of diagenesis in the basal Belly River sandstone, Pembina I-Pool, Alberta. *Journal of Sedimentary Petrology*, **56**, 78–88.

—— 1987. Mineralogical and oxygen-isotope studies of clastic diagenesis: implications for fluid flow in sedimentary basins. *In*: Hitchon, B. (ed.). *Hydrogeology of Sedimentary Basins*. Proceedings of the Third Annual Canadian/American Conference on Hydrogeology, National Water Well Association (in press).

—— & Ayalon, A. 1986. Oxygen-isotope studies of diagenesis in clastic rocks from the Viking and Belly River Formations, Alberta (abstract). *Terra Cognita*, **6**, 109

Magara, K. 1976. Thickness of removed sedimentary rocks, paleopore pressure, and paleotemperature, southwestern part of western Canada Basin. *American Association of Petroleum Geologists Bulletin*, **60**, 554–65.

Majorowicz, J. A. & Jessop, A. M. 1981. Regional heat flow patterns in the western Canadian sedimentary basin. *Tectonophysics*, **74**, 209–38.

——, Rahman, M., Jones, F. W. & McMillan, N. J. 1985. The paleogeothermal and present thermal regimes of the Alberta basin and their significance for petroleum occurrences. *Bulletin of Canadian Petroleum Geology*, **33**, 12–21.

Nurkowski, J. R. 1984. Coal quality, coal rank variation and its relation to reconstructed overburden, Upper Cretaceous and Tertiary Plains coals, Alberta, Canada. *American Association of Petroleum Geologists Bulletin*, **68**, 285–95.

O'Neil, J. R., Clayton, R. N. & Mayeda, T. K. 1969. Oxygen isotope fractionation in divalent metal carbonates. *Journal of Chemical Physics*, **51**, 5547–58.

Piel, K. 1971. Palynology of Oligocene sediments of central British Columbia. *Canadian Journal of Botany*, **49**, 1885–920.

Postma, D. 1982. Pyrite and siderite formation in brackish and freshwater swamp sediments. *American Journal of Science*, **282**, 1151–83.

Robb, G. A. 1985. Sedimentology and diagenesis of the Viking Formation, Garrington Oil Field, south-central Alberta. *MSc thesis*. University of Alberta, Edmonton, Canada.

Rosenbaum, J. & Sheppard, S. M. F. 1986. An isotopic study of siderites, dolomites and ankerites at high temperatures. *Geochimica et Cosmochimica Acta*, **50**, 1147–50.

Savin, S. M. & Epstein, S. 1970a. The oxygen and hydrogen isotope geochemistry of clay minerals. *Geochimica et Cosmochimica Acta*, **34**, 25–42.

—— & —— 1970b. The oxygen and hydrogen isotope geochemistry of ocean sediments and shales. *Geochimica et Cosmochimica Acta*, **34**, 43–63.

—— & Yeh, H. 1981. Stable isotopes in ocean sediments. *In*: Emiliani, C. (ed.). *The Sea, volume 7, The Oceanic Lithosphere*, pp. 1521–54. Wiley, New York.

Schwartz, F. W., Muehlenbachs, K. & Chorley, D. W. 1981. Flow-system controls of the chemical evolution of groundwater. *Journal of Hydrology*, **54**, 225–43.

Sharma, T. & Clayton, R. N. 1965. Measurements of O^{18}/O^{16} ratios of total oxygen of carbonates. *Geochimica et Cosmochimica Acta*, **29**, 1347–53.

Slipper, S. E. 1918. Viking gas field, structure of area. *Geological Survey of Canada, Summary Department 1917*, **Part C**, 8c.

Suchecki, R. K. & Land, L. S. 1983. Isotopic geochemistry of burial-metamorphosed volganogenic sediments, Great Valley sequence, northern California. *Geochimica et Cosmochimica Acta*, **47**, 1487–99.

Syers, J. K., Chapman, S. L., Jackson, M. L., Rex, R. W. & Clayton, R. N. 1968. Quartz isolation from rocks, sediments and soils for determination of oxygen isotopes. *Geochimica et Cosmochimica Acta*, **32**, 1022–5

Taylor, H. P. Jr 1974. The application of oxygen and hydrogen isotope studies to problems of hydrothermal alteration and ore deposition. *Economic Geology*, **69**, 843–83.

Taylor, R. S., Mathews, W. H. & Kupsch, W. O. 1964. Tertiary. *In*: McCrossan, R. G. & Glaister, R. P. (eds). *Geological History of Western Canada*, pp. 190–4. Alberta Society of Petroleum Geologists.

Thomas, M. B. & Oliver, T. A. 1979. Depth-porosity relationships in the Viking and Cardium Formations of central Alberta. *Bulletin of Canadian Petroleum Geology*, **27**, 209–28.

Tilley, B. J. & Longstaffe, F. J. 1986. $\delta^{18}O$ results for silicate and carbonate cements from Lower Cretaceous sedimentary rocks in the Alberta Deep Basin, Canada (abstract). *Terra Cognita*, **6**, 107.

Weimer, R. J. 1983. Relation of unconformities, tectonics, and sea level changes Cretaceous of the Denver Basin and adjacent areas. *In*: Reynolds, M. W. & Dolly, E. D. (eds). *Mesozoic Paleogeography of West-central United States*, pp. 359–76. Rocky Mountain Section, Society of Economic Paleontologists and Mineralogists.

Wenner, D. B. & Taylor, H. P. Jr 1971. Temperatures of serpentinization of ultramafic rocks based upon O^{18}/O^{16} fractionation between coexisting serpentine and magnetite. *Contributions of Mineralogy and Petrology,* **32**, 165–85.

WILLIAMS, G. D. & STELCK, C. R. 1975. Speculations on the Cretaceous paleogeography of North America. *In:* CALDWELL, W. G. E. (ed.). *The Cretaceous System in the Western Interior of North America,* pp. 1–7. Geological Association of Canada, Special Paper, **13**.

WILSON, M. D. & PITTMAN, E. D. 1977. Authigenic clays in sandstones: recognition and influence on reservoir properties and paleoenvironmental analysis. *Journal of Sedimentary Petrology,* **476**, 3–31.

YEH, H. & SAVIN, S. M. 1976. The extent of oxygen isotope exchange between clay minerals and seawater. *Geochimica et Cosmochimica Acta,* **40**, 743–8.

—— & —— 1977. Mechanism of burial metamorphism of argillaceous sediments: 3. O-isotope evidence. *Geological Society of America Bulletin,* **88**, 1321–30.

FRED J. LONGSTAFFE, Department of Geology, University of Western Ontario, London, Ontario N6A 5B7, Canada.

AVNER AYALON, Geological Survey of Israel, 30 Malchei Israel Street, Jerusalem 95501, Israel.

Secondary porosity in hydrocarbon-bearing transgressive sandstones on an unstable Lower Palaeozoic continental shelf, Welsh Borderland

J. Parnell

SUMMARY: Coarse sandstones were deposited during each of four transgressions across the shelf of the Lower Palaeozoic back-arc Welsh Basin. Hydrocarbon shows record porosity in the deposits of the Cambrian, Arenig, Caradoc and Llandovery transgressions. Hydrocarbons migrated from two sources: from Lower Palaeozoic source rocks within the basin, and later from Carboniferous rocks.

Diagenesis and reservoir potential is related to the mineralogy of the sandstones, which reflects a balance between reworking of shelf sediment and an input from the eroded substrate. The substrate included Precambrian volcanic and plutonic rocks formed at the southern margin of the Iapetus Ocean. Quartz arenites are quartz-cemented or exhibit thick pore-linings of clay which restrict permeability. They have only minor reservoir potential. Arkoses do not exhibit macroporosity but have effective microporosity in the immediate vicinity of source rocks. Subarkoses have optimal reservoir potential, consisting of primary porosity about non-quartz grains (feldspars, glauconite) and secondary (intragranular) porosity in feldspars and lithics. The secondary porosity formed by leaching of clay alteration products in feldspars. Subarkoses retained sufficient framework stability to preserve the secondary porosity, whereas arkoses experienced widespread alteration and framework collapse. Limited secondary porosity after calcite dissolution may be due to surficial weathering of the Lower Palaeozoic rocks during Devonian/Lower Carboniferous exposure.

An active continental margin presents a favourable setting for marine transgressive events (Swift 1968). For example, several thick transgressive sandstones are recorded in the Cretaceous sequence of the Pacific coast of North America (Bourgeois 1980). Rises in sea-level in the fore-arc region or marginal basin of a destructive plate boundary are recorded by transgressive stratigraphy across the adjacent foreland. The transgressive sediments tend to be relatively coarse siliciclastic deposits, whilst fine-grained sediments are deposited in the outer shelf regions where an increase in water depth accompanies the transgression. Outer shelf mudrocks deposited during transgressive events can contain substantial proportions of organic matter, dominated by types I and II kerogen (Demaison & Moore 1980). In favourable circumstances, hydrocarbons generated from a basinal/outer shelf mudrock sequence could be reservoired in shelf sandstones (Galloway & Hobday 1983). The shelf mudstones may function as reservoir seals as well as source rocks. Brenner (1978) has described significant petroleum accumulations in shelf sandstones, preserved by updip stratigraphic pinchouts with a shale seal, from the late Cretaceous Sussex Sandstone of Wyoming.

However, another feature of unstable continental margins is that volcanic material may make a major contribution to the siliciclastic sediment budget. Volcanic lithic fragments, generally feldspathic, are highly reactive and susceptible to hydration reactions (Surdam & Boles 1979). The diagenetic products (clays, zeolites, carbonates, oxides) tend to be less dense and therefore occupy a greater volume to the detriment of porosity. The reservoir potential of such rocks is therefore generally poor (Nagtegaal 1978, Galloway 1979). Nevertheless, there is potential for generation of secondary porosity by leaching of volcanic grains if sufficient quartz is present to provide a stable framework (e.g. Loucks *et al.* 1979), and some volcaniclastic reservoirs are known (reviewed by Seeman & Scherer 1984). Petrographic studies of the sediment in back-arc and fore-arc basins show them to be mixtures of immature rock fragments and reworked quartz-rich sediment (e.g. Ingersoll & Suczek 1979). The relative proportions of these components will influence the framework stability of the sediment.

Transgressive deposits in the Lower Palaeozoic Welsh Basin have been examined to determine whether they function/have functioned as hydrocarbon reservoirs. The Welsh Basin is a particularly interesting case for study because hydrocarbon migration and entrapment can be envisaged from two distinct sources: (i) maturation of Lower Palaeozoic source rocks,

From: MARSHALL, J. D. (ed.), 1987, *Diagenesis of Sedimentary Sequences*, Geological Society Special Publication No. 36, pp. 297–312.

i.e. a source within the basin, (ii) migration from Carboniferous reservoir rocks (containing hydrocarbons of Carboniferous origin) into uplifted Lower Palaeozoic rocks, i.e. a younger external source (further details in Parnell 1987).

Only transgressive shelf sandstones are discussed here. The basinal deposits include extensive turbidite sandstones, many of which are petrographically greywackes. No porosity was observed in the basinal sandstones.

The Welsh Basin

The Lower Palaeozoic Welsh Basin evolved as a back-arc basin at the southern continental margin of the Iapetus Ocean. A comprehensive review of tectonic setting, stratigraphy and palaeogeography is given by Bassett (1980). Towards the end of the Precambrian, the margin was a site of very active subsidence, plutonism, volcanism and deformation. The Lower Palaeozoic rocks accumulated on a basement of schists, gneisses, low grade metasediments (Longmyndian) and a variety of igneous rocks dominated by the calc-alkaline products of volcanic arcs (Thorpe *et al.* 1984). In the Welsh Borderland the basin shallowed to a shelf sea of varying width on the western side of the Midland Platform (Fig. 1). The sea transgressed across the shelf on numerous occasions, most importantly during the Lower/Middle Cambrian, Arenig, lower Carodoc and Llandovery (Fig. 2). The shelf sediments are generally shallow-water siliciclastic and carbonate deposits. In the basin, sedimentation was more continuous, in the form of pelagic muds/silts and turbiditic sands.

The basinal sediments include substantial thicknesses of organic-rich sediment. Significant black shale deposition also occurred in outer shelf areas following transgressive events, particularly during and after the Caradoc and Llandovery transgressions (Leggett 1978, 1980). The basinal sediments experienced mild metamorphism to greenschist facies (Bevins & Rowbotham 1983), and consequently organic-rich beds are overmature. Sediments on the inner shelf have not experienced this degree of metamorphism. Kelling (1978) estimates that the shelf succession reaches 5 km thickness, whilst up to 15 km accumulated in the basin. Accordingly, the organic-rich beds of the inner shelf may have some residual hydrocarbon potential, particularly for gas.

Transgressive sandstone deposits

Transgressive sandstones are exposed where they onlap against basement rocks. Precambrian basement rocks outcrop in two main regions today, forming the topographic highs of the Longmynd and the Malvern Hills (Fig. 1). These regions were also topographic highs during the Lower Palaeozoic (Bridges 1975). The outcrops of Cambrian, Arenig and Caradoc rocks are all in the two regions of basement outcrop. A map of the Longmynd region (Fig. 3) shows successive onlap of Cambrian, Caradoc, Llandovery and Carboniferous rocks on to the Precambrian, with unconformities between each. Llandovery rocks are more widely exposed (Fig. 4).

The mineralogy of the transgressive sandstones reflects a balance between reworking of sediment on the shelf and an input of new sediment as the shelf substrate was eroded during transgression. The balance has an important bearing on the mineralogical and textural maturity (and hence diagenesis) of the sediments. Whilst reworking enhanced the quartz content of the sands, erosion of the substrate created a supply of lithic clasts, many of which were susceptable to alteration. Sandstones of each transgressive episode contain abundant well-rounded quartz grains. In addition to shelf reworking, a contribution from aeolian reworking is suggested by records of dreikanters in the Cambrian and Caradocian. Unstable lithic material from the substrate included volcanic, plutonic and mudrock clasts. In particular, Uriconian (Precambrian) basic to acid volcanic clasts occur in many of the coarse beds of the northern Welsh Borderland. An Ordovician volcanic basement in the western part of the region also supplied unstable material.

Hydrocarbon shows

Hydrocarbon shows have been recorded in the deposits of all four transgressive episodes. Details of occurrences in eight districts (Fig. 4) are summarized in Table 1. The hydrocarbon shows in Cambrian sandstones are described and discussed by Parnell (1987).

Trecoed Beds sandstones occupy an ideal setting for a hydrocarbon reservoir from Lower Palaeozoic source rocks. They were deposited on the outer shelf of the basin (James 1983) resting unconformably on Ordovician black shales and overlain by Upper Llandovery/Wenlock black shales (Jones 1949). They are overlapped by the overlying shales in a shorewards direction, i.e. the shales form a stratigraphic seal as well as a potential source rock (Fig. 5). The hydrocarbon shows in the Folly Sandstone occur in the same region, underlying bituminous Wenlockian limestones. The Lower Cambrian sandstones of the Malvern Hills are hydrocarbon-bearing where they are faulted against Upper Cambrian

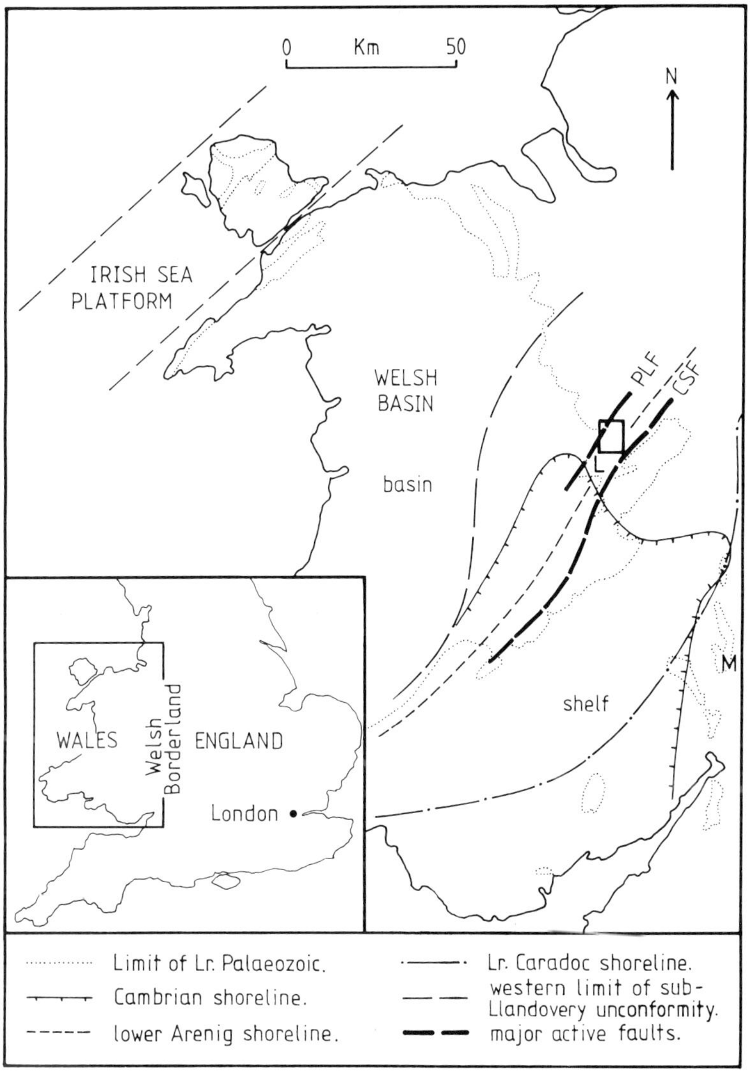

FIG. 1. Location of Welsh Basin. Shoreline changed through time (Cambrian after Bassett, 1980, Arenig and Caradoc after Anderton *et al.* 1978). L, M = Longmynd and Malvern Hills, regions of basement outcrop which are/were topographic highs. PLF = Pontesford–Linley Fault, CSF = Church Stretton Fault. Box indicates region of Fig. 3.

black shales (Parnell 1987). Both of these instances probably also had a Lower Palaeozoic source.

In the northern part of the Welsh Borderlands, the shows in the Cambrian of Shropshire, Stiperstones Quartzite, Hoar Edge Grit and Bog Quartzite probably had a Carboniferous source. They occur close to unconformably overlying Upper Carboniferous Sandstones (Figs 3 and 4) that are locally impregnated with oil. Organic geochemical studies confirm that shows in Lower Palaeozoic and Carboniferous rocks in Shropshire had the same source (Robinson *et al.* 1986). The Black Grit is similarly near the Carboniferous outcrop (Fig. 3), but is overlain by dark Caradocian shales which are intruded by a dolerite sill (Wedd 1932). Elsewhere in the Welsh Borderlands hydrocarbons were generated from Lower Palaeozoic shales by the heat of igneous intrusions (Parnell 1983), and a Lower Palaeozoic source for the Black Grit hydrocarbons is likely.

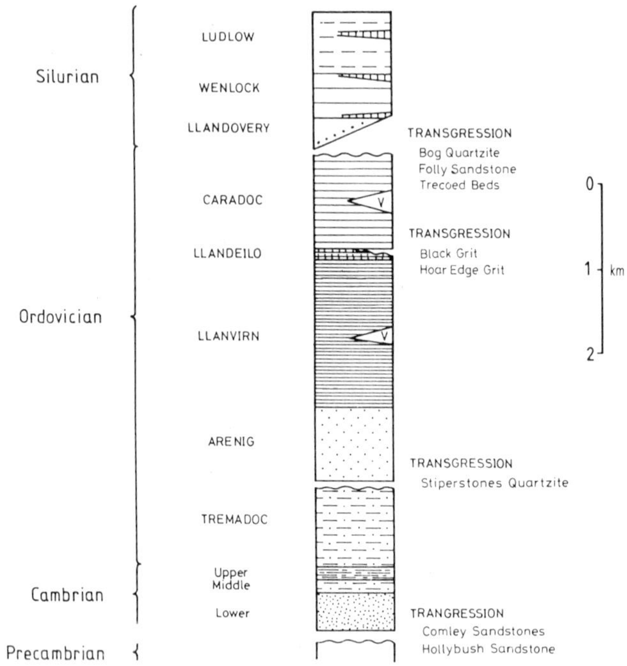

FIG. 2. Generalized stratigraphic succession for the Welsh Borderland (from Earp & Hains 1971), showing main transgressive episodes. Named stratigraphic units are hydrocarbon-bearing transgressive sandstones. V = episodes of volcanic activity.

Sample analysis

Sandstone samples, from surface outcrop, were examined in thin section. Many samples are from breached hydrocarbon traps and contain reservoir bitumen/pyrobitumen. Samples not plugged with hydrocarbons were impregnated with blue resin. Percentage porosity in hydrocarbon-bearing sandstones, recorded in Table 1, is porosity now occupied by hydrocarbon. These porosity values are therefore not exaggerated by surface weathering. Clay mineralogy was examined and identified using scanning electron microscopy with an EDAX facility. Analyses were supported by X-ray diffraction analyses.

Sandstone diagenesis

The relative proportions of quartz and unstable grains is fundamental to the reservoir properties of the Welsh Borderland sandstones. Accordingly, quartz arenites, subarkoses and sublitharenites, and arkoses and litharenites (Fig. 6, classified according to Folk 1974) are discussed separately. The samples plotted are medium to coarse-grained sandstones, with the exception of samples from the Hollybush Sandstone and Black Grit, which are fine-grained hydrocarbon-bearing arkoses and litharenites.

The composition of Cambrian and Llandovery sandstones clearly differs between the northern and southern Welsh Borderlands. The sandstones in the south are less mature. The proportion of acid volcanic rocks in the northern basement is higher, contributing quartzose grains to the sandstones there. In Llandovery times, the sedimentary environments were different with a predominance of marine reworked sands in the north, but with a contribution from regressive sands in the south (Bridges 1975).

Quartz arenites

Quartz arenites were collected from the Cambrian sandstones of Shropshire, the Caradoc Hoar Edge Grit and open marine Llandovery sandstones. The quartz arenites are all medium to coarse sandstones. They generally show less than 5% porosity. Porosity has been reduced by: (i) quartz overgrowths (Fig. 7a) which may completely fill the pore-space, particularly in the Cambrian sandstones; (ii) pore-filling kaolinite, particularly in the Hoar Edge Grit, where quartz arenites are interbedded with arkoses and subarkoses.

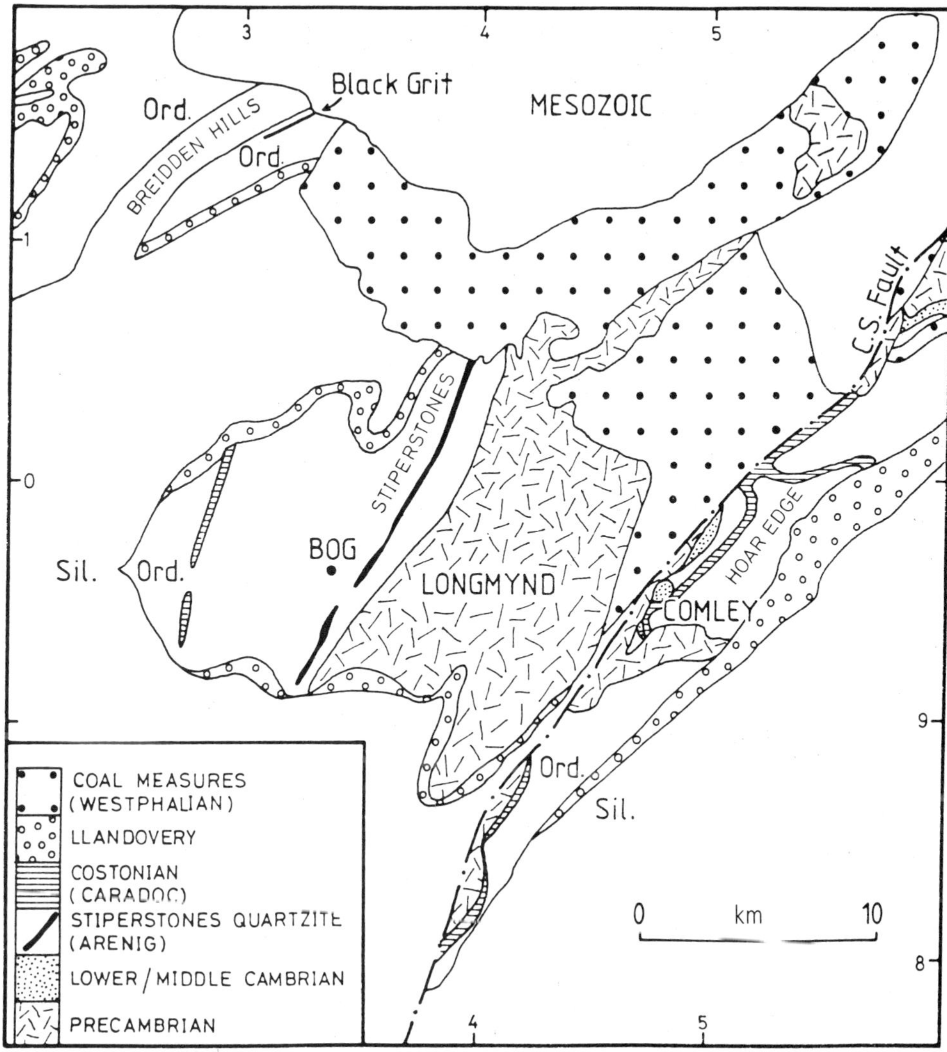

FIG. 3. Outcrops of transgressive deposits about the Longmynd topographic high, Shropshire. Cambrian, Caradoc, Llandovery and Carboniferous sandstones successively onlap the Precambrian basement, with unconformities between each.

Scanning electron microscopy shows that a thin grain coating of clay (variably illite or chlorite) predates the quartz overgrowths. At Rubery, the easternmost and possibly least buried exposure, cementation is incomplete and the sandstone locally has 15–20% porosity. However, very well developed pore-linings of smectite/mixed-layer clay (Fig. 7b) severely limit permeability.

Hydrocarbon shows

Limited hydrocarbon shows occur in quartz arenites, in: (i) Cambrian sandstones in Shropshire in residual primary porosity (Fig. 7a); (ii) the Hoar Edge Grit, in which very minor traces (less than 1%) occur in kaolinite-cemented quartz arenites and subarkoses.

Subarkoses or sublitharenites

The majority of sandstone samples can be classified as subarkoses/sublitharenites. Glauconite is an important component of the Cambrian sandstones and a minor component of Caradocian and Llandovery sandstones.

In many cases the subarkoses/sublitharenites exhibit significant porosity due to the preserva-

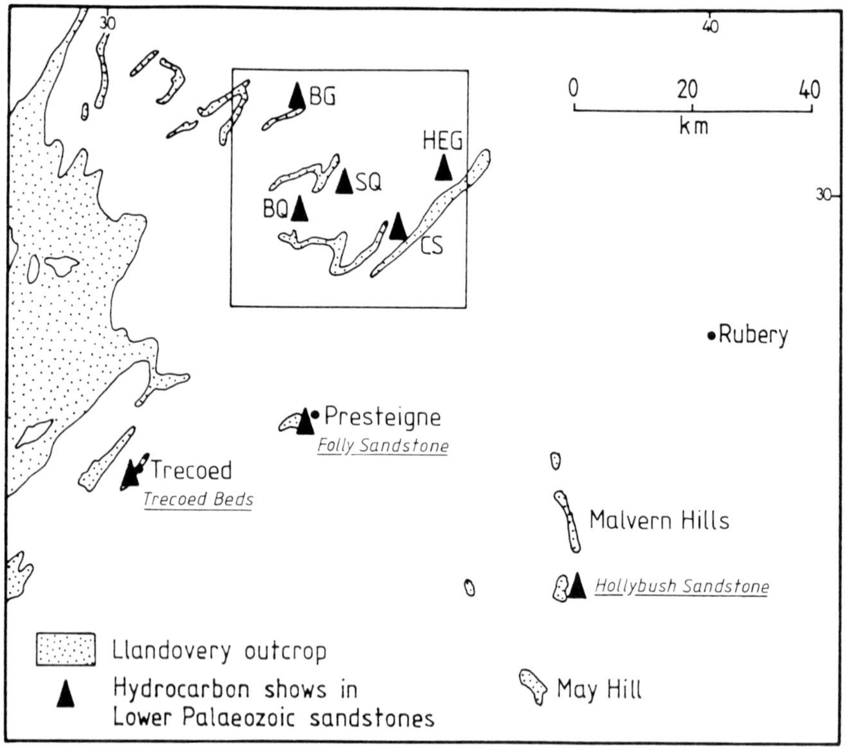

FIG. 4. Llandovery outcrops in Welsh Borderland, and locations of hydrocarbon shows in Lower Palaeozoic sandstones. BG = Black Grit (Caradoc), BQ = Bog Quartzite (Llandovery), CS = Comley Sandstones (Cambrian), HEG = Hoar Edge Grit (Caradoc), SQ = Stiperstones Quartzite (Arenig). Box indicates region of Fig. 3.

tion of residual primary porosity and the leaching of unstable grains. Porosity is particularly associated with feldspar and glauconite grains because: (i) overgrowths are not developed about glauconite and only rarely about feldspar, (ii) glauconite grains develop fissures during glauconitization (Odin & Matter 1981), (iii) alteration and leaching of feldspars gave rise to significant intragranular porosity.

The honeycombed nature of the feldspar grains is strong evidence for post-depositional leaching as they would rapidly disintegrate during sediment transport. Microprobe analysis of altered feldspars in Cambrian sandstones shows that sodic portions of perthitic intergrowths have been preferentially altered and leached. In addition to the leaching of feldspar grains, the leaching of volcanic and mudrock clasts and calcite bioclasts gave rise to porosity of up to 20%. The lithic clasts show varying stages of replacement by kaolinite, illite or chlorite. The replacive clays were locally leached to yield pore space. There is a gradation from feldspar or lithic grains with patches of replacive clay through grains with pores irregularly lined with clay to grains exhibiting 'clean' porosity.

The mineralogy of the clay mineral replacements is broadly related to the mineralogy of the original grains. Kaolinite replacements occur within feldspar grains, illitization is prominent in feldspars and mudrock and volcanic clasts and chlorite is a common replacement in basic volcanic clasts (see Burley *et al.* 1985). Early clay precipitation on grain surfaces pre-dated quartz overgrowths. Where the clay layer was thin (not more than a few microns) overgrowths were readily developed. Thicker pore-lining clays apparently prevented quartz overgrowth. Thick pore-linings are intact in rocks showing evidence for the leaching of clays from altered feldspars. Pore-lining clays include illite and mixed layer illite/smectite, but are most commonly iron-rich chlorite. In several districts, particularly in the southern Welsh Borderlands, the chlorites are stained red by haematite/goethite due to alteration of the chlorite or later percolation of iron-rich fluids draining the Precambrian basement rocks. The iron oxides severely limit microporo-

TABLE 1. *Occurrences of hydrocarbon-bearing transgressive sandstones, Welsh Borderland*

District	Grid ref.	Transgression	Stratigraphic unit	Petrography	Hydrocarbon occurrence	Max. porosity (%)
1. Comley	SO 4895, 4896	Cambrian	Lower and Upper Comley Sandstone	Quartz arenite, subarkose	Inter/intra-granular	12 (H)
2. Malvern Hills	SO 7636	Cambrian	Hollybush Sandstone	Arkose	Microporosity	*
3. Stiperstones	SJ 3803, 3904, 3905	Arenig	Stiperstones Quartzite	Sublitharenite	Inter/intra-granular	17 (10H + 7P)
4. Hoar Edge	SO 4996	Caradoc	Hoar Edge Grit	Quartz arenite, subarkose	Intergranular	19 (P)
5. Breidden Hills	SJ 3114, 3315	Caradoc	Black Grit	Arkose	Microporosity	*
6. Bog	SO 3498	Llandovery	Bog Quartzite	Sublitharenite	Mouldic porosity	12 (P)
7. Presteigne	SO 3162, 3163	Llandovery	Folly Sandstone	Subarkose	?Accretionary nodules	*
8. Trecoed	SO 0555, 0556	Llandovery	Trecoed Beds	Subarkose	Inter/intra-granular	13 (H)

* Porosity not determinable in hydrocarbon-impregnated clay matrix or in accretionary (replacive) nodules. H = porosity now occupied by hydrocarbon, P = extant porosity.

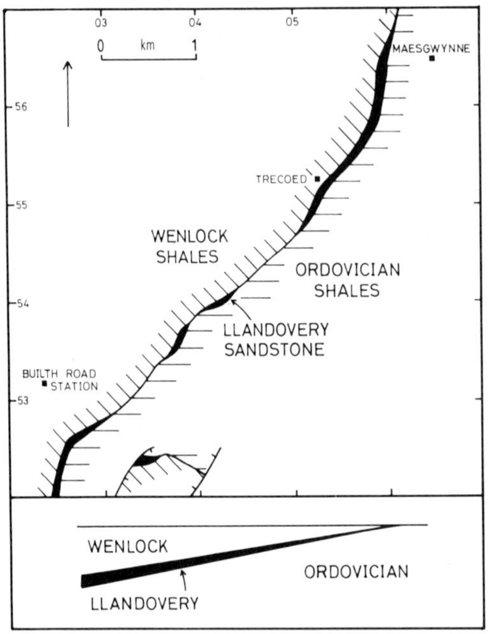

FIG. 5. Outcrop of outer shelf Llandovery sandstone (Trecoed Beds) near Builth Wells. Sandstones are sandwiched between Ordovician and Wenlock black shales, and overstepped by the latter. Shales acted as a seal to hydrocarbons in the sandstone.

sity within the pore-linings, and in some Llandovery rocks in the Malvern Hills they completely fill the remaining pore space. Some sandstones in which cementation by quartz was incomplete contain pore-bridging kaolinite. For example, Hoar Edge Grit samples can show a sequence of illite grain-coating, quartz overgrowth and pore-filling kaolinite.

Calcite is present in three forms: detrital, cement and replacement. Calcite bioclasts are usually represented by moulds, but calcite is preserved in fresh exposures such as road cuttings and borehole samples. The mouldic porosity is clearly not deeply penetrating. Sandstones above and below inner shelf limestones were susceptible to early carbonate cementation. Sandstone immediately adjacent to limestone is still calcite-cemented. The transition from calcite-cemented sandstone to quartz-cemented sandstones is usually sharp and there are no traces of calcite cement pre-dating overgrowths in the quartzose sandstones. It can be assumed that calcite cementation was limited to a zone adjacent to the limestones. The Hoar Edge Grit is the only sandstone under consideration which contains numerous limestone bands and traces of calcite cement. Many Hoar Edge Grit subarkoses do not exhibit quartz overgrowths but are cemented by authigenic kaolinite. Their minus-cement porosity is abnormally high (up to 25%) and features including corroded grains, inhomogeneity of grains and over-sized pores suggest pre-kaolinite secondary porosity. Kaolinite-cemented sandstone grades laterally into calcite-cemented sandstone.

Hydrocarbon shows

Hydrocarbon occurrences in the Comley Sandstones, the Stiperstones Quartzite and the Trecoed Beds all depend on combined primary and secondary porosity. Hydrocarbons fill the primary pores associated with grains which do not exhibit overgrowths, as noted above (Fig. 7c).

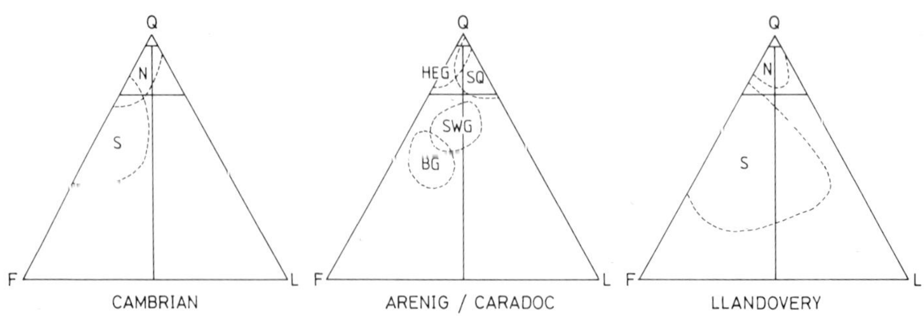

FIG. 6. Compositional field of sandstones from transgressive deposits, Welsh Borderland. Subdivision into quartz arenites, subarkoses, sublitharenites, arkoses and litharenites according to Folk (1974). End members Q = monocrystalline and polycrystalline quartz including chert, F = total feldspars, L = lithics excepting quartzose but including calcite bioclasts and glauconite. Interstitial material not counted. Plagioclase/total feldspar ratio varies from 0.2 to 1.0. N, S = northern and southern Welsh Borderland, BG = Black Grit, HEG = Hoar Edge Grit, SQ = Stiperstones Quartzite, SWG = Spy Wood Grit. Samples are medium/coarse-grained, except fine-grained Cambrian (south) and Black Grit.

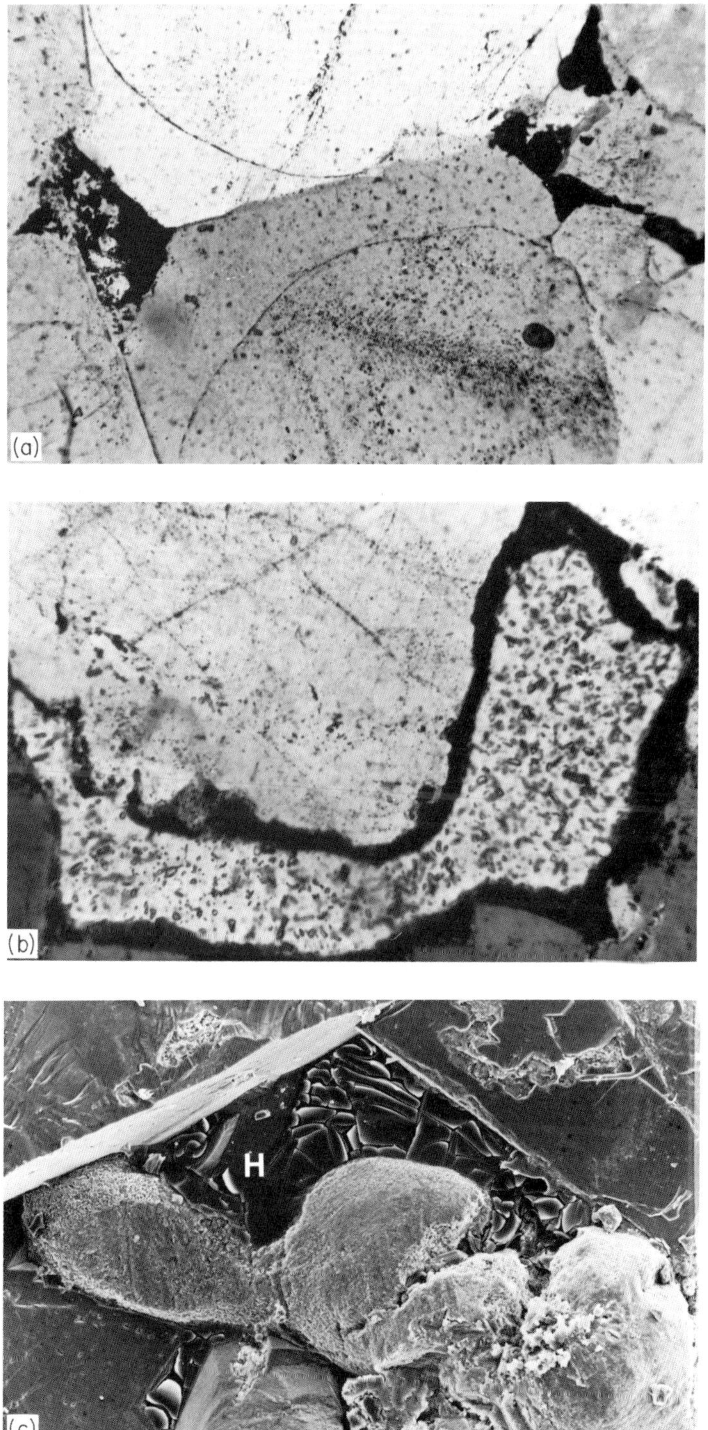

FIG. 7. (a) Residual primary porosity between quartz overgrowths, infilled with hydrocarbon (black), Cambrian Sandstone, Comley (field width 0.8 mm). (b) Pore-lining of mixed layer clay impregnated with iron oxides, Llandovery quartz arenite, Rubery. Pore is filled with mounting resin (field width 1.2 mm). (c) Glauconite grain with hydrocarbon in shrinkage fissure and in adjacent primary porosity, Cambrian sandstone, Comley (SEM, field width 0.25 mm).

Fissuring in glauconite grains in the Comley Sandstones, is commonly occupied by hydrocarbon (Fig. 7c and Parnell 1987). Hydrocarbons fill micropores between individual crystals in the boxwork fabric of glauconite grain surfaces (see Odin & Matter 1981). Hydrocarbon filling of secondary pores within feldspars is also widespread in the Cambrian sandstones (Fig. 8a) and in the Llandovery Trecoed Beds (Fig. 8b). This secondary porosity appears to be after leaching of clays rather than calcite dissolution. The gradation from replacement clay to clean pores in feldspar grains shows that secondary porosity is due to leaching of replacive clay rather than direct leaching of feldspars. Feldspar grains in the Cambrian sandstones exhibit alteration to chlorite and illite in quartz-cemented beds, but are little altered in calcite-cemented beds. Secondary porosity, and thence hydrocarbons, is limited to the quartz-cemented beds.

Fifty samples of Middle Cambrian sandstone (Upper Comley Group) from Comley, Shropshire, were point-counted for quartz (percentage of grains) and hydrocarbon (percentage of volume) (Fig. 9). The results show low hydrocarbon contents for quartz arenites, but significant values for quartz grain contents 75 to 90%. Few samples were available with quartz percentages less than 75% and none yielded hydrocarbons.

The role of alteration to clay minerals is most obvious in the Stiperstones Quartzite, in which feldspar grains and feldspathic volcanic clasts are extensively altered to illite. The illite has been subsequently leached out and the secondary porosity filled with hydrocarbon (Fig. 8c and Parnell 1987). In the most extreme cases clasts of up to 1.5 cm width were completely leached away to leave cavities supported by the surrounding framework of quartzose clasts and cement.

Hydrocarbons occupying secondary porosity after dissolution of calcite occur in the Bog Quartzite, where they fill surficial mouldic porosity (Whittard 1932), and as very minor traces in the Hoar Edge Grit within kaolinitic cement.

Arkoses/litharenites

Arkoses were collected from Cambrian sandstones in the Malvern Hills, the Caradoc Black Grit and Llandovery sandstones in the southern Welsh Borderlands. The Cambrian and Caradoc arkoses include fine-grained sandstones. Labile grains such as volcanic and mudrock clasts were deformed during compaction and squeezed between rigid grains to reduce porosity. Compaction was also accommodated by mechanical fracturing of feldspars. The compaction postdated quartz overgrowth and the precipitation of pore-lining clays. The arkoses are substantially cemented and replaced by clay minerals, particularly chlorite.

Hydrocarbon shows

Hydrocarbons impregnate arkosic sandstones in the Black Grit and Hollybush Sandstone. The 6 m Black Grit is thoroughly impregnated with hydrocarbon at localities 2 km distant along-strike. The Hollybush Sandstone is patchily hydrocarbon-bearing at both extremes of a 1.2 km length outcrop, but only where faulted against black shales. Sandstone 100 m from the fault is petrographically similar but not hydrocarbon-bearing (Fig. 10a,b). Grains appear to float in a hydrocarbon cement (Figs 8d and 10a) because the hydrocarbons 'stain' the clay matrix. Backscattered electron imagery of polished surfaces show that the hydrocarbons fill micropores (*sensu lato*; 5–20 μm pore size) in chloritic clays. Hydrocarbons also fill pores in altered feldspars (Fig. 8d).

Arkosic rocks elsewhere in the Welsh Borderlands are *not* impregnated with hydrocarbon.

Discussion

Diagenesis and reservoir potential are clearly related to lithological composition. Figure 11 summarizes the main diagenetic changes experienced by a quartz arenite, subarkose and arkose during progressive burial and then uplift to the surface. The consequences of early cementation by calcite are also figured. Hydrocarbon migration is indicated as following the leaching of replacive clays and calcite cement (Fig. 11).

Importance of secondary porosity

Schmidt & McDonald (1979a) distinguish five textural origins of secondary porosity in sandstones; fracturing, shrinkage, dissolution of sedimentary material, dissolution of authigenic cement and dissolution of authigenic replacement. All five types are recognized and play a role in hydrocarbon distribution in the Welsh Borderlands. Fracture-bound hydrocarbons are widespread and are discussed elsewhere (Bath *et al.* 1986). Hydrocarbon shows occupying shrinkage fissures in glauconite grains and secondary porosity after dissolution of calcite cement, calcite bioclasts and clay replacements have all been recorded above.

In the Comley Sandstones, Stiperstones Quartzite and Trecoed Beds, where hydrocarbons are relatively abundant, they predominantly occupy

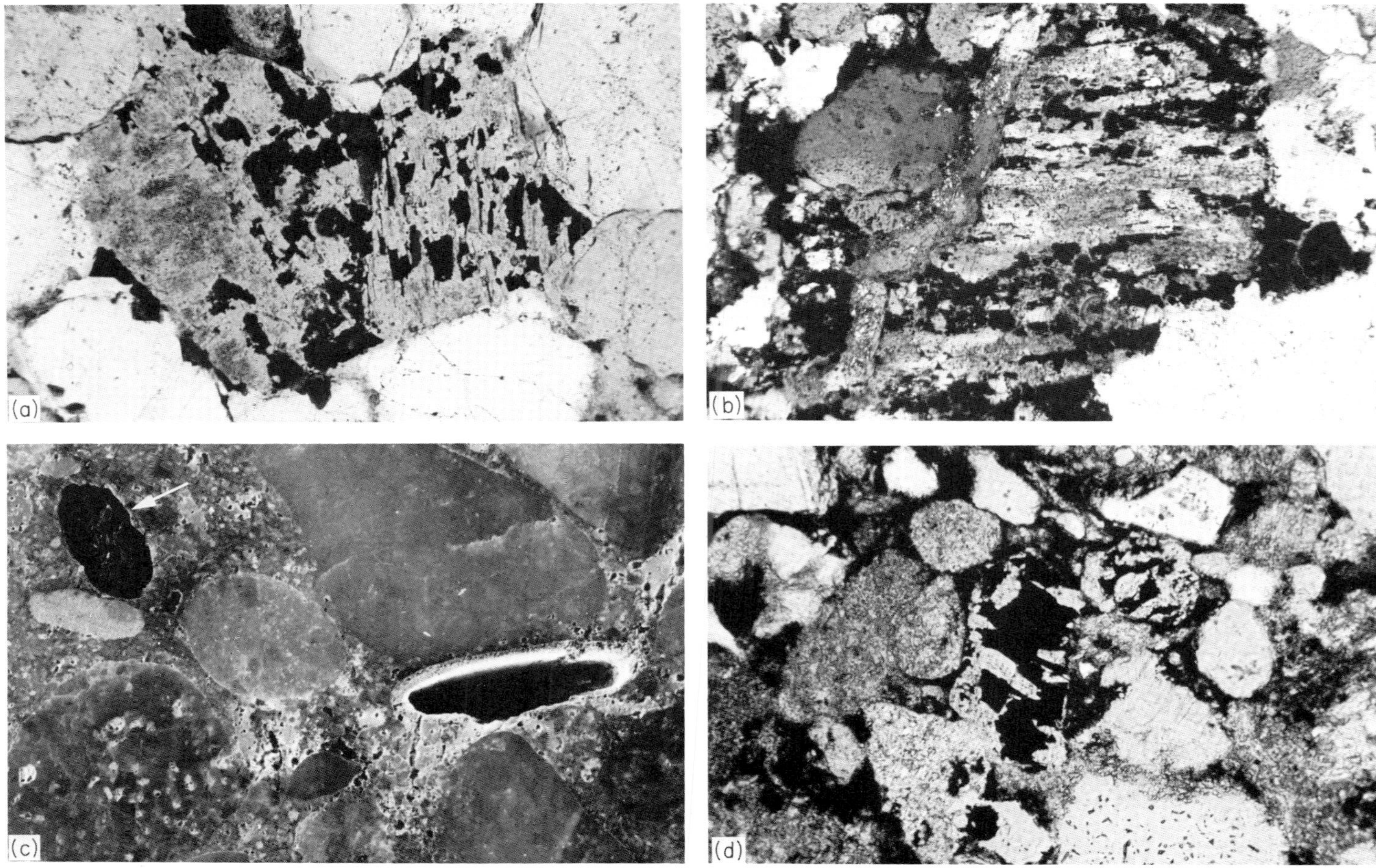

FIG. 8. Secondary porosity within feldspars, filled with hydrocarbon (black). (a) Alkali feldspar grains, Cambrian Sandstone, Comley (field width 1.3 mm). (b) Alkali feldspar grain, Llandovery Trecoed Beds (field width 1.3 mm). (c) Altered and leached clasts, Arenig Stiperstone Quartzite. One clast mould (arrowed) now filled with hydrocarbon. Original clasts were feldspathic volcanic (field width 30 mm). (d) Porous alkali feldspar grain (centre) and other grains apparently floating in a hydrocarbon cement, Caradoc Black Grit (field width 1.3 mm).

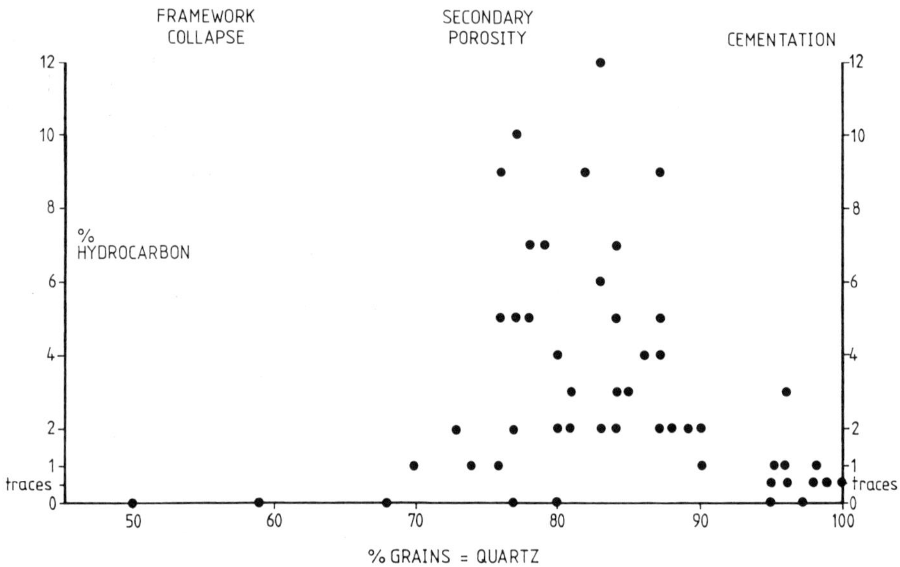

FIG. 9. Hydrocarbon content of Cambrian sandstone (Upper Comley Group) samples, Comley, Shropshire. Fifty samples were point-counted for quartz as a percentage of grains and hydrocarbon as a total volume percentage. Maximum hydrocarbon content is in subarkoses, reflecting secondary porosity in feldspars preserved by framework (Fig. 8a). Quartz arenites (Fig. 7a) and arkoses show low porosity.

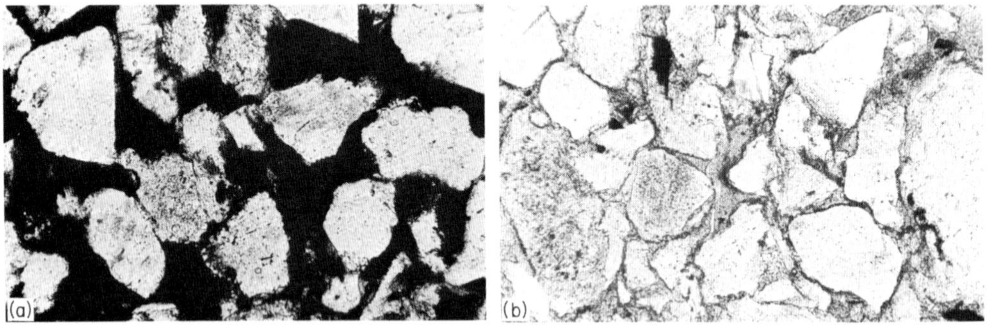

FIG. 10. (a) Fine-grained arkosic Cambrian Hollybush Sandstone, with hydrocarbon-impregnated matrix (field width 2.4 mm), which passes laterally into (b) sandstone with chloritic matrix not impregnated with hydrocarbon (field width 2.4 mm).

secondary porosity. Primary porosity is limited because of quartz overgrowth, or pore-lining chlorite or illite. Hydrocarbons have not been observed within pores with thick clay-linings, but do occur rarely within kaolinite cement. Although the porosity of the chlorite-bearing sandstones is often higher because quartz overgrowth has been inhibited, the pore-lining restricted hydrocarbon migration and the porosity is isolated and therefore not effective. Pore-filling kaolinite clearly reduces porosity but still exhibits an effective microporosity because permeability between pores is not so restricted.

Primary porosity in arkoses

Primary porosity is subordinate in medium/coarse-grained sandstone, but may be the main mode of porosity in the two fine-grained sandstones which bear hydrocarbons. The microporosity measured in chlorite/hydrocarbon mixtures does not exceed 20 μm width. This is of the magnitude expected for primary intra-cement pores, whilst secondary intra-cement pores often exceed this size (Schmidt & McDonald 1979b). Although the sandstones are substantially replaced and cemented by chloritic clay, which is detrimental to porosity, hydrocarbon impregna-

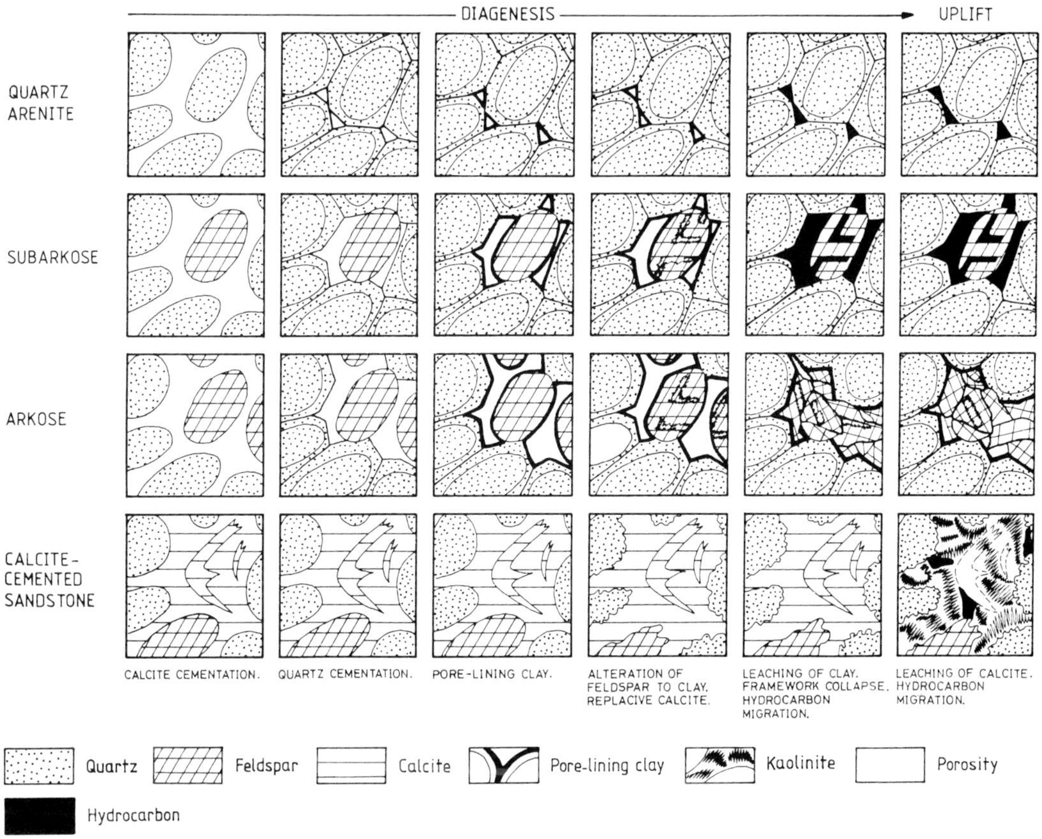

FIG. 11. Diagenetic changes experienced by sandstones of different composition, through progressive burial and uplift. Hydrocarbon migration into secondary porosity follows leaching of replacive clays and calcite cement.

tion is accounted for by the very close proximity of the source rock horizons (see above).

Framework stability

Secondary porosity in sandstones is dependent on the alteration of grains and cement, but it is not enhanced by any resultant mechanical compaction. For secondary porosity to be preserved, there must be a balance between leaching of grains and retaining sufficient stability to prevent pore collapse. The importance of sandstone framework stability to porosity and permeability has been discussed by Nagtegaal (1978), who showed that quartz frameworks are the most stable. Sandstones containing abundant lithic clasts are the least stable because they experience plastic deformation of mechanically weak clasts. In sandstones free of allochthonous cement, Nagtegaal (1978) concluded that 'pure quartz arenites have the optimal potential of preserving high porosity and permeability during burial diagenesis'. However, in the Lower Palaeozoic sandstones of the Welsh Borderlands, where cementation is significant, the optimal reservoir potential is developed in sandstones containing a proportion of non-quartz grains.

The cementation in quartz arenites left only minor primary porosity and negligible secondary porosity. Arkosic sandstones have poor porosity except where in immediate contact with the source rock. The optimum reservoir potential was in the subarkoses, where secondary porosity in feldspars is combined with a framework stability conferred by quartz grains and their cementing overgrowths. Clearly there must be a limit to the porosity which can be supported before the framework collapses and porosity is destroyed. As the proportion of feldspars and lithic fragments becomes greater, mechanical deformation, collapse/precipitation of cementing clays becomes increasingly important and the reservoir potential progressively lower. The importance of the non-quartz component to reservoir potential is demonstrated by the data in Fig. 9.

Significance of marginal basin setting

Volcanic and plutonic rocks formed at the plate margin contributed material to the sandstones. Whilst products of Ordovician volcanism contributed to the Llandovery sediments, volcanic and plutonic basement rocks formed during the late Precambrian evolution of the margin were a more important source. Clasts from Malvernian, Uriconian and Longmyndian basement are widely recognized. The time-scale of volcanicity, erosion and diagenesis was much longer than reported in other studies of marginal basin sandstones (e.g. Davies *et al.* 1979, Galloway 1979, Surdam & Boles 1979, Vessell & Davies 1981). High heat flow in back-arc basins normally accelerates diagenesis (Lee & Klein 1986). Where erosion of volcanic rocks made a substantial contribution to sediments they are arkoses and litharenites and reservoir quality is negligible as expected. However, without an input from volcanic and plutonic rocks, successive transgressions would have reworked sediment almost entirely from pre-existing shelf deposits to yield abundant quartz arenites. This occurred in regions where no volcanic or plutonic basement was exposed, e.g. the Llandovery transgression reworked sediments of the Cambrian transgression at Rubery and the Caradoc transgression in east Shropshire. Where sediments incorporated a moderate input of unstable grains from volcanic and plutonic sources, the sediment composition was such as to have some reservoir potential as discussed above.

Generation of secondary porosity

The generation of secondary porosity requires a fluid to leach away unstable components. The setting of the Welsh Borderland shelf sandstones updip from a basin of organic-rich mudrocks invites speculation that large volumes of fluid migrating from the compacting mudrocks might pass through the shelf sandstones and leach them. The 20 m thick Trecoed Beds sandstones pass laterally into over 1000 m Llandovery mudrocks within 10 km distance (Ziegler *et al.* 1968). Extensive secondary porosity in these beds was developed prior to the migration of hydrocarbons from the Lower Palaeozoic black shales. Schmidt & McDonald (1979a) suggest that secondary porosity is created particularly by acidic fluids produced by CO_2 generation from organic-rich sediments prior to liquid hydrocarbon generation. However, Giles & Marshall (1986) report that organic-rich shales are unlikely to yield enough CO_2 to generate secondary porosity. Furthermore, whilst feldspars are leached in the Trecoed Beds calcite bioclasts are still preserved. Acidic fluids would have leached the calcite. Similar selective leaching of feldspars relative to calcite is recorded by Heald & Larese (1973). The origin of the secondary porosity is therefore not clear.

Shropshire: meteoric fluids

The hydrocarbon shows from Shropshire occupy secondary porosity after both feldspar (clay) dissolution *and* calcite dissolution. Calcite is *preserved* in the Trecoed Beds, so a different mechanism is needed to explain its leaching at Bog and in the Hoar Edge Grit. The calcite mouldic porosity recorded above at Bog appears to be surficial and yet contains hydrocarbons, so at least some of the surficial porosity was not caused by recent exposure. Given that the Bog hydrocarbons probably migrated from Carboniferous rocks, the likely scenario is a post-Lower Palaeozoic exposure of the sandstones, surficial leaching of calcite to yield secondary porosity, burial beneath Carboniferous sediments and then ultimately hydrocarbon migration into the Llandovery sandstones. The exposure and weathering of Lower Palaeozoic rocks under a semi-arid Devonian/Lower Carboniferous climate would have facilitated secondary porosity development. In the Hoar Edge Grit, evidence for pre-kaolinite secondary porosity and the gradation from kaolinite cement to calcite cement suggest that a calcite cement in the sandstones was similarly leached and replaced by kaolinite. Acidic meteoric waters would account for both calcite dissolution and kaolinite precipitation. Calcite dissolution by an acidic fluid could cause the pH of the fluid to rise to the point where it is supersaturated and the mineral kaolinite is precipitated (Curtis 1983). Feldspar dissolution in the Shropshire sandstones could also be a product of freshwater leaching.

Reservations about freshwater leaching and clay precipitation have been expressed by some workers on the basis that fluids tend to migrate upwards in sedimentary basins (Burley *et al.* 1985). However, the Carboniferous sandstones were deposited on a terrane of considerable relief. The distribution of hydrocarbons in the region (Bath *et al.* 1986) suggests that migration was updip from the Carboniferous into uplifted topographically *higher* Lower Palaeozoic rocks. Hydrocarbon seeps are still active in Carboniferous sandstones in Shropshire (Bath *et al.* 1986), so it is important to emphasize that hydrocarbon migration into the Lower Palaeozoic sandstones was not recent. Hydrocarbons have been recorded in a borehole through Cambrian sandstones at over 50 m depth below drift (Parnell 1987). Some hydrocarbon-bearing Lower Palaeozoic rocks occur at distances over 10 km from the nearest Carboniferous outcrop. Recent or sub-recent seeps have left viscous hydrocarbon residues unlike the brittle solids found in Lower Palaeozoic sandstones. These aspects do not absolutely rule out a sub-recent origin, but petrographic evidence is more conclusive. Hydrocarbons in the Welsh Borderlands of suspected Lower Palaeozoic origin have high reflectivity, low H/C ratio and low solubility in organic solvents, all indicative of high thermal maturity. They are clearly distinguishable from hydrocarbons of younger origin which have lower reflectivity, higher H/C ratio and ready solubility (Bath *et al.* 1986, Parnell 1987). The hydrocarbons from the Trecoed Beds, Hollybush Sandstone and the Black Grit exhibit one or more of the features of high thermal maturity. Hydrocarbons do not migrate in this state and high maturity reflects their emplacement for a geologically very long period. The hydrocarbons were generated during the Lower Palaeozoic (Bath *et al.* 1986), and hence secondary porosity in the Trecoed Beds was generated before exposure to meteoric fluids.

Hydrocarbons of probable Carboniferous origin were also emplaced in the geological past. In samples where some pore space is isolated by a thick pore-lining of clay, hydrocarbons do not occur in the isolated pore space despite the fact that sub-recent weathering has broken the pore-linings. The hydrocarbons were therefore emplaced pre-weathering. Furthermore, some hydrocarbons occupying fracture porosity below the sub-Carboniferous unconformity are coeval with calcite veins (Parnell 1983).

Conclusion

The deposits of each of the four main transgressions in the Welsh Basin locally contain hydrocarbons. The composition of the sandstones was important to reservoir potential, as their porosity is predominantly secondary, after alteration and leaching of feldspars in subarkoses. Alteration in arkoses was too extensive to allow the preservation of a framework around secondary porosity. Precambrian volcanic and plutonic rocks which formed at the unstable southern margin of the Iapetus Ocean were a source of unstable grains, particularly feldspars. However the grains were made available by long-term erosion of basement rocks, rather than direct fall-out of volcanic detritus or penecontemporaneous reworking of marginal basin volcanic and plutonic rocks.

The secondary porosity was generated by *fluids* from two sources; fluids within the basin and fluids which post-dated the basin history. Limited secondary porosity, after dissolution of calcite, was generated by the meteoric fluids. The porosity was occupied by *hydrocarbons* from two sources: organic-rich rocks within the basin and source rocks which were deposited at a much later stage. Hydrocarbon migration within the basin occurred as the basin subsided during the Lower Palaeozoic, *before* secondary porosity was generated by meteoric fluids. Hydrocarbon migration from Carboniferous rocks could take advantage of secondary porosity of both origins.

ACKNOWLEDGMENTS: The manuscript benefited considerably from reviews by J. D. Marshall, G. Strong and E. K. Walton. E. Lawson and E. McKelvey provided valuable technical support. P. Eakin kindly supplied Fig. 7(c). Fieldwork was supported by the Queen's University of Belfast and NERC grant F60/A1/64.

References

ANDERTON, R., BRIDGES, P., LEEDER, M. & SELLWOOD, B. 1978. *A Dynamic Stratigraphy of the British Isles*. Allen & Unwin, London.

BASSETT, M. G. 1980. The Caledonides of Wales, the Welsh Borderland and Central England. *In*: OWEN, T. R. (ed). *United Kingdom: Introduction to General Geology and Guides to Excursions 002, 005, 093 & 151*. Institute of Geological Sciences, London.

BATH, A. H., BRASSELL, S. C., EGLINTON, G., HILL, R. I., HOOKER, P. J., O'NIONS, R. K., PARNELL, J., ROBINSON, N. & SPIRO, B. 1986. Deep source gases and hydrocarbons in the U.K. crust. *Report of the Fluid Processes Unit, British Geological Survey*, No. 86/2.

BEVINS, R. E. & ROWBOTHAM, G. 1983. Low-grade metamorphism within the Welsh sector of the paratectonic Caledonides. *Geological Journal*, **18**, 141–67.

BOURGEOIS, J. 1980. A transgressive shelf sequence exhibiting hummocky stratification: the Cape Sebastian Sandstone (Upper Cretaceous), Southwestern Oregon. *Journal of Sedimentary Petrology*, **50**, 681–702.

BRENNER, R. L. 1978. Sussex Sandstone of Wyoming—example of Cretaceous offshore sedimentation. *American Association of Petroleum Geologists Bulletin*, **62**, 181–200.

BRIDGES, P. H. 1975. The transgression of a hard substrate shelf: the Llandovery (lower Silurian) of the Welsh Borderland. *Journal of Sedimentary Petrology*, **45**, 79–94.

BURLEY, S. D., KANTOROWICZ, J. D. & WAUGH, B. 1985. Clastic diagenesis. *In*: BRENCHLEY, P. J. & WILLIAMS, B. P. J. (eds). *Sedimentology: Recent Developments and Applied Aspects*, pp. 189–226. Geological Society of London Special Publication, 18. Blackwell Scientific Publications, Oxford.

CURTIS, C. D. 1983. Geochemical studies on development and destruction of secondary porosity. *In*: BROOKS, J. (ed.). *Petroleum Geochemistry and Exploration of Europe*, pp. 113–25. Geological Society of London Special Publication, 12. Blackwell Scientific Publications, Oxford.

DAVIES, D. K., ALMON, W. R., BONIS, S. B. & HUNTER, B. E. 1979. Deposition and diagenesis of Tertiary–Holocene volcaniclastics, Guatemala. *Society of Paleontologists and Mineralogists Special Publication*, **26**, 281–306.

DEMAISON, G. J. & MOORE, G. T. 1980. Anoxic environments and oil source bed genesis. *Bulletin of the American Association of Petroleum Geologists*, **64**, 1179–209.

EARP, J. R. & HAINS, B. A. 1971. *British Regional Geology: the Welsh Borderland*. HMSO, London.

FOLK, R. L. 1974. *Petrology of Sedimentary Rocks*. Hemphills, Austin, Texas.

GALLOWAY, W. E. 1979. Diagenetic control of reservoir quality in arc-derived sandstones: implications for petroleum exploration. *Society of Economic Paleontologists and Mineralogists Special Publication*, **26**, 251–62.

—— & HOBDAY, D. K. 1983. *Terrigenous Clastic Depositional Systems*. Springer-Verlag, New York.

GILES, M. R. & MARSHALL, J. D. 1986. Constraints on the development of secondary porosity in the subsurface: re-evaluation of processes. *Marine and Petroleum Geology*, **3**, 243–55.

HEALD, M. T. & LARESE, R. E. 1973. The significance of the solution of feldspar in porosity development. *Journal of Sedimentary Petrology*, **49**, 1217–28.

INGERSOLL, R. V. & SUCZEK, C. A. 1979. Petrology and provenance of Neogene sand from Nicobar and Bengal fans, DSDP sites 211 and 218. *Journal of Sedimentary Petrology*, **49**, 1217–28.

JAMES, D. M. D. 1983. Observations and speculations on the northeast Towy "axis", mid-Wales. *Geological Journal*, **18**, 283–96.

JONES, O. T. 1949. Geology of the Llandovery district Part 2: the northern area. *Quarterly Journal of the Geological Society of London*, **105**, 43–64.

KELLING, G. 1978. The paratectonic Caledonides of mainland Britain. *Geological Survey of Canada Paper*, 78–13, 89–95.

LEE, Y. I. & KLEIN, G. DE V. 1986. Diagenesis of sandstones in the back-arc basins of the Western Pacific Ocean. *Sedimentology*, **33**, 651–75.

LEGGETT, J. K. 1978. Eustacy and pelagic regimes in the Iapetus Ocean during the Ordovician and Silurian. *Earth and Planetary Science Letters*, **41**, 163–9.

—— 1980. British Lower Palaeozoic black shales and their palaeo-oceanographic significance. *Journal of the Geological Society of London*, **137**, 139–56.

LOUCKS, R. G., DODGE, M. M. & GALLOWAY, W. E. 1979. Importance of secondary leached porosity in lower Tertiary sandstones reservoir along The Texas Gulf Coast. *Transactions of the Gulf Coast Association Geological Society*, **29**, 164–71.

NAGTEGAAL, P. J. C. 1978. Sandstone-framework instability as a function of burial diagenesis. *Journal of the Geological Society of London*, **135**, 101–5.

ODIN, G. S. & MATTER, A. 1981. De glauconiarum origine. *Sedimentology*, **28**, 611–41.

PARNELL, J. 1983. The distribution of hydrocarbon minerals in the Welsh Borderlands and ajacent areas. *Geological Journal*, **18**, 129–39.

—— 1987. The occurrence of hydrocarbons in Cambrian sandstones of the Welsh Borderland. *Geological Journal*, **22**.

ROBINSON, N., EGLINTON, G., BRASSELL, S. C., GOWER, A. P. & PARNELL, J. 1986. Hydrocarbon compositions of bitumens associated with igneous intrusions and hydrothermal deposits in Britain. *Organic Geochemistry*, **10**, 145–52.

SCHMIDT, V. & MCDONALD, D. A. 1979a. The role of secondary porosity in the course of sandstone diagenesis. *Society of Economic Paleontologists and Mineralogists Special Publication*, **26**, 175–207.

—— & —— 1979b. Texture and recognition of secondary porosity in sandstones. *Society of Economic Paleontologists and Mineralogists Special Publication*, **26**, 209–25.

SEEMANN, U. & SCHERER, M. 1984. Volcaniclastics as potential hydrocarbon reservoirs. *Clay Minerals*, **9**, 457–70.

SURDAM, R. C. & BOLES, J. R. 1979. Diagenesis of volcanic sandstones. *Society of Economic Paleontologists and Mineralogists Special Publication*, **26**, 227–42.

SWIFT, D. J. P. 1968. Coastal erosion and transgressive stratigraphy. *Journal of Geology*, **76**, 444–56.

THORPE, R. S., BECKINSALE, R. D., PATCHETT, P. J., PIPER, J. D. A., DAVIES, G. R. & EVANS, J. A. 1984. Crustal growth and late Precambrian–early Palaeozoic plate tectonic evolution of England and Wales. *Journal of the Geological Society of London*, **141**, 521–36.

VESSELL, R. K. & DAVIES, D. K. 1981. Nonmarine sedimentation in an active fore arc basin. *Society of Economic Paleontologists and Mineralogists Special Publication*, **31**, 31–45.

WEDD, C. B. 1982. Notes on the Ordovician rocks of Bausley, Montgomeryshire. *Summary of Progress, Geological Survey of Great Britain*, 1931 (Part 2), pp. 49–55.

WHITTARD, W. F. 1932. The stratigraphy of the Valentian rocks of Shropshire: the Longmynd-Shelve and Breidden outcrops. *Quarterly Journal of the Geological Society of London*, **88**, 859–902.

ZIEGLER, A. M., COCKS, L. R. M. & MCKERROW, W. S. 1968. The Llandovery transgression of the Welsh Borderland. *Palaeontology*, **11**, 736–82.

JOHN PARNELL, Department of Geology, The Queen's University of Belfast, Belfast BT7 1NN, Ireland.

Carbonate cements in clastic reservoir rocks from offshore Norway—relationships between isotopic composition, textural development and burial depth

G. C. Saigal & K. Bjørlykke

SUMMARY: Many Jurassic and Lower Cretaceous sandstone reservoir rocks from offshore Norway contain carbonate-cemented intervals which typically make up 1–10% of the total sequence. Being nearly impermeable to fluid flow, their distribution may strongly influence reservoir properties. It is therefore important to understand the genesis of these carbonate cements.

Studies of the Jurassic and Lower Cretaceous reservoir rocks from the Haltenbanken and Viking Graben area show that the sandstones contain dominantly two types of calcite cements: (i) microsparry and sparry calcite (Type I) and (ii) poikilotopic calcite (Type II). Type I calcite with $\delta^{18}O$ values between −4.5 and −8.4‰ PDB precipitated relatively rapidly at shallow burial and at temperatures between 15 and 40°C, possibly from porewater of meteoric origin. Type II calcite with $\delta^{18}O$ values between −10 and −12‰ PDB have precipitated later at temperatures corresponding to 50–60°C (at ~1.5 km burial depth). Fluid inclusion data also indicate that Type II calcites have crystallized at temperatures between 56 and 68°C.

Silicate grains are often corroded by Type I calcite while grains contained in Type II calcite are uncorroded, suggesting that the porewater was in near equilibrium with the main silicate phases (quartz and feldspar) during the precipitation of Type II calcite.

The degree of supersaturation in the porewater with respect to low Mg calcite probably decreased with increasing burial depth as sources of more soluble carbonate (aragonite, high Mg calcite) were exhausted, causing slower precipitation rates and increased crystal size. The Type II (poikilotopic) calcite is thought to have formed at the expense of only slightly more soluble low Mg calcite crystals.

The fact that the Type II calcite shows nearly the same $\delta^{18}O$ value (−10 to −12‰ PDB) in sandstones ranging in present burial depth from 1.6 to 5 km suggests that this type of carbonate cementation probably took place early (at temperatures less than 60–70°C), and that very little dissolution and reprecipitation occurred at higher temperatures (greater burial depth).

Carbonate cemented intervals in reservoirs form impermeable barriers to fluid flow during hydrocarbon production. It is therefore important to try to understand how these form diagenetically. Only then can predictions about the lateral distribution of carbonate cemented layers encountered in the cores be made. The first question to ask is: what is the origin of carbonate cements in sandstones? There are three possible sources of carbonate cements: (1) Biogenic carbonate within the sandstone sequence—deposited as scattered fossils or enriched in lag deposits (coquinas). (2) Early marine carbonate cement precipitated during periods of slow sedimentation. (3) Carbonate cement introduced from outside the sandstone sequence.

Very little carbonate may be precipitated from porewaters without the dissolution of other carbonate minerals in the sequence. Although carbon is released from maturing kerogen as CO_2, or organic acids, the amount of carbonate that can precipitate is limited by the availability of cations such as Ca^{++}, Mg^{++} and Fe^{++}. Unless significant amounts of calcium are released from non-carbonate minerals, very little 'new calcite' will precipitate. The Jurassic and Lower Cretaceous sandstones examined in the present study from the Norwegian continental shelf, contain very little plagioclase or other possible sources of Ca^{++} from non-carbonate minerals.

The calcite cements formed during deep burial in the sandstones therefore represent redistributed carbonate which was buried with the sandstone and adjacent shales. The initial carbonate distribution is facies controlled, and the later redistribution is caused by dissolution and reprecipitation of essentially the same volume of carbonate. Bjørkum (1984), in a study of the Jurassic reservoir sandstones from the Oseberg field, argued that the precipitation of calcite must have taken place in a semi-closed system. He observed high Sr/Ca ratios in the poikilotopic

From: MARSHALL, J. D. (ed.), 1987, *Diagenesis of Sedimentary Sequences*, Geological Society Special Publication No. 36, pp. 313–324.

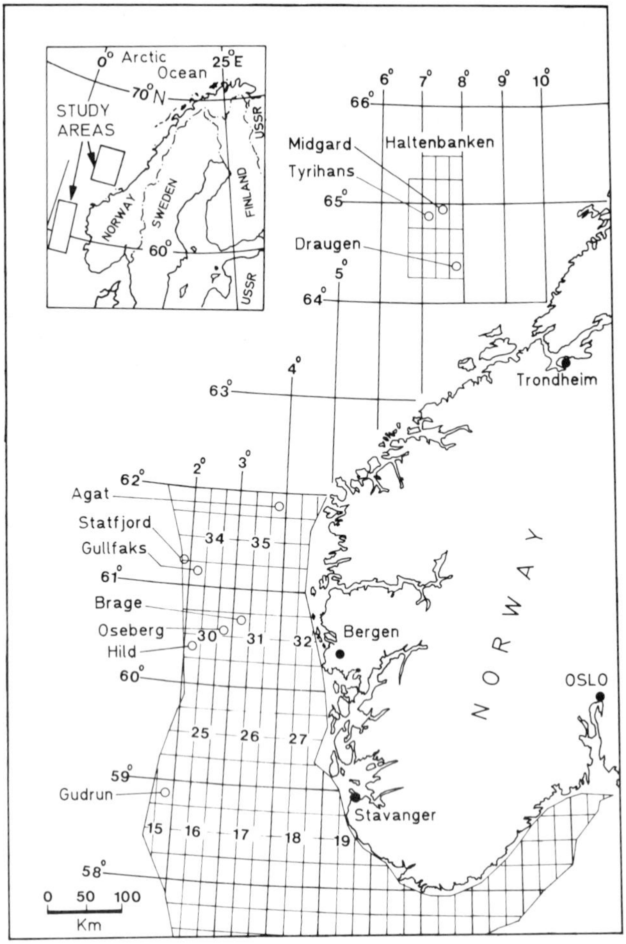

FIG. 1. Map showing wells used in the present study. The inset shows the location of study areas from offshore Norway.

calcite and also in samples of the formation water. He suggested that precipitation of poikilotopic calcite took place in porewater that was not very mobile and that the Sr/Ca ratio of the porewater would build up in the process.

In the present study we have compared isotopic data with textural analysis of the different types of calcite cements present in the Jurassic (Draugen, Midgard, Gullfaks, Brage and Gudrun) and Lower Cretaceous (Agat) clastic reservoir rocks from offshore Norway (Fig. 1).

Petrology

The average bulk composition and detrital mineralogy of Jurassic sandstones is partly based on the work of Bjørlykke *et al.* (1986). This is also supplemented by point-count analyses of 73 samples examined during this study. The framework grain mineralogy and average bulk composition of Lower Cretaceous sandstones is based on the point-count analyses of 82 thin sections. Approximately 250 points per slide were counted.

The Jurassic sandstones are feldspathic and micaceous, and range from fine to coarse sand. The grains are subrounded to subangular and poorly sorted. The feldspar content is typically between 10 and 20%. Potassium feldspar is nearly always dominant over plagioclase. Quartz content varies between 40 and 50%, while mica constitutes 5–10% of the total rock. More quartzitic facies occur locally. Cements and porosity make up the remaining proportion of

FIG. 2. Photomicrograph showing sparry (S) calcite (Type I) cement often replacing detrital grains; the arrow shows relict of a quartz grain. Other grains seen in this picture are of quartz. Scale bar = 200 μm.

the rock volume. Poikilotopic calcite and diagenetic kaolinite are the most common cements, and are found in variable quantities in these sandstones.

The Lower Cretaceous sandstones are of fine to coarse grain size. The grains are subrounded to subangular and poorly sorted. The sandstones contain 35–55% quartz (mono and polycrystalline), 6–15% feldspars (dominantly K-feldspars with some plagioclase), 3–8% rock fragments (dominantly quartzitic with occasional volcanic) and 5–10% micas (mainly biotite with some muscovite). Cements and porosity make up the remaining proportion of the rock volume. Authigenic clay minerals (kaolinite, chlorite and illite), quartz overgrowths (rare to common forming euhedral outlines) and calcite (dominantly poikilotopic with some microsparry to sparry) are the common cements found in these sandstones.

Calcite textures

The sandstones examined show dominantly two types of calcite cement: (i) microsparry and sparry calcite and (ii) poikilotopic calcite. Samples may contain either one type of calcite cement (pure), or a mixture of both in varying proportions (impure).

Type I.

Microsparry and sparry calcite constitutes 28–47% of the total rock and its crystal size varies between <10 and 200 μm. The sandstones are normally grain supported (quartz and feldspar grains) but may also show a floating texture. The contacts between calcite crystals are commonly sharp and straight but sutured (microstylolitic) contacts are also seen. Microstylolitic contacts are often more common in samples showing floating texture. Remains of shell debris and/or ghosts of shell fragments are enclosed in sparry calcite cement in some samples. The calcite cement commonly partially replaces detrital grains (Fig. 2), and the floating texture can, to some extent, be attributed to similar replacement of detritals by calcite cement. Under cathodoluminescence the calcite cement shows patches of more and less bright luminescence. Individual sparry calcite crystals show uniform luminescence without any zoning. SEM studies show that calcite commonly occurs with authigenic kaolinite as pore filling cement (Fig. 3). The relative timings of calcite and kaolinite formation in these pores is not clear. Microsparry and sparry calcite is hereafter referred to as Type I calcite.

Type II.

Poikilotopic calcite is ferroan in composition and makes up between 5 and 45% of the sandstones examined. The size of poikilotopic crystals varies between approximately 300 and 3000 μm. These crystals enclose a few to several detrital grains (Fig. 4). The contact between the calcite crystals and detrital grains is usually

FIG. 3. SEM photograph showing sparry calcite (S) and kaolinite filling a pore. Scale bar = 10 μm.

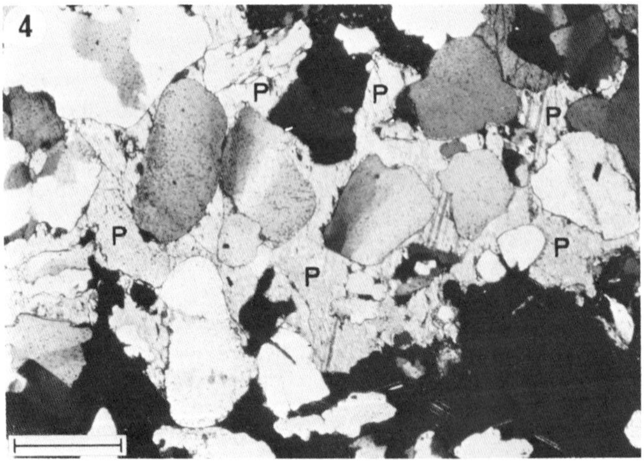

FIG. 4. Typical poikilotopic calcite (Type II) crystal (P) enclosing several quartz grains. Note that the detrital grains are not corroded by poikilotopic calcite. Scale bar = 500 μm.

sharp and, unlike sparry calcite, it (poikilotopic calcite) seldom shows any significant replacement of detrital silicate grains (Fig. 4). Like sparry calcite, the contacts between poikilotopic calcite crystals are either straight and sharp or sutured (microstylolitic). Microstylolitic type contacts are more common in samples showing a floating texture. Under cathodoluminescence poikilotopic calcite crystals commonly appear to be made up of more and less bright laths oriented parallel to each other. Often individual poikilotopic calcite crystals also show uniform luminescence without any zoning or laths. SEM studies show that detrital grains (including quartz and feldspar overgrowths) and authigenic kaolinite are commonly enclosed within poikilotopic calcite crystals. Kaolinite occurring as vermicular stacks of pseudo-hexagonal crystals and discrete booklets is often partly or almost completely engulfed in poikilotopic calcite crystals (Fig. 5). Poikilotopic calcite is hereafter referred to as Type II calcite.

Fluid inclusions

Fluid inclusions present in Type II calcite cement (poikilotopic) have been studied using a Linkam TH 600 stage. Generally, fluid inclusions are rare and < 10 μm in diameter. Most of the inclusions have a very tiny vapour bubble (high liquid to vapour ratio) and are globular or

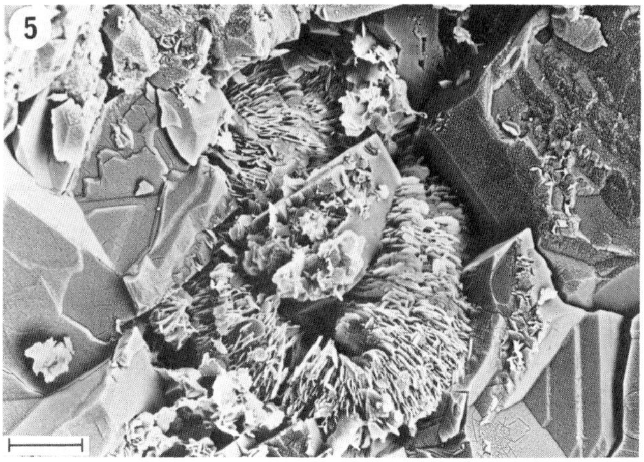

FIG. 5. SEM photograph showing poikilotopic calcite engulfing vermicular stacks of kaolinite. Scale bar = 20 μm.

elongate rounded in shape. It was not possible to record the temperature of melting of the first or last ice crystal but the temperature of homogenization (Th) was recorded from a total of 19 fluid inclusions. Out of these, eight inclusions were present in the Jurassic sandstones of the Draugen field while 11 were present in the Lower Cretaceous sandstones of the Agat field. The present burial depth of these Jurassic and Lower Cretaceous samples is about 1680 and 3600 m, respectively. The Th measurements were made on a doubly polished, approximately 70 μm thick slice. The range of measured Th values is from 56 to 68°C but the majority of inclusions (from both Jurassic and Lower Cretaceous sandstone samples) show values between 64 and 68°C. Similar values have been recorded by Olav Walderhaug in poikilotopic calcite from the Haltenbanken area (personal communication).

Stable isotope geochemistry

On the basis of petrological studies, samples containing different types of calcite cement were selectively chosen and analysed for stable isotope compositions. Organic-free, powdered calcite cement from 24 samples of Jurassic sandstones and 19 samples of Lower Cretaceous sandstones have been analysed for carbon and oxygen isotopes. The results of the analyses are given in Tables 1 and 2. Petrographic and XRD studies confirmed that calcite cement comprised >90% of the carbonate material in all the samples. Carbon dioxide was extracted from samples by reacting with phosphoric acid using the procedure described by McCrea (1950).

The $\delta^{18}O$ values of Type I calcite cement in Jurassic and Lower Cretaceous sandstones range from −4.5 to −8.4‰ PDB, whereas $\delta^{18}O$ values of Type II calcite cement vary between −9.2 and −12.1‰ PDB (Tables 1 and 2). Samples containing a mixture of Types I and II calcite show intermediate values.

The $\delta^{13}C$ values of Type I calcite cement present in the Jurassic sandstones of the Haltenbanken, Brage and Gudrun areas range from −0.9 to −15.6‰ PDB, whereas $\delta^{13}C$ values of Type II and mixed calcite samples vary between −2.7 and −34.4‰ PDB (Table 1). In the Jurassic sandstones of the Gullfaks area, $\delta^{13}C$ values of Type I calcite cement range from +5.3 to +12.6‰ PDB, whereas $\delta^{13}C$ values of Type II and mixed calcite samples are +9.8 and +5.8‰ PDB, respectively (Table 1). In the Lower Cretaceous sandstones (Agat), $\delta^{13}C$ values of Type I calcite range from −5.1 to 1.8‰ PDB, whereas $\delta^{13}C$ values of Type II and mixed calcite samples vary between −10.5 and 0.6‰ PDB (Table 2).

Discussion

Stable isotope analyses of calcite cements present in the Jurassic and Lower Cretaceous sandstones show that Type II cements consisting of coarse poikilotopic calcite crystals are characterized by the most negative $\delta^{18}O$ values (−9.2 to −12.1‰ PDB) and that the finer sized Type I calcite have more positive $\delta^{18}O$ values (−4.5 to −8.4‰ PDB, Fig. 6). The most negative $\delta^{18}O$ values (−11.1 to −12.1‰ PBD, for Type II calcite) are found in the shallower Draugen field (in the

TABLE 1. $\delta^{18}O$ and $\delta^{13}C$ for calcite cement types present in Jurassic sandstones

Field	Well and sample no.	Depth (m)	Calcite type*	Calcite %	Average calcite crystal size (μm)	$\delta^{18}O$ (PDB)	$\delta^{13}C$ (PDB)
Draugen	6407/9-1						
	33	1669.7	I	8	120	−8.37	−15.58
	18	1670.1	I+II	I—28	I—100	−9.57	−15.03
				II—5	II—450		
	12	1676.3	II	50	1400	−12.10	−18.51
	7	1681.2	II	45	2000	−10.36	−34.36
	44	1682.3	II	43	1500	−11.14	−32.93
	3	1666.3	I+II	I—18	I—180	−10.34	−15.94
				II—12	II—650		
	38	1679.9	II	42	1750	−10.57	−31.40
Midgard	6407/2-1						
	15	3088.9	I	45	<15	−5.1	−11.7
	43	3495.6	I	42	<20	−4.8	−0.9
Midgard	6407/2-2						
	69	2501.6	II	45	1000	−10.4	−27.2
	65	2510.1	II	47	800	−9.8	−21.1
Gullfaks	34/10-3						
	10	1930.9	I	55	100	−6.24	12.56
	11	1942.1	I	40	200	−6.38	11.33
	34/10-4						
	12	1852.00	I+II	I—28	I—150	−5.08	5.81
				II—10	II—400		
	13	1863.70	I	40	200	−5.80	5.33
	14	2351.30	II	40	2500	−9.22	9.84
Brage	31/4-2						
	7	2222.76	I	45	150	−4.52	−8.55
	8	2494.30	II+I	II—20	II—1200	−11.46	−9.11
				I—15	I-200		
Gudrun	15/3-3						
	1	4262.95	II	42	2200	−9.61	−2.69
	2	4268.10	II	40	1600	−10.87	−6.40
	3	4268.28	II	35	1300	−10.45	−3.66
	4	4269.20	I+II	I—15	I—150	−11.53	−5.77
				II—7	II—1200		
	5	5000.70	II	14	1500	−11.57	−8.89
	15/9-3						
	6	3523.40	II	45	1100	-9.91	−8.20

* I—microsparry and sparry calcite, II—poikilotopic calcite.

Haltenbanken region) at about 1680 m, which at present is at its maximum depth of burial, and has a bottom hole temperature (at 1700 m) of about 60°C. The Haltenbanken region experienced rapid subsidence and sedimentation in late Tertiary and Quaternary time (Ellenor & Mozetic 1986), totalling about 1000 m, and at present is exposed to the maximum burial depth and temperature yet experienced. The bottom hole temperature (60°C) can therefore be taken to be the maximum temperature of crystallization for Type II calcite. Hence, the most negative $\delta^{18}O$ value (−12‰ PDB) of Type II calcite and maximum temperature of crystallization (60°C) together suggest that the heaviest water from which Type II calcite in the Draugen field could have precipitated had $\delta^{18}O = -4‰$ (SMOW) (Bjørlykke *et al.* 1986 and Fig. 7). Therefore, Type II calcite in the Draugen field either precipitated from porewater near present reservoir conditions or from more negative (less than −4‰ $\delta^{18}O$ SMOW) formation water at lower (<60°C) temperatures. Precipitation from porewater with a more negative oxygen isotope composition would imply that it took place during meteoric water flushing.

However, since Type II calcite can be shown to form late and clearly post-date the formation

TABLE 2. $\delta^{18}O$ *and* $\delta^{13}C$ *for calcite cement types present in Lower Cretaceous sandstones*

Field	Well and sample no.	Depth (m)	Calcite type*	Calcite %	Average calcite crystal size (μm)	$\delta^{18}O$ (PDB)	$\delta^{13}C$ (PDB)
Agat	35/3-2						
	7	3614.1	II	9	2000	−11.75	−2.14
	11	3638.5	I	45	<20	−6.16	−0.56
	13	3698.3	II	25	1750	−11.50	−10.48
	19	3950.0	II	10	1500	−11.38	−10.26
	22	4008.7	I	47	200	−8.27	−5.14
	35/3-4						
	1	3403.3	I	47	<20	−5.42	1.55
	5	3448.7	I	45	150	−6.10	1.46
	10	3470.1	I	40	80	−6.73	−0.015
	15	3500.3	II	20	2000	−9.61	−7.54
	35/3-5†						
	1	3230.2	I	38	200	−8.03	−0.26
	8	3240.3	II + I	II—20	II—1200	−7.60	−1.86
				I—13	I—220		
	11	3253	I + II	I—6	I—50	−8.65	−0.93
				II—3	II—700		
	14	3255.2	II	35	2250	−9.37	−0.91
	18	3265.2	I + II	I—30	I—150	−7.55	−0.12
				II—15	II—1150		
	23	3272.1	II	5	1100	−11.78	−1.13
	29	3282.1	II	8	900	−10.81	−1.26
	31	3284.6	I + II	I—30	I—250	−7.93	0.65
				II—10	II—800		
	34	3290.4	I	48	<20	−4.80	1.81
	37	3292.2	I + II	I—8	I—200	−7.72	0.10
				II—4	II—800		

* I—microsparry and sparry calcite, II—poikilotopic calcite.
† 35/3-5—dry well.

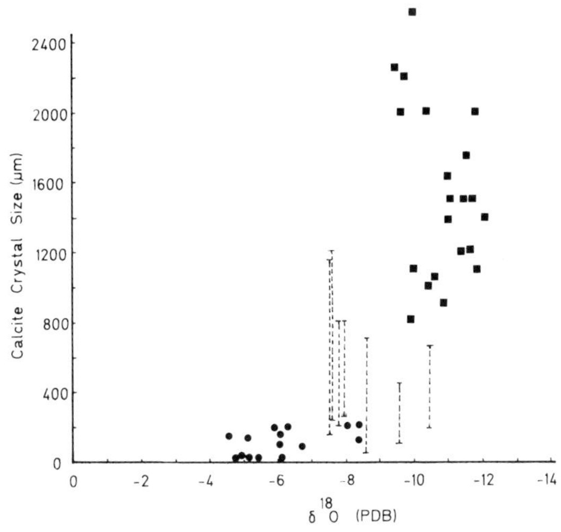

FIG. 6. Plot of oxygen isotope ratio against calcite crystal size for all samples from Jurassic and Lower Cretaceous sandstones. Circles and squares represent Types I and II calcite, respectively. Broken lines represent impure samples containing both Types I and II calcite cements.

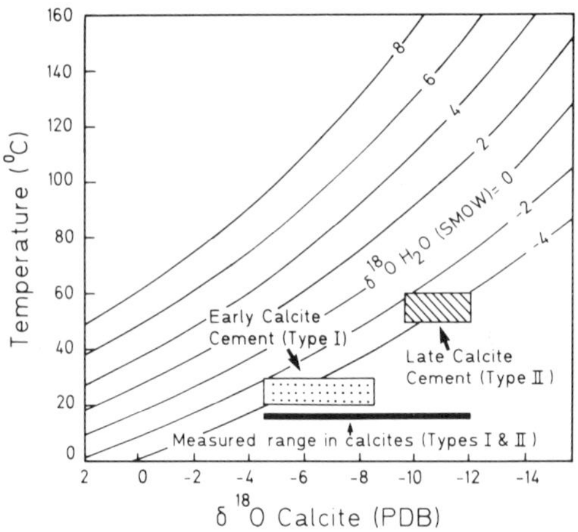

FIG. 7. Temperature dependence of the calcite-water isotopic equilibrium (after Bjørlykke *et al.* 1986). The dark bar represents the observed range of calcite cements (Types I and II) for Jurassic and Lower Cretaceous sandstones examined in this study.

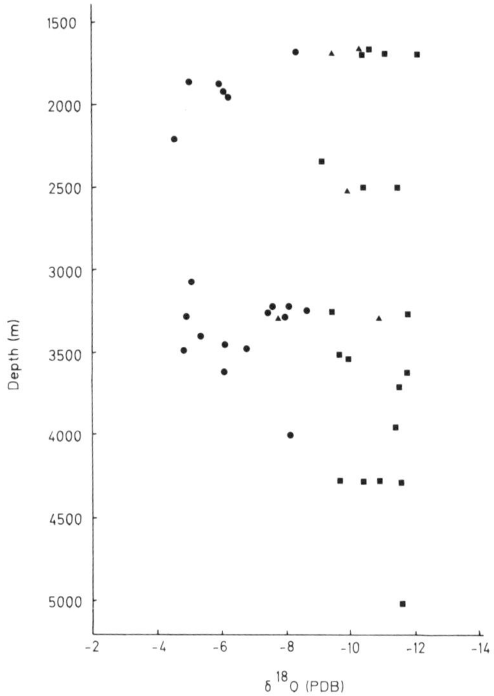

FIG. 8. Plot of oxygen isotope ratio versus depth for all samples from Jurassic and Lower Cretaceous sandstones. Circles and squares represent Types I and II calcite, respectively. Triangles represent impure samples containing both Types I and II calcite cements. Note that Type II calcite crystals (squares) show similar $\delta^{18}O$ values (−10 to −12‰ PDB) at all depths between 1600 and 5000 m.

of kaolinite (Fig. 5) we consider the Type II calcite to have formed near present-day conditions and may continue to form at present right below the oil/water contact. It is probable that most of the Type I calcite formed during meteoric water flushing at lower temperatures. Oxygen isotope values for Type I calcite as negative as −8‰ (PDB) would at 20–30°C imply porewater compositions of about −4 to −6‰ (SMOW) (Fig. 7).

In the deeper wells of the Haltenbanken (4300 m) and Hild fields (4–5 km) analyses of formation water gives more positive values: $\delta^{18}O = +1.3$ and +2.5 to +3.5‰ SMOW, respectively (Bjørlykke *et al.* 1986 and Lønøy *et al.* 1986). The porewater probably experienced $\delta^{18}O$ enrichment by isotopic exchange with minerals during burial diagenesis (Craig *et al.* 1956). The rate at which porewater becomes more positive at depth does not quite correspond to the temperature-dependent fractionation of $\delta^{18}O$ between porewater and calcite. For example, in the Haltenbanken area the present formation water with $\delta^{18}O = +1.3$‰ SMOW at 4300 m gives a maximum temperature of crystallization for poikilotopic calcite ($\delta^{18}O = -11.75$‰ PDB) of about 95°C while the bottom hole temperature is nearly 120°C. Similarly, in the Hild field the formation water with $\delta^{18}O = +3.5$‰ SMOW gives a maximum temperature of crystallization for poikilotopic calcite ($\delta^{18}O = -11$ to −14‰ PDB) of about 120°C (Lønøy *et al.* 1986) while the bottom hole temperature is nearly 150°C. If the Type II calcite formed at shallower depth then isotopic composition of the formation water would have been more negative and the calculated temperature of crystallization would have been lower.

Calcite cemented sandstones buried to a much greater depth (4–5 km) analysed in this study do not show $\delta^{18}O$ values below −12‰ PDB (Fig. 8), indicating that the formation of Type II calcite was probably limited to a narrow temperature range (50–60°C, assuming $\delta^{18}O$ of porewater to be −4‰ SMOW) (Figs 9 and 10). Irwin & Hurst (1983), analysing carbonate cement from Jurassic sandstones from the North Sea, also found that poikilotopic ferroan calcite had $\delta^{18}O$ values mostly between −10 and −12‰ PDB. Similar values for calcite cement were observed by Lønøy *et al.* (1986) from the Hild field which has been buried relatively deeply (4–5 km) in the Viking Graben. In the Oseberg field, however, the calcite cements have $\delta^{18}O$ values of about −14 to −15‰ PDB (Bjørkum 1984). We do not understand the reason why the poikilotopic calcite cement in sandstones from the Hild and

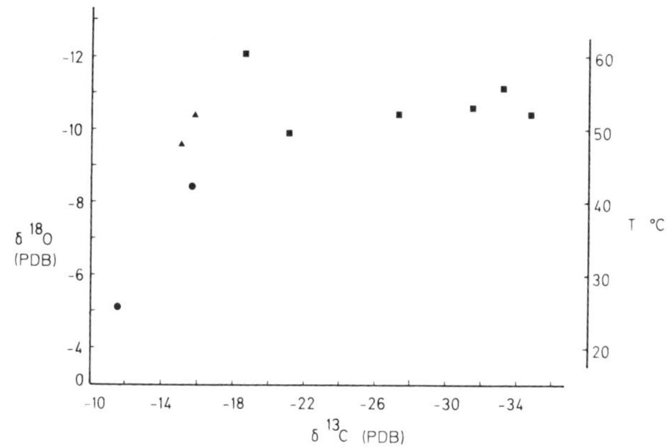

FIG. 9. Plot of oxygen isotope ratio against carbon isotope ratio for Haltenbanken samples. Circles and squares represent Types I and II calcite, respectively. Triangles represent impure samples containing both Types I and II calcite cements.

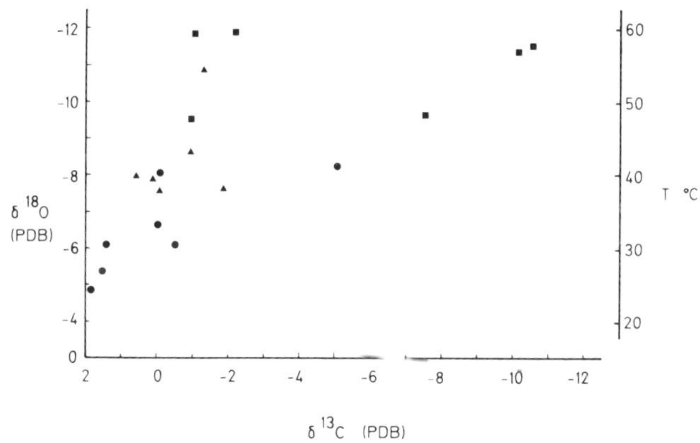

FIG. 10. Plot of oxygen isotope ratio against carbon isotope ratio for Agat samples. Circles and squares represent Types I and II calcite, respectively. Triangles represent impure samples containing both Types I and II calcite cements. The temperature scale has been determined using $\delta^{18}O$ of water $= -4‰$ (SMOW).

Oseberg fields have more negative $\delta^{18}O$ values than the samples analysed in the present study.

The temperature and burial depth we infer from the $\delta^{18}O$ value of the Type II calcite cement depend on the porewater composition at the time of crystallization. Earlier work and available data of present-day formation water show that the porewater becomes isotopically heavier with increasing burial (Craig *et al.* 1956, Bjørkum 1984, Bjørlykke *et al.* 1986 and Lønøy *et al.* 1986). The fact that the oxygen isotope composition of Type II (poikilotopic) calcite cement does not become more negative with increasing burial can theoretically be explained in two ways:

(i) The Type II calcite precipitated at shallow depths (1500–1700 m) did not dissolve, and remained stable on burial.

(ii) The Type II calcite did dissolve and reprecipitated at greater burial depths but the porewater composition changed towards more positive values in such a way that the $\delta^{18}O$ values of the calcite precipitated remained the same despite the increasing temperature.

Alternative (ii) seems less probable as the published and available preliminary data on the porewater compositions indicate that the $\delta^{18}O$

versus depth gradient is not sufficiently steep to produce calcite cement with the same $\delta^{18}O$ values at different depths. Also the fluid inclusion data indicate that crystallization of Type II calcite took place at temperatures between 56 and 68°C (between approximately 1.5 and 2 km burial depth). The negative $\delta^{18}O$ values found in the porewater of shallow reservoirs show that meteoric water was trapped in the structures and remained below the oil-water contact. This trapped meteoric water was not flushed by deeper compactional water which would have had more positive $\delta^{18}O$ values. The initial $\delta^{18}O$ composition of the trapped meteoric water may also depend on when the structures were sealed off from meteoric water influx. The $\delta^{18}O$ composition of the trapped waters may also change during burial and hence produce carbonate cements having different $\delta^{18}O$ values than those formed from otherwise open porewaters at the same temperatures.

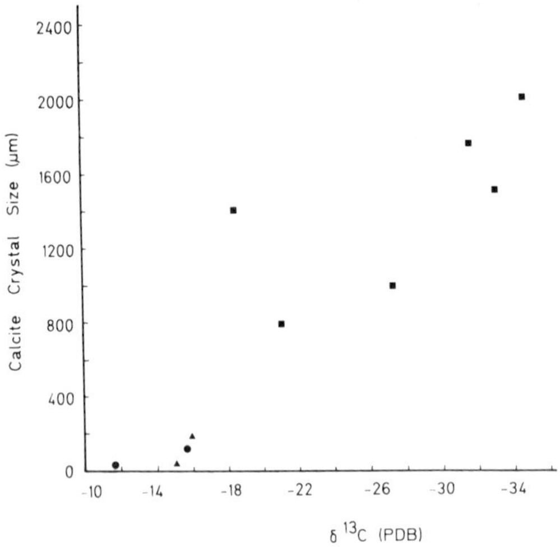

FIG. 11. Plot of carbon isotope ratio against calcite crystal size for Haltenbanken samples. Circles and squares represent Types I and II calcite, respectively. Triangles represent impure samples containing both Types I and II calcite cements.

Samples of Type I calcite with smaller crystals (<200 μm) all have $\delta^{18}O$ values that are significantly more positive (−4.5 to −8.4‰ PDB) indicating that they have formed at somewhat lower temperatures, between 15 and 40°C (assuming $\delta^{18}O$ of formation water = −4 to −6‰ SMOW). Within this range the carbonate cement with the smallest crystals tends to have the most positive $\delta^{18}O$ value (−4.5‰ PDB) and are therefore probably formed at lower temperatures (about 15°C).

The Type I calcite cements from the Gullfaks area with $\delta^{13}C$ values between +5.3 and +12.6‰ (PDB) and $\delta^{18}O$ values of about −6‰ PDB (Table 1) were produced in a bacterial fermentation zone (Irwin & Hurst 1983). Type II calcite cement from the same well (Gullfaks, sample no. 14) shows a similar $\delta^{13}C$ composition (9.84‰ PDB) but a comparatively lighter $\delta^{18}O$ value, −9.22‰ PDB. This $\delta^{18}O$ value indicates formation at a somewhat higher temperature, probably below the bacterial fermentation zone. The positive $\delta^{13}C$ value in this sample of Type II

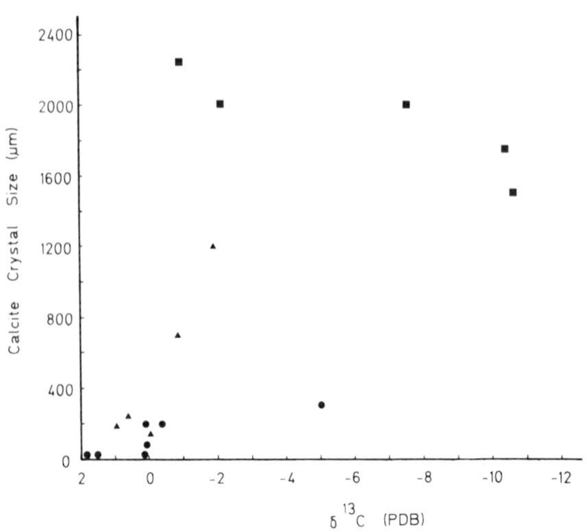

FIG. 12. Plot of carbon isotope ratio against calcite crystal size for Agat samples. Circles and squares represent Types I and II calcite, respectively. Triangles represent impure samples containing both Types I and II calcite cements.

calcite may have been inherited from dissolving calcite precipitated in the bacterial fermentation zone.

Type II calcite from the Haltenbanken and Agat areas shows comparatively lighter carbon ($\delta^{13}C$ from −0.9 to −34.4‰ PDB, Tables 1 and 2 and Figs 11 and 12), suggesting an influence from organically derived carbon (CO_2) derived from the source rock (Irwin *et al.* 1977) probably at the time of oil migration. The Type II calcite cement from the Haltenbanken area with $\delta^{13}C$ of about −34‰ PDB can not have come from decarboxylation, but may have formed by dissolution of calcite precipitated at very shallow depth under the influence of oxidation of methane (Irwin & Hurst 1983). In the dry well of the Agat region (35/3-5), there is less influence from negative carbon in the Type II calcite cement than in wells with hydrocarbons (Table 2). Thus it is suggested that the early carbonate cement (Type I) was dissolved by formation water rich in organic acids associated with oil migration, and that simultaneous precipitation of Type II calcite took place. The difference in solubility of the Types I and II calcite was probably very small, producing a very low degree of supersaturation and a very slow precipitation of coarse poikilotopic calcite (Type II). Similar relationships are also found in the Jurassic sandstones of the Oseberg field (Bjørkum & Bjørlykke, personal communication).

During the formation of Type II calcite, the porewater was in near equilibrium with silicate phases as there is little evidence of their corrosion by the Type II cement. However, in the finer sized Type I calcite-cemented sandstone, there is evidence of corrosion of feldspar and quartz. Type I calcite cement is believed to have formed by relatively rapid precipitation at numerous nucleation sites from solution with a higher (i.e. more than Type II calcite precipitating solutions) degree of supersaturation. This highly super-saturated porewater was probably produced due to dissolution of more soluble phases such as aragonite and possibly also high Mg calcite. Corrosion of silicate grains by Type I calcite cement suggests that porewater from which Type I calcite precipitated was not in equilibrium with silicate phases. Burley & Kantorowicz (1986) also observed that finer carbonate crystals are more corrosive than coarsely crystalline spar or poikilotopic carbonates.

Conclusions

Carbonate cement in the Jurassic and Lower Cretaceous clastic reservoir rocks from offshore Norway is composed of two main types:

(1) Microsparry and sparry calcite cement (Type I) with crystals smaller than 0.2 mm and $\delta^{18}O$ values between −4.5 and −8.4‰ PDB. The least negative values are associated with finer crystalline cement indicating precipitation at low temperatures (about 15°C) and shallow burial, possibly in porewater of meteoric origin. The coarser crystals have more negative $\delta^{18}O$ values and may have precipitated at temperatures up to 40°C.

(2) Poikilotopic calcite cement (Type II) with $\delta^{18}O$ values between −9.2 and −12.1‰ PDB may have precipitated at about 1.5 km of burial depth from isotopically light porewater ($\delta^{18}O = -4$‰ SMOW) at temperatures corresponding to 50–60°C.

The fluid inclusion data indicate that crystallization of Type II calcite took place at temperatures between 56 and 68°C. This includes the deeply buried samples that are at present at much greater temperatures.

In our study, sandstones buried to greater depths (4–5 km) show no isotopic evidence of late carbonate cementation.

The Type I calcite cements probably precipitated relatively rapidly from porewater supersaturated by dissolution of more soluble phases such as aragonite and possibly also high Mg calcite. This cementation was associated with corrosion of silicate minerals.

The Type II calcite cements precipitated slowly from porewater, a process controlled by the dissolution of the only slightly more soluble low Mg calcite phases. The Type II calcite cements do not seem to have corroded silicate minerals, suggesting that the porewater then was in near equilibrium with the silicate minerals.

ACKNOWLEDGMENTS: This research has been supported by VISTA, a research cooperation between the Norwegian Academy of Science and Letters and Den Norske Stats Oljeselskap (Statoil). Support from Norwegian Research Councils (NAVF and NTNF) is also appreciated. G. C. Saigal acknowledges the receipt of post-doctoral fellowship from NTNF and is grateful to the Oil and Natural Gas Commission of India for granting leave of absence. We also thank Arco, Saga, Shell, BP, Norsk Hydro and Conoco for their cooperation in providing samples used in this study. The isotopic analyses have been carried out in the Mass Spectrographic Laboratory at the Department of Geology, University of Bergen. Fluid inclusion studies were carried out in the Department of Geology, St Andrews University, Scotland and we thank Judy Kinnaird for her help during this work. Thanks are also due to Berit Løken Berg for her help with SEM studies. We thank Eli B. Bjerke for typing the manuscript.

Constructive reviews by S. Rainey, J. Gluyas, D. Stow and J. D. Marshall were very beneficial to the final manuscript.

References

BJØRKUM, P. A. 1984. Et studium av de mellomjurassiske sandsteiner: kjerne 30/6-7 Oseberg, med særlig vekt på diagenese. Hovedoppgave. *MSc Thesis*. University of Bergen.

BJØRLYKKE, K., AAGAARD, P., DYPVIK, H., HASTINGS, D. S. & HARPER, A. S. 1986. Diagenesis and reservoir properties of Jurassic sandstones from the Haltenbanken area, offshore mid Norway. *In*: SPENCER, A. M. *et al.* (eds). *Habitat of Hydrocarbons on the Norwegian Continental Shelf*, pp. 275–86. Graham & Trotman, London.

BURLEY, S. D. & KANTOROWICZ, J. D. 1986. Thin section and S.E.M. textural criteria for the recognition of cement-dissolution porosity in sandstones. *Sedimentology*, **33**, 587–604.

CRAIG, H., BOATO, G. & WHITE, D. E. 1956. *Isotopic Geochemistry of Thermal Waters*, pp. 29–38. National Academy of Sciences, National Research Council Publication, 400.

ELLENOR, D. W. & MOZETIC, A. 1986. The Draugen oil discovery. *In*: SPENCER, A. M. *et al.* (eds). *Habitat of Hydrocarbons on the Norwegian Continental Shelf*, pp. 313–6. Graham & Trotman, London.

IRWIN, H., CURTIS, C. & COLEMAN, M. 1977. Isotopic evidence for source of diagenetic carbonates formed during burial of organic-rich sediments. *Nature*, **269**, 209–13.

—— & HURST, A. 1983. Application of geochemistry to sandstone reservoir studies. *In*: BROOKS, J. (ed.). *Petroleum Geochemistry and Exploration of Europe*, pp. 127–46. Geological Society of London Special Publication, 12. Blackwell Scientific Publications, Oxford.

LØNØY, A., AKSELSEN, J. & RØNNING, K. 1986. Diagenesis of a deeply buried sandstone reservoir: Hild field northern north sea. *Clay Minerals*, **21**, 497–511.

MCCREA, J. M. 1950. On the isotope chemistry of carbonates and a paleotemperature scale. *Journal of Chemical Physics*, **18**, 849–57.

GIRISH C. SAIGAL, KDM Institute of Petroleum Exploration, Oil and Natural Gas Commission, Kaulagarh Road, Dehradun—248 195, India. Present address: Institutt for geologi, Universitetet i Oslo, Oslo, Norway.

KNUT BJØRLYKKE, Institutt for geologi, Universitetet i Oslo, Oslo, Norway.

Aspects of the diagenesis of the Sherwood Sandstones of the Wessex Basin and their influence on reservoir characteristics

G. E. Strong & A. E. Milodowski

SUMMARY: The Sherwood Sandstone Group sandstones of the Wessex Basin are important potential reservoirs for both hydrocarbons and geothermal brines. Analyses of conglomerate, sandstone and siltstone samples from outcrop and deep boreholes indicate a complex diagenetic history that has had important positive and negative effects on the reservoir properties. Over most of the basin, early diagenesis (eodiagenesis) is dominated by calcrete development but, towards the basin centre, non-ferroan dolomite associated with early evaporitic sulphate cements that appear to have been deposited in an inland sabkha or playa environment are important. Later diagenesis (mesodiagenesis) is characterized by selective framework grain dissolution of feldspars, and anhydrite cementation and subsequent dissolution—processes that have locally yielded significant secondary porosity. However, in some cases this secondary porosity has been destroyed by the precipitation of late manganiferous ferroan calcite and ferroan dolomite. Quartz cements are locally important. Rocks near the present-day outcrop have been affected by weathering processes (telodiagenesis) that have resulted in decalcification, and the precipitation of kaolinite, illite and iron hydroxides. The key factors determining the permeability of the sandstones are the original grain size and degree of sorting, the precipitation and subsequent removal of anhydrite, and the extent of early and late carbonate cementation, framework grain dissolution and overgrowth.

Petrological investigations of the Sherwood Sandstone Group sandstones of the Wessex Basin were undertaken to identify primary and diagenetic features for correlation with hydraulic properties, as part of the investigation of the geothermal potential of the basin by the British Geological Survey for the Department of Energy (Milodowski *et al.* 1986). The diagenesis of the Sherwood Sandstone in the Wessex Basin has been discussed previously (Lott & Strong 1982, Burley 1984, Knox *et al.* 1984, Burley & Kantorowicz 1986), based on material from outcrop and from the Winterborne Kingston and Marchwood boreholes. For this study, material from a wider number of borehole localities was available for analysis, resulting in significant new information regarding the diagenesis of this formation. Samples were taken from the outcrop at the western margin of the basin (east Devon to west Somerset) and from cores and drill-chippings from deep boreholes at Southampton (Marchwood and Western Esplanade), Seaborough, Nettlecombe, Winterborne Kingston, Wytch Farm, Shrewton and two confidential wells. Figure 1 shows the isopachytes of the Sherwood Sandstone Group for the Wessex Basin and locations of non-confidential boreholes. In total, 122 samples were examined by petrological microscope, cathodoluminescence (CL), and scanning electron microscope (SEM) using backscattered electron imaging techniques (BSI). The samples were examined as polished thin sections and where possible they were prepared from core plugs that had been selected for determination of porosity and permeability values. The samples were also studied by SEM as rough stubs suitable for analysis by secondary electron imaging. SEM identification of minerals was aided by qualitative energy-dispersive X-ray microanalysis (EDS). Selected samples were also analysed by electron microprobe and by X-ray diffraction analysis (XRD).

Regional setting

The lower Triassic Sherwood Sandstone Group (Warrington *et al.* 1980) in the Wessex Basin consists mainly of fluvial sediments dominated by feldspathic sandstones, but lithologies range from conglomerates (lithic and intraformational) to ferruginous mudstones and siltstones. At outcrop in Devon and west Somerset, the Sherwood Sandstone consists of a lower conglomeratic unit (Budleigh Salterton Pebble Beds) overlain by thick cross-bedded sandstones (Otter Sandstone). In east Devon the sediments are only weakly cemented with patchy calcite cements, and the pebble beds are dominated by pebbles and cobbles of quartz and quartzite with subordinate slate and igneous rocks. North of Wellington the sediments become increasingly calcareous and well cemented and the Budleigh Salterton Pebble Beds in particular contain so

From: MARSHALL, J. D. (ed.), 1987, *Diagenesis of Sedimentary Sequences*, Geological Society Special Publication No. 36, pp. 325–337.

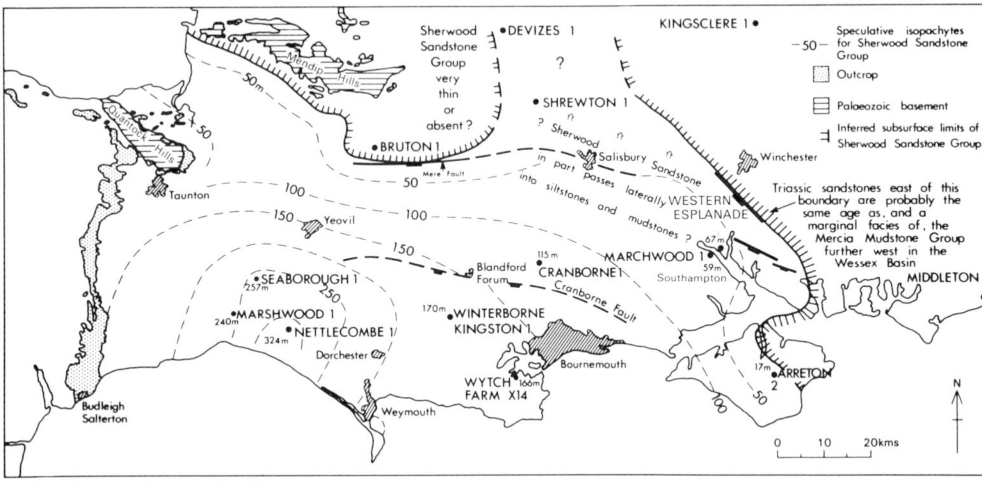

FIG. 1. Isopachytes for the Sherwood Sandstone Group of the Wessex Basin (after Milodowski *et al.* 1986).

much detrital carbonate that locally the deposits have been worked for lime. The limestone pebbles and cobbles have been identified as Devonian and Carboniferous limestones (Edmonds & Williams 1985) and are presumably derived from the nearby Quantock Hills or Mendip areas. In the subsurface this division into a conglomeratic lower unit and a sandy upper unit can be recognized, although the conglomeratic unit is not present everywhere, probably because of erosion.

In general the sequence consists of a series of fining-upward cycles about 1–5 m thick; each cycle has an erosional base, above which a conglomerate grades upwards through coarse cross-bedded sandstone into laminated siltstone and mudstone. Concretionary calcites may be developed towards the tops of cycles, producing mottled and irregularly nodular horizons that are interpreted as calcretes. Many of the carbonate clasts seen in the intraformational conglomerates closely resemble these calcretes in microfabric and are considered to be derived penecontemporaneously by erosion of earlier material. The sediments may be cemented by carbonate (generally calcite, but dolomite and ferroan dolomite may occur as replacement cements), sulphate (mainly anhydrite, minor gypsum), or less commonly silica, K-feldspar or albite. At outcrop limonitic cements are also important.

Diagenesis

The sediments have been subject to diagenetic processes which have modified both the detrital grains and the pore space. The diagenetic regimes encountered can be classified, following Schmidt & McDonald (1979a), according to whether diagenetic reactions are controlled by near-surface environments prior to burial (eodiagenesis), by conditions prevailing during burial (mesodiagenesis) or by near-surface processes following uplift of the sediments (telodiagenesis). In general, eodiagenetic and early mesodiagenetic changes have tended to reduce the primary intergranular pore space through cementation, syntaxial grain overgrowths, early compaction and pressure solution. Later mesodiagenetic processes caused the removal of earlier cement and the dissolution of some framework grains, resulting in the production of a secondary porosity which has had a major influence on the aquifer properties of the sediments. In some cases, however, the secondary porosity has been in turn reduced by the precipitation of late carbonates. Telodiagenesis affects only the sediments near the outcrop where the influx of meteoric water (low pH) has locally resulted in the dissolution of carbonate cements (particularly in the southern part of the outcrop), dissolution of feldspar grains and precipitation of kaolinite and illite. An outline of the temporal relationships of the diagenetic events is shown in Table 1, and the distribution of authigenic phases is given in Table 2.

Eodiagenesis

Iron oxide surface veneers

The earliest diagenetic product preserved appears to be a fine veneer of reddish-brown or black iron oxide on many grain surfaces which is ubiquitous throughout the area of study. SEM

TABLE 1. *Temporal relationships of the diagenetic events in the Sherwood Sandstone of the Wessex Basin*

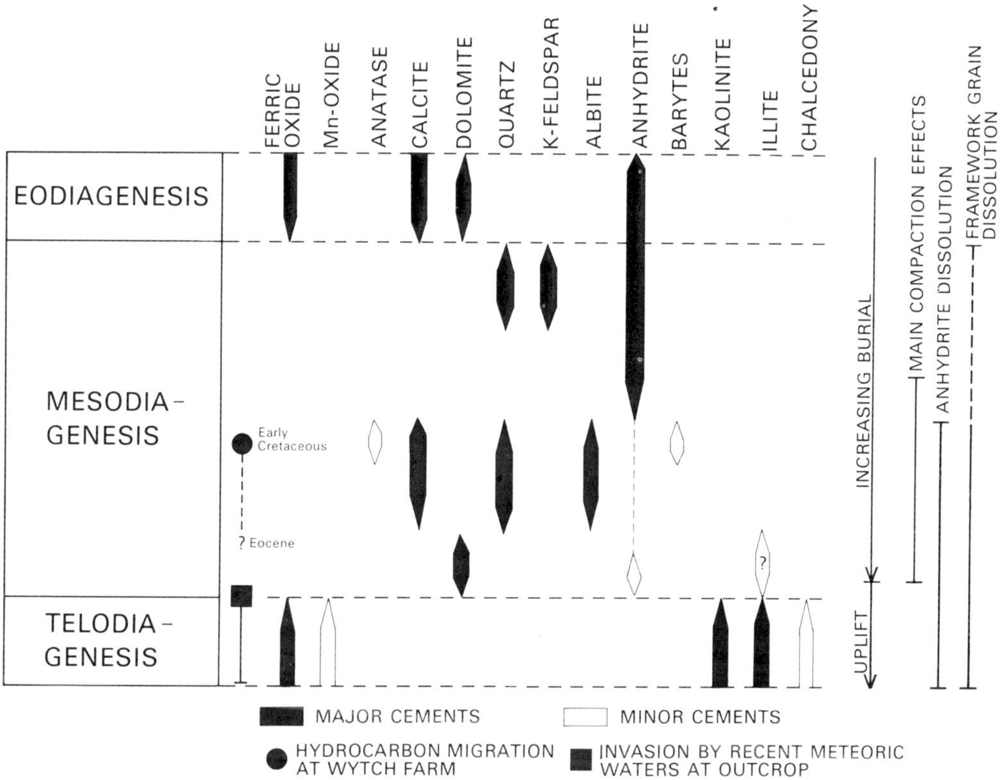

TABLE 2. *Distribution of authigenic phases in the Sherwood Sandstone of the Wessex Basin*

LOCALITY	He	At	Ba	An (I)	An (II)	Ct(I)	Ct(II)	Do(I)	Do (II)	Qz(I)	Qz (II)	Ch	Kf	Ab	Ka	Il	Mn	Fe
W. SOMERSET OUTCROP	●	●	●			●	●			●		●	●		●		●	●
E. DEVON OUTCROP	●	●	●			●	●			●		●	●		●	●		●
MARCHWOOD	●	●	●			●	●		●	●			●			●		
WESTERN ESPLANADE	●	●	●		●	●	●		●	●			●			●		
SHREWTON	●	●	●		●	●	●		●	●			●					
WYTCH FARM	●	●	●		●	●	●			●			●					
WINTERBORNE KINGSTON	●	●	●	●	●		●	●	●	●	●		●	●				
NETTLECOMBE, SEABOROUGH	●			●				●		●								

He: hematite, At: anatase, Ba: baryte, An(I): eodiagenetic anhydrite, An(II): mesodiagenetic anhydrite, Ct(I): early calcite, Ct(II): later calcite, Do(I): early non-ferroan dolomite, Do(II): later ferroan dolomite, Qz(I): early quartz, Qz(II): later quartz, Ch: chalcedony, Kf: K-feldspar, Ab: albite, Ka: kaolinite, Il: illite, Mn: Mn-oxide/hydroxide, Fe: late Fe-oxide/hydroxide

observations reveal that this veneer is composed of tiny (1 μm or less) plates, probably of haematite, and its absence at points of grain contact indicates that it formed after deposition. Some ferromagnesian grains have been replaced by fine haematite; in particular biotites commonly exhibit haematite plates developed along cleavage planes (Fig. 2a). Some interstitial clays

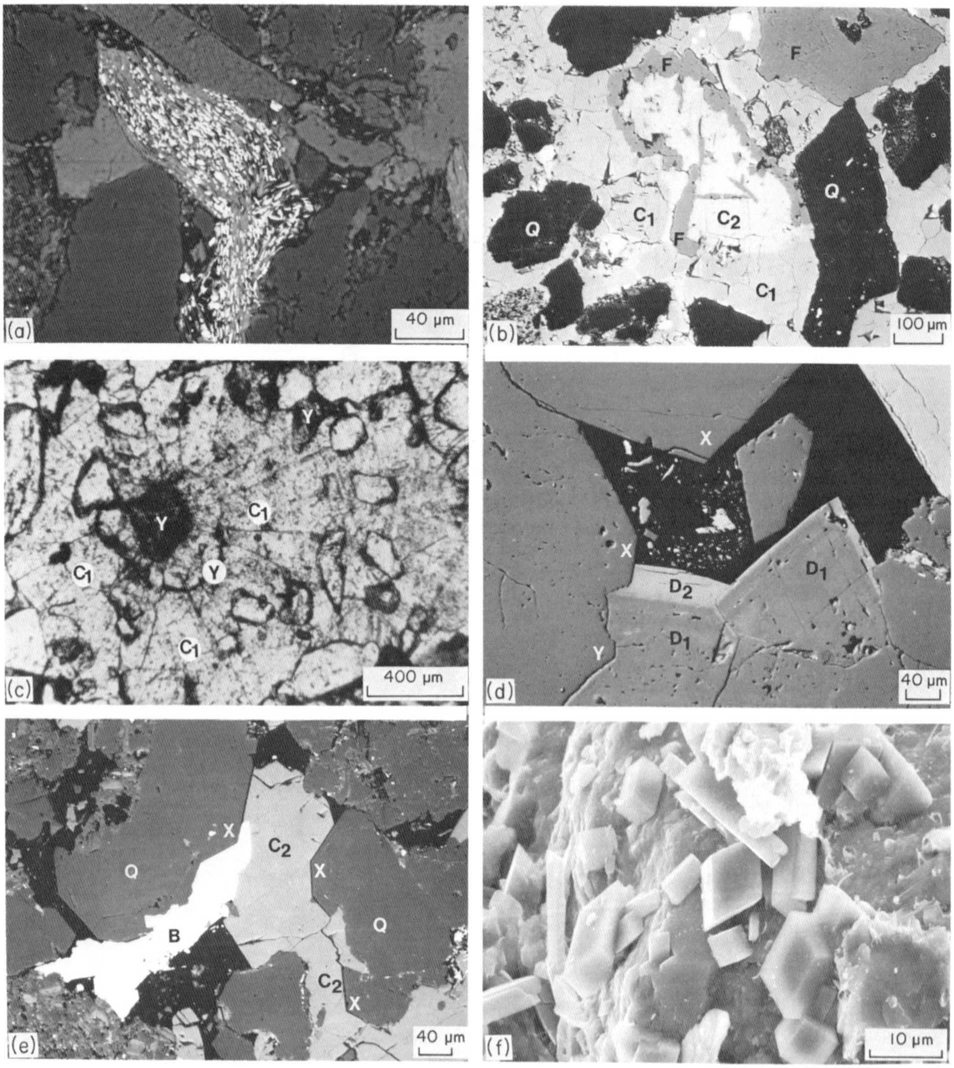

FIG. 2. (a) Haematite plates (bright) replacing biotite grain deformed by compaction. Marchwood borehole, 1680 m. (b) Open framework sandstone cemented by early calcite (C_1) with later calcite (C_2) filling remaining porosity in K-feldspar grain. Quartz (Q), K-feldspar (F). Marchwood borehole, 1695 m. (c) Stellate calcite (C_1) with early iron oxide films (Y) and expanded grain framework. Road cutting, Langford Budville (NGR ST 1235 2218). (d) Euhedral terminations of early non-ferroan dolomite (D_1) into open pore space. Quartz overgrowth absent at (Y), but present at (X) against later ferroan dolomite (D_2) and open pore space. Winterborne Kingston borehole, 2429 m. (e) Euhedral quartz overgrowths (X) developed into pore space, and later enclosed by barytes cement (B) and late calcite (C_2). Both barytes and calcite show some replacement of adjacent quartz. Marchwood borehole, 1696 m. (f) Authigenic K-feldspar ('adularia') coating grain surface of detrital K-feldspar. Marchwood borehole, 1686 m. (a, b, d, e)—SEM backscattered electron micrographs; (c)—plane polarized light; (f)—SEM secondary micrograph.

also show partial replacement by haematite. These features are characteristic of eodiagenesis in present-day sediments in hot arid or semi-arid environments (e.g. Walker *et al.* 1978).

Early calcite cementation

Early cementation by calcite provides a major cement in many parts of the Sherwood Sandstone. Optically, most of the calcite appears as neomorphic near-equigranular xenotopic spar, although calcite intraclasts, some rim cements and neomorphic stellate or radial-fibrous structures (cf. Bathurst 1976, p. 484) can be seen (Fig. 2c). Stellate or radial-fibrous calcite is well developed in the Otter Sandstone at outcrop and in samples from Marchwood and Western Esplanade. It is coarser than the rim cement and often disrupts the surrounding detrital grain fabric (indicating growth under minimal overburden), and some stellate structures show concentric fine banding by iron oxide (goethite or haematite) which presumably precipitated during intermittent calcite growth in a manner similar to the haematite coatings described previously. SEM, EDS and cathodoluminescence analyses reveal that the early calcites occur as non-ferroan, non-luminescent or faintly luminescent (often with fine, bright multiple zoning) rims around detrital grains. They are commonly best developed around carbonate intraclasts (Fig. 3a), which appear to have acted as nucleation sites. In many cases these cements appear to have been irregularly and repeatedly fractured producing a micropseudo-breccia texture with the microfractures filled by calcite of slightly different luminescence. Early calcite cementation is usually patchy and cemented areas may coalesce into nodular horizons, but in many cases it is completely pore-filling, producing a rock with an uncompacted 'open' framework texture (Fig. 2b). These early cement fabrics are typical of calcretes. We have found no evidence of dissolution of either detrital carbonate intraclasts ('cornstones') or early calcite cements (except at outcrop where affected by telodiagenesis). Early rim cements very commonly show euhedral terminations into open pore space, thereby precluding calcite dissolution as a mechanism for secondary porosity generation away from the outcrop.

Early dolomite

In the Winterborne Kingston borehole early calcite appears to be absent and an early (pre-quartz overgrowth) non-ferroan dolomite replaces detrital cornstone and siltstone clasts. This dolomite is found only in the lower cored section (2421.44–2436.65 m), where it is the major cement phase in cemented horizons. It occurs as idiotopic rhombs completely replacing the intraclasts, and locally it replaces quartz and feldspar. Limited overgrowths of dolomite may penetrate into adjacent pore-space, causing localized nodular cementation and inhibition of authigenic quartz and feldspar overgrowths (Fig. 2d). In open pore-space the dolomite exhibits euhedral crystal faces and is commonly overgrown by syntaxial late ferroan dolomite. Early dolomite cut by luminescent late calcite-filled fractures was also noted in a confidential borehole 15 km west of Winterborne Kingston. The early dolomite is believed to have precipitated before significant burial of the sediment, for dolomite-cemented sandstones show little evidence of compaction. The early dolomite appears to be restricted to sediments from boreholes near the basin depocentre (e.g. Winterborne Kingston, Nettlecombe, Seaborough).

Early anhydrite

Some siltstones and mudstones from boreholes at Nettlecombe, Seaborough and a confidential borehole to the NW of Winterborne Kingston contain anhydrite that occurs as felted masses of subparallel or radiating laths or irregular nodules. The anhydrite has clearly disrupted and expanded the detrital fabric of the sediments, and it is usually intimately associated with nodular non-ferroan dolomite, suggesting a near-surface evaporitic origin. The textures are similar to those of sabkha sediments described by Shearman (1966), Kinsman (1969) and Dean *et al.* (1975). Anhydritic siltstones and mudstones dominate the lower parts of the Sherwood Sandstone Group sequences immediately above the ?Budleigh Salterton Pebble Beds of the Nettlecombe 1 and Seaborough 1 boreholes, located close to the basin depocentre, and gypsiferous sandstones and shales are recorded above ?Budleigh Salterton Pebble Beds in the Marchwood 1 borehole.

Mesodiagenesis

Burial history

The burial history of the Sherwood Sandstone in the Wessex Basin is discussed in detail by Milodowski *et al.* (1986). At Winterborne Kingston, maximum depths of about 2.5 km may have been reached at the end of the Jurassic, and also at the end of the Cretaceous. Maximum depths for Marchwood and Southampton (Western Esplanade) are estimated at 1.8 km (early Oligocene), and Wytch Farm about 1.9 km (end of Cretaceous).

Early quartz and K-feldspar syntaxial overgrowths

Authigenic overgrowths of quartz and K-feldspar on detrital grains developed extensively into pore-spaces unoccupied by early carbonates (Figs 2e and 4b) throughout the Wessex Basin. In some sandstones these quartz overgrowths have produced pore-filling cements (e.g. Winterborne Kingston, Wytch Farm). The apparently 'compacted' textures of some of these sandstones, as seen under the optical microscope, are shown by SEM analysis to be the result of mutual interference of overgrowth surfaces. Overgrowths tend to be inhibited by early ferruginous or clay coatings, and in the more argillaceous sediments K-feldspar with rhombic adularia morphology (Fig. 2f) forms in preference to syntaxial overgrowths. Microprobe analyses show that authigenic K-feldspar is essentially pure, whereas the detrital grains may contain significant Na^+ or Ba^{2+}. Doubly terminated quartz crystals are sometimes developed around silt grains resting in open pores. Quartz and K-feldspar authigenesis appears to have begun before significant burial compaction, for quartz overgrowths are commonly found compacted into less competent grains (e.g. carbonate clasts, micas and mudstone clasts).

Anhydrite cementation

In general, anhydrite cementation seems to have occurred relatively early during mesodiagenesis because the anhydrite-cemented sandstones generally lack features indicative of compaction (Fig. 4a). The mesodiagenetic anhydrite is usually pore-filling and poikilotopic, and does not show any features indicating near-surface precipitation (expanded fabrics, felted lath or nodular habits, etc.). Its relationships to other diagenetic phases is complex and probably due to repeated dissolution and reprecipitation of earlier anhydrite. On account of its high solubility, remobilization of anhydrite is common during diagenesis (e.g. Holliday 1970, 1973, Mossop & Shearman 1973). There is a variation in the timing of quartz and feldspar overgrowth precipitation with respect to anhydrite cementation, which in some cases inhibited authigenic quartz and feldspar but in other cases, even within the same borehole (e.g. Winterborne Kingston), anhydrite post-dates overgrowths. Both quartz and feldspar grains and overgrowths locally display signs of etching and embayment against anhydrite. At Winterborne Kingston anhydrite also occurs in compacted sandstones and siltstones where it post-dates compaction effects (e.g. it fills fractures) and encloses late ferroan dolomite. Some feldspar dissolution sites may be completely filled by anhydrite (Fig. 4a), leaving only the more stable pure K-feldspar overgrowths. This has created a grain-supported fabric with anomalously oversized areas occupied by anhydrite cement—a fabric typical of framework grain dissolution or replacement (Schmidt & McDonald 1979b). Anhydrite is a major cement in sandstones in the Wytch Farm, Winterborne Kingston and Shrewton boreholes, and in other boreholes in the Winterborne Kingston area. It occurs as a minor cement in Western Esplanade, and in a confidential well between Southampton and Salisbury, but was not found in samples from Marchwood or the outcrop. The source of the mesodiagenetic anhydrite is presently unresolved. Early eodiagenetic anhydrite and gypsum that formed in the basin depocentre may have provided sulphates for mobilization during burial and diagenesis. The Permian Aylesbeare Group (Smith *et al.* 1974) beneath the Sherwood Sandstone, and the Mercia Mudstone Group overlying the sequence, are both anhydritic and potential sources of sulphate.

Compaction, framework grain dissolution

Compaction has affected most sandstones that remained uncemented by early calcite, dolomite or anhydrite. Those most affected are normally argillaceous. Deformation of clasts (especially of micas and mudstone clasts), pressure-welding and grain point-contact fracturing occurred. Cemented frameworks remained relatively unaffected. The dissolution of framework grains occurred after the main effects of compaction, since skeletal remnants and overgrowths of partially dissolved grains, particularly feldspar overgrowths, remain preserved (e.g. Figs 2b and 4a). Microperthites, plagioclases and K-feldspars are the grains most susceptible to grain dissolution, particularly the sodic laminae of microperthites, but quartz and unstable heavy minerals such as garnet (Morton 1982), magne-

FIG. 3. (a) Cathodoluminescence micrograph showing early non-luminescent calcite fringes with euhedral terminations (C_1) around cornstone clast (X) with brightly luminescent late calcite (C_2) filling remaining porosity. Quartz (Q) shows dull luminescence, K-feldspar (F) blue. Scale bar = 400 μm. Marchwood borehole, 1694 m. (b) Micrograph showing poikilotopic calcite partly filling pore space in sandstone and invading stress fracture in K-feldspar grain (crossed polarizers). Scale bar = 200 μm. Western Esplanade, 1752 m.
(c) Cathodoluminescence micrograph of the same area as (b) showing early non-luminescent calcite (C_1) seeded on a detrital calcite intraclast (X), with later luminescent calcite (C_2). Calcite C_2 post-dates the feldspar overgrowths at (Y), fills stress fractures at (Z), post-dating compaction. Calcite C_2 has euhedral terminations into open pore space (P). Quartz (Q), K-feldspar (F). Scale bar = 200 μm. Western Esplanade, 1752 m.

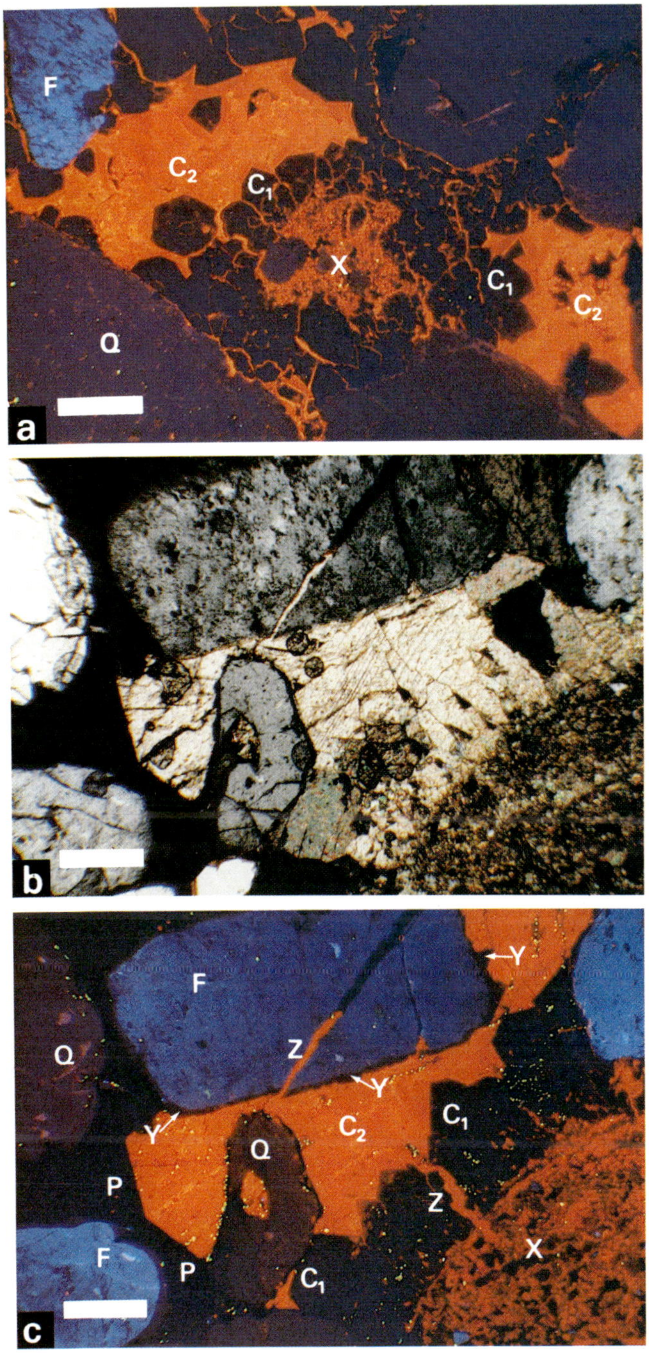
F
C_2
C_1
X
C_1
C_2
Q
a
b
Y
F
Q
Z
Y
Y
C_2
C_1
Q
P
Z
X
F
P
C_1
c

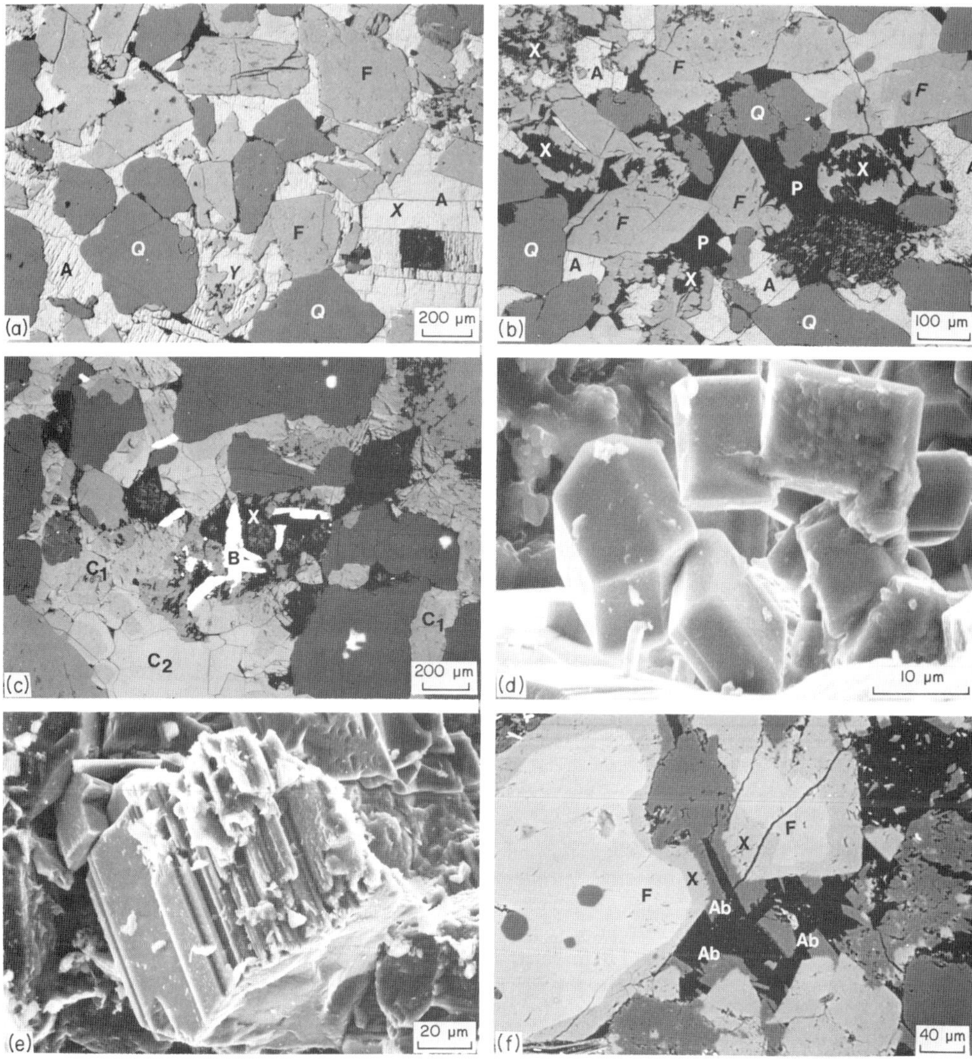

FIG. 4. (a) Anhydrite cemented sandstone showing minimal compaction. Anhydride (A) occupies oversized pores (X) and infills K-feldspar dissolution sites (Y). Quartz (Q), K-feldspar (F). Wytch Farm X14 borehole, 1598 m. (b) Anhydrite cemented sandstone showing some cement-dissolution porosity (P), extensive framework grain dissolution (X), residual anhydrite (A). Quartz (Q), K-feldspar (F). Wytch Farm X14 borehole, 1600 m. (c) Barytes laths (B) occupying K-feldspar dissolution site (X), in sandstone largely cemented by late calcite (C_2) and some early calcite (C_1). Marchwood borehole, 1694 m. (d) Euhedral authigenic anatase in open pore space. Wytch Farm X14 borehole, 1600 m. (e) Deeply corroded anhydrite in open pore space. Wytch Farm X14 borehole, 1636 m. (f) Detrital Ba-rich K-feldspars (F) with early pure K-feldspar overgrowths (X), and later albite overgrowth (Ab). Winterborne Kingston borehole, 2318 m. (a, b, c, f)—SEM backscattered electron micrographs; (d, e)—SEM secondary electron micrographs.

tite and ilmenite are also affected. Some local collapse of partly dissolved grains was noted, particularly where grain dissolution has been locally extensive. Although this secondary compaction may locally reduce porosity, overall framework grain dissolution may account for up to 10% of the porosity in some samples. Barium (detected by probe analysis in many K-feldspars) and titanium (in ilmenites and other heavy minerals) liberated by grain dissolution were rapidly reprecipitated as authigenic barytes (Figs 2e and 4c) and anatase (Fig. 4d). The barytes commonly occupies K-feldspar grain-dissolution sites and, less commonly, plagioclase grain-dissolution sites. Anatase is commonly seen resting on early calcite surfaces but enclosed within later calcite. Although common, barytes and anatase are volumetrically insignificant as cements.

Development of cement dissolution porosity

In samples from Wytch Farm, Winterborne Kingston and Western Esplanade there is strong evidence for secondary porosity arising from the dissolution of anhydrite cements. In these wells anhydrite is strongly etched and corroded against open pore space (Fig. 4b, e). In some samples gypsum occurs as haloes around remnants of anhydrite cement. Framework grain dissolution appears to be closely associated with anhydrite removal and areas devoid of anhydrite cement usually exhibit a greater degree of framework grain dissolution. Most of the cement dissolution porosity appears to have developed relatively late during mesodiagenesis as it is largely unaffected by compaction. At Wytch Farm there appears to be a relationship between the presence of hydrocarbon and the extent of anhydrite dissolution. Anhydrite is most abundant in the oil-saturated sandstones, whereas it occurs to a much lesser extent in oil-free sandstones. In both cases the anhydrite is etched and corroded but it would appear that hydrocarbon invasion may have suspended further anhydrite dissolution in the now oil-saturated sandstones.

Late quartz and albite

A late mesodiagenetic generation of quartz and albite overgrowths was noted in some samples from the Winterborne Kingston borehole. Quartz is developed as an overgrowth around grains with pressure-solution grain boundaries and it also heals fractured grains. Albite similarly occurs as overgrowths on K-feldspar grains, both on external grain surfaces and intragranular dissolution surfaces. The albite rims commonly develop on earlier K-feldspar overgrowths (Fig. 4f) or fill stress fractures cutting both grains and early overgrowths (Fig. 5a). In places the albite overgrowths have coalesced and filled pore spaces, forming a rigid framework cement. This generation of quartz and albite clearly post-dates the main effects of compaction, but appears to precede the late dolomite. The common association of authigenic albite with sites of feldspar dissolution suggests that dissolution of detrital plagioclase and K-feldspar may locally have supplied ions for albite authigenesis and that this process may have occurred concomitantly with framework grain dissolution.

Late calcite

Late calcite is characterized by bright orange luminescence under CL analysis and was identified by microprobe and SEM-EDS to be manganese-bearing ferroan calcite. It is a major cement at outcrop and in the Marchwood and Western Esplanade wells, but it was identified only in minor amounts in Winterborne Kingston (upper cored section only, 2315.96–2334.11 m) and Wytch Farm. These calcites are generally poikilotopic and pore-filling (Figs 2e, 3b, c and 4c) although they also commonly replace framework grains or fill framework grain dissolution sites (Figs 2b and 5b). Stress fractures in framework grains may also be filled by late calcite (Fig. 3c). The late calcite cements fill minor fractures in dolomite-cemented sandstone in a well near Winterborne Kingston and in early calcite-cemented sandstone in Marchwood. Intense veining by late calcite at Wolston Quarry (NGR ST 0945 4015) in the Budleigh Salterton Pebble Beds at the west Somerset outcrop appears to be associated with local faulting. These cements are clearly a late mesodiagenetic event since they cross-cut other phases. They post-date quartz and feldspar overgrowths which they commonly etch or replace (Figs 2e and 5c) and they fill or partly fill the remaining intergranular space not occupied by earlier cements. In open pore-spaces late calcites show uncorroded euhedral terminations (Figs 2e and 5b). The relationships between anhydrite dissolution, framework grain dissolution and late calcite cementation are complex, and spatio-temporal overlap of all three processes is evident. Samples in which anhydrite dissolution clearly preceded the precipitation of late calcite have varying degrees of calcite cement (e.g. upper cored section of Winterborne Kingston, Western Esplanade, Wytch Farm and outcrop), whereas the continued presence of anhydrite appears to have inhibited late calcite in some sandstones (e.g. lower cored section of Winterborne Kingston, Wytch Farm).

Oxygen isotope analyses of late calcites in the Wessex Basin indicate precipitation at a range of burial temperatures up to 70°C (Bath *et al.* this volume) if a basinal brine with similar isotopic composition to the present is assumed. This temperature is close to present-day temperatures but could have been achieved since maximum burial (i.e. from late Cretaceous). However, these deep burial calcites are not in carbon isotopic equilibrium with the present formation fluid (Knox *et al.* 1984, Bath *et al.* this volume) and must therefore pre-date the present hydrogeological regime.

Late dolomite

A late replacive ferroan dolomite occurs in wells in the Winterborne Kingston area, Shrewton, and between Southampton and Salisbury. The

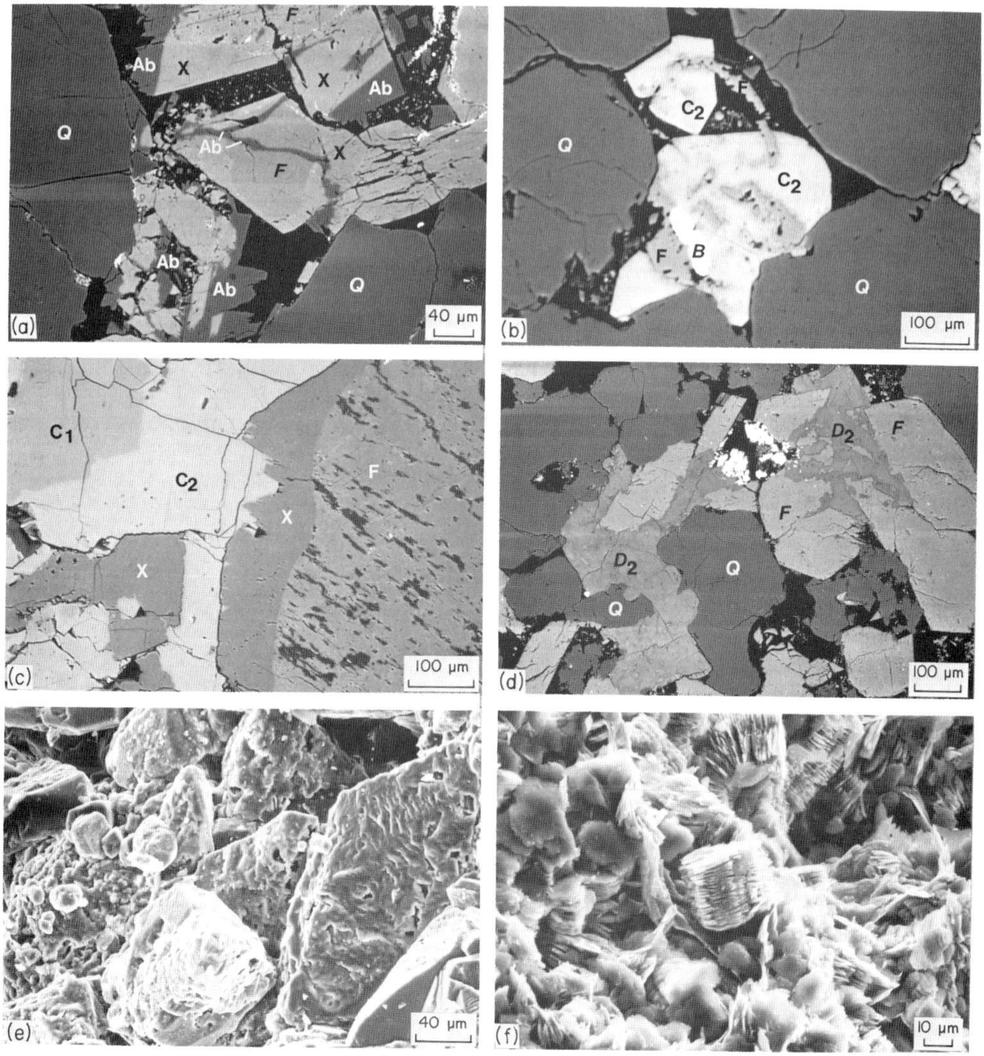

FIG. 5. (a) Compacted sandstone framework of detrital quartz (Q) and K-feldspar (F) with authigenic overgrowths (X). Late authigenic albite forms overgrowths on K-feldspar and infills stress fractures (Ab). Winterborne Kingston borehole, 2318 m. (b) Euhedral late calcite (C_2) occupying secondary porosity in feldspar dissolution site. Minor barytes (B). K-feldspar (F), Quartz (Q). Marchwood borehole, 1694 m. (c) Detrital perthitic K-feldspar (F) with authigenic overgrowth (X) corroded by late calcite (C_2). Early calcite (C_1). Marchwood borehole, 1694 m. (d) Detrital K-feldspar (F) and quartz (Q), with late ferroan dolomite (D_2) replacing framework grains. Winterborne Kingston borehole, 2316 m. (e) Corroded scalenohedral calcite lining pore space in sandstone. Budleigh Salterton Pebble Beds, Holywell Quarry (NGR ST 1269 2703). (f) Pore-filling and grain-replacive books of authigenic kaolinite, seen in intimate association with ragged flakes of authigenic illite. Otter Sandstone, Budleigh Salterton (SY 060 816). (a, b, c, d)—SEM backscattered electron micrographs; (e, f)—SEM secondary electron micrographs.

late dolomite is best developed in the Winterborne Kingston borehole, occurring as euhedral overgrowths on earlier dolomite (Fig. 2d), as a non-selective replacement of framework grains and locally as poikilotopic cements (Fig. 5d). Small isolated rhombs of late ferroan dolomite are occasionally seen on quartz and feldspar overgrowths from Marchwood and Western

Esplanade core samples. In the Winterborne Kingston borehole, some late anhydrite occurs which post-dates the ferroan dolomite.

Fibrous illite

In the porous zones of some sandstones in the upper unit of the Marchwood sequence very minor amounts of filamentous illite were noted. The illite clearly post-dates quartz overgrowths but the relationship to other authigenic phases is uncertain. It is associated with concentrations of detrital clay and clay coatings on grains, and locally appears to be replacive. Most of the illite collapsed on to pore surfaces as a result of core-drying (cf. McHardy *et al.* 1982).

Telodiagenesis

The diagenetic mineralogy of the rocks from outcrops in Somerset and Devon differs considerably from that of the sandstones sampled from the boreholes, presumably because of the direct influence of recent weathering processes. The most obvious effects are the dissolution of calcite cements (Fig. 5e), framework grain dissolution, the precipitation of kaolinite (Fig. 5f) commonly occupying sites of feldspar dissolution, and the precipitation of minor amorphous or chalcedonic silica on detrital quartz grain surfaces. Flakey illite presumably also of authigenic origin may be intimately associated with kaolinite in these rocks. Some pore walls may also be lined with spongy precipitates of manganese oxides or hydroxides. Framework grain dissolution is considerably more extensive at outcrop compared with the deep boreholes. However, it is not possible to distinguish telodiagenetic from mesodiagenetic dissolution sites except where the presence of late calcite in the secondary pores indicates early features. Telodiagenetic effects are most obvious in the outcrop south of Wellington where the sediments may have originally had less calcite cementation. In many parts of the east Devon outcrop hard concretionary bands of limonite or goethite are developed near the surface ('iron pans'). No anhydrite or gypsum was found in outcrop samples but, as stated before, their presence in wells 30 km to the east suggests that they may have been originally present as mesodiagenetic cements which have subsequently been removed by meteoric groundwaters. The groundwaters in the Sherwood Sandstone at outcrop are saturated with respect to both kaolinite and illite and indicate that K-feldspar alteration and dissolution is occurring under the present telodiagenetic regime (Walton 1981).

Discussion

Effects of diagenesis on reservoir properties

The reservoir properties of the Sherwood Sandstone in the Wessex Basin have been described in detail previously (Burgess *et al.* 1981, Burgess 1982, Downing *et al.* 1982, Knox *et al.* 1984, Milodowski *et al.* 1986) and hence are only briefly discussed here. The main factors that determine the intergranular permeability of the Sherwood Sandstone at the present time are the sedimentological characteristics (grain size, sorting) and the extent of cementation and grain dissolution. In general, the sediments were deposited in a series of upward-fining fluvial cycles (Henson 1970, Lott & Strong 1982). The mudstones, siltstones and muddy sandstones forming the upper parts of these cycles are considered to represent overbank flood deposits, and vertical accretion deposits of temporary lakes and back-levee swamps. They possess low porosities (5–10%) and permeabilities generally less than 1 mD. The primary porosity in these lithologies has been reduced by compaction and also in some cases by the replacement or cementation of the matrix by calcite. The moderately sorted grainstones, representing lateral accretion deposits at point or channel bars, possess the best reservoir characteristics, but porosities vary considerably from <5–28% and permeabilities range from 1 mD to about 7 D. Some sandstones from the east Devon outcrop may have porosities up to 35% and permeabilities up to 9 D (Milodowski *et al.* 1986). Conglomerates, representing channel bedload deposits, also show considerable variation in porosity and permeability depending mainly on the degree of cementation of the sandy matrix. Knox *et al.* (1984) established an inverse relationship between the extent of calcite cementation and the permeabilities in sandstones of the Marchwood well, with slight modifications due to compaction and detrital clay contents. This is also true elsewhere in the basin, although the effects of other cements (anhydrite, dolomite) are significant.

An important feature of the Sherwood Sandstones throughout the Wessex Basin is the occurrence of discrete high permeability zones that have porosities ranging from 18 to 28% and permeabilities generally ranging from 100 mD to about 6 D. They form significant aquifers at Marchwood, Western Esplanade, Winterborne Kingston and Wytch Farm and may account for as much as 95% of the total transmissivity of the formation in the Marchwood and Western Esplanade wells (Milodowski *et al.* 1986). They have typically only minor or absent early and late carbonates, extensive framework grain dissolution of feldspars (which may account for as much as 10% of the total porosity of some samples), and an extensive secondary porosity due to cement dissolution inferred here to be after anhydrite. In contrast, sandstones that have extensive calcrete development, or have been affected significantly by late calcite (possibly due to early anhydrite removal), are tightly cemented and offer little or no aquifer potential. Knox *et al.* (1984) considered that the porosity of these permeable horizons is mainly primary porosity enhanced by framework grain dissolution and that it largely escaped calcite cementation. However, their study was limited to material from the Marchwood sequence, where anhydrite is absent. Burley (1984), and Burley & Kantorowicz (1986) have suggested that the secondary porosity in the highly permeable zones may be due to calcite dissolution, but the ubiquitous presence of well preserved detrital cornstone clasts and euhedral terminations into open pore-space of both early and late calcites would appear to preclude this mechanism for the Wessex Basin sandstones, apart from those affected by telodiagenesis near the outcrop. We consider that much of the secondary porosity in the highly permeable bands is due to the late removal of anhydrite, which has protected pore-space from the effects of compaction and has prevented extensive precipitation of late calcite in these zones. The subsequent dissolution of the anhydrite thereby rejuvenated considerable primary and secondary porosity. Although anhydrite has not been reported from the Marchwood well, its presence at Western Esplanade and elsewhere in the subcrop suggests that it may well have been present at Marchwood, and in rocks now at outcrop. There is no evidence at present to suggest that the highly permeable zones are laterally extensive and hence correlatable. In a variable fluviatile sequence this seems highly unlikely. However, it is possible that they may be hydraulically interconnected via fissures, faults etc.

Anhydrite removal at Wytch Farm appears to have been restricted by the migration of hydrocarbons into the sandstones. Oil migration at Wytch Farm is believed to have begun during the early Cretaceous and could have continued until the Alpine (Eocene) uplift (Colter & Havard 1981). The relatively limited extent of anhydrite corrosion in the oil-saturated sandstones appears to indicate only minor dissolution by an initial water flush preceding hydrocarbon accumulation. More extensive dissolution of anhydrite followed only in sandstones below the oil–water contact.

As noted earlier, parts of the Sherwood Sandstone Group sequences in the basin depocentre contain eodiagenetic (near-surface) anhydrite which may represent an inland sabkha or playa environment of deposition with minimal development of calcretes. In areas marginal to the playa environment some calcretes have developed that have been later replaced by early non-ferroan dolomite (e.g. lower part of the Winterborne Kingston sequence). Near the basin margins, however, calcretes appear to have been more extensively developed, but there is no evidence of early dolomitization. Isotopic analyses of the early dolomite (Bath *et al.* this volume) indicate that this dolomitization occurred near the surface under highly evaporitic conditions. Dolomitized calcretes have been described as an intermediate lithofacies between calcretes and playa deposits (gypsites) in Australian inland palaeodrainage systems (Arakel & McConchie 1982). They consider that the dolomitization occurs in zones of fluctuating groundwater level where repeated solution and reprecipitation of the calcrete have taken place.

The exceptionally high porosities and permeabilities in sandstones of the east Devon outcrop are due to extensive decalcification and feldspar grain dissolution. Outcrop groundwater chemistries indicate that these processes (together with the precipitation of authigenic kaolinite and illite) are occurring at the present time (Walton 1981). However, it is unlikely that calcite dissolution extends very far into the subcrop since (according to Walton 1981) the groundwaters rapidly achieve calcite saturation with depth. Any anhydrite originally present in the rocks now at outcrop would have dissolved under the present telodiagenetic regime. The extent of anhydrite dissolution eastwards into the basin is unknown, although anhydrite is present in deep boreholes about 30 km east of the outcrop.

ACKNOWLEDGMENTS: This work formed part of the BGS Low Enthalpy Geothermal Energy Project funded by the Department of Energy. The authors would like to thank colleagues of the British Geological Survey for support and encouragement, in particular the Project Manager, Dr R. A. Downing, and Dr S. Holloway for discussions relating to the regional geology. We are grateful to Drs N. A. Chapman, R. W. O'B Knox and G. Warrington for critically reading the manuscript, and to Dr M. Scherer and an anonymous reviewer for further constructive comments. This paper is published by permission of the Director, British Geological Survey, (NERC). Figures and illustrations are NERC copyright reserved.

References

ARAKEL, A. V. & MCCONCHIE, D. 1982. Classification and genesis of calcrete and gypsite lithofacies in paleodrainage systems of inland Australia and their relationship to carnotite mineralization. *Journal of Sedimentary Petrology*, **52**, 1149–70.

BATH, A. H., MILODOWSKI, A. E. & SPIRO, B. 1987. Diagenesis of carbonate cements in Permo–Triassic sandstones in the Wessex and East Yorkshire–Lincolnshire Basins, UK: a stable isotope study. *This volume.*

BATHURST, R. G. C. 1976. *Carbonate Sediments and their Diagenesis.* Developments in Sedimentology 12. Elsevier, Amsterdam.

BURLEY, S. D. 1984. Patterns of diagenesis in the Sherwood Sandstone Group (Triassic), United Kingdom. *Clay Minerals*, **19**, 403–40.

—— & KANTOROWICZ, J. D. 1986. Thin section and S.E.M. criteria for the recognition of cement-dissolution porosity in sandstones. *Sedimentology*, **33**, 587–604.

BURGESS, W. G. 1982, Hydraulic characteristics of the Triassic Sherwood Sandstone and the Lower Jurassic Bridport Sands intervals, as derived from drill stem tests, geophysical logs and laboratory tests. *In*: RHYS, G. H., LOTT, G. K. & CALVER, M. A. (eds). *The Winterborne Kingston Borehole, Dorset, England*, pp. 164–75. Report of the Institute of Geological Sciences No. 81/3.

—— BURLEY, A. J., DOWNING, R. A., EDMUNDS, W. M. & PRICE, M. 1981. The Marchwood Geothermal borehole—a preliminary assessment of the resource. *Investigation of the Geothermal Potential of the United Kingdom.* Institute of Geological Sciences.

COLTER, V. S. & HAVARD, D. J. 1981. The Wytch Farm Oil Field, Dorset. *In: Petroleum Geology of the Continental Shelf of North-West Europe*, pp. 494-503. Institute of Petroleum, London.

DEAN, W. E., DAVIS, G. R. & ANDERSON, R. Y. 1975. Sedimentological significance of nodular and laminated anhydrite. *Geology*, **3**, 367–72.

DOWNING, R. A., ALLEN, D. J., BURGESS, W. G. & EDMUNDS, W. M. 1982. The Southampton (Western Esplanade) Geothermal Well—a preliminary assessment of the resource. *Investigation of the Geothermal Potential of the United Kingdom.* Institute of Geological Sciences.

EDMONDS, E. A. & WILLIAMS, B. J. 1985. The geology of the country around Taunton and the Quantock Hills. *Memoirs of the British Geological Survey Sheet*, 295.

HENSON, M. R. 1970. The Triassic rocks of South Devon. *Proceedings of the Ussher Society*, **2**, 172.

HOLLIDAY, D. W. 1970. The petrology of secondary gypsum rocks: a review. *Journal of Sedimentary Petrology*, **40/2**, 734–44.

—— 1973. Early diagenesis in nodular anhydrite rocks. *Transactions of the Institute of Mining and Metallurgy*, **82**, 81–4.

KINSMAN, D. J. J. 1969. Modes of formation, sedimentary associations and diagenetic features of shallow water and supratidal evaporites. *Bulletin of the American Association of Petroleum Geologists*, **53**, 830–40.

KNOX, R. W. O'B., BURGESS, W. G., WILSON, K. S. & BATH, A. H. 1984. Diagenetic influences on reservoir properties of the Sherwood Sandstone (Triassic) in the Marchwood Geothermal borehole, Southampton, U.K. *Clay Minerals*, **19**, 441–56.

LOTT, G. K. & STRONG, G. E. 1982. The petrology and petrography of the Sherwood Sandstone (?Middle Triassic) of the Winterborne Kingston borehole, Dorset. *In:* RHYS, G. H., LOTT, G. K. & CALVER, M. A. (eds). *The Winterborne Kingston Borehole, Dorset, England*, pp. 135–42. Report of the Institute of Geological Sciences No. 81/3.

MCHARDY, W. J., WILSON, M. J. & TAIT, J. M. 1982. Electron microscope and X-ray diffraction studies of filamentous illite clay from sandstones of the Magnus Field. *Clay Minerals*, **17**, 23–39.

MILODOWSKI, A. E., STRONG, G. E., WILSON, K. S., ALLEN, D. J., HOLLOWAY, S. & BATH, A. H. 1986. Diagenetic influences on the aquifer properties of the Sherwood Sandstone in the Wessex Basin. *Investigation of the Geothermal Potential of the UK.* British Geological Survey.

MORTON, A. C. 1982. Heavy minerals from the sandstones of the Winterborne Kingston borehold, Dorset. *In:* RHYS, G. H., LOTT, G. K. & CALVER, M. A. (eds). *The Winterborne Kingston Borehole, Dorset, England.* Report of the Institute of Geological Sciences No. 81/3.

MOSSOP, G. D. & SHEARMAN, D. J. 1973. Origins of secondary gypsum rocks. *Transactions of the Institute of Mining and Metallurgy*, **82**, 147–54.

SCHMIDT, V. & MCDONALD, D. A. 1979a. The role of secondary porosity in the course of sandstone diagenesis. *In:* SCHOLLE, P. A. & SCHLUGER, P. R. (eds). *Aspects of Diagenesis*, pp. 175–207. Society of Economic Paleontologists and Mineralogists Special Publication No. 26.

—— &—— 1979b. Texture recognition of secondary porosities in sandstones. *In:* SCHOLLE, P. A. & SCHLUGER, P. R. (eds). *Aspects of Diagenesis*, pp. 209–25. Society of Economic Paleontologists and Mineralogists Special Publication No. 26.

SHEARMAN, D. J. 1966. Origin of marine evaporites by diagenesis. *Transactions of the Institute of Mining and Metallurgy*, **75**, 208–15.

SMITH, D. B., BRUNSTROM, R. G. W., MANNING, P. I., SIMPSON, S. & SHOTTON, F. W. 1974. A correlation of Permian rocks in the British Isles. *Journal of the Geological Society of London*, **130**, 1–45.

WALKER, T. R., WAUGH, B. & GRONE, A. J. 1978. Diagenesis in first-cycle desert alluvium of Cenozoic age, southwestern United States and northwestern Mexico. *Bulletin of the Geological Society of America*, **89**, 19–32.

WALTON, N. R. G. 1981. A detailed investigation of the geochemistry of groundwaters from the Triassic sandstone aquifer of south-west England. *Report of the Institute of Geological Sciences* No. 81/5.

WARRINGTON, G., AUDLEY-CHARLES, M. G., ELLIOTT, R. E., EVANS, W. B., IVEMEY-COOK, H. C., KENT, P. E., ROBINSON, P. L., SHOTTON, F. W. & TAYLOR, F. M. 1980. A correlation of Triassic rocks in the British Isles. *Geological Society of London Special Report* No. 13.

G. E. STRONG & A. E. MILODOWSKI, British Geological Survey, Keyworth, Nottingham NG12 5GG, UK.

Deposition and diagenesis in an extensional basin: the Corallian Formation (Jurassic) near Oxford, England

C. B. de Wet

SUMMARY: The mixed carbonate-siliclastic Corallian Formation was deposited on horsts and in grabens which were produced during Jurassic extensional tectonics. The high areas became sites for sandstone and carbonate deposition in shallow subtidal environments while clays and allochthonous debris accumulated in the depressions. Tectonic activity, perhaps confined to single fault blocks, produced depositional variation distinct to each area.

The basin has been subdivided into four major east–west trending fault blocks. The most northerly of these is the Oxford Shallows, interpreted as a small accretionary-margin slope. The depositional facies and diagenesis of this area are discussed relative to the basin as a whole.

The diagenesis of the Corallian sediments has been divided into three phases of cementation: pre-burial, shallow burial and deep burial. Phase 1 spars include marine precipitates and non-ferroan, neomorphic and drusy spars. Phase 2 cements are ferroan, precipitating during shallow burial, including an intermediate ferroan (mauve) spar which may represent development of a redox front. Phase 3 spar post-dates vuggy dissolution which probably took place during Barremian uplift. The cement precipitated during the ensuing burial phase.

Geochemistry and stable isotope data show an overall fluid evolution path for the first two spars, although Sr^{2+} values in replaced corals reveal variability in the degree of meteoric flushing. The chemistry of the late spars suggests low water/rock ratios, where local dissolution provided the cations for the cement. The oxygen isotope values suggest, however, that meteoric and connate waters were mixed at slightly elevated temperatures.

The Oxfordian Corallian Formation is a mixed carbonate-siliclastic sequence found at outcrop and in the subsurface of southern England. The study area has traditionally been divided into the Hampshire Basin to the west and the Wealden Basin to the east, but as recent structural surveys (Stoneley 1982, Whittaker 1985) and this study demonstrate, east and west are depositionally and diagenetically linked, forming the Wessex Basin (Fig. 1). Lithofacies range from shale to sandstone to oolite and packstone (Fig. 2) with preserved thicknesses varying from 4 m on the edge of the basin to over 100 m near its centre. Previous studies of the Corallian Formation have explained the lithologic variability in terms of shallowing-upward cycles (Talbot 1973, Brookfield 1973, 1978, Wright 1981 and Chowdhury 1982). Many depositional models have been proposed, including an analogy with the present Gulf of Mexico (Brookfield 1977) and a carbonate regime like the Bahamas (Chowdhury 1980).

In this study, diagenesis, coupled with a recent structural synthesis (Chadwick 1986), has elucidated the basin's fluid pathways and has led to a new depositional model. This paper discusses recent developments in the understanding of the basin's structural history, proposes a new depositional hypothesis and examines the diagenesis in the Oxford area as it relates to the whole basin.

Structure

Chadwick (1986) at the British Geological Survey produced a synthesis of the basin's structural evolution based on seismic and borehole data. He proposes that the Wessex Basin experienced episodic extensional tectonics producing a series of grabens and fault block highs. Chadwick invokes McKenzie's (1978) model of rifting, followed by subsidence, to explain the repeated pattern of shallow water deposits being followed by laterally widespread deeper water sediments. The Corallian Formation is one of the units deposited during an extensional phase. Extension produced highs where predominantly carbonates and coarse clastics were deposited and intervening lows that collected clays and allochthonous debris (Fig. 3). These highs experienced local tectonic activity, moving them above and below sea-level independently of movements elsewhere in the basin. This helps to explain much of the lithologic variation, and difficulty in correlation, found north–south across the area.

Reports in the literature on other extensional regimes are revealing similar patterns to that seen in the Wessex Basin. Profiles across China's Jianghan and Bohai Bay basins, which are of roughly the same dimensions as the Wessex

From: MARSHALL, J. D. (ed.), 1987, *Diagenesis of Sedimentary Sequences*, Geological Society Special Publication No. 36, pp. 339–353.

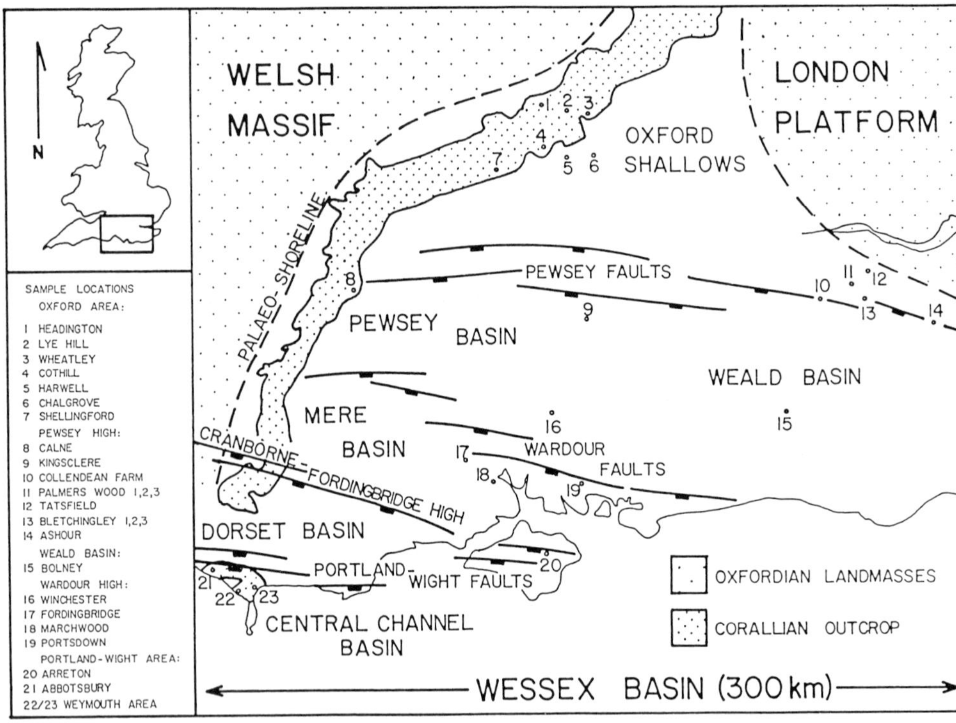

FIG. 1. Map of southern England showing location of the Corallian Formation outcrop and borehole sites used in this study. Major east–west trending faults and basins dissect the Wessex Basin.

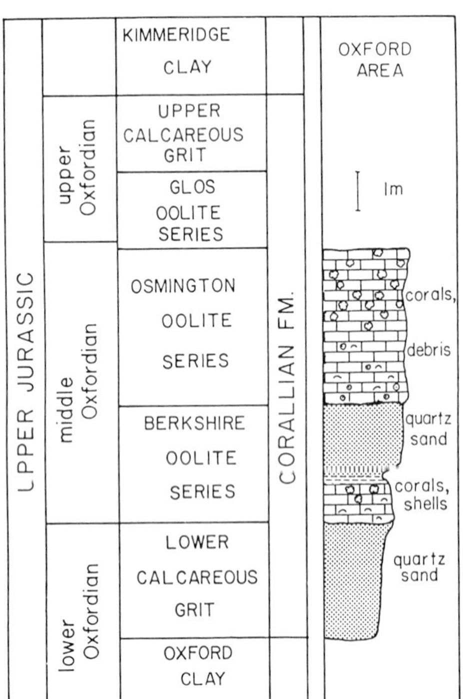

FIG. 2. Corallian Formation stratigraphy and Oxford area lithological log.

basin, show an almost identical arrangement of tilted fault blocks stepping down into the basin's interior (Hefu 1986). While at present on the South China and Red Sea fault-bounded highs, carbonates and coarse clastics are accumulating and thicker deposits of argillaceous and allochthonous debris are filling in the lows (Ru & Pigott 1986).

Observations: Oxford Shallows

An examination of one structural high, the Oxford Shallows block, reveals many of the depositional and diagenetic features predicted by the proposed model. The Oxford Shallows is the most northerly section within the Wessex Basin and between two land masses, the London Platform and the Welsh Massif.

Depositional environments

At outcrop the Formation (maximum 10 m thick) consists of a predominantly cross-bedded sandstone (exposed in quarries at Shellingford and Cothill, NGR SU 960 430, SU 980 450) followed by coral patch reefs and inter-reef shelly debris. The sandstones are well-sorted quartz sands with minor feldspar, containing occasional flaser

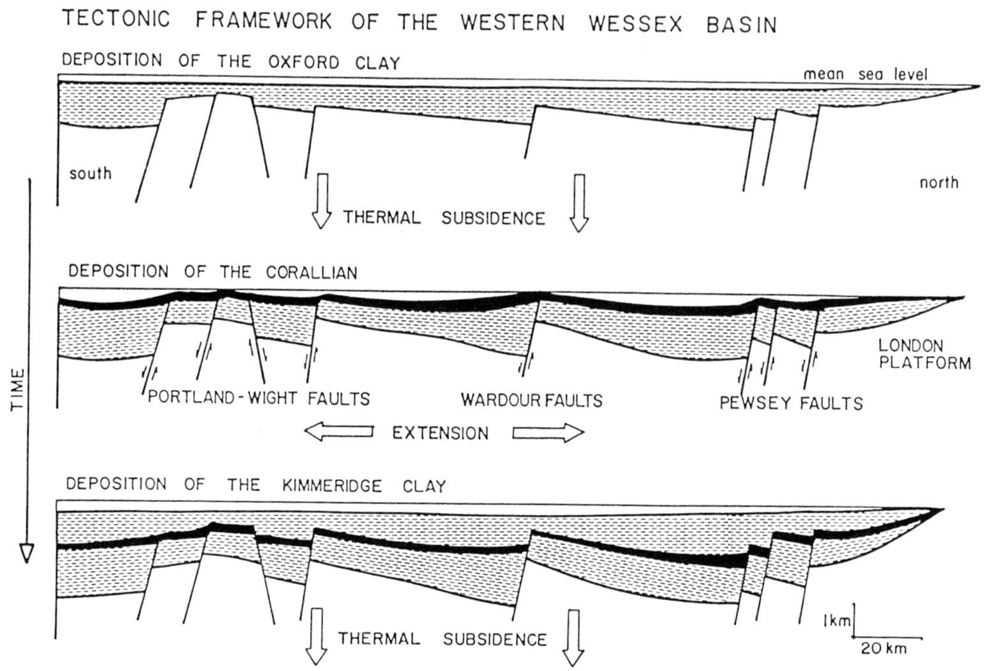

FIG. 3. Tectonic model for the western Wessex Basin (after Chadwick 1986). The Corallian Formation is bounded by the Oxford and Kimmeridge Clay Formations. The Corallian Formation rocks onlap the London Platform to the north. Equivalent deposits are reported from Normandy to the south (Fürsich 1975).

bedding and mud filled burrows. Shell fragments and *Rhaxella* sponge spicule moulds are locally abundant.

The overlying limestone contains cross-laminated sands of heavily micritized shell and coral fragments (Fig. 4a), pisoids, intraclasts, oncoids and ooids. Well preserved *Girvanella* oncoids are present plus oncoids showing multiple stages of encrustation and reworking (Fig. 4b). Echinoids, bivalves, and *in situ* corals comprise most of the patch reef assemblage (Sellwood 1978). Disarticulation of *Rhaxella* sponges liberated round opaline spicules which often form ooid nuclei or cement together to form grapestone clasts. Rounded intraclasts containing neomorphic coral remains with non-ferroan and ferroan cements are common. Patchily distributed within the grainstones are micrite clumps, which form meniscuses and bridges between grains partially occluding porosity (Fig. 4c).

The sediments thicken to 40 m in the Harwell 4 and Chalgrove boreholes to the east. The sequence at Harwell 4 consists of laminated mudstones which coarsen upward into bioturbated siltstones. These are interrupted by a 10 cm thick, graded, cross-laminated fine sandstone bed with an erosional base and abrupt upper contact. Terrigeneous sands are gradually superseded by wackestones and packstones containing ooids and pisoids in a micritic-clay matrix. At the top of the Corallian there are reworked coral fragments and shell fragments. The Chalgrove borehole contains mudstones and wackestones which are extensively bioturbated, with bivalve debris concentrated in four main horizons.

Diagenetic features

The coarse-grained deposits which formed on the block crests acted as conduits for diagenetic fluid flow. The Corallian's stratigraphic position between two thick and impermeable clay formations, the Oxford and Kimmeridge Clays, enabled it to function as an aquifer, transporting meteoric water in east–west conduits, isolated by the surrounding intra and interformational clays.

Lithification is divided into three phases; preburial, shallow burial and deep burial. Phase 1 includes marine precipitates, vadose features, non-ferroan neomorphic and drusy spars. The phase 2 cements are ferroan and precipitated under reducing conditions, probably accompanying shallow burial. Phase 3 deep burial stylolites are associated with a blocky spar that has a patchy response to staining (Dickson 1966) and a characteristic mottled cathodoluminescence (CL) pattern. The Corallian Formation also contains subordinate amounts of dolomite,

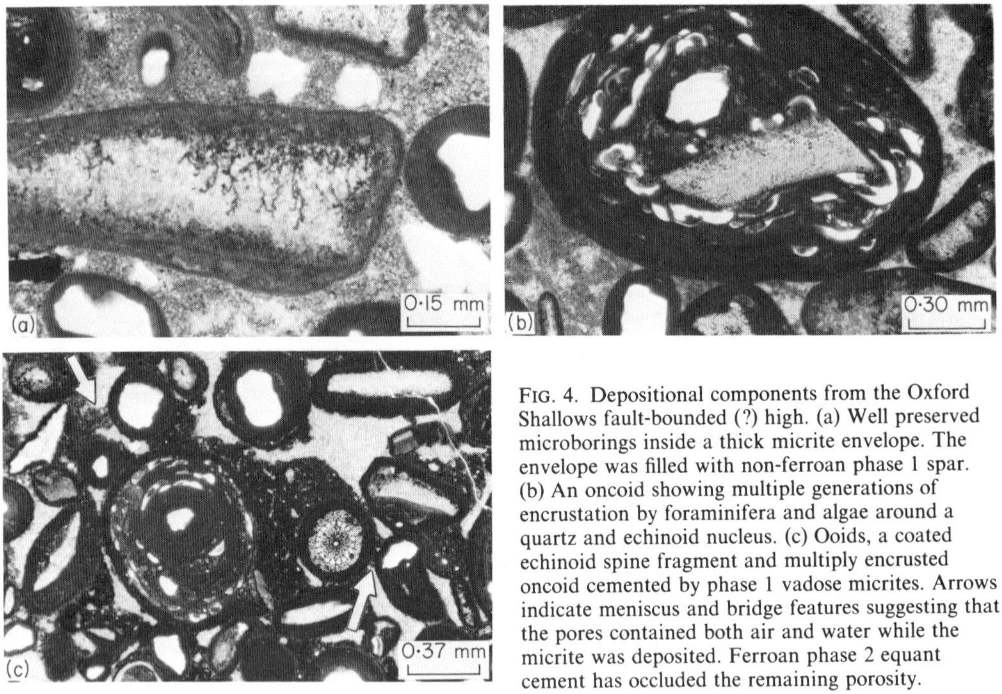

FIG. 4. Depositional components from the Oxford Shallows fault-bounded (?) high. (a) Well preserved microborings inside a thick micrite envelope. The envelope was filled with non-ferroan phase 1 spar. (b) An oncoid showing multiple generations of encrustation by foraminifera and algae around a quartz and echinoid nucleus. (c) Ooids, a coated echinoid spine fragment and multiply encrusted oncoid cemented by phase 1 vadose micrites. Arrows indicate meniscus and bridge features suggesting that the pores contained both air and water while the micrite was deposited. Ferroan phase 2 equant cement has occluded the remaining porosity.

siderite, silica and hydrocarbon not discussed here.

Basin burial curves are often used to delineate diagenetic trends. They may, however, present an oversimplified view as local permeability barriers, water-table fluctuations and mineralogic transformations create complications only distantly related to tectonic movements. As this study and many others (Moore & Druckman 1981, Saller 1986) demonstrate, simple burial curves may belie the complications of diagenesis.

FIG. 5. Cements from diagenetic phases 1, 2 and 3. All samples were stained according to Dickson (1966) except D, which was polished for cathodoluminescence. (a) Serpulid encrustations, micrite envelopes and an echinoid spine fragment showing phase 1 and 2 cements. LMC discrete peloids, p, may have had a HMC precursor, representing marine cements that precipitated during phase 1. Non-ferroan phase 1 spar is particularly well developed on the echinoderm as a syntaxial overgrowth (arrow 1). It precipitated from oxidizing and/or iron-poor meteoric water. The first ferroan (mauve) spar that precipitated during phase 2 is also syntaxial to the echinoderm spine (arrow 2a). It may represent the development of a redox front. The second ferroan spar, 2b, also precipitated during shallow burial from iron-rich, reducing formation water. The pale blue to the right of the photograph is dyed impregnating plastic. (b) Early diagenetic cements. Phase 1 cements are represented by peloids, p, now LMC. They probably had a HMC marine precursor. Spar 1 (1) is a non-ferroan syntaxial overgrowth on part of an echinoderm shell wall, e. The field of view is inside an echinoid test cavity that was partially filled by micrite envelopes and shell debris. Phase 2 cements are represented by 2a and 2b. 2a represents the first ferroan (mauve) spar precipitated during shallow burial. 2b was subsequently precipitated in the remaining pores as the formation water became reducing and iron-rich. (c) Phase 1, 2 and 3 spars are shown. Phase 1 non-ferroan equant spar forms isopachous rims around ooids (arrow 1) and micrite envelopes. Phase 1 spar is present inside some micrite envelopes, indicating that aragonite dissolution preceded its precipitation. Phase 2 precipitation (2) is represented by ferroan spar that occluded the remaining porosity. Phase 3 dissolution, probably the result of undersaturated Barremian meteoric water entering the Corallian Formation preferentially dissolved a coral(?) fragment. Blocky phase 3 spar (3) precipitated in the void. (d) Photomicrograph showing phase 1 non-ferroan equant spar (1) and a micrite-filled coral mould which has been partially dissolved by phase 3 fluids. Phase 3 (3) crystals contain relics and inclusions of the previous fabric. (e) Cathodoluminescence photomicrograph showing cements from phases 1 and 3. Phase 1 spar (1) is dully zoned adjacent to the brightly luminescent substrate. The zones become more brightly luminescent towards the pore centre. The remaining porosity is filled by phase 3 (3) blocky spar which exhibits a characteristic mottled luminescence.

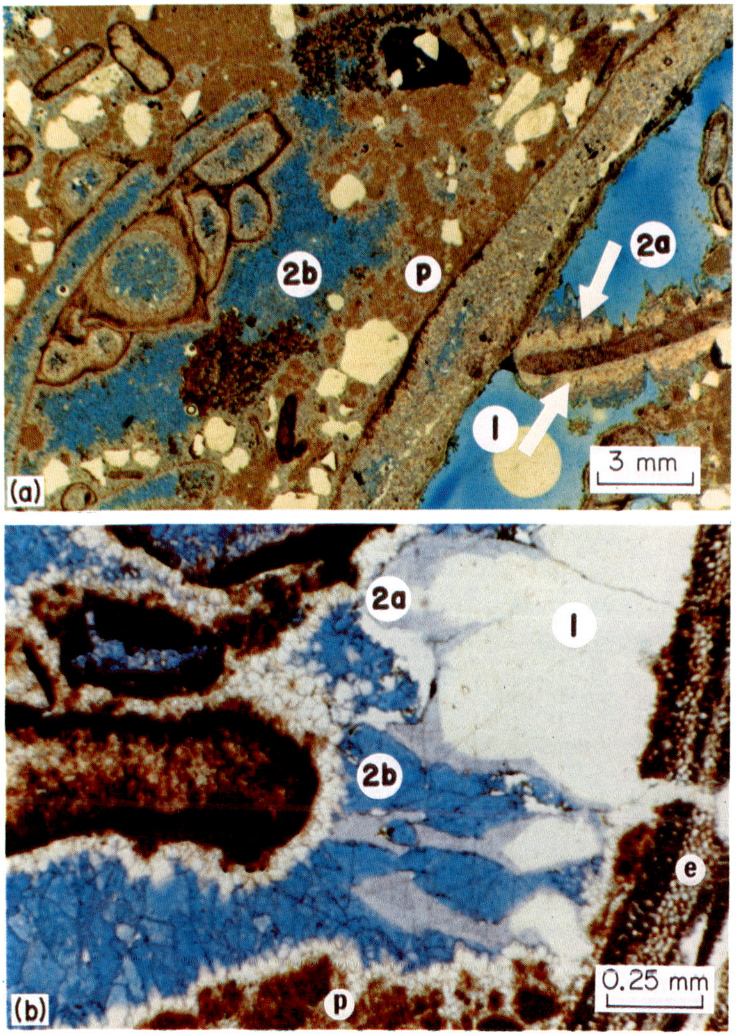
2b
p
2a
1
3 mm
(a)
2a
1
2b
e
0.25 mm
p
(b)

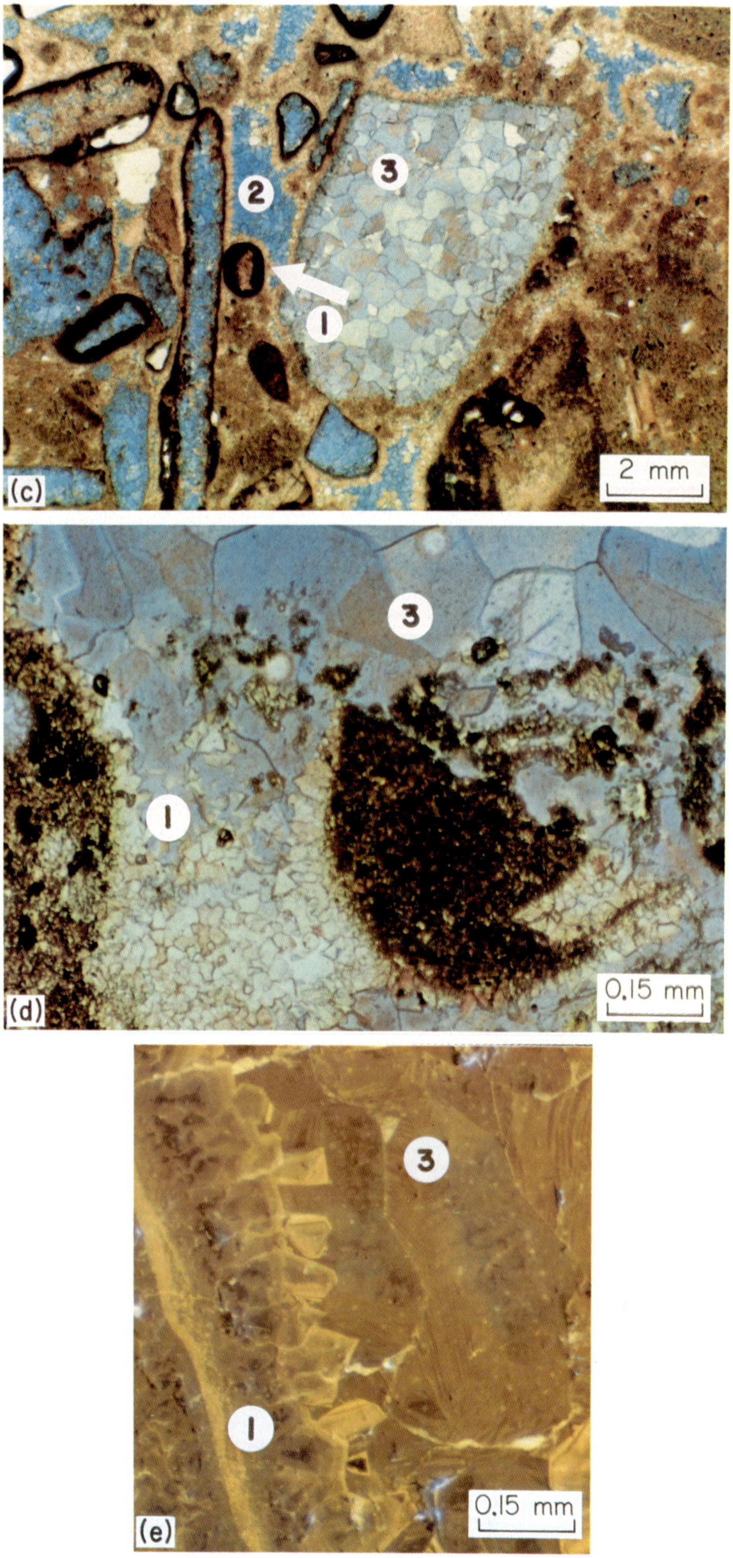

2
3
1
2 mm
(c)
3
1
0.15 mm
(d)
3
1
0.15 mm
(e)

Phase 1 cements

Phase 1 cements are most common in carbonate horizons. Micrite cemented grapestone and limeclasts constitute about 5% of the carbonate allochems in the grainstones. The earliest pore-occluding cements are columnar fringing spars. These are particularly common on prismatic bivalve fragments that escaped micritization (Fig. 6a). Peloidal micrites are usually found inside protected cavities (Fig. 5a, b), however, silt-filled fractures, micritic meniscuses between grains and clotted muds are locally common (see Fig. 4c). Extensive mouldic porosity, up to 50%, was created by the dissolution of aragonite, leaving open micrite envelopes (Fig. 6b). This fragile framework was stabilized by neomorphic replacement and the precipitation of the first non-ferroan druse spars in primary and dissolution pores (Fig. 5c). The cements are dog-tooth to equant, usually isopachous, and line both the interior and exterior of micrite envelopes and coral chambers. Under CL most of the phase 1 spars begin with a non-luminescent zone followed by brightly luminescent and non-luminescent subzones (Fig. 5e).

Phase 2 cement

The first ferroan (mauve) cement is particularly well developed as syntaxial overgrowths on echinoderms (Fig. 5a, b). Although volumetrically rare the spar is both spatially and chemically transitional between phase 1 non-ferroan and later phase 2 ferroan cements. Under CL the overgrowths show complex zones of non-luminescent and brightly luminescent bands. The second phase 2 spar precipitated during shallow burial as an equant or granular ferroan cement (Fig. 5a, c). It is the final pore-occluding cement in the grainstones and is the first and often only spar in the stratigraphically lower clastic horizons. *Rhaxella* sponge spicule moulds are almost exclusively filled with this ferroan spar. The cement post-dates minor compaction, filling small fractures, broken micrite envelopes and spalled ooid jackets (Fig. 6c). The crystals are usually dark under CL, but may show three to five dull-luminescing zones not detected by staining.

Phase 3 cement

Phase 3 is the only spar to show a consistent crystal morphology, stain and CL response across the entire basin. The crystals are blocky, often poikilotopic, and are up to 1 mm across. They have a patchy iron distribution based on staining (Fig. 5c, d) and a mottled CL response (Fig. 5e). The most striking aspect of the spar is its association with dissolution cavities (Fig. 6d). Though it post-dates all previous spars, it occurs in non fabric-selective vugs and fractures. The contact with pre-existing fabrics is marked by corroded crystal faces, and ghosts and relics of precursor crystals (Fig. 5d).

Interpretations: Oxford Shallows

Depositional environment

The sandstones at outcrop have been interpreted as representing deltaic input (Brookfield 1973), or nearshore shoals and beaches (Talbot 1973), possibly derived by the reworking of an older sandstone (Triassic?) from topographically subdued landmasses to the east and west. Marine shells and *Rhaxella* spicule moulds concur with a nearshore marine setting.

The presence of the carbonate units indicates a diversion of the terrigeneous supply, allowing pisoids and oncoids to form. Photic zone conditions supported blue-green algal oncoids and scleractinian corals. Micritization, boring, rounding, and encrusting indicate slow sedimentation rates on the high. Patch reefs developed on stabilized sandstone shoals near the escarpment where currents were most active (Headington, NGR SP 030 430, Shellingford, Cothill). The abundance of *Rhaxella* spicule moulds attests to quiet water conditions in the back-reef and lagoon (McKerrow 1978). Intraclasts containing equant cements in coral vugs imply that the marine sediments were subaerially exposed. The presence of both non-ferroan and ferroan equant cements within the clasts suggests that the spars precipitated in a well-developed meteoric system. Additional evidence of exposure includes micritic clots and muds forming meniscuses between grains. These features suggest that small fluctuations in sea-level resulted in subaerial exposure and the development of a phreatic lens below the shoal tops.

Eastward, across the slope break, the much-thickened facies at Harwell 4 represents deeper water deposition (Milodowski & Wilmot 1984). The sediments mirror the shallow water equivalents, progressing from terrigeneous clastics, including a storm deposit or small turbidite (the 10 cm thick erosional sand bed), to pisoids and ooids, through to shelly debris and coral fragments. All of the allochems are allochthonous, swept off the platform and encased in clay and micrite. The finer grained, more uniform sediments of Chalgrove represent the deeper subtidal zone, more distal to the high (Stanley & Moore 1983).

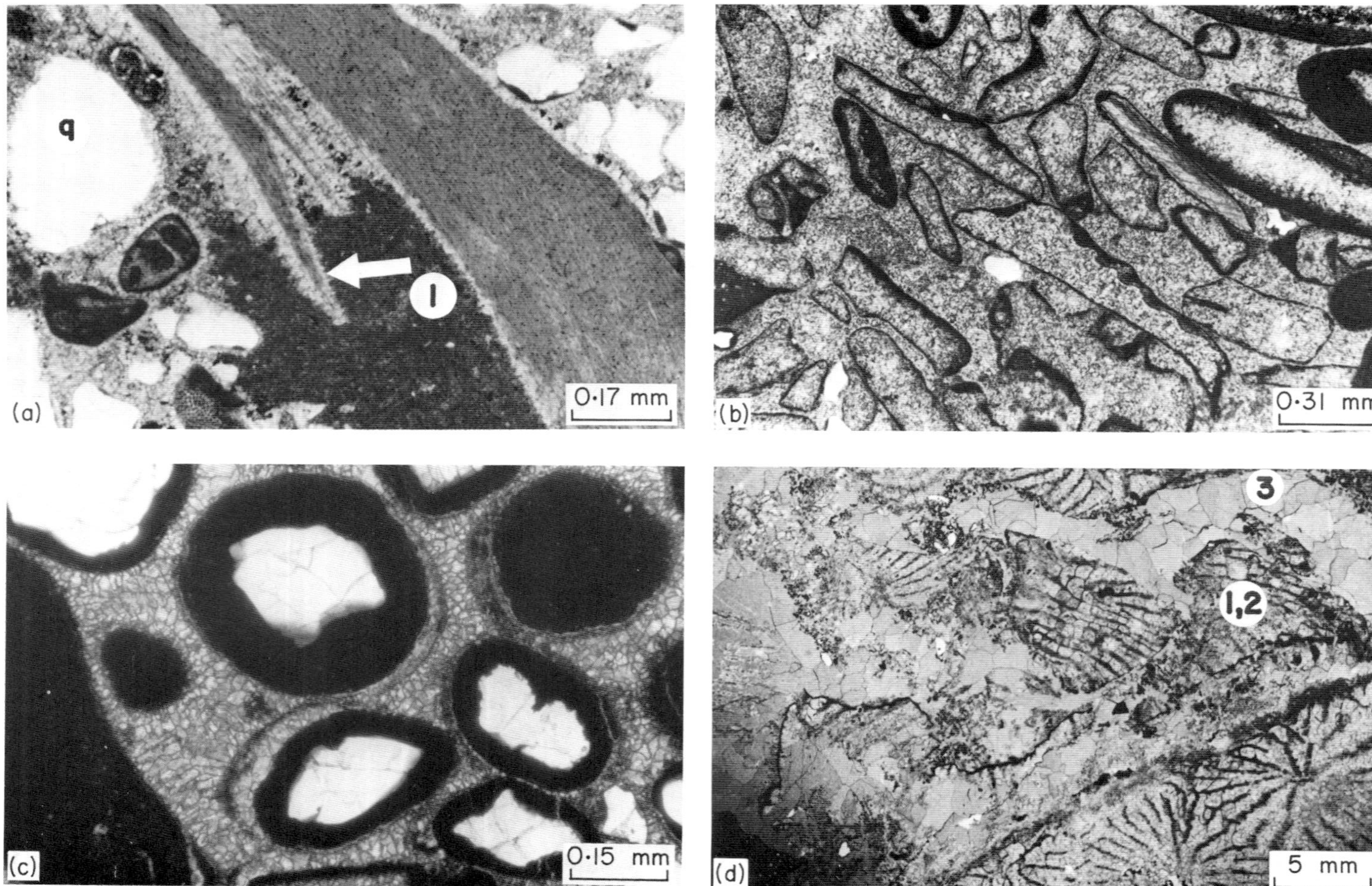

FIG. 6. Diagenetic cements and fabrics. (a) Phase 1 non-ferroan isopachous columnar cement (1) on a bivalve fragment. The cement appears to have displaced the muds around it, implying early precipitation, probably in the marine phreatic environment. Quartz sand (q) and silt grains are common. (b) Photomicrograph illustrating aragonite dissolution prior to the precipitation of the phase 2 ferroan cement which fills both the primary and secondary pores. (c) Minor compaction during phase 2 shallow burial spalled ooid cortices. Phase 2 ferroan cement fills all porosity. (d) Photomicrograph showing that dissolution of previously stabilized allochems and cements phase 1 and 2 (1,2), preceded the precipitation of phase 3 cement (3). This implies that phase 3 waters were

Diagenesis

Phase 1 cements

Micritic marine cements bind together the grapestone and intraclasts, while peloidal precipitates are preserved in cavities. These spherical peloids, often surrounded by an interlocking crystal mosaic, are distinct from clotted micrites which occur in depositional layers and burrows. According to Macintyre (1985), spherical shape and isolated formation site are two criteria useful for distinguishing chemically precipitated peloids from biologically induced pellets. Corallian peloids meeting these criteria are interpreted as being originally magnesium calcite precipitated as a cement. Other possible marine cements include columnar fringes around ooids and on some bivalve fragments. They are presumed to have had a high magnesium calcite (HMC) or aragonite precursor based on their columnar habit and relation to the substrate. The crystals in Fig. 6(a) have displaced the surrounding micritic matrix indicating that the muds must have been soft at the time of growth. This implies that precipitation commenced before compaction could solidify the sediments. It is possible that the sediment was introduced after the growth of the cement, however because of its geopetal distribution this would be physically the more difficult alternative. Talbot (1971) and Chowdhury (1982) interpret the columnar spars on ooids as originally HMC or aragonite marine phreatic cements. Neither the distinct peloids nor these fringes could be separated for chemical analysis.

Silt-filled fractures, small collapse breccias, clotted muds and micritic meniscus cements indicate local subaerial exposure, presumably on shoal tops. These vadose features may represent pedogenic alteration (Martin, Wilkinson & Lohmann 1986), or beachrock cementation of limited duration.

Common aragonite dissolution attests to the presence of a well developed, initially undersaturated meteoric water system (Longman 1980, Moldovanyi & Lohmann 1984). As the oxidizing ground waters became saturated with calcite they precipitated the non-ferroan druse and neomorphic spars (Talbot 1971, Chowdhury 1982). As these cements are non-ferroan and non-luminescent the initial cement probably lacked iron or manganese. Over time, however, sufficient Mn^{2+} was incorporated to promote luminescence. The cement remained non-ferroan, suggesting that Mn^{2+} may be activating luminescence regardless of iron content. It should be noted that other factors, such as rate of precipitation and trace element concentration, may also affect luminescence (ten Have & Heijen 1985).

Phase 2 cement

Through shallow burial the meteoric waters gradually became more reducing and/or iron-rich, precipitating the first ferroan (mauve-staining) spar. This cement's intermediate iron values and its position between phase 1 and the later phase 2 ferroan cements provide a record of transition as the redox front developed and progressed through the sediments. This phase was short-lived as most of the cement crystals are too small to be observed except at Cothill (Fig. 5) and in the more rapidly precipitating echinoid overgrowths.

As burial continued, marine waters mixed with, and were displaced by, reducing meteoric fluids. This was accompanied by minor grain breakage and overpacking. The precipitation of ferroan granualar cement occluded the remaining porosity and prevented further compaction. Talbot (1971) states that this ferroan cement is a deep burial precipitate, however, it is cut by stylolites and pre-dates hydrocarbon migration. Non-fabric-selective solution voids filled with the youngest cement cut across the ferroan crystals. Therefore the ferroan spar must have formed prior to Barremian uplift and dissolution (see below).

The Lower Calcareous Grit lacks phase 1 cements because a minimum of shelly material was available for dissolution and reprecipitation. At outcrop the sandstone is cemented by phase 2 ferroan spar, but in the Harwell 4 borehole there are friable decalcified sands in discrect beds and along fractures, suggesting dissolution of the original ferroan cement (Milodowski & Wilmot 1984). Abundant intra- and interformational clays may have provided the iron for the ferroan cement, for even at shallow burial depths iron may be mobilized from iron hydroxides (Emery, this volume).

Phase 3 cement

This spar post-dates all others, cross-cutting pre-existing fabrics in vugs and fractures. Its blocky, often poikilotopic, ferroan crystals are typical of burial precipitates (Scholle & Halley 1985). However, its association with solution and its patchy response to staining are more enigmatic. Dickson & Coleman (1980), and Meyers & Lohmann (1978) report patchy staining and CL response, Dickson & Coleman attributing it to homogenization of formerly zoned cements while Meyers & Lohmann suggest that corrosion during the stabilization of HMC to LMC left inclusions and relicts. In the Corallian, previously stabilized allochems and cements are dissolved and phase 3 calcite is precipitated. In vugs where the host rock is predominantly non-ferroan, the new spar contains little iron, while vugs in ferroan hosts contain more iron-rich cement. This suggests that local chemistries dictated the composition of the precipitate even though the uniformity of the CL response indicates a homogeneous source. The stable isotopes and geochemistry discussed in the next section help to resolve this apparent contradiction.

To what environment should these phase 3 cements be ascribed? Aggressive, undersaturated solutions capable of vuggy dissolution are commonly attributed to active meteoric circulation or to brines preceding hydrocarbon migration. In the Corallian the phase 3 spars clearly post-date some initial burial, but only the centre of the basin reached burial depths sufficient for hydrocarbon generation. However, Barremian uplift exposed the Formation's margins (Fig. 7) as there are Corallian clasts in the Cretaceous Faringdon Sponge Gravel (Talbot 1971) and in the Lower Greensand Formation (Kelly 1985). This period of exposure at the surface would almost certainly have allowed undersaturated meteoric waters to enter the Formation and be driven basinward by gravity flow. During the ensuing phase of deep burial, the trapped meteoric water would have mixed with connate water that was being expelled from clays and slowly spread throughout the basin. At present, basinal fluid flow is still converging on the Corallian in the Harwell 4 borehole (J. Black, personal communication, 1987) due to cross-formational flow from the Oxford and Kimmeridge clays. Therefore it is suggested that such a mix of meteoric and formation waters produced the phase 3 spars.

Summary

The three cementation phases represent residence time in distinct diagenetic environments. Phase 1 cements precipitated in surface marine, vadose and freshwater phreatic environments. Phase 2 cements formed in reducing fluids during initial burial. Minor compaction effects, such as spalled ooid cortices indicate that cementation was not complete before burial. Phase 3 involved tectonic uplift during the Barremian, allowing meteoric recharge through the Corallian Formation, causing vuggy dissolution. Phase 3 cements precipitated during the ensuing deep burial (Table 1).

Geochemistry

Trace elements

Within the carbonate system, analysing for cations whose distribution coefficients are less than and greater than unity may be useful in determining the amount of diagenetic alteration a given component has experienced (Moore & Druckman 1981, Veizer 1983, Majid & Veizer 1986). By comparing Mg^{2+} and Sr^{2+} (with distribution coefficients of 0.013–0.06 and 0.13 respectively, values from Veizer 1983) with Mn^{2+} and Fe^{2+} (with distribution coefficients of 15–30 and 10–20, Veizer 1983) in specific cement phases, general trends in the 'openness' of the fluid system may become apparent. Phase 1–3 cements from the Corallian Formation were separated and analysed on a Phillips 50 MHz inductively coupled argon plasma spectrometer (1–3% precision).

As magnesium and iron have opposing distribution coefficients relative to 1 they should act antithetically during mineralogic transforma-

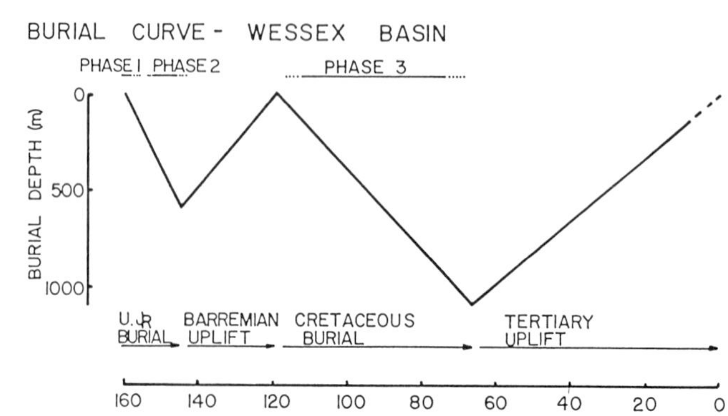

FIG. 7. General burial curve for the western Wessex Basin indicating the relative timing of cementation phases 1–3. Barremian uplift exposed the Formation's western and northern margins to meteoric recharge.

TABLE 1. *Carbonate cement summary*

PHASE	CL RESPONSE	STAIN RESPONSE	DESCRIPTION	DIAGENETIC ENVIRONMENT
1	non-luminescent to bright	brownish	micritic peloids, grapestone, intraclasts	marine phreatic
	non-luminescent to bright	non-ferroan	isopachous fibrous	marine phreatic
	zoned dull to bright	non-ferroan	isopachous dogtooth, equant, microspar	oxidizing meteoric phreatic nearsurface
2	complex zoning	ferroan (mauve)	syntaxial overgrowths equant	mildly reducing meteoric phreatic
	non-luminescent or dully zoned	ferroan	equant, granular	reducing meteoric+connate shallow burial
3	dull with dark patches	variable	blocky, poikilotopic	Cretaceous meteoric+ connate deep burial

tions involving a fluid phase. Partitioning theory predicts that Mg^{2+} will be preferentially retained in the fluid while Fe^{2+} will become incorporated into the crystal (Pingitore 1982). Therefore, decreases in Mg^{2+} and increases in Fe^{2+} from an original marine value should indicate increasing diagenetic alteration. The Corallian cement sequence data on phase 1 and 2 cements shows a steady decrease in Mg^{2+} away from a marine value, as Fe^{2+} increases (Fig. 8a). This is the expected result assuming aragonite and HMC, both containing seawater-derived Mg^{2+}, stabilized to LMC in a relatively Mg-poor freshwater system. The iron content rises due to the reduction of Fe^{3+} to Fe^{2+} and its incorporation into calcite during shallow burial. Mobile iron may have been derived from primary glauconite, chamositic ooids and berthierine found to the south (Talbot 1973), suggesting that collodial iron complexes were available for reduction even during shallow burial.

The phase 3 cements exhibit variable magnesium concentrations and a decrease in iron relative to phase 2 cements, probably reflecting a multi-sourced origin. Barremian meteoric water dissolved previously stabilized phase 1 and 2 cements, creating a fluid with a new composition. The phase 3 cement that precipitated from this fluid contained similar concentrations of Mg^{2+} and less Fe^{2+} than the adjacent phase 2 crystals. The majority of the samples analysed exhibit this small chemical gradient, suggesting that the bulk solution began to approach equilibrium with the new fluid. In the example where the phase 3 spar chemically matches the host phase 1 cement, local dissolution and reprecipitation without chemical homogenization with the new fluid must have occurred. In this case, bulk solution disequilibrium (Veizer 1983) was

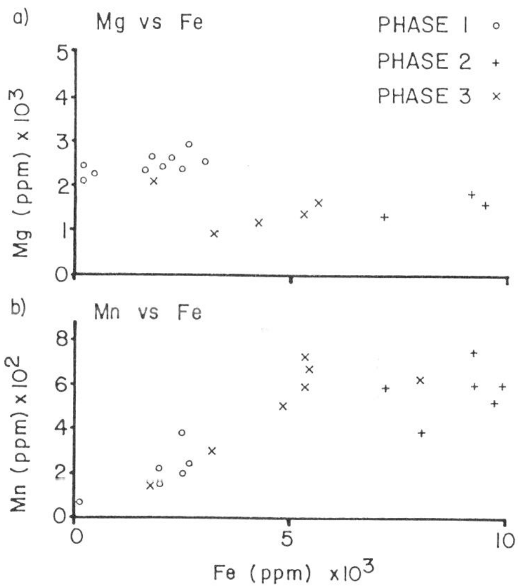

FIG. 8. Geochemical data for phase 1, 2 and 3 cements from the Corallian Formation, Oxford Shallows area. (a) Plot of magnesium against iron. Phase 1 cements are depleted in Mg^{2+} relative to present seawater (16300–75400 ppm) and are enriched in Fe^{2+} (2–39 ppm, marine calcite values from Veizer 1983). Phase 2 spars show a significant increase in iron concentration accompanied by continuing magnesium depletion. Phase 3 cements contain intermediate iron concentrations and variable magnesium values. (b) Plot of magnesium against iron. Phase 1 cements are depleted in both cations relative to phase 2 and 3 cements. Phase 2 spars contain the highest concentrations of manganese and iron, indicating that cations were being supplied from source sediments, probably intra- and interformational clays, into reducing waters. Data from the phase 3 cements show two groups; one containing Mn^{2+}-rich samples and a second containing two Mn^{2+}-depleted samples.

probably the dominant mechanism for cement growth.

Mn^{3+} is more readily reduced to Mn^{2+} than Fe^{3+} is reduced to Fe^{2+}. Therefore the incorporation of Mn^{2+} into calcite should precede the incorporation of Fe^{2+}. Corallian Formation phase 1 cements luminesce brightly under cathodoluminescence excitation although they have a non-ferroan response to staining. In Fig. 8(b), the most depleted phase 1 spar contains negligible iron but 80 ppm Mn^{2+}. This may reflect the initial development of the redox front.

The trend from phase 1 to phase 2 cements in Fig. 8(b) is consistent with a shallow burial model, demonstrating that initially oxygenated, and/or iron and manganese-poor marine and meteoric waters became progressively reduced and enriched in both Fe^{2+} and Mn^{2+}. The iron and manganese were probably derived from colloids and hydroxides.

In Fig. 8(b), the phase 3 cements form Mn-rich and Mn-poor categories. The Mn-rich group is depleted in Fe^{2+} compared to the phase 2 cements, suggesting that Fe^{2+}-depleted Barremian meteoric water mixed with dissolving phase 2 calcites, thus diluting the overall iron concentration. The slight increase in Mn^{2+} may account for the similarity of CL response in almost all of the phase 3 spars. Although staining showed a variable iron content, the pattern of luminescence and intensity are uniform in all of the blocky phase 3 spars. This indicates that Mn^{2+} predominates over Fe^{2+} in dicating CL response. If the phase 3 waters were only slightly reducing or a redox potential fluctuated then Mn^{2+} would be more likely to be incorporated into the precipitating calcite than would Fe^{2+}. As the Oxford Shallows are adjacent to the area exposed during Barremian uplift, it is likely that phase 3 waters were only midly reducing when the sequence was reburied during the rest of the Cretaceous. The phase 3 cement would have precipitated before the fluids became fully reduced. An alternative explanation could be that there was an increase in the concentration of Mn^{2+} to the system. This would suggest that the phase 3 cement precipitated during maximum burial as compaction released Mn^{2+} from adjacent clay horizons. In this environment iron would also be released, but authigenic clays and pyrite could have incorporated it (Milodowski & Wilmot 1984), leaving the calcite relatively depleted.

The two Mn-poor samples in Fig. 8(b) contain Mn^{2+} and Fe^{2+} concentrations that are very similar to the host phase 1 cements. As in the previous discussion of Mg^{2+} and Fe^{2+} in phase 3 cements, this may imply that localized dissolution and reprecipitation has occurred, without chemical equilibration with the bulk solution.

Strontium has been widely used as a tracer for evaluating diagenetic alteration, particularly in corals (Pingitore 1982, Pingitore & Eastman 1986, Yoshioka *et al.* 1986) since it is readily incorporated into the aragonite structure. However, the original coral structure may have a significant effect on the amount of alteration experienced during stabilization (Walter 1985, Constantz 1986). Talbot (1971) discussed differences in textural preservation relative to coral species within the Corallian Formation. Good textural preservation, accompanied by high Sr^{2+} values, may indicate areas where the dominant mechanism for cation movement was diffusion rather than mass transport (Pingitore 1982). In this study, neomorphosed *Theocosmilia* corals have Sr^{2+} values ranging from 950 to 400 ppm (Fig. 9), suggesting differing amounts of flushing by phase 1 meteoric waters. All of the phase 1 calcites have relatively high Sr^{2+} concentrations and low Mn^{2+} values indicative of early stabilization and precipitation in an oxygenated and/or Mn^{2+} poor environment.

The cements that precipitated during shallow burial are depleted in Sr^{2+} and enriched in Mn^{2+} relative to the phase 1 spars. This is the expected result assuming that strontium's distribution coefficient is less than one and manganese's is greater than one. During burial, Mn^{3+} became reduced to Mn^{2+} and, as predicted by partioning theory, was substituted into the calcite lattice while Sr^{2+} remained in the fluid. The most enriched phase 2 cement contains 1200 ppm Mn^{2+} and only 300 ppm Sr^{2+} (Fig. 9).

Phase 3 cements are generally intermediate between phases 1 and 2, a consistent trend seen

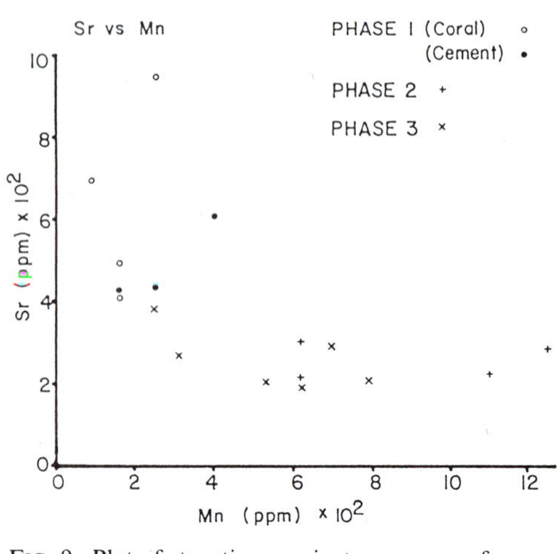

FIG. 9. Plot of strontium against manganese for phase 1, 2 and 3 cements from the Corallian Formation, Oxford Shallows area.

in all of the geochemical plots. This probably reflects mixing between cations introduced in Barremian waters and cations released during phase 3 dissolution. In Fig. 9, one sample contains Sr^{2+} and Mn^{2+} concentrations that closely match phase 1 values, indicating incomplete homogenization with the bulk solution relative to other phase 3 cements.

Stable isotopes

Stable isotope compositions were measured on individual cements collected from serial thin sections and rock chips. Analysis of CO_2 was carried out using a VG SIRA 12 mass spectrometer. Each run of 10 samples contained two internal standards of Lincolnshire Limestone. Precision was better than 0.1 ml^{-1}. Corrected results (after Craig 1957) are expressed relative to the PDB standard.

In Fig. 10, the carbon isotopic compositions indicate a gradual evolution in phase 1 and 2 cements from values near to Jurassic marine compositions (Hudson 1977, Walls *et al.* 1979, Marshall & Ashton 1980), through to values with more negative $\delta^{13}C$ probably caused by the incorporation of organogenic carbon during shallow burial. Phase 1 cements have 'marine' carbon and oxygen isotopic compositions (Hudson 1977). Using the average $\delta^{18}O$ of -2 ‰ palaeotemperature determinations were made using the equation of Shackelton & Kennett (1975):

$$T°C = 16.9 - 4.38\ (\delta c - \delta w) + 0.10\ (\delta c - \delta w)^2$$

(where δc is the measured isotopic value of CO_2 and δw is the isotopic composition of CO_2 in equilibrium with formation water). Assuming a value of $w = -1.2$ for Jurassic sea water (Marshall & Ashton 1980) this gives a palaeotemperature of 20.5°C for the average phase 1 cements. This is a reasonable estimate of seawater temperature. It is slightly lower than Marshall & Ashton's (1980) estimate of 22–25°C from British Jurassic oysters, but is higher than values calculated from their marine cements and ooids. The $\delta^{18}O$ values for phase 2 cements are similar to those for phase 1 but the $\delta^{13}C$ values are about 3 ‰ 'lighter'. As discussed above the petrographic and geochemical evidence suggest that phase 2 cements formed under the influence of meteoric water. The measured carbon isotopic values are consistent with minor input from light (photosynthetically-fixed) soil-derived CO_2 but the oxygen values would normally be expected to show more negative 'freshwater influenced'

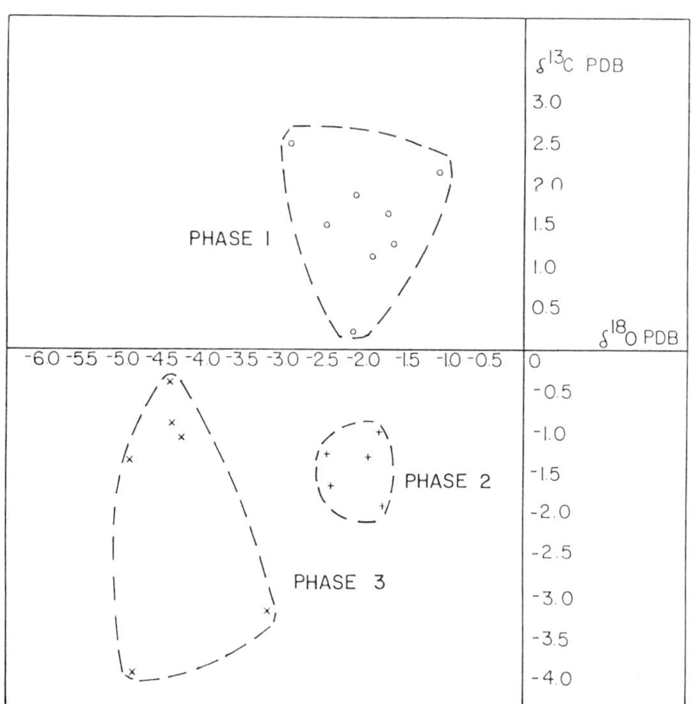

FIG. 10. Plot of $\delta^{13}C$ PDB against $\delta^{18}O$ PDB. Stable isotope data for cement phases 1, 2 and 3 from the Corallian Formation, Oxford Shallows area.

values. This apparent discrepancy may be explained by suggesting that in the context of the Corallian limestones the 'fresh' water may not have been isotopically very different from sea water. Near-coastal rainfall in low latitudes has an isotopic composition which approaches local sea water (Anderson & Arthur 1983). In the proposed tectonic setting for the Corallian the fresh water involved in phase 2 cements would have returned as rain without considerable transport as vapour, thereby limiting fractionation. It is conceivable that it could have been virtually indistinguishable, in terms of $\delta^{18}O$, from the sea water particularly if any marginal depletion in $\delta^{18}O$ from the meteoric water cycle was compensated for by an opposite marginal enrichment by local evaporation on the exposed highs.

The difference in $\delta^{18}O$ between phases 1 and 2 and phase 3 cements is about 2.5 ‰. If precipitation again took place from water with a constant oxygen isotopic composition the temperature would have to have increased by 12.8°C. The Wessex Basin has a geothermal gradient of 30°C km^{-1} and in the Marchwood borehole waters in the Corallian Formation are presently at 40°C at 850 m depth (Burgess *et al.* 1981). However, the isotopic change in the calcites could be due, at least in part, to isotopic changes in the precipitating water. As argued above, the phase 3 cements probably formed in response to the introduction of Barremian meteoric water in a groundwater system. Such water would almost certainly have had a more negative $\delta^{18}O$. The $\delta^{18}O$ values in the cements may therefore reflect both temperature effects and changing isotopic composition of the porewater. Most of the phase 3 samples have $\delta^{13}C$ values which lie between −0.4 and −1.4 ‰; however, in two samples, the carbon values are significantly lower at −3.3 and −3.9 ‰ PDB. These carbon values probably reflect minor input from a source of isotopically light organogenic carbon into a pore fluid whose carbon values were dominated by the dissolution of marine carbonate. Possible sources might include thermal breakdown of organic matter ('decarboxylation') or surface derived carbon from the recharge area. The latter is unlikely except for the limestones adjacent to the recharge area and soil zones so thermal breakdown is the preferred alternative for the phase 3 cements.

Discussion

The depositional environment of the Corallian sediments near Oxford consisted of a rimmed accretionary shelf adjacent to the Welsh Massif. The model presented in Fig. 11 relies on Chadwick's (1986) tectonic interpretation for the Wessex basin and predicts lateral variation in the north–south direction. Facies belts trend east–west and seismic profiles show sediment thickening dramatically in the troughs (R. Benmore, personal communication, 1987). The steeper, generally south-facing, fault-controlled slope break marks an abrupt facies change from predominantly clastic and carbonate sands to predominantly clays. Allochthonous debris, produced by storms and gravity sliding, collected in the muds at the base of the slope. The gentler, usually north-facing slope is characterized by lagoonal micrites and sands, grading into clays.

The Wessex basin remained dissected by east–west trending faults throughout its early diagenetic history; phases 1 and 2. Corridors of coarse-grained sediments transported fluids from marginal meteoric recharge sites along fault block highs into the basin. During phase 2 shallow burial meteoric water became reducing and/or iron and manganese-rich as colloids and hydroxides released cations. Local facies distribution produced permeability differences that influenced phase 1 and 2 cementation. The low permeability wackestones and packstones at Harwell contain very little phase 1 cement relative to the grainstones on the Oxford high. The highly permeable sandstone beds both on the Oxford Shallows high and off it, across the slope break at Harwell, are almost exclusively cemented by phase 2 spar or are decalcified. As the fault highs were surrounded by both inter- and intraformational clays they were essentially isolated from each other during diagenetic phases 1 and 2. The cements that precipitated during that time can be correlated by CL internally within a block, but not across to an adjacent high. This is unlike the broad cement correlations that can be made in more structurally uniform basins (Meyers 1978).

It is only the phase 3 spar, which probably precipitated during the deepest burial phase, that has a basin-wide CL pattern. The chemistry of the late spars is generally intermediate between the two preceding phases of cementation, suggesting that introduced Barremian meteoric water was not in equilibrium with the host rock. Initially the waters were undersaturated, creating secondary porosity as non-fabric selective vugs and fractures which attest to this disequilibrium. Phase 3 cements precipitated in the voids as the fluids became supersaturated. Cations were supplied both from the introduced water and from the dissolution of previously stabilized allochems and cements. Hundreds of pore volumes of fluid must have passed through the voids, yet relics of precursor fabrics and phase 3

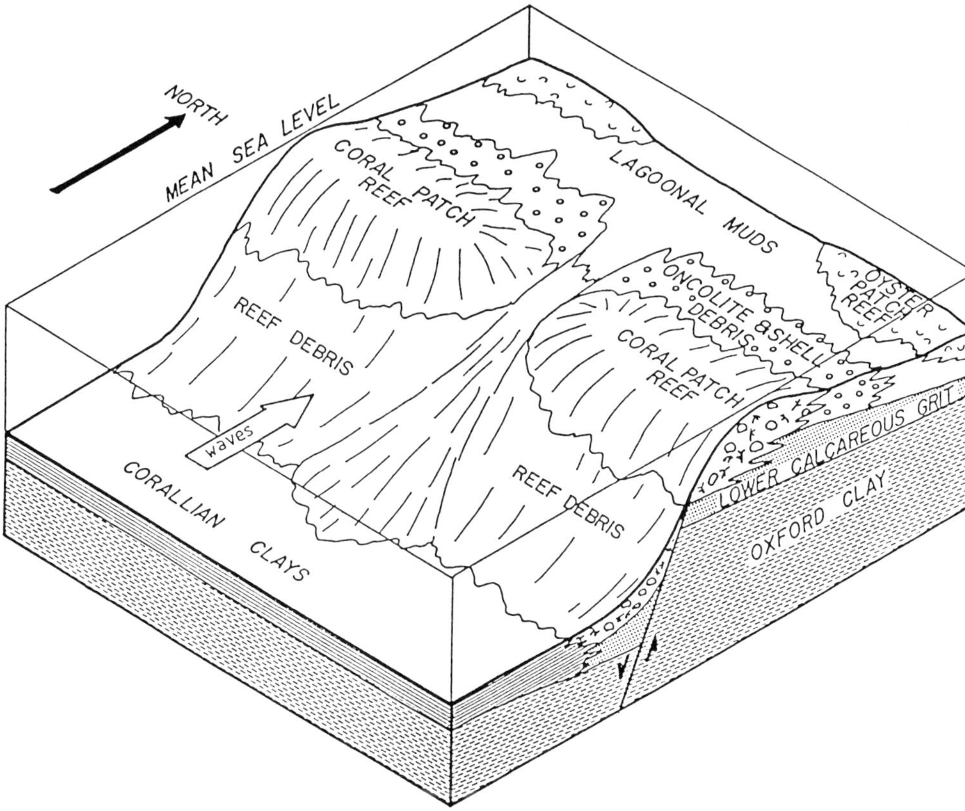

FIG. 11. Proposed depositional model and facies distribution along block crests and fault-bounded slopes for the Corallian Formation within the Wessex basin.

cement chemistries that almost perfectly match the host rock suggest that in particular cavities diffusion was the dominant mass transfer mechanism. In the majority of phase 3 cements the chemistry is internally more uniform, implying increased homogenization with the bulk solution via mass transport.

Conclusions

(1) A depositional model, consistent with, and dependent on, the internal structure of the Wessex Basin, is proposed. Fault-bounded blocks step down into the basin, trending longitudinally east–west across it. Syndepositional tectonics regulate local relative sea-level independently over each block, producing variable facies distributions. In the Oxford Shallows area moderate to high energy currents create sand shoals and permit patch reef development in a shallow subtidal shelf environment.

(2) Early diagensis (cements 1 and 2) was strongly influenced by depositional trends. Marine cements, probably HMC or aragonite, grew as fringes, peloids and grapestone micrites in areas of moderate circulation. Muds were deposited in meniscuses and bridges between grains during shoal top exposure. During burial, waters moved along conduits created by coarser grained sandstones and grainstones in a freshwater phreatic system. The fluids were initially oxidizing but became reducing and iron-rich during shallow burial. Early Cretaceous regional uplift exposed Corallian margins, admitting undersaturated meteoric water and initiating phase 3 vuggy dissolution.

(3) Geochemical and stable isotopic data show fluid evolution in cements 1 and 2, with Mn^{2+} and Fe^{2+} increasing and Sr^{2+} and Mg^{2+} decreasing through time. The late burial spar 3 occurs throughout the basin and has a depleted oxygen isotopic value suggesting increased temperature as fluids sourced from connate water were expelled from clays. In some examples the spar contains ionic concentrations which almost

match those of the host rock, in these cases local dissolution and reprecipitation in a relatively closed system have dictated the new spar's chemical composition.

ACKNOWLEDGMENTS: I wish to thank the staff at Royal Holloway and Bedford New College for running the ICP samples, and Dr J. D. Marshall and H. P. Attenborough at the University of Liverpool for allowing me to run the stable isotopes there. (The stable isotope laboratory is supported by NERC Research Grant GR3/5339 to Dr Marshall.) I have benefitted greatly from discussions with Dr J. A. D. Dickson. Drs M. Talbot, P. Gutteridge, J. A. D. Dickson and J. D. Marshall made useful suggestions to improve the manuscript. The British Geological Survey provided the Harwell and Chalgrove borehole samples and a grant from Sigma Xi Research contributed towards field expenses. Texaco (Houston) has kindly supported laboratory expenses for the Cambridge carbonate group.

References

ANDERSON, T. F. & ARTHUR, M. A. 1983. Stable isotopes of oxygen and carbon and their application to sedimentologic and paleoenvironmental problems. *In: Stable Isotopes in Sedimentary Geology*, pp. 1-1–151. Society of Economic Paleontologists and Mineralogists, Dallas, Short Course, no. 10.

BROOKFIELD, M. 1973. Palaeogeography of the upper Oxfordian and lower Kimmeridgian (Jurassic) in Britain. *Palaeogeography, Palaeoclimatology and Palaeoecology*, **14**, 137–67.

—— 1977. The use of coarse fraction analysis in palaeonvironmental interpretation: upper Oxfordian and lower Kimmeridgian (Jurassic) sediments in Dorset, England. *Sedimentary Geology*, **20**, 249–80.

—— 1978. The lithostratigraphy in the upper Oxfordian and lower Kimmeridgian beds of south Dorset, England. *Proceedings of the Geologists' Association*, **89**, 1–32.

BURGESS, W. G. *et al.* 1981. *The Marchwood Geothermal Borehole—a Preliminary Assessment of the Resource.* Institute of Geological Sciences, Geophysics and Hydrology Division.

CHADWICK, R. A. 1986. Extension tectonics in the Wessex Basin, Southern England. *Journal of the Geological Society of London*, **143**, 465–88.

CHOWDHURY, A. 1980. Geochemistry and sedimentology of the Corallian sediments of southern England. *Unpublished PhD Thesis*, p. 579. University of Southampton, England.

—— 1982. Variations in diagenetic environments of the Coral Rag and underlying oolite of the Osmington Oolite Formation (Upper Oxfordian) of the Oxford-Berkshire area, U.K., and their implications. *Sedimentary Geology*, **32**, 247–63.

CONSTANTZ, B. R. 1986. The primary surface area of corals and variations in their susceptibility to diagenesis. In: SCHROEDER, J. H. & PURSER, B. H. (eds). *Reef Diagenesis*, pp. 53-76. Springer-Verlag, Berlin.

CRAIG, H. 1957. Isotopic standards for carbon and oxygen and correction factors for mass-spectrometric analysis of carbon dioxide. *Geochimica et Cosmochimica Acta*, **12**, 133–49.

DICKSON, J. A. D. 1966. Carbonate identification and genesis as revealed by staining. *Journal of Sedimentary Petrology*, **36**, 491-505.

—— & COLEMAN, M. L. 1980. Changes in carbon and oxygen isotope composition during limestone diagenesis. *Sedimentology*, **27**, 107–18.

FÜRSICH, F. T. 1975. Trace fossils as environmental indicators in the Corallian of England and Normandy. *Lethaia*, **8**, 151–72.

HEFU, L 1986. Geodynamic scenario and structural styles of Mesozoic and Cenozoic basins in China. *Bulletin of the American Association of Petroleum Geologists*, **70**, 377–95.

HUDSON, J. D. 1977. Stable isotopes and limestone lithification. *Journal of the Geological Society of London*, **133**, 637–60.

KELLY, S. R. A. 1985. Jurassic and Cretaceous deposits of Upware, Cambridgeshire, *Bulletin of the Geological Society of Norfolk*, **35**, 3–38.

LONGMAN, M. W. 1980. Carbonate diagenetic textures from near-surface diagenetic environments. *Bulletin of the American Association of Petroleum Geologists*, **64**, 461–87.

MACINTYRE, I. G. 1985. Submarine cements—the peloidal question. *In:* SCHNEIDERMANN, N. & HARRIS, P. (eds). *Carbonate Cements*, pp. 3-16. Society of Economic Paleontologists and Mineralogists Special Publication no. 36.

MAJID, A. H. & VEIZER, J. 1986. Deposition and chemical diagenesis of Tertiary carbonates, Kirkuk oil field, Iraq. *Bulletin of the American Association of Petroleum Geologists*, **70**, 898–913.

MARSHALL, J. D. & ASHTON, M. 1980. Isotopic and trace element evidence for submarine lithification of hardgrounds in the Jurassic of eastern England. *Sedimentology*, **27**, 271–89.

MARTIN, G., WILKINSON, B. & LOHMANN, K. C. 1986. The role of skeletal porosity in aragonite neomorphism—*Strombus* and *Montastrea* from the Pleistocene Key Largo Limestone, Florida. *Journal of Sedimentary Petrology*, **56**, 194–203.

MCKENZIE, D. P. 1978. Some remarks on the development of sedimentary basins. *Earth and Planetary Science Letters*, **40**, 25–32.

MCKERROW, W. S. (ed.) 1978. *The Ecology of Fossils.* Duckworth & Co., London.

MEYERS, W. J. 1978. Carbonate cements: their regional distribution and interpretation in Mississippian limestones of southwestern New Mexico. *Sedimentology*, **25**, 371–400.

—— & LOHMANN, K. C. 1978. Microdolomite-rich syntaxial cements: proposed meteoric-marine mixing zone phreatic cements from Mississippian limestones, New Mexico. *Journal of Sedimentary Petrology*, **48**, 475–88.

MILODOWSKI, A. E. & WILMOT, R. D. 1984. Diagenesis, porosity and permeability in the Corallian

Beds (Upper Oxfordian) from the Harwell Research Site, South Oxfordshire, U.K. *Clay Minerals*, **19**, 323–41.

MOLDOVANYI, E. & LOHMANN, K. C. 1984. Isotopic and petrographic record of phreatic diagenesis: Lower Cretaceous Sligo and Cupido Formations. *Journal of Sedimentary Petrology*, **54**, 972–85.

MOORE, C. H. & DRUCKMAN, Y. 1981. Burial diagenesis and porosity evolution, Upper Jurassic Smackover, Arkansas and Louisiana. *Bulletin of the American Association of Petroleum Geologists*, **65**, 597–628.

PINGITORE, N. E. 1982. The role of diffusion during carbonate diagenesis. *Journal of Sedimentary Petrology*, **52**, 27–39.

—— & EASTMAN, M. P. 1986. The coprecipitation of Sr^{2+} with calcite at 25°C and 1 atm. *Geochimica et Cosmochimica Acta*, **50**, 2195–203.

RU, K. & PIGGOTT, J. 1986. Episodic rifting and subsidence in the South China Sea. *Bulletin of the American Association of Petroleum Geologists*, **70**, 1136–55.

SALLER, A. H. 1986. Radiaxial calcite in Lower Miocene strata, subsurface Enewetak Atoll. *Journal of Sedimentary Petrology*, **56**, 743–62.

SCHOLLE, P. A. & HALLEY, R. B. 1985. Burial diagenesis: out of sight, out of mind! *In:* SCHNEIDERMANN, N. & HARRIS, P. (eds). *Carbonate Cements*, pp. 309-34. Society of Economic Paleontologists and Mineralogists Special Publication no. 36.

SELLWOOD, B. 1978. The Jurassic. *In:* MCKERROW, W. S. (ed). *The Ecology of Fossils*, pp. 204-79. Duckworth & Co., London.

STANLEY, D. J. & MOORE, G. T. 1983. *The Shelfbreak: Critical Interface on Continental Margins*. Society of Economic Paleontologists and Mineralogists Special Publication no. 33.

STONELEY, R. 1982. The structural development of the Wessex Basin. *Journal of the Geological Society of London*, **139**, 543–54.

TALBOT, M. R. 1971. Calcite cements in the Corallian Beds (Upper Oxfordian) of southern England. *Journal of Sedimentary Petrology*, **41**, 261–73.

—— 1973. Major sedimentary cycles in the Corallian Beds (Oxfordian) of southern England. *Palaeogeography, Palaeoclimatology and Palaeoecology*, **14**, 293–317.

TEN HAVE, T. & HEIJEN, W. 1985. Cathodoluminescence activation and zonation in carbonate rocks: an experimental approach. *Geologie en Mijnbouw*, **64**, 297–310.

VEIZER, J. 1983. Chemical diagenesis of carbonates: theory and application of trace element technique. *In: Stable Isotopes in Sedimentary Geology*, pp. 3-1–100. Short Course Society of Economic Paleontologists and Mineralogists, Dallas, no. 10.

WALLS, R. A., MOUNTJOY, E. W. & FRITZ, P. 1979. Isotopic composition and diagenetic history of carbonate cements in Devonian Golden Spike reef, Alberta, Canada. *Bulletin of the Geological Society of America*, **90**, 963–82.

WALTER, L. 1985. Relative reactivity of skeletal carbonates during dissolution: implications for diagenesis. *In:* SCHNEIDERMANN, N. & HARRIS, P. (eds). *Carbonate Cements*, pp. 3–16. Society of Economic Paleontologists and Mineralogists Special Publication no. 36.

WHITTAKER, A. (ed). 1985. *Atlas of Onshore Sedimentary Basins in England and Wales: Post-Carboniferous Tectonics and Stratigraphy*. Blackie, Glasgow.

WRIGHT, J. K. 1981. The Corallian rocks of north Dorset. *Proceedings of the Geologists' Association*, **92**, 17–32.

YOSHIOKA, S., OHDE, S., KITANO, Y. & NANAMORI, K. 1986. Behavior of magnesium and strontium during the transformation of coral aragonite to calcite in aquatic environments. *Marine Chemistry*, **18**, 35–48.

C. B. DE WET, Department of Earth Sciences, Downing Street, University of Cambridge, Cambridge CB2 3EQ, UK.

Index

Pages on which figures appear are printed in *italic*, and those with tables in **bold**